Strength of Materials

Dr P. Purushothama Raj
Dean
Sri Aravindar Engineering College
Villupuram

Dr V. Ramasamy
Dean
Adhiparasakthi Engineering College
Melmaruvathur

Editor—Acquisitions: Sandhya Jayadev
Associate Editor—Production: M. R. Ramesh

ISBN 978-81-317-6854-9

First Impression

Published by Pearson India Education Services Pvt. Ltd, CIN: U72200TN2005PTC057128

Head Office: 1st Floor, Berger Tower, Plot No. C-001A/2, Sector 16B, Noida – 201 301,Uttar Pradesh, India.

Registered Office: Featherlite, 'The Address' 5th Floor, Survey No 203/10B, 200 Ft MMRD Road, Zamin Pallavaram, Chennai - 600044
Website: in.pearson.com; Email:companysecretary.india@pearson.com

Digitally printed in India by Trinity Academy for Corporate Training Ltd, New Delhi in the year of 2025.

Contents

Preface

Strength of materials is a basic engineering subject that is to be understood by any one who is concerned with the strength and physical performance of structures. This is a subject of study dealing with deformable solid bodies. This subject comprises of fundamental concepts of stress–strain, deformations, strain energy and load-carrying capacity of certain structures. It forms a core subject in the curriculum of civil engineering, mechanical engineering, production engineering, aeronautical engineering, automobile engineering, chemical engineering and electrical engineering.

This book has been written for the benefit of students who have commited to understand the elements of strength of materials. The elementary concepts of this subject form the basis for the design and analysis of a large variety of mechanical, electrical, chemical and civil engineering structures.

The subject has been presented systematically to maintain a logical continuity with adequate tables and illustrations so as to make the concepts self-explanatory. Salient points of each chapter are presented at the chapter-end to facilitate easy recapitulation.

In addition, numerical problems are solved at the end of each section to consolidate the knowledge and concepts studied under theory. From the university examination point of view, university questions of previous years are solved.

The core content has been written after consulting numerous text books, manuals and codes, for which the authors express their sincere gratitude and such books are duly listed in the reference. This being a textbook, the authors do not claim any originality.

Constructive suggestions and opinions are invited from faculty members and students for further improvement of the book.

Dr P. Purushothama Raj

Dr V. Ramasamy

About the Authors

Dr P. Purushothama Raj is currently Dean of Sri Aravindar Engineering College, Tamil Nadu. He obtained his B. E. Degree from University of Madras, M.E. Degree from Indian Institute of Technology, Roorkee (UP) and PhD from Bangalore University.

Dr Raj was awarded Bangalore University Silver-Jubilee Independence Day Award for Best Lecturer, Mysore University Best Research Award and Best Paper Awards from Institution of Engineers (India) and Bangalore University. He was a post-doctoral advanced research fellow for two years at National Research Council of Canada and involved in Geothermal Studies at Division of Seismology and Geothermal Studies at Ottawa, Canada. He was a visiting research fellow at University of California, Berkeley, USA, and involved in Earth Dam Design.

Dr Raj was Dean, College Development Council, University of Madras and he was involved in the guidance of all affiliated colleges in the field of learning, teaching, research, infrastructure, autonomy, student amenities and interaction with University Departments. He has published about 45 papers and 10 books.

Dr Raj's biographic notes are included in the Reference Volumes under the following headings: Reference Asia – Asia's Who's Who of men and women of achievements, Asian/American – Who's Who and Asia/Pacific – who's who, published by Rifacimento International, New Delhi.

He served as Director and Principal, VRS College of Engineering and Technology, Director and Principal, Adhiparasakthi Engineering College, Principal and Head of Civil Engineering, Pondicherry Engineering College, and Reader and Lecturer at Bangalore University.

Dr V. Ramasamy is currently Professor of Civil Engineering and Dean of Adhiparasakthi Engineering College. He obtained his B. E. Degree from University of Madras, M. E. Degree from Annamalai University and PhD from VIT University.

Dr Ramasamy has been working in Adhiparasakthi Engineering College for over 17 years. He has published five papers in journals and presented 20 papers in conferences and symposia.

He is a member of Indian Society of Technical Education and Indian Concrete Institute.

He has published five books and guiding five students for PhD.

1

Stress–Strain Behaviour of Solids

LEARNING OBJECTIVES

1.1 INTRODUCTION

Various structures like buildings, bridges, steel frames, and so on and machines like cranes, ships, aeroplanes, and so on are found to consist of numerous parts or members connected together. These members are connected in such a way so as to perform a specific function and to withstand externally applied loads.

Applied loads on a member may effect an *axial tension* or *oval* or *axial compression* or *bending* or a *torsion* (i.e. twist) or a combination of one or more of these. These four basic types of loading of a member are frequently encountered in both structural and machine design problems. They essentially constitute the principal subject matter of strength of materials.

1.2 CONTINUUM

Molecules within substances are in constant motion and collide with each other. The number of molecules involved in a substance is immense. In engineering problems, we are not interested in the motion of individual particles but in the overall or bulk property although the property arises from its molecular structure.

If the microscopic nature of matter is not directly considered but the properties of a substance are assumed to exist as a continuous distribution of mass, then this idealized substance is called a continuum. This hypothetical continuum concept emphasizes that the conditions at a point are the average of a very large number of molecules surrounding that point within a reasonable distance. Thus, in a continuum the gross effects of the actions of the molecules and the atoms are conveyed by the concepts of density, pressure, temperature, etc., which are of interest in engineering applications.

1.2.1 Rigid Solid Body

A rigid solid body may be defined as a body in which the relative positions of any two arbitrary points is invariant under any condition. That is, a body may be considered as a rigid body if the deformation between its parts is negligible or there is no change in the linear or angular dimensions during the course of its analysis.

1.2.2 Deformable Solid Body

A deformable solid body may be defined as a body which may change its linear or angular dimensions or deform under certain conditions. Deformation may be due to application of force, pressure, torque, temperature, etc. However, such applications may be temporary, permanent, instantaneous or continuous.

It is true that all the solid bodies in nature are essentially deformable to some extent. However, the magnitude of deformation may be insignificant. For example, the deformation of rubber may be appreciable, whereas the deformation of a rock may be negligible.

The study of deformable bodies is involved in the determination of strength, stiffness and stability of a body, under varying loading conditions, by experimental or analytical methods, which along with other properties are discussed in the next section.

1.3 EXTERNAL AND INTERNAL FORCES

1.3.1 External Forces

Force may be considered as a measure of interaction between two systems. The interaction may be of mechanical, electrical, electromagnetic or chemical in origin. In this book, the interaction by mechanical origin alone is discussed. Mechanical interaction may be through contact with other members or due to the effects of fluid pressure, gravity or inertia.

A system (body) in a specified origin surrounded by other systems is considered. The body can be isolated from the actions of all the outside systems by representing such actions as a set of forces. These forces are denoted as external forces. Figure 1.1 represents such an isolated body and $\vec{E}_1, \vec{E}_2, \vec{E}_3, \ldots \vec{E}_{n-1}, \vec{E}_n$ are the external forces acting on the system.

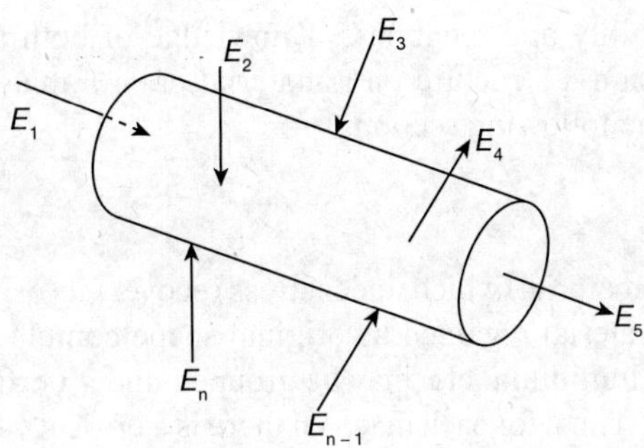

Fig. 1.1 Isolated body subjected to external forces.

External forces comprise of both body and surface forces. *Body forces* are distributed over the entire volume of the body and are expressed as force per unit volume. Examples of body forces include gravitational forces, electrostatic forces, inertia forces, magnetic forces, etc.

Surface forces are the forces which act through a contact surface on the boundary of the body and are expressed as force per unit area. Examples of surface forces are hydrostatic pressure, surface tension, friction, etc.

External forces may also be classified as static, quasi-static or dynamic based on the time of load application. Static forces are those which do not change their magnitude, direction or point of application with respect to time. Quasi-static forces are those forces which may vary at a small rate, whereas dynamic forces are those forces which change with time and may vary at a fast rate.

1.3.2 Internal Forces

In a solid at a stable equilibrium, molecules are spaced at a specific space when the net force is zero. To resist any change in spacing of the molecules, they have to either move closer, i.e., causing an attraction, or move apart, i.e., causing a repulsion. This fundamental concept applies very well to a system which consists of a very large number of molecules.

Thus, in a system, at the time of building up of internal resistance to balance the external forces, molecules do change their mutual positions resulting in a change of spacing between them. This process effects deformation in the system. Such forces of interaction between molecules which are brought into action to resist deformation (caused due to external forces) are called internal forces. Considering a system consisting of several bodies, it is observed that the internal forces not only arise in the interior of all the bodies but also between the bodies interacting in the system.

1.4 PROPERTIES OF SOLID MATERIALS

Engineering materials have different properties, viz., physical, mechanical, chemical, thermal, electrical, etc. Among these properties, physical and mechanical properties are very important for construction of materials.

Physical properties include density, porosity, water absorption, thermal conductivity and permeability, whereas mechanical properties are elasticity, plasticity, strength, abrasion, hardness, ductility, brittleness, malleability and toughness. Knowledge of both the physical and mechanical properties are needed to design a structure on a material or of a material. Some of the important properties are discussed in the following sections.

1.4.1 Elasticity

Elasticity is the property of a material, which under stress recovers its original shape after the removal of the external load. If the material regained its original shape completely, then it is said to be *perfectly elastic*. Steel, copper, aluminium, etc. may be grouped under perfectly elastic materials within certain limits of deformation. Thus, for each material there is a critical value of load, generally known as *elastic limit*, which marks the partial breakdown of elasticity. Loading the material beyond this point leads to permanent deformation.

1.4.2 Plasticity

Plasticity is the property of a material by which a strained material retains the deformed position even after the removal of the external load which caused the deformation. Under large loads or forces, most of the materials become plastic.

1.4.3 Ductility

Ductility is the property of a material by which it can be drawn into a wire by external forces. Thus, a ductile material can withstand large deformation before failure. During the process of extension, a ductile material may show a certain degree of elasticity together with a degree of plasticity to considerable extent. Source of the ductile materials include copper, aluminium, gold, etc.

1.4.4 Brittleness

Brittleness is the property of a material by which it is not capable of undergoing a significant deformation due to the application of an external load but breaks or ruptures suddenly. This is the most undesirable property of a construction material. Some of the brittle materials include glass, porcelain, etc.

1.4.5 Malleability

Malleability is the property of a material by which it can be uniformly lengthened or widened by hammering or rolling without rupture. A malleable material possesses a high degree of plasticity. This property has a wide use in forging, hot rolling, drop stamping, etc. Some of the malleable materials include wrought iron, copper, mild steel, etc.

1.4.6 Strength

Strength is the property of a material determined by the maximum stress that the material can withstand prior to failure. Based on the nature of loading and the nature of stress the strength is defined. There is no unique value which can define strength in all cases.

For example, an adequately designed structural member or machine element is not expected to fail under normal operating conditions. This is ensured when the material of the member is strong enough to withstand the forces imposed on it.

1.4.7 Hardness

Hardness of a material is the ability of a material to resist penetration by a hard material or object. The hardest material is diamond and the one with least hardness is talcum.

1.4.8 Toughness

Toughness is the property of a material which enables the material to absorb energy without fracture. This is a very useful property of a material which is applicable in cyclic or instantaneous loading.

1.4.9 Stiffness

Stiffness is the property that enables a material to withstand high stress without large deformation. Stiffness of a material depends on its elastic property.

For example, in a structural member or machine element large deformations are undesirable and the material should be stiff enough to withstand the load.

1.4.10 Stability

Stability is the overall property of a member made out of a material to maintain the overall equilibrium preventing complete collapse.

For example, a component member made out of a particular material is just long enough to prevent a buckling when subjected to a force acting along its axis.

1.5 STRESSES AND STRAINS

When a body is acted upon by external forces or loads, internal forces develop and the body is said to be in a state of stress. Considering the conditions of equilibrium, the resultant of internal forces must be equal to the resultant of external forces.

Thus, if P is the external force acting over a body of area of cross section A, then the force per unit area, p is given as

$$p = \frac{P}{A} \tag{1.1}$$

This force per unit area is called the *intensity of stress* or for brevity, the *stress*. Thus, stress can be defined as the force of resistance per unit area offered by a body against deformation. The unit of stress in SI Units is N/mm^2.

The ratio of change of dimension, due to an application of external load, to the original dimension is called *strain*. Strain is a non-dimensional quantity.

1.5.1 Tensile Stress and Strain

When an external force causes an elongation of a body in the direction of loading, then the force is called as a *tensile force* (P_t) and the corresponding stress is called *tensile stress* (p_t) (Figure 1.2), i.e.,

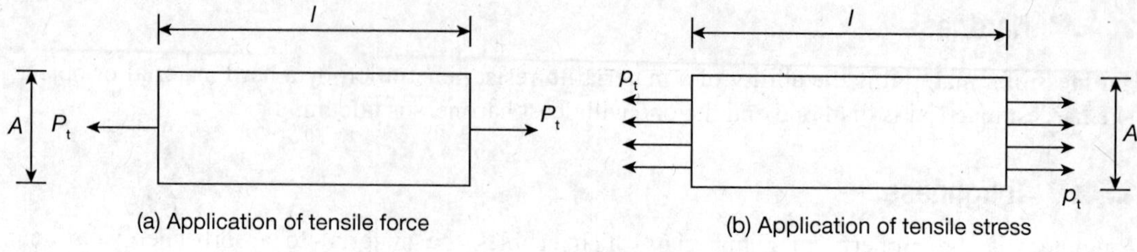

Fig. 1.2 Tensile force and stress.

$$\text{Tensile stress } (p_t) = \frac{\text{Tensile force}}{\text{Area of cross section}}$$

i.e.,
$$p_t = \frac{P_t}{A} \tag{1.2}$$

If the elongation or increase in length δl is caused due to external force over the original length, l, then

$$\text{Tensile strain } (e_t) = \frac{\text{Increase in length}}{\text{Original length}}$$

$$e_t = \frac{\delta l}{l} \tag{1.3}$$

Both the tensile stress and tensile strain are considered as positive.

1.5.2 Compressive Stress and Strain

When an external force causes a shortening of a body in the direction of loading, then the force is called as a compressive force (P_c) and the corresponding stress is called as compressive stress (p_c), i.e.,

$$\text{Compressive stress } (p_c) = \frac{\text{Compressive force}}{\text{Area of cross section}}$$

$$p_c = \frac{P_c}{A} \tag{1.4}$$

If the shortening or decrease in length δl is caused due to external force over the original length, l, then

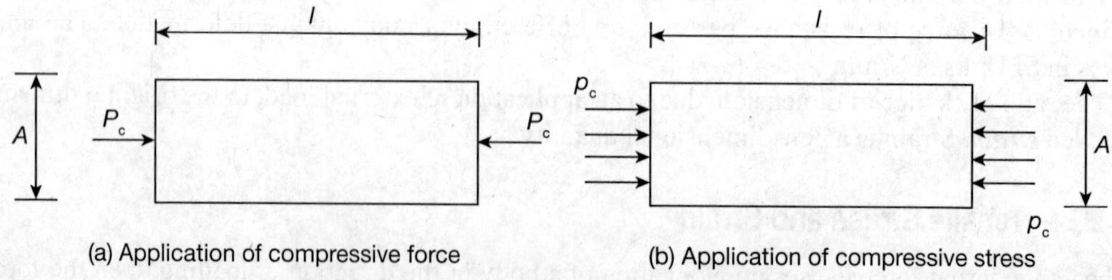

Fig. 1.3 Compressive force and stress.

$$\text{Compressive strain } (e_c) = \frac{\text{Decrease in length}}{\text{Original length}}$$

$$e_c = \frac{\delta l}{l} \tag{1.5}$$

Both the compressive stress and the compressive strain are considered negative. Tensile and compressive stresses are also referred to as direct stresses.

1.5.3 Shear Stress and Strain

When an external force is acting along the section of a body, then the force is termed as shear force (Q) and the corresponding stress is called the shear stress (q)

$$\text{Shear stress } (q) = \frac{\text{Shear force}}{\text{Area under shear}}$$

$$q = \frac{Q}{A} \tag{1.6}$$

The block ABCD is considered. Figure 1.4(a) shows the position of block before displacement. Let the top surface be displaced transversely while the bottom surface is fixed. Let the new position of AB be A_1B_1 (Figure 1.4(b)), i.e., the block ABCD has distorted to a position A_1B_1CD through the angle θ. Then

$$\text{Shear strain } (e_s) = \frac{\text{Transverse displacement}}{\text{Distance from the fixed surface}}$$

i.e.,
$$\text{Shear strain } (e_s) = \frac{\delta l}{h} = \tan\theta$$

i.e.,
$$(e_s) = \theta \text{ since } \theta \text{ is very small} \tag{1.7}$$

Thus, the shear strain may be defined as the angle in radius through which a body is distorted by a shear force.

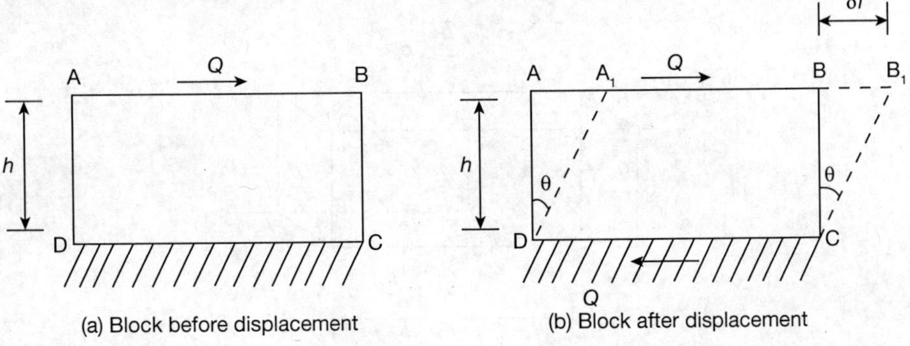

Fig. 1.4 Shear force and displacement.

1.5.4 Volumetric Strain

Volumetric strain is defined as the change in volume δV of an elastic body (of volume, V) due to external forces in unit original volume.

Then

$$\text{Volumetric strain } (e_v) = \frac{\text{Change in volume}}{\text{Original volume}}$$

i.e.,
$$e_v = \frac{\delta V}{V} \tag{1.8}$$

A rectangular bar of dimensions $l \times b \times d$ subjected to triaxial system of stresses be considered (Figure 1.5).

Original volume, $V = l \times b \times d$

Change in volume, $\delta V = \text{(Final volume)} - \text{(Original volume)}$

$$= (l \times b \times d + b \times d \times \delta l - l \times b \times \delta d - l \times d \times \delta b) - l \times b \times d$$

where δl, δb and δd are the changes in the respective dimensions.

Change in volume, $\delta V = b \times d \times \delta l - l \times b \times \delta d - l \times d \times \delta b$

$\therefore$ Volumetric strain,
$$e_v = \frac{\delta V}{V}$$
$$= \frac{\delta l}{l} - \frac{\delta d}{d} - \frac{\delta b}{b}$$

i.e.,
$$e_v = e_x + e_y + e_z$$

where e_x, e_y and e_z are the strains in three directions.

Let the stresses in three mutually perpendicular directions be p_x, p_y and p_z, then

$$e_x = \frac{p_x}{E} - \frac{p_y}{mE} - \frac{p_z}{mE}$$

$$e_y = \frac{p_y}{E} - \frac{p_x}{mE} - \frac{p_z}{mE}$$

$$e_z = \frac{p_z}{E} - \frac{p_x}{mE} - \frac{p_y}{mE}$$

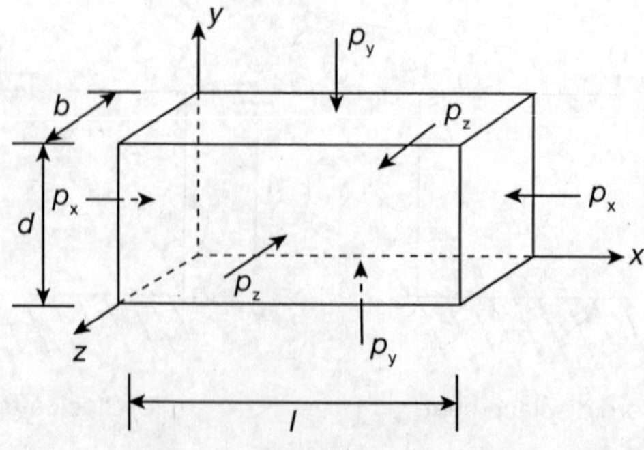

Fig. 1.5 Triaxial system of stresses.

where $1/m = \mu$, Poisson's ratio (discussed elsewhere). Adding all strains

$$e_x + e_y + e_z = \frac{1}{E}\left(p_x + p_y + p_z\right) - \frac{2}{mE}\left(p_x + p_y + p_z\right)$$

$$= \frac{1}{E}\left(p_x + p_y + p_z\right)\left(1 - \frac{2}{m}\right)$$

Volumetric strain = Total strain

i.e.,
$$\frac{\delta V}{V} = \frac{p_x + p_y + p_z}{E}\left(1 - \frac{2}{m}\right) \tag{1.9}$$

For uniaxial stress system

$$\left.\begin{aligned}
&\left.\begin{array}{l}\text{in } x\text{-direction}\\ (p_y = p_z = 0)\end{array}\right\} \rightarrow \quad \frac{\delta V}{V} = \frac{p_x}{E}\left(1 - \frac{2}{m}\right)\\
&\left.\begin{array}{l}\text{in } y\text{-direction}\\ (p_x = p_z = 0)\end{array}\right\} \rightarrow \quad \frac{\delta V}{V} = \frac{p_y}{E}\left(1 - \frac{2}{m}\right)\\
&\left.\begin{array}{l}\text{in } z\text{-direction}\\ (p_x = p_y = 0)\end{array}\right\} \rightarrow \quad \frac{\delta V}{V} = \frac{p_z}{E}\left(1 - \frac{2}{m}\right)
\end{aligned}\right\} \tag{1.10}$$

1.5.5 Lateral Strain

When a body is subjected to a single vertical force acting along its axis, it not only deforms in the direction of loading (i.e., longitudinally) but also in the lateral direction. Under a tensile force, the lateral dimension decreases while under compressive force it increases. Hence,

$$\text{Lateral strain} = \frac{\text{Change in lateral dimension}}{\text{Original lateral dimension}}$$

1.5.6 Poisson's Ratio

Poisson's ratio is the ratio of the lateral strain to the axial strain (strain in the direction of loading) which is a constant for a material when the material is stressed within the elastic limit. This ratio is denoted as μ or $1/m$. It varies from 0.25 to 0.33 or in other words the value of m varies from 3 to 4.

1.6 HOOKE'S LAW

It has been well established that many structural materials up to a certain load level exhibit a direct proportionality of strain with the stress produced in it. This is usually stated as stress is proportional to strain.

Thus, Hooke's law is defined as the stress induced in a material is proportional to the strain within the elastic limit. This law is applicable both for tension and compression conditions.

This law is valid within certain limits of stress for most ferrous alloys and also with sufficient accuracy to other engineering materials such as timber, concrete and non-ferrous alloys.

1.7 STRESS–STRAIN DIAGRAM

The behaviour of a material at different stages of loading is represented by a diagram called as stress–strain diagram. A stress–strain diagram of ductile material like mild steel is a conventional diagram to understand fully the load-deformation behaviour of a material.

A ductile material subjected to a tensile test passes through different stages before reaching the failure stage. Figure 1.6 shows a typical stress–strain diagram obtained from a tension test conducted on a mild steel bar.

It may be observed that in the initial stages of loading (from the origin O to the point A), stress is proportional to strain which is depicted by a straight line. In this region, the material regains its original shape after the applied load is completely released and accordingly the point A is referred to as the *proportionality limit* or *limit of proportionality.*

From A to B, the strain is not proportional to stress but the material retains its original shape after the complete removal of the load. The stress corresponding to the point B is called the *elastic limit.* The portion OAB represents a straight line curve, the strain produced is proportional to the stress that produced it, i.e., it obeyed Hooke's law.

Beyond the point B, the deformation is not recoverable; hence, the material experiences a permanent deformation called *permanent set.* It is the beginning of a *semiplastic stage*, i.e., the yielding stage and the strain is very large even for a small increase of stress.

In the diagram, C′ and C represent upper and lower yield points, respectively. The stress corresponding to the load at the point C is called the *yield stress.* The portion CD represents *plastic yielding*, i.e., it is the strain which occurs after the yield point with no increase in stress.

The stressing of the material beyond the point D causes a substantially large strain for a very small stress increase. The point E corresponds to *maximum stress* or *ultimate stress.* From the point D up to

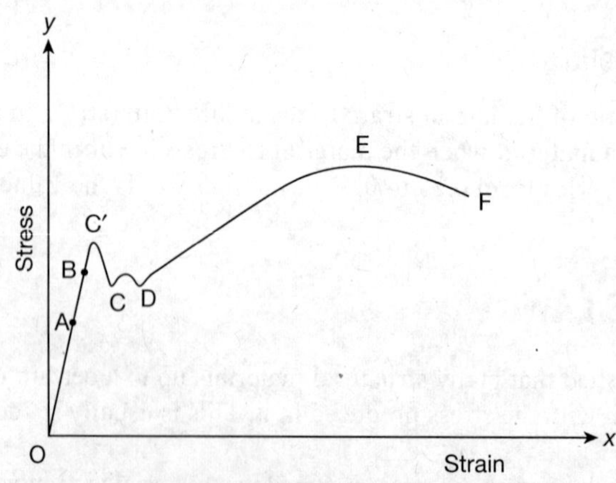

Fig. 1.6 Typical stress–strain diagram.

point E, the cross-sectional area of the bar decreases in proportion with increase in length and at the point E a waist is formed in the bar.

Beyond the point E, the extension continues without any increase in stress and the material breaks at the stress corresponding to point F. This stress is referred to as *breaking stress* which is less than the ultimate stress.

1.8 ALLOWABLE STRESS

Most of the materials begin to exhibit some non-linearity between stress and strain when loaded beyond a certain limit. With this non-linear behaviour of the materials, the problem of assessing deformation and distribution of internal forces in structural members and machine element become much more complicated compared with a material with linear stress–strain behaviour. To have the structure in an elastic condition and to remove the possibility of permanent set, it is usual practice to keep the stress in the material in the elastic range such that permanent set is not caused.

With the knowledge of the limit of proportionality, the yield point and the ultimate stress of the material, it is possible to fix for a given field problem, the magnitude of the stress which may be considered as a safe stress. This stress is referred to as the *allowable stress* or *working stress.*

While deciding the allowable stress, the following factors have to be considered apart from the material of which the member is made up of:

(i) the type of loading, viz., static or dynamic,
(ii) the nature of load application, viz., repetitive or normal,
(iii) the consequence of the failure of the member and
(iv) the accuracy of the theoretical analysis.

Allowable stress is obtained by dividing the ultimate stress or the yield stress of the material by a suitable factor called the *factor of safety.* It is the practice to mention whether the ultimate stress or yield stress is adopted in arriving at the allowable stress.

In the case of structural steel, it is logical to take the yield stress as the basis because beyond this a considerable permanent set may develop which is not permissible in engineering structures. For brittle materials such as cast iron, concrete, stone, wood, etc., the ultimate stress is taken as a basis for determining the working stress. Generally, a factor of safety of two will give a conservative value for a static loading and larger factors of safety may have to be used in cases of suddenly applied loads or variable loads.

SOLVED PROBLEM 1.1

A rod of length 300 mm and diameter 20 mm is subjected to an axial force of 20 kN. The modulus of elasticity of the material is 2×10^5 N/mm^2 and Poisson's ratio is 0.28. Calculate the changes in the dimension of the rod and the volume of the rod.

Solution:

Given data: Length of rod, $l = 300$ mm, Poisson's ratio = 0.28, diameter of rod, $d = 20$ mm and $E = 2 \times 10^5$ N/mm^2.

Area of rod, $A = \dfrac{\pi \times 20^2}{4} = 314.16 \text{ mm}^2$

Change in length of the rod, $\delta l = \dfrac{PL}{AE}$

$$\delta l = \frac{20 \times 1,000 \times 300}{314.16 \times 2 \times 10^5} = 0.0955 \text{ mm}$$

Longitudinal strain, $e = \dfrac{0.0955}{300} = 0.000318$

Poisson's ratio, $\mu = \dfrac{\text{Lateral strain}}{\text{Longitudinal strain}}$

$\therefore$ Lateral strain, $e_1 = 0.28 \times 0.000318 = 0.000089$

Change in the volume of the rod $= \dfrac{\pi}{4} \times (20 - 0.00178)^2 \times (300 - 0.0955)$

$= 94,201.008 \text{ mm}^3$

Net change in volume $= (314.16 \times 300) - 94,201.008$

$= 46.99 \text{ mm}^3$

SOLVED PROBLEM 1.2

A steel rod 50 mm in diameter and 3 m long is subject to a tension of 100 kN. Up to what length the bar should be bored centrally so that the total extension of the rod will increase by 25% under the same pull, the bore being 30 mm in diameter?

$E = 200$ GP.

Solution:

Given data: Diameter of rod, $d = 50$ mm, length of rod, $l = 3$ m, tension, $P = 100$ kN and $E = 200$ GP.

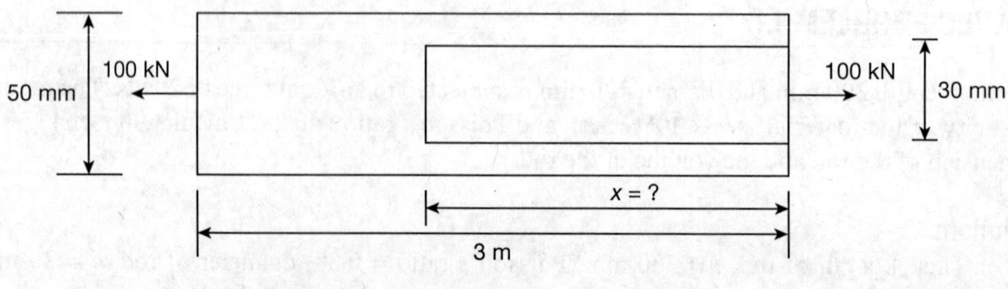

Fig. 1.7

Stress in solid rod, $p = \dfrac{100 \times 1,000}{\dfrac{\pi}{4} \times 50^2}$

$= 50.92\ \text{N/mm}^2$

Elongation of solid rod, $\delta l = \dfrac{p \times l}{E} = \dfrac{50.92 \times 3,000}{200 \times 10^3}$

$= 0.76\ \text{mm}$

Elongation after the rod is bored $= 1.25 \times 0.76 = 0.95\ \text{mm}$

Let the bar be bored to a length x

Area at the reduced section $= \dfrac{\pi}{4}\left(50^2 - 30^2\right)$

$= 1,256.64\ \text{mm}^2$

Stress in the reduced section, $p' = \dfrac{100 \times 1,000}{1,256.64} = 79.57\ \text{N/mm}^2$

∴ Elongation of the rod $= \dfrac{p\left(3,000 - x\right)}{E} + \dfrac{p'x}{E} = 0.95\ \text{mm}$

$$= \frac{50.92(3,000 - x)}{200 \times 10^3} + \frac{79.57x}{200 \times 10^3} = 0.95$$

Solving $x = 1,299.83\ \text{mm}$

∴ The rod has to be bored up to a length of 1,299.83 mm

SOLVED PROBLEM 1.3

A brass bar, having a cross-sectional area of 900 mm² is subjected to axial forces as shown in Figure 1.8. Find the total elongation of the bar. Take $E = 1 \times 10^5\ \text{N/mm}^2$.

Solution:

Given data: Area of bar = 90 mm² and $E = 1 \times 10^5\ \text{N/mm}^2$.

Free-body diagram of the bar is drawn as given below (Figure 1.9).

Let δl_1, δl_2 and δl_3 be the changes in lengths of *PQ*, *QR* and *RS*, respectively:

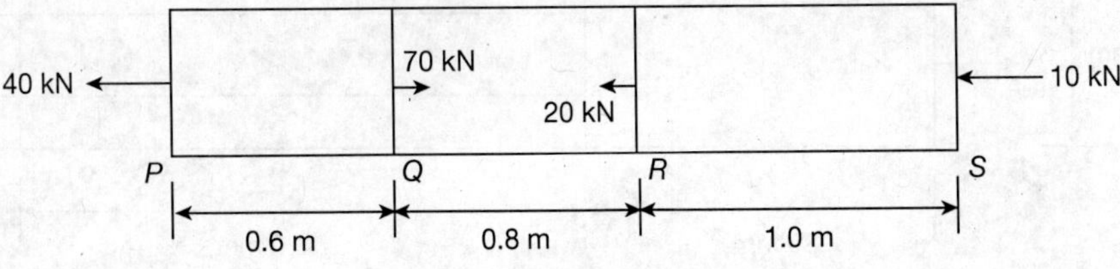

Fig. 1.8

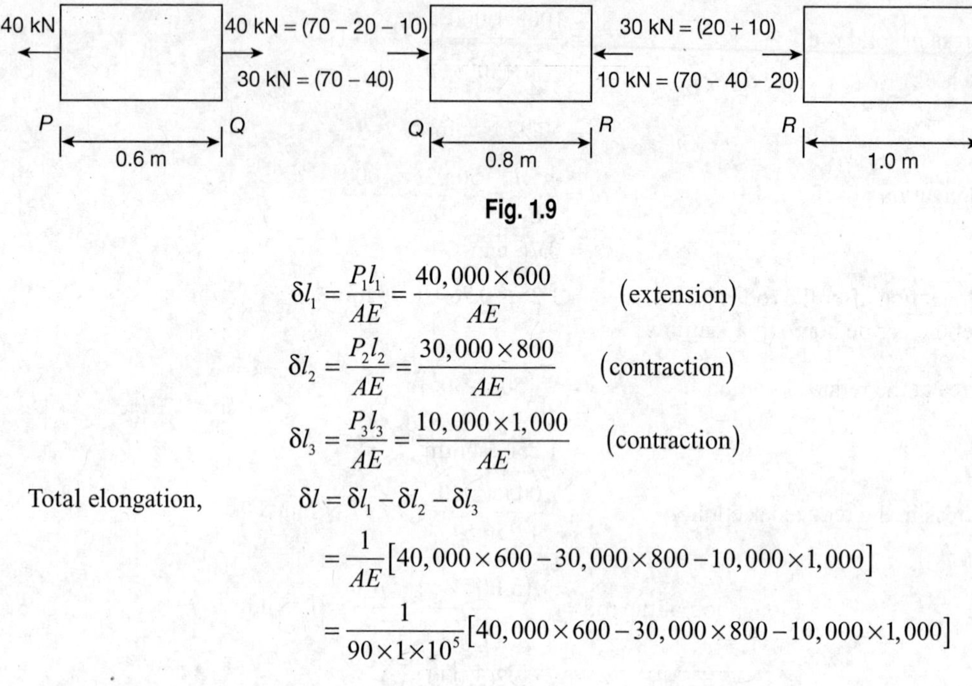

Fig. 1.9

$$\delta l_1 = \frac{P_1 l_1}{AE} = \frac{40{,}000 \times 600}{AE} \quad \text{(extension)}$$

$$\delta l_2 = \frac{P_2 l_2}{AE} = \frac{30{,}000 \times 800}{AE} \quad \text{(contraction)}$$

$$\delta l_3 = \frac{P_3 l_3}{AE} = \frac{10{,}000 \times 1{,}000}{AE} \quad \text{(contraction)}$$

Total elongation,

$$\delta l = \delta l_1 - \delta l_2 - \delta l_3$$

$$= \frac{1}{AE}[40{,}000 \times 600 - 30{,}000 \times 800 - 10{,}000 \times 1{,}000]$$

$$= \frac{1}{90 \times 1 \times 10^5}[40{,}000 \times 600 - 30{,}000 \times 800 - 10{,}000 \times 1{,}000]$$

$$= -1.1 \text{ mm}$$

Negative sign indicates that there will be a decrease in the length of the bar.

SOLVED PROBLEM 1.4

Determine the total strain in a bar made up of 40 mm diameter solid for a length of 80 mm and a hollow circular section of outer diameter 40 mm and inner diameter of 20 mm for a length of 120 mm. Take $E = 200$ kN/mm^2. The axial load is 80 kN. What will be the strain if the bar is solid section for the full length.

Solution:

Given data: $E = 200$ kN/mm^2 and axial load = 80 kN.

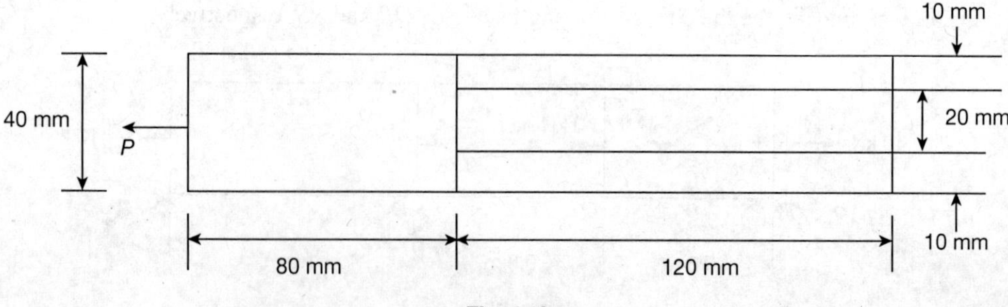

Fig. 1.10

Total change in the length of the bar will be equal to the sum of the changes in sections due to the axial force P.

Total elongation δl will be the sum of individual elongations, i.e.,

$$\delta l = \delta l_1 + \delta l_2 = \frac{P}{E}\left(\frac{l_1}{A_1} + \frac{l_2}{A_2}\right)$$

Here, $l_1 = 80$ mm and $l_2 = 120$ mm

$$A_1 = \frac{\pi}{4}\left(40^2\right) = 1{,}256.6 \text{ mm}^2$$

$$A_2 = \frac{\pi}{4}\left(40^2 - 20^2\right) = 942.48 \text{ mm}^2$$

$$\text{Total strain in hollow tube} = \frac{80 \times 1{,}000}{200 \times 1{,}000}\left(\frac{80}{1{,}256.6} + \frac{120}{942.48}\right)$$

$$= 0.0764 \text{ mm}$$

$$\text{Total strain with solid tube} = \frac{80 \times 1{,}000}{200 \times 1{,}000}\left(\frac{80}{1{,}256.6} + \frac{120}{1{,}256.6}\right)$$

$$= 0.06366 \text{ mm.}$$

1.9 ELASTIC CONSTANTS

Different types of stresses and their corresponding strains within elastic limit are related which are referred to as elastic constants. The three types of elastic constants are:

1. Modulus of elasticity or Young's modulus (E),
2. Bulk modulus (K) and
3. Modulus of rigidity or shear modulus (M, C or G).

1.9.1 Modulus of Elasticity or Young's Modulus

Ratio of stress to strain within the elastic limit is a constant which is defined by Hooke as the Modulus of elasticity or Young's modulus (E):

i.e.,
$$E = \frac{\text{Stress}}{\text{Strain}}$$

i.e.,
$$E = \frac{p}{e} \tag{1.11}$$

where p is the stress due to tension or compression within elastic limit and e is the corresponding strain. Since e is nondimensional, the unit for modulus of elasticity is that of stress, i.e., N/mm^2. Substituting for $p = P/A$ and $e = \delta l/l$, we have the expression for change of length as

$$\delta l = \frac{Pl}{AE} \tag{1.12}$$

1.9.2 Bulk Modulus

When a material is subjected to three direct stresses along the three mutually perpendicular directions with equal intensity (Figure 1.11), then the ratio of the direct stress to the volumetric strain is denoted as the bulk modulus, *K*.

$$K = \frac{\text{Direct stress}}{\text{Volumetric strain}}$$

i.e.,

$$K = \frac{p}{e_v}$$

i.e.,

$$K = \frac{p}{(\delta V / V)} \tag{1.13}$$

As the volumetric strain is nondimensional, the unit of bulk modulus is that of stress, i.e., N/mm^2.

1.9.3 Modulus of Rigidity or Shear Modulus

This is the ratio of shear stress to shear strain. When the material under the elastic limit

i.e.,

$$G = \frac{\text{Shear stress}}{\text{Shear strain}}$$

i.e.,

$$G = \frac{q}{e_\lambda} = \frac{q}{\theta} \tag{1.14}$$

As the shear is nondimensional, the unit of modulus of rigidity is that of stress, i.e., N/mm^2.

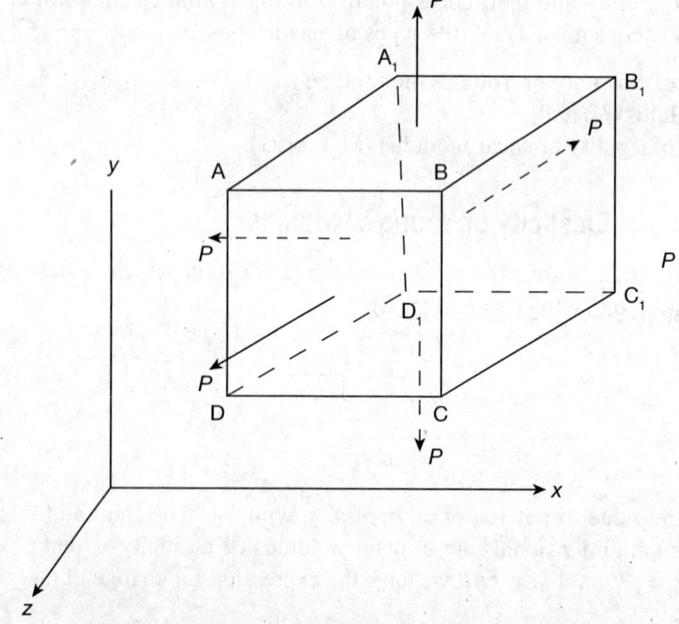

Fig. 1.11 Equal directed stresses on all the faces of a body.

1.9.4 Relation Between Elastic Constants

1. Relation between *E* and *G*

A cube subjected to a shear stress of q on all sides in considered. In Figure 1.12, one face of cube of length l is shown before and after distortion.

Due to the shear stresses, the cube undergoes distortion such that the diagonal BD and AC undergo distortion. The shear stress q causes a shear strain, i.e., the diagonal BD is distorted to B_1 D.

$$\text{Strain of DB} = \frac{DB_1 - DB}{DB} = \frac{B_1B_2}{DB}$$

$$= \frac{BB_1 \cos 45°}{BC \sec 45°} = \frac{BB_1}{2BC}$$

i.e.,
$$e = \frac{\theta}{2} = \frac{q}{2G} \tag{1.15}$$

Thus, the linear strain of the diagonal BD is half of the shear strain and tensile in nature. This also holds good for the diagonal AC.

The shear stresses acting on the sides AB, CD, CB and AD cause a tensile stress on the diagonal BD and a compressive stress on the diagonal AC. Thus, the tensile strain on diagonal BD due to tensile stress on the diagonal BD is q/E and the tensile strain on the diagonal BD due to compressive stress on the diagonal AC is q/mE.

Combined strain on the diagonal is given as

$$e = \frac{q}{E} + \frac{q}{mE} = \frac{q}{E}\left(1 + \frac{1}{m}\right)$$

i.e.,
$$e = \frac{q}{E}\left(\frac{m+1}{m}\right) \tag{1.16}$$

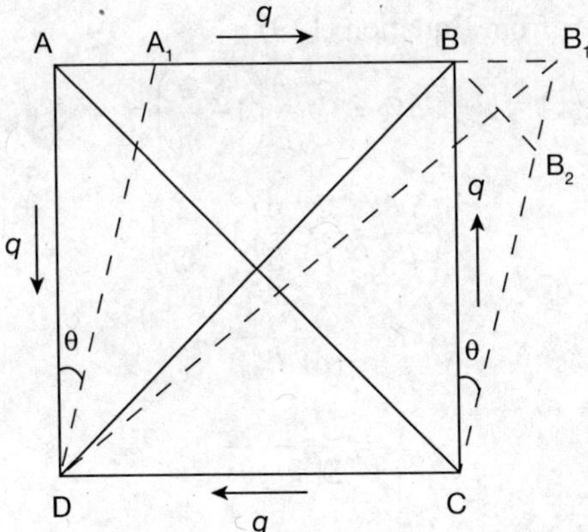

Fig. 1.12 Shear displacement.

Equating Equations (1.15) and (1.16)

$$\frac{q}{2G} = \frac{q}{E}\left(\frac{m+1}{m}\right)$$

i.e.,
$$G = \frac{mE}{2(m+1)} \tag{1.17}$$

i.e.,
$$E = 2G\left(\frac{m+1}{m}\right) \tag{1.18}$$

2. Relation between *E* and *K*

A cube of side length l subjected to three mutually perpendicular tensile stresses of equal intensity is considered (Figure 1.7).

Consequent to stresses acting on all the faces of the cube, every side undergoes deformation due to all the stresses. Thus, the side AB undergoes the following deformations:

(i) a tensile strain of p/E due to the stress on the faces BB_1C_1C and AA_1D_1D;
(ii) a compressive lateral strain of p/mE due to the stress on the faces AA_1B_1B and DD_1C_1C and
(iii) a compressive lateral strain of p/mE due to the stress on the faces ABCD and $A_1B_1C_1D_1$.

Therefore, the net tensile strain of the side AB is given as

$$e = \frac{\delta l}{l} = \frac{p}{E} - \frac{p}{mE} - \frac{p}{mE}$$
$$e = \frac{p}{E}\left(1 - \frac{2}{m}\right) \tag{1.19}$$

Original volume of the cube $V = l^3$

Differentiating V with respect to l, we have

$$\frac{\delta V}{\delta l} = 3l^2$$

i.e.,
$$\delta V = 3l^2 \delta l$$

Substituting the value of δl from Equation (1.19)

$$\delta V = 3l^2 \frac{p}{E}\left(1 - \frac{2}{m}\right)$$

∴
$$\frac{\delta V}{V} = \frac{3l^2}{l^3}\frac{p}{E}\left(1 - \frac{2}{m}\right)$$

$$K = \frac{p}{(\delta V/V)} \tag{1.20}$$

$$K = \frac{mE}{3(m-2)}$$

or
$$E = 3K\left(\frac{m-2}{m}\right)$$

or
$$E = 3K\left(1 - \frac{2}{m}\right) \tag{1.21}$$

3. Relation between *E*, *K* and *G*

Considering Equations (1.17) and (1.20) and eliminating *m*, we get

$$E = \frac{9KG}{G + 3K} \tag{1.22}$$

Similarly, considering Equations (1.18) and (1.21) and eliminating *E* we get

$$\mu = \frac{1}{m} = \frac{3K - 2G}{6K + 2G} \tag{1.23}$$

SOLVED PROBLEM 1.5

A bar of 30 mm diameter is subjected to a pull of 60 kN. The measured extension of the gauge length of 200 mm is 0.09 mm and the change in diameter is 0.0039 mm. Calculate the Poisson's ratio and the value of the three modules.

Solution:

Given data: Diameter of bar = 30 mm, pull P = 60 kN, gauge length l = 200 mm, extension δl = 0.09 mm and change in diameter, δd = 0.0039 mm.

Applied stress $= \dfrac{60 \times 1{,}000}{(\pi \times 30^2)/4} = 84.88 \text{ N/mm}^2$

Axial strain $= \dfrac{0.09}{200} = 0.00045$

Lateral strain $= \dfrac{0.0039}{30} = 0.00013$

Poisson's ratio, $\mu = \dfrac{0.00013}{0.00045} = 0.289$

Modulus of elasticity, $E = \dfrac{84.88}{0.00045} = 1.89 \times 10^5 \text{ N/mm}^2$

Modulus of rigidity, $G = \dfrac{mE}{2(m+1)}$

$$= \frac{(1/0.289) \times 1.89 \times 10^5}{2[(1/0.289) + 1]} = 0.73 \times 10^5 \text{ N/mm}^2$$

Bulk modulus, $K = \dfrac{mE}{3(m-2)}$

$$= \frac{(1/0.289) \times 1.89 \times 10^5}{3[(1/0.289) - 2]} = 1.49 \times 10^5 \text{ N/mm}^2$$

SOLVED PROBLEM 1.6

Calculate the modulus of rigidity and bulk modulus of the cylindrical bar of diameter 25 mm and of length 1.5 m, if the longitudinal strain in the bar during a tensile test is four times the lateral strain. $E = 1.5 \times 10^5$ MPa.

Solution:

Given data: Diameter of bar, $d = 25$ mm, $\mu = 0.25$, length of bar, $l = 1.5$ m and $E = 1.5 \times 10^5$ MPa.

$$\text{Longitudinal strain} = 4 \times (\text{Lateral strain})$$

Poisson's ratio,

$$\mu = \frac{\text{Lateral strain}}{\text{Longitudinal strain}} = \frac{1}{4}$$

$$\mu = 0.25$$

i.e.,

$$\frac{1}{m} = 0.25$$

Now,

Using the relationship,

$$E = 2G\left(1 + \frac{1}{m}\right)$$

$$1.5 \times 10^5 = 2G\,(1 + 0.25)$$

Rearranging and solving,

$$G = 6 \times 10^4 \text{ N/mm}^2$$

Using the relationship,

$$E = 3K\left(1 - \frac{2}{m}\right)$$

$$1.5 \times 10^5 = 3K(1 - 2 \times 0.25)$$

Solving for K

$$K = 1 \times 10^5 \text{ N/mm}^2$$

SOLVED PROBLEM 1.7

A rectangular block that is 350 mm long, 100 mm wide and 80 mm thick is subjected to axial load as follows: 50 kN tensile force in the direction of length, 100 kN compressive force in the direction of its thickness and 60 kN tensile force in the direction of its width. Determine (i) change in volume, (ii) bulk modulus and (iii) modulus of rigidity. Take $E = 2 \times 10^5$ N/mm^2 and Poisson's ratio = 0.25.

Solution:

Given data: Length = 350 mm, width = 100 mm, thickness = 80 mm, tensile force = 50 kN, compressive force = 100 kN, tensile force = 60 kN, $E = 2 \times 10^5$ N/mm^2 and Poisson's ratio = 0.25.
As we have discussed earlier, tensile stress and strain are considered as positive and compressive stress and strain are considered as negative.

$p_x = 50$ kN acting over an area of 100 mm × 80 mm
$p_y = 100$ kN acting over an area of 300 mm × 100 mm
$p_z = 60$ kN acting over an area of 300 mm × 80 mm

$$\therefore \quad \text{Stress in } x\text{-direction} = \frac{50 \times 10^3}{100 \times 80} = 6.25 \text{ N/mm}^2$$

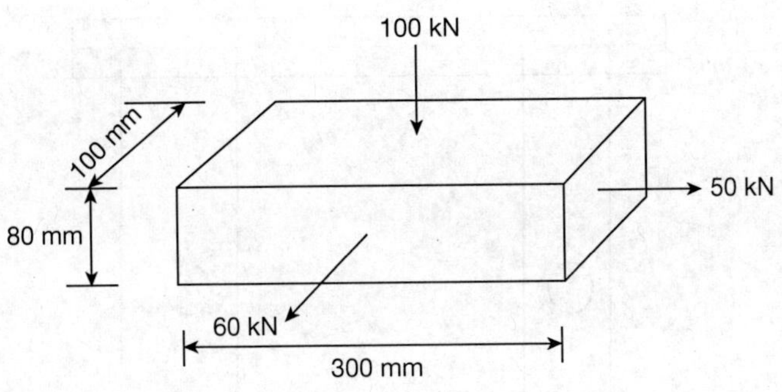

Fig. 1.13

Stress in y-direction $= -\dfrac{100\times 10^3}{300\times 100} = -3.33\ \text{N/mm}^2$

Stress in z-direction $= \dfrac{60\times 10^3}{300\times 80} = 2.5\ \text{N/mm}^2$

Using $\dfrac{\delta V}{V} = \dfrac{1}{E}(p_x - p_y + p_z)\left(1-\dfrac{2}{m}\right)$

$$= \frac{1}{2\times 10^5}(6.25-3.33+2.5)(1-2\times 0.25)$$

$$= 1.355\times 10^{-5}$$

(i) Change in volume $= 1.355\times 10^{-5}\times(300\times 100\times 80)$

$$= 32.52\ \text{mm}^3$$

(ii) Bulk modulus $E = 3K\left(1-\dfrac{2}{m}\right)$

i.e., $2\times 10^5 = 3K(1-2\times 0.25)$

Rearranging, $K = 1.33\times 10^5\ \text{N/mm}^2$

(iii) Modulus of rigidity $E = 2G\left(1+\dfrac{1}{m}\right)$

i.e., $2\times 10^5 = 2G\,(1+1\times 0.25)$

Rearranging, $G = 0.8\times 10^5\ \text{N/mm}^2$

1.10 COMPOSITE SECTIONS

Composite sections may form by combination of two or more bars of equal lengths but of different material rigidity and fixity so as to act as one unit (Figure 1.14). In composite bars, the following factors are considered:

(i) the change in length in each bar and the corresponding strain is equal and
(ii) the external load is shared by the bars based on their material property (i.e., on the modulus of elasticity).

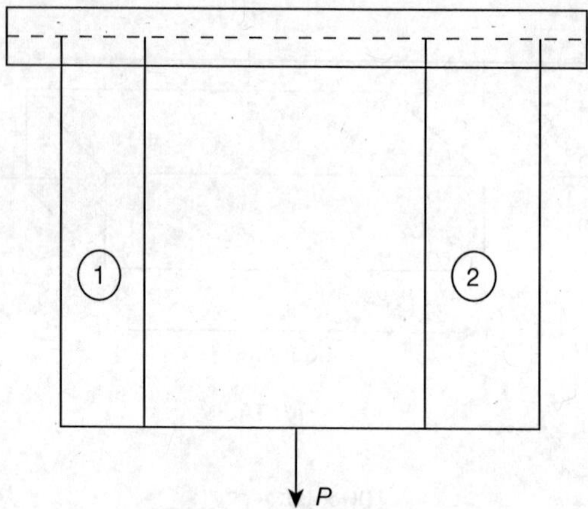

Fig. 1.14 Composite bars.

Considering Figure 1.14, the load P is shared by the bar ① as P_1 and bar ② as P_2. Thus,

$$P = P_1 + P_2 \tag{1.24}$$

Stresses in bars ① and ② are, respectively, as $p_1 = \frac{P_1}{A_1}$ and $p_2 = \frac{P_2}{A_2}$

i.e.,
$$P_1 = p_1 A_1 \tag{1.25}$$
and
$$P_2 = p_2 A_2 \tag{1.26}$$

where A_1 and A_2 are the areas of cross section of bar ① and bar ②, respectively.

Substituting for P_1 and P_2 in Equation (1.24)

$$P = p_1 A_1 + p_2 A_2 \tag{1.27}$$

Strain in bar ①, e_1 and in bar ②, e_2 are

$e_1 = \frac{p_1}{E_1}$ and $e_2 = \frac{p_2}{E_2}$

where E_1 and E_2 are modulus of elasticity of bar ① and bar ②, respectively.

As strain is equal in both the bars, $e_1 = e_2$

i.e.,
$$\frac{p_1}{E_1} = \frac{p_2}{E_2} \tag{1.28}$$

Thus, the load and the stress in bar ① and bar ② can be found out by using Equations (1.25), (1.26) and (1.28).

SOLVED PROBLEM 1.8

A reinforced concrete column 300 mm × 300 mm is reinforced with eight steel rods with a total area of 1,820 mm^2. The column carries an axial load of 400 kN. If the modulus of elasticity of steel is 18 times as that of concrete, find the loads carried by concrete and steel. Find also the adhesive force between the steel and concrete.

Solution:

Given data: Area of column = 300 mm × 300 mm, area of steel = 1,820 mm² and axial load = 400 kN.

Reinforced cement concrete column is a monolithic rigid structure with steel rods embedded in a particular pattern in a mass concrete. Hence, the strain produced is same in the concrete and the steel.

Area of column $= 300 \times 300 = 90{,}000 \text{ mm}^2$

Area of steel rods, $A_s = 1{,}820 \text{ mm}^2$

Area of concrete, $A_c = 90{,}000 - 1{,}820 = 88{,}180 \text{ mm}^2$

The load is shared by both steel and concrete based on the stress and the areas of cross section of each.

$$p_s A_s + p_c A_c = 400 \times 1{,}000 \qquad \text{(i)}$$

where, p_s and p_c are stresses on steel and concrete, respectively.

For equal strain condition, $\dfrac{p_s}{E_s} = \dfrac{p_c}{E_c}$

$$\therefore \quad p_s = \frac{E_s}{E_c} p_c = 18 p_c$$

Substituting for p_s in Equation (i)

$$18 p_c (1{,}820) + p_c \times 88{,}180 = 400{,}000$$

Solving for p_c, $p_c = 3.307 \text{ N/mm}^2$

Then $p_s = 18 \times 3.307 = 59.53 \text{ N/mm}^2$

Average compressive stress $= \dfrac{400{,}000}{90{,}000} = 4.44 \text{ N/mm}^2$

Average load carried by concrete $= 4.44 \times 88{,}180$

$= 391{,}911.1$ N

Actual load carried by concrete $= 3.307 \times 88{,}180$

$= 29{,}166.6$ N

Load carried by steel $= 400{,}000 - 291{,}677$

$= 10{,}283.34$ N

Adhesive force between concrete and steel surface $= 391{,}911.1 - 29{,}166.6$

$= 392{,}744.5$ N.

SOLVED PROBLEM 1.9

A solid steel cylinder 500 mm long and 70 mm in diameter is placed inside an aluminium cylinder having 75 mm inner diameter and 100 mm outer diameter. The aluminium cylinder is 0.15 mm longer than the steel bar. An axial load of 600 kN is applied to the bar and cylinder through rigid cover plates at both ends. Find the stresses and deformations developed in the steel bar and aluminium tube. Assume $E_s = 220$ kN/mm^2 and $E_a = 70$ kN/mm^2.

Solution:

Given data: Steel bar length = 500 mm, diameter of steel bar = 70 mm, diameter of aluminium = 75 mm, $E_s = 220$ kN/mm^2 and $E_a = 70$ kN/mm^2.

The aluminium cylinder is 0.15 mm longer than the steel cylinder.

The cylinder has to be compressed by 0.15 mm (i.e., $\delta l = 0.15$ mm), and the corresponding load has to be found.

$$\delta l = \frac{PC}{AE}$$

$$\therefore \quad P = \frac{EA\delta l}{l} = \frac{70\times 10^3 \times \frac{\pi}{4}(100^2 - 75^2)\times 0.15}{(500+0.15)}$$

$$P = 72{,}136.8 \text{ N}$$

At this stage, the load shared by the aluminium and steel cylinders will be

$$600{,}000 - 72{,}136.8 = 527{,}863.2 \text{ N}$$

Both the cylinders are of the same length and are compressed by the same amount. Then,

Strain in steel cylinder = Strain in aluminium cylinder.

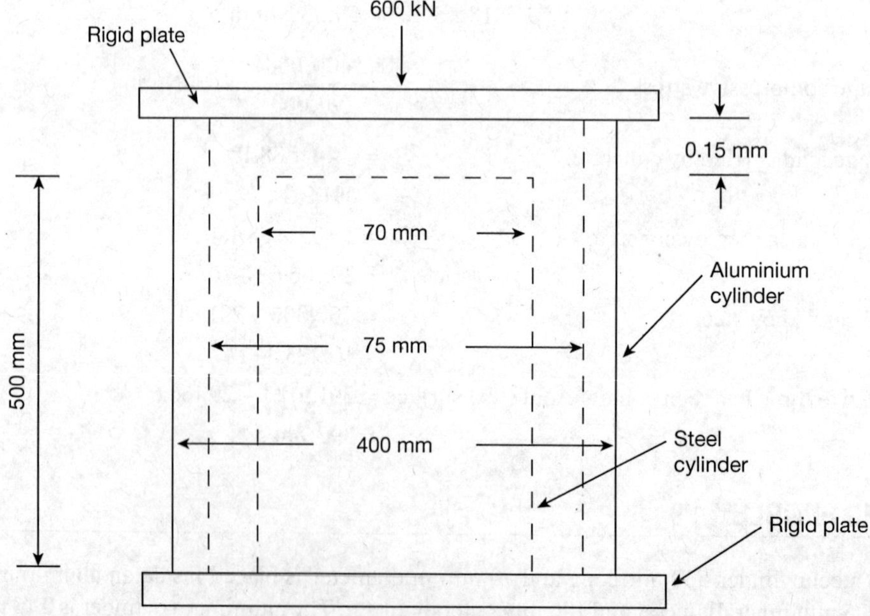

Fig. 1.15

i.e.,
$$\frac{p_s}{E_s} = \frac{p_a}{E_a}$$

or
$$p_s = \frac{E_s}{E_a} p_s = \frac{220 \times 10^3}{70 \times 10^3} p_a$$
$$p_s = \frac{22}{7} p_a$$

where p_s and p_a are stress on steel and concrete, respectively.

The load is taken by both steel and aluminium based on the stress and the area of cross section.

$$p_s A_s + p_a A_a = 527{,}863.2 \text{ N}$$

i.e.,
$$\frac{22}{7} p_a A_s + p_a A_a = 527{,}863.2$$
$$p_a = \frac{527{,}863.2}{\frac{22}{7} A_s + A_a}$$

i.e.,
$$= \frac{527{,}863.2}{\frac{22}{7} \times \frac{\pi}{4} \times 70^2 + \frac{\pi}{4}(100^2 - 75^2)}$$

i.e.,
$$p_a = 33.987 \text{ N/mm}^2$$

and
$$p_s = 33.987 \times \frac{22}{7} = 106.82 \text{ N/mm}^2$$

Stress in aluminium cylinder due to load 72,136.8 N

$$= \frac{72{,}136.8}{\frac{\pi}{4}(100^2 - 75^2)}$$
$$= 20.99 \text{ N/mm}^2$$

Total stress in aluminium cylinder $= 33.987 + 20.99$

$$= 54.980 \text{ N/mm}^2$$

SOLVED PROBLEM 1.10

Two copper rods and one steel rod together support a load as shown in Figure 1.16. If the stresses in copper and steel are not to exceed 60 MPa and 120 MPa, respectively, find the safe load that can be supported. Young's modulus for steel is twice that of copper.

Solution:

Given data: Allowable stress in copper = 60 MPa, allowable stress in steel = 120 MPa, copper rod section = 25 mm × 25 mm, steel rod section = 30 mm × 30 mm and $E_s = 2E_c$.

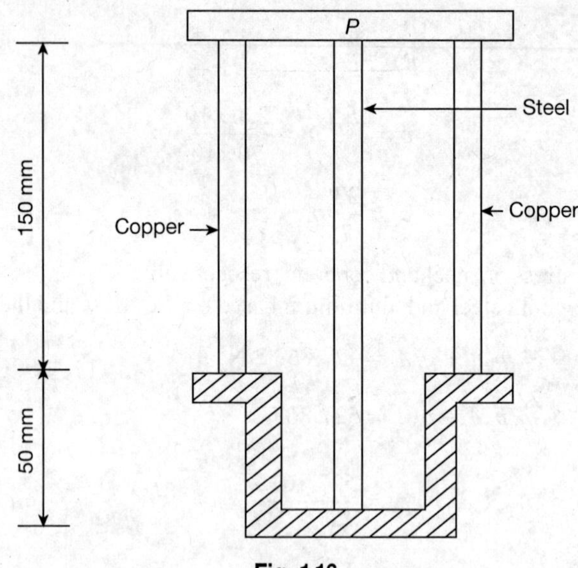

Fig. 1.16

As each rod will be compressed to the same extent, decrease in length of steel

= Decrease in length of copper rods

(Strain in steel rod) × (Length of steel rod) = (Strain in copper rod) × (Length of copper rod)

i.e.,
$$e_s l_s = e_c l_c$$

or
$$\frac{e_s}{e_c} = \frac{l_c}{l_s} = \frac{150}{200} = 0.75$$

Stress in steel = (Strain in steel) $\times E_s$

i.e.,
$$p_s = e_s \times E_s$$

Stress in copper = (Strain in copper) $\times E_c$

$$p_c = e_c \times E_c$$

$$\therefore \quad \frac{p_s}{p_c} = \frac{l_s \times E_s}{l_c \times E_c}$$

$$= 0.75 \times 2 = 1.5$$

Suppose steel is permitted to reach its allowable stress of 120×10^6 N/mm^2, the corresponding stress in copper, $p_c = \frac{120 \times 10^6}{1.5} = 80 \times 10^6$ m/mm^2 which exceeds its allowable stress of 60×10^6 N/mm^2.

Therefore, let copper be allowed to reach its allowable value of 60×10^6 N/mm^2, then the corresponding stress in steel will be $1.5 \times 60 \times 10^6$ N/mm^2, i.e., 90×10^6 N/mm^2 which is less than the allowable stress in steel of 120×10^6 N/mm^2.

$$\therefore \quad \text{Total load} = p_s A_s + p_c A_c$$

$$= 90\times10^6 \times 0.03\times0.03+60\times10^6 \times0.025\times0.025$$

$$= 118,500 \text{ N}$$

$$= 118.5 \text{ kN}$$

SOLVED PROBLEM 1.11

Two vertical rods, one of steel and the other of copper, are each rigidly fixed at the top and are 50 cm apart. Diameter and length of each rod are 20 mm and 4 m, respectively. A cross bar fixed to the rods at the lower ends carries a load of 5 kN, such that the cross bar remains horizontal even after loading. Find the tension in each rod and the position of the load on the bar. $E_s = 2\times10^5$ N/mm^2 and $E_{cu} = 1\times10^5$ N/mm^2.

Solution:

Given data: Spacing of rods = 50 cm, diameter of rod = 20 mm, length of rod = 4 m, load, P = 5 kN, $E_s = 2\times10^5$ N/mm^2 and $E_{cu} = 1\times10^5$ N/mm^2.

$$\text{Area of steel rod} = \frac{\pi}{4}\times20^2 = 314.15 \text{ mm}^2$$

$$\text{Area of copper rod} = \frac{\pi}{4}\times20^2 = 314.15 \text{ mm}^2$$

As the cross bar remains horizontal, the extensions of the steel and copper rods are equal. Further, as these rods have the same length, the strains in these rods are equal.

Strain in steel = Strain in copper

$$\frac{P_s}{E_s} = \frac{p_{cu}}{E_{cu}}$$

i.e.,

$$\frac{p_s}{2\times10^5} = \frac{p_{cu}}{1\times10^5}$$

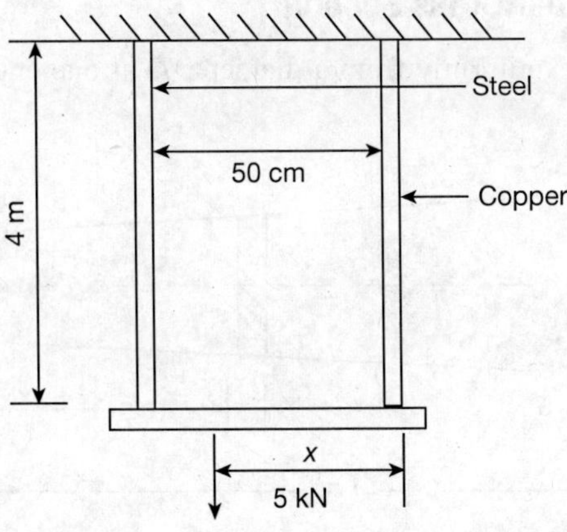

Fig. 1.17

$$\therefore \quad p_s = 2p_{cu}$$

Total load = (Load on steel rod) + (Load on copper rod)

$$5{,}000 = p_s A_s + p_{cu} A_{cu}$$

i.e., $p_s \times 314.15 + p_{cu} \times 314.15 = 5{,}000$

i.e., $2p_{cu} \times 314.15 + p_{cu} \times 314.15 = 5{,}000$

$$p_{cu} = 5.305 \text{ N/mm}^2$$

$$\therefore \quad p_s = 2 \times 5.305 = 10.610 \text{ N/mm}^2$$

Load carried by steel rod $= p_s \times A_s = 10.61 \times 314.15$

$= 3{,}333$ N

Load carried by copper rod $= 5{,}000 - 3{,}333$

$= 1{,}667$ N

Let the load 5 kN be at a position x from the copper rod

$$5{,}000 \times x = p_s \times 50$$

$$x = 33.33 \text{ cm}$$

Position of the load on the bar from the copper rod = 33.33 cm.

1.11 BARS OF VARYING CROSS SECTIONS

In practice, one may come across with bars of varying cross sections (e.g., circular or rectangular). In such cases, it is usual to assume that the load is distributed over the cross section as inversely proportional to the area.

1.11.1 Bars of Circular Cross Section

A bar of length l tapering uniformly from a diameter D at one end to a diameter d at the other is considered (Figure 1.18).

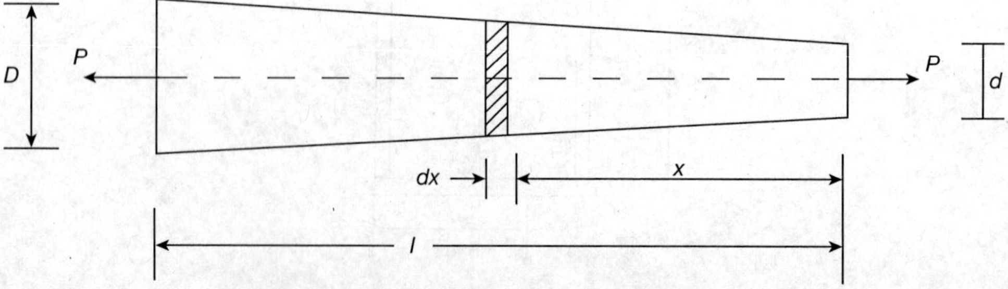

Fig. 1.18 Bar with varying circular cross section.

A small strip of length dx at a distance x from the small-diameter end is considered. Diameter at this level is

$$D_x = d + \frac{(D-d)x}{l} \tag{1.29}$$

Extension of this strip due to the load P acting at the end is given as

$$\delta(dx) = \frac{Pdx}{\left(\frac{\pi D_x^2}{4}\right)E} \tag{1.30}$$

Considering the whole rod, the extension

$$\delta l = \int_0^l \delta(dx) = \int_0^l \frac{4Pdx}{\pi D_x^2 E}$$

i.e.,

$$\delta l = \int_0^l \frac{4Pdx}{\pi\left[d + \frac{(D-d)x}{l}\right]^2 E}$$

$$= -\frac{l}{D-d}\frac{4P}{\pi E}\left[\frac{l}{d+(D-d)\frac{x}{l}}\right]_0^l$$

i.e.,

$$\delta l = \frac{4Pl}{\pi DdE} \tag{1.31}$$

1.11.2 Bars of Rectangular Cross Section

Consider a bar of constant thickness t and uniformly tapering in width B from one end to the width b at the other end (Figure 1.19).

A small strip of length dx at a distance x from the small-width end is considered. Width at this level is

$$B_x = b + \frac{(B-b)x}{l}$$

Area of strip,

$$A_x = \left[b + \frac{(B-b)x}{l}\right]t$$

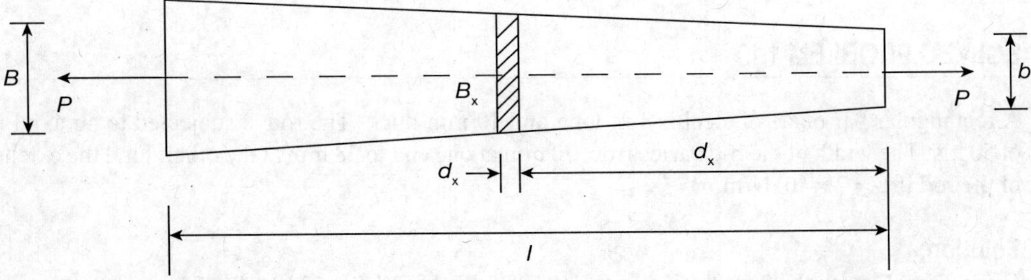

Fig. 1.19 Bar of varying rectangular cross section.

Extension of this strip due to the load P acting at the end is given as

$$\delta(dx) = \frac{Pdx}{A_x E} \tag{1.32}$$

Considering the entire length

$$\delta l = \int_0^l \delta(dx) = \int_0^l \frac{Pdx}{A_x E}$$

$$= \int_0^l \frac{Pdx}{\left[b + \frac{B-b}{l}x\right] tE}$$

Integrating

$$\delta l = \frac{P}{tE}\left(\frac{l}{B-b}\right)\left[\log_e\left(b + \frac{B-b}{l}x\right)\right]_0^l$$

i.e.,

$$\delta l = \frac{Pl}{(B-b)tE}\log_e \frac{B}{b} \tag{1.33}$$

SOLVED PROBLEM 1.12

A circular steel rod tapers uniformly from 4 cm diameter to 1.5 cm diameter in a length of 40 cm. How much the bar is elongated under an axial pull of 40 kN. $E = 2 \times 10^5$ N/mm^2.

Solution:

Given data: $D = 4$ cm, $d = 1.5$ cm, pull, $P = 40$ kN and $E = 2.0 \times 10^5$ N/mm^2.

Elongation, $$\delta l = \frac{4Pl}{\pi DdE}$$

Substituting the values, we have

$$\delta l = \frac{4 \times 40 \times 1{,}000 \times 400}{\pi \times 4 \times 1.5 \times 2.0 \times 10^5}$$

$$\delta l = 0.17 \text{ m}$$

SOLVED PROBLEM 1.13

A rectangular bar made of steel is 3 m long and 10 mm thick. The rod is subjected to an axial load of 50 kN. The width of the rod varies from 70 mm at one end to 28 mm at the other. Find the extension of the rod if $E = 2 \times 10^5$ N/mm^2.

Solution:

Given data: Length, $l = 3$ m, thickness, $t = 10$ mm, axial load, $P = 50$ kN, $B = 70$ mm, $b = 28$ mm and $E = 2 \times 10^5$ N/mm^2.

Extension of the rod, $$\delta l = \frac{Pl}{(B-b)tE}\log_e \frac{B}{b}$$

$$\delta l = \frac{50\times 1,000\times 3\times 1,000}{(70-28)\times 10\times 2\times 10^5}\log_e \frac{70}{28}$$

$$\delta l = 1.785\times 0.916$$

$$\delta l = 1.635 \text{ mm}$$

SOLVED PROBLEM 1.14

An axial pull of 40 kN is acting on a bar consisting of three sections of lengths 300 mm, 250 mm and 200 mm and of diameters 20 mm, 40 mm and 50 mm, respectively. Find (i) the stress in each section and (ii) total extension of the bar. $E = 2 \times 10^5$ N/mm^2.

Solution:

Given data: Diameters, 20 mm/40 mm/50 mm, $E = 2 \times 10^5$ N/mm^2, length, 300 mm/250 mm/200 mm and pull, $P = 40$ kN.

In the case of bars of varying cross sections acted upon by an axial tensile or compressive force P, the total elongation of the bar will be equal to sum of elongation or contraction of each section.

Thus, the total elongation, $\delta l = \delta l_1 + \delta l_2 + \delta l_3$ where δl_1, δl_2 and δl_3 are the elongation of respective sections.

Now, $$A_1 = \frac{\pi}{4}\times 20^2 = 314.16 \text{ mm}^2$$

$$A_2 = \frac{\pi}{4}\times 40^2 = 1,256.64 \text{ mm}^2$$

$$A_3 = \frac{\pi}{4}\times 50^2 = 1,963.50 \text{ mm}^2$$

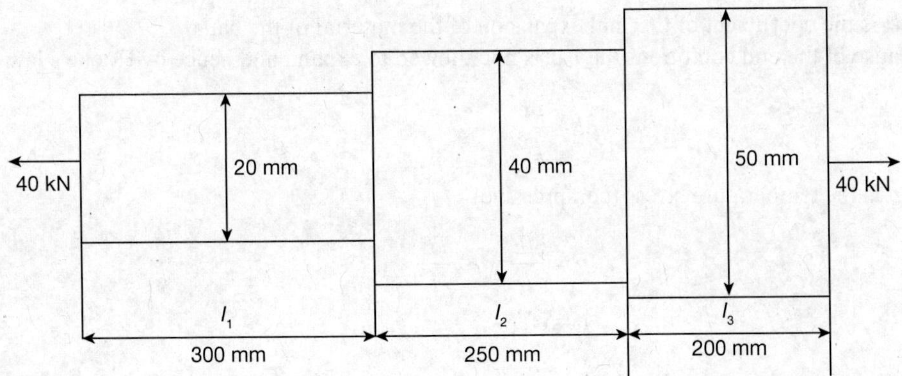

Fig. 1.20

Stress in Each Section

$$\text{Stress in 20 mm bar } = \frac{40 \times 1{,}000}{314.16} = 127.32 \text{ N/mm}^2$$

$$\text{Stress in 40 mm bar } = \frac{40 \times 1{,}000}{1{,}256.64} = 31.84 \text{ N/mm}^2$$

$$\text{Stress in 50 mm bar } = \frac{40 \times 1{,}000}{1{,}963.50} = 20.37 \text{ N/mm}^2$$

Total extension

$$\delta l = \frac{P}{E}\left(\frac{l_1}{A_1} + \frac{l_2}{A_2} + \frac{l_3}{A_3}\right)$$

$$\delta l = \frac{40 \times 1{,}000}{2 \times 10^5}\left(\frac{300}{314.16} + \frac{250}{1{,}256.64} + \frac{200}{1{,}963.50}\right)$$

$$\delta l = 0.251 \text{ mm}$$

1.12 TEMPERATURE STRESSES

Increase or decrease of temperature of a free body causes the body to expand or contract and no stresses are induced. However, if the deformation of the body is constrained, some stresses are induced in the body, and such developed stresses are called temperature stresses which may be tensile or compressive based on either the contraction is prevented or extension is prevented.

A bar whose ends are fixed to rigid supports, so that the expansion is prevented, is considered. Let the length of the bar be l subjected to an increase in temperature $T°$. The expansion of the bar will be

$$\delta l = l\alpha T$$

where α is the coefficient of thermal expansion of the material of the bar.

Because of the end conditions the bar is not allowed to expand and hence by Hooke's law

$$\delta l = \frac{pl}{E}$$

where p is the temperature stress (compressive)

$$\therefore p = \frac{\delta l E}{l} = l\alpha \frac{TE}{l}$$

i.e.,

$$p = \alpha TE \tag{1.34}$$

Let there be a situation wherein the end yields by an amount δe. Then, the amount of expansion prevented is only $(\delta l - \delta e)$:

$$\therefore p = (\delta l - \delta e)\frac{E}{l} \tag{1.35}$$

Figure 1.21 shows a composite bar consisting of two materials, say B and S, subjected to an increase in temperature $T°$. If the members are allowed to expand, no stress will be induced in the members. As the bars are fixed at one end, both the bars as a whole expand by the same amount. Let the coefficient of linear expansion of bar B be more than that of bar S. Then, bar B will expand more than bar S.

When the bars are fixed and not allowed to expand, the expansion of the composite bar as a whole will be less than bar B and more than that of bar S. Hence, the stress induced in bar B will be compressive but the stress in bar S will be tensile. Further, the expansion of one material will be a combination of free expansion of the material plus the expansion or contraction due to the other.

Let A_b, p_b, e_b, α_b and E_b be the area, stress, strain, coefficient of linear expansion and Young's modulus of bar B, respectively.

Similarly, let A_s, p_s, e_s, α_s and E_s be for bar S.

Let δl be the expansion of the composite bar.

Load on bar, B = $p_b A_b$

Load on bar, S = $p_s A_s$

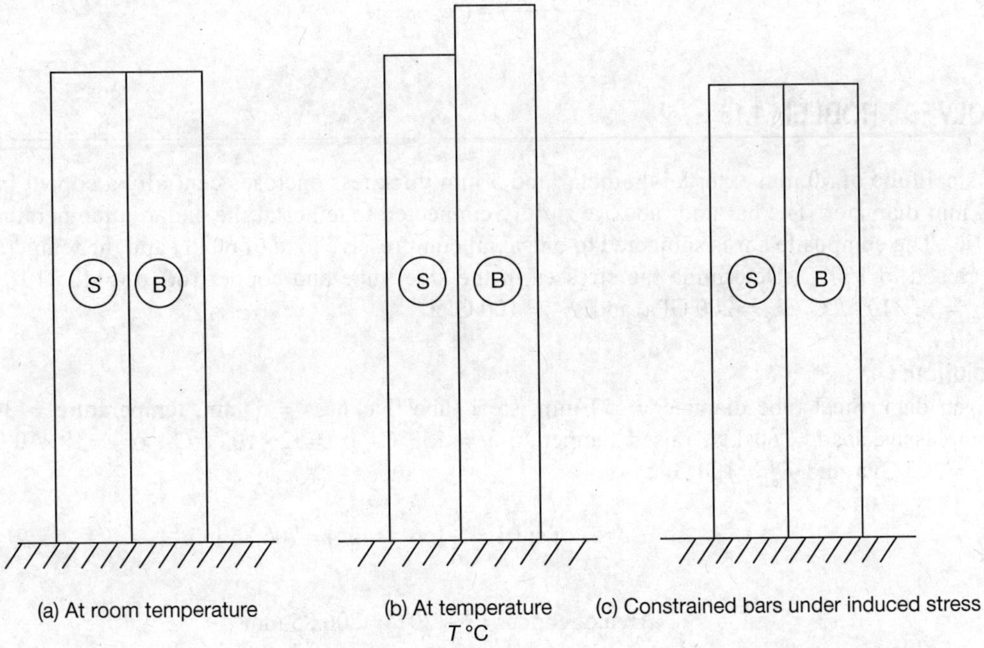

Fig. 1.21 Bars at different temperature conditions.

S – Bar of material, S
B – Bar of material, B

Load on bar B = Load on bar S.

i.e., $$p_b A_b = p_s A_s \tag{1.36}$$

For equilibrium, the contraction in bar B should be equal to the expansion in bar S. In other words, the actual expansion of bar S and bar B is equal.

Actual expansion of bar S = (Free expansion of bar S) + (Expansion due to tensile stress in bar S) (1.37)

$$\text{Actual expansion of bar B} = \alpha_s Tl + \frac{p_s}{E_s} l$$

i.e., (Free expansion of bar B) – (Contraction due to compressive stress induced by bar B)

$$= \alpha_s Tl - \frac{p_b}{E_b} l \tag{1.38}$$

Equating Equations (1.37) and (1.38) we have

$$\alpha_s T + \frac{p_s}{E_s} = \alpha_s T - \frac{p_b}{E_b} \tag{1.39}$$

i.e., $$\frac{p_s}{E_s} + \frac{p_b}{E_b} = (\alpha_b - \alpha_s)T \tag{1.40}$$

SOLVED PROBLEM 1.15

A steel tube of 50 mm external diameter and 5 mm thickness encloses centrally a copper bar of 30 mm diameter. The bar and tube are rigidly connected together at the end at a temperature of 30°C. The composite bar is subjected to one axial compressive load of 60 kN and the temperature is raised to 150°C. Determine the stresses in the steel tube and copper rod $\alpha_\lambda = 12\times10^{-6}/°C$, $\alpha_{cu} = 18\times10^{-6}/°C$, $E_s = 200$ GPa and $E_{cu} = 100$ GPa.

Solution:

Given data: Steel tube diameter = 50 mm, steel tube thickness = 5 mm, temperature = 30°C, compressive load = 60 kN, raised temperature = 150°C, $\alpha_s = 12\times10^{-6}/°C$, $\alpha_{cu} = 18\times10^{-6}/°C$ $E_s = 200$ GPa and $E_{cu} = 100$ GPa.

$$\text{Area of steel} = \frac{\pi}{4}\left(50^2 - 40^2\right) = 706.85 \text{ mm}^2$$

$$\text{Area of copper} = \frac{\pi}{4}(30)^2 = 706.85 \text{ mm}^2$$

Temperature stress due to rise in temperature

When the temperature rises, both copper and steel bars expand. Since the coefficient of thermal expansion of copper is more than that of the steel, the copper rod shall expand more than the steel tube.

As the two bars are joined together, the copper rod will be subjected to compressive stress and the steel rod will be subjected to tensile stress.

Let p_s and p_{cu} be the stresses in steel and copper, respectively.

$$\text{Total tension in steel} = \text{Total compression in copper}$$

i.e.,
$$p_s A_s = p_{cu} A_{cu}$$

i.e.,
$$p_s \times 706.5 = p_{cu} \times 706.5$$

i.e.,
$$p_s = p_{cu}$$

Actual expansion of steel = Actual expansion of copper.

(Free expansion of steel) + (Expansion due to tensile stress) = (Free expansion of copper) – (Contraction due to compressive stress).

i.e.,
$$\alpha_s .T.L + \frac{p_s}{E_s} \times L = \alpha_{cu} \times T \times L - \frac{p_{cu}}{E_{cu}} \times L$$

i.e.,
$$\alpha_s T.L + \frac{p_s}{E_s} = \alpha_{cu} T - \frac{p_{cu}}{E_{cu}}$$

Substituting the respective values

$$12 \times 10^{-6} \times 120 + \frac{p_s}{E_s} = 18 \times 10^{-6} \times 120 - \frac{p_{cu}}{E_{cu}}$$

As $p_s = p_{cu}$

$$12 \times 10^{-6} \times 120 + \frac{p_s}{2 \times 10^5} = 18 \times 10^{-6} \times 120 - \frac{p_s}{100 \times 10^3}$$

Solving, $p_s = 48$ N/mm^2 (tensile)

and $p_{cu} = 48$ N/mm^2 (compressive)

Stresses due to external compressive load

Let p'_s and p'_{cu} be the stresses due to external loading in steel and copper, respectively.

Then
$$\frac{p'_s}{E_s} = \frac{p'_{cu}}{E_{cu}}$$

i.e.,
$$\frac{p'_s}{2 \times 10^5} = \frac{p'_{cu}}{1 \times 10^5}$$

i.e.,
$$p'_s = 2p'_{cu}$$

But

$$\text{(Load on steel)} + \text{(Load on copper)} = \text{Total load}$$

$$p'_s A_s + p'_{cu} A_{cu} = 60 \times 10^3$$

i.e.,
$$2' p'_{cu} \times 706.5 + p'_{cu} 706.5 = 60 \times 10^3$$

Solving $p'_{cu} = 28.3 \text{ N/mm}^2 \text{(compressive)}$

$p'_s = 56.6 \text{ N/mm}^2 \text{(compressive)}$

Final Stresses

Stress in copper = 48 + 28.3 = 76.3 N/mm^2 (compressive)

Stress in steel = 56.6 – 48 = 8.6 N/mm^2 (compressive)

SOLVED PROBLEM 1.16

A steel rod of 50 mm diameter and 6 m length is connected to two grips and the rod is maintained at a temperature of 100°C. Determine the stress and pull exerted when the temperature falls to 20°C, if (i) the ends do not yield, and (ii) the end yield by 0.15 cm. Take $E = 2 \times 10^5$ and $\alpha = 12\times10^{-6}/°\text{C}$.

Solution:

Given data: Diameter of rod = 50 mm, length of rod = 6 m, constant temperature = 100°C, $E = 2 \times 10^5$, and $\alpha = 12\times10^{-6}/°\text{C}$.

Area of the rod, $A = \frac{\pi}{4}\times 50^2 = 1{,}963 \text{ mm}^2$

Fall of temperature, $T = 100 - 20 = 80°\text{C}$.

When the ends do not yield

$$\text{Temperature stress} = \alpha \times T \times E$$
$$= 12 \times 10^{-6} \times 80 \times 2 \times 10^5 = 192$$
$$= 192 \text{ N/mm}^2 \text{ (tensile)}$$

$$\text{Pull in the rod} = \text{Stress} \times \text{Area}$$
$$= 192 \times 1{,}963$$
$$= 376{,}896 \text{ N}$$

When the ends yield by 0.15 cm

$$\text{Temperature stress} = \frac{\alpha TL - \delta l}{L}\times E$$
$$= \left(\frac{12\times10^{-6}\times 80\times 6{,}000 - 1.5}{6{,}000}\right)\times 2\times10^5$$
$$= \left(\frac{5.76 - 1.5}{6{,}000}\right)\times 2\times10^3$$
$$= 142 \text{ N/mm}^2$$

$$\text{Pull in the rod} = 142 \times 1{,}963$$
$$= 278{,}746 \text{ N}$$

SOLVED PROBLEM 1.17

The stepped bar shown in Figure 1.22 is stress free at 25°C. Determine the stresses in steel and copper if the temperature of the bar is raised to 55°C. A_{cu} = 100 mm², $A_s = 200$ mm², $E_s = 2\times10^5$ MPa, $E_{cu} = 1\times10^5$ MPa, $\alpha_\lambda = 12\times10^{-6}$ /°C and $\alpha_{cu} = 18\times10^{-6}$ /°C.

Solution:

Given data: Constant temperature = 25°C, $A_{cu} = 100$ mm², $A_s = 200$ mm², $E_s = 2\times10^5$ MPa, $E_{cu} = 1\times10^5$ MPa, $\alpha_s = 12\times10^{-6}$ /°C, $\alpha_{cu} = 18\times10^{-6}$ /°C and raised temperature = 55°C.

If the temperature of a bar fixed between two rigid supports is changed, then stresses are developed in the bar due to the prevention of free expansion or contraction of the bar.

Due to increase in temperature of the bar, the bar will expand and exert a pressure on the wall as the expansion is prevented. This prevention will exert an equal and opposite pressure on the bar.

Here, for the given condition because of rise in temperature the two portions will be in compression as the temperature tends to expand the bar.

Temperature rise, $T = 55 - 25 = 30$°C.

Elongation in copper $= \alpha_{cu}\, T \times l_{cu}$

$= 18 \times 10^{-6} \times 30 \times 1{,}500 = 0.81$ mm

Elongation in steel $= \alpha_s + l_s$

$= 12 \times 10^{-6} \times 30 \times 1{,}000 = 0.36$ mm

For equilibrium, $f_{cu}A_{cu} = f_s A_s$

i.e., $f_{cu} \times 100 = f_s \times 200$

$f_{cu} = 2f_s$

Strain in copper $= \dfrac{f_{cu}}{E_{cu}}$

Strain in steel $= \dfrac{f_s}{E_s}$

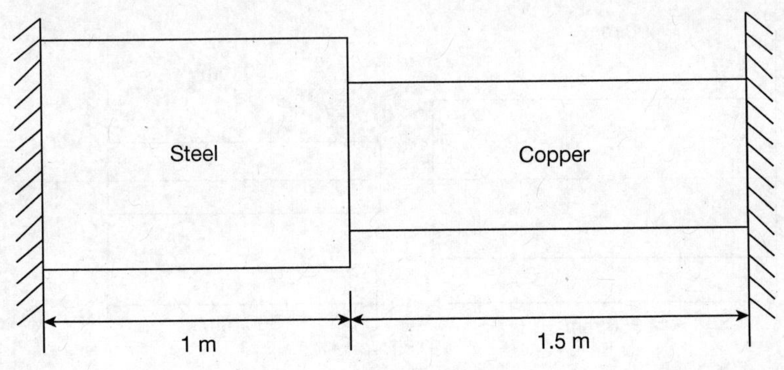

Fig. 1.22

Contraction in copper $= \dfrac{f_{cu}}{E_{cu}} \times l_{cu}$

Contraction in steel $= \dfrac{f_s}{E_s} \times l_s$

Total contraction due to prevention of free expansion = 0.81 + 0.36 mm
= 1.17 mm

$$\therefore \quad \frac{f_{cu}}{E_{cu}} \times l_{cu} + \frac{f_s}{E_s} \times l_s = 1.17$$

$$\frac{f_{cu}}{E_{cu}} \times l_{cu} + \frac{f_{cu}}{2E_s} \times l_s = 1.17$$

Substituting

$$\frac{f_{cu}}{1 \times 10^5} \times 1,500 + \frac{f_{cu}}{2 \times 2 \times 10^5} \times 1,000 = 1.17$$

Solving for f_{cu},

$$f_{cu} = 66.85 \text{ N/mm}^2$$

Stress in steel, $f_s = \dfrac{66.85}{2} = 33.43 \text{ N/mm}^2$

SOLVED PROBLEM 1.18

At a room temperature of 20°C, a gap of 0.6 m exists between the ends of rods A and B as shown in Figure 1.23. Both the rods A and B are aluminium. Area of cross section of rod A is 400 mm² and that of rod B is 200 mm². Determine the stresses in rods A and B if temperature of both rods is raised by 100°C. $E_{al} = 70$ GPa and $\alpha_{al} = 23 \times 10^{-6}$/°C.

Solution:

Given data: Room temperature = 20°C, area of rod A = 400 mm², area of rod B = 200 mm².

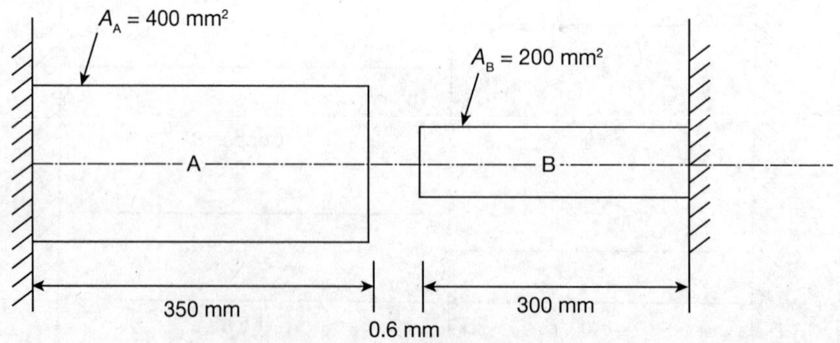

Fig. 1.23

Raised temperature = 100°C, $E_{al} = 70$ GPa and $\alpha_{al} = 23\times10^{-6}$/°C.

Change in temperature, $(T_2 - T_1) = 100 - 20 = 80$°C. If both the rods are free to expand, then the free expansion in both the rods $= (350 + 300) \times 80 \times 23 \times 10^{-6}$
$= 1.196$ mm

But the gap between the rods is only 0.6 mm.

∴ Free expansion = 1.196 – 0.6 = 0.596 mm which will produce compressive force in rods A and B.

For equilibrium

$$f_A \times A_A = f_B \times A_B$$
$$f_A \times 400 = f_B \times 200$$
$$f_A = 0.5 f_B$$

Change in length (due to compressive force) in rod $A = \frac{f_A}{E} \times l_A$

Change in length in rod $B = \frac{f_B}{E} \times l_B$

Total change in length,

$$\frac{f_A}{E} \times 350 + \frac{f_B}{E} \times 300 = 0.596 \text{ mm}$$

i.e.,
$$\frac{0.5 f_B}{E} \times 350 + \frac{f_B}{E} \times 300 = 0.596 \text{ mm}$$

Solving, $f_B = 87.83$ N/mm^2 and $f_A = 43.92$ N/mm^2.

SOLVED UNIVERSITY QUESTIONS

SOLVED PROBLEM 1.19

Determine the value of Young's modulus and Poisson's ratio of a metallic bar of length 300 mm, width 40 mm and depth 40 mm, when the bar is subjected to an axial compressive force of 400 kN, the decrease in length is given as 0.75 mm and increase in width as 0.03 mm (Anna Univ., June 2006, CE).

Solution:

Given data: Length, l = 300 mm, width, b = 40 mm, depth, d = 40 mm, decrease in length, $\delta l = 0.75$ mm, load, $P = 400$ kN and increase in width, $\delta b = 0.03$ mm.

Young's modulus,
$$E = \frac{\text{Stress}}{\text{Strain}} = \frac{Pl}{A\delta l}$$
$$= \frac{400\times10^3\times300}{(40\times40)\times0.75} = 1\times10^5 \text{ N/mm}^2$$

$$\text{Linear or longitudinal strain} = \frac{\delta l}{l} = \frac{0.75}{300} = 0.0025$$

$$\text{Lateral strain} = \frac{\delta b}{b} = \frac{0.03}{40} = 0.00075$$

$$\text{Poisson's ratio} = \frac{\text{Lateral strain}}{\text{Linear strain}}$$

$$\therefore \quad = \frac{0.00075}{0.0025} = 0.30$$

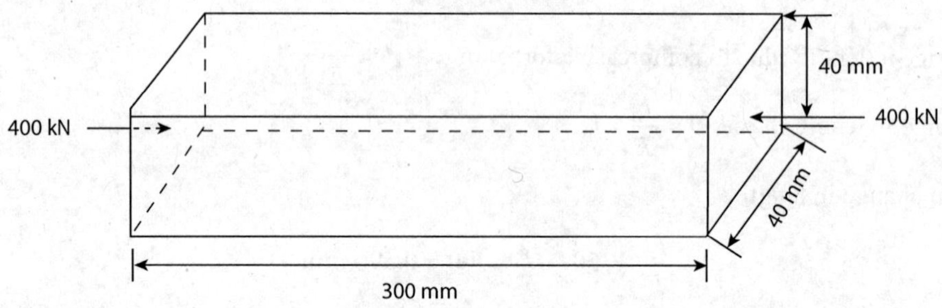

Fig. 1.24

SOLVED PROBLEM 1.20

A bar 30 mm in diameter was subjected to a tensile load of 54 kN and measured extension on 300 mm gauge length was 0.12 mm and change in diameter was 0.00366 mm. Calculate the ratio and the Poisson's values of three elastic moduli (Anna Univ., May 2006, ME).

Solution:

Given data: Diameter, $d = 30$ mm, load, $P = 54$ kN, length, $l = 300$ mm, change in length, $\delta l = 0.112$ mm and change in diameter, $\delta d = 0.00366$ mm.

$$\text{Poisson's ratio} = \frac{\text{Lateral strain}}{\text{Linear strain}}$$

$$\text{Lateral strain} = \frac{\delta b}{b} \text{ or } \frac{\delta d}{d}$$

$$= \frac{0.00366}{30} = 1.22 \times 10^{-4}$$

$$\text{Longitudinal strain} = \frac{\delta l}{l} = \frac{0.112}{300}$$

$$= 3.73 \times 10^{-4}$$

$\therefore$ Poisson's ratio $= \dfrac{1.22\times10^{-4}}{4.04\times10^{-4}} = 0.30$

Young's modulus, $E = \dfrac{\text{Tensile stress}}{\text{Tensile strain}}$

$$= \frac{(54\times1,000)/\left(\frac{\pi}{4}\times30^2\right)}{3.73\times10^{-4}}$$

$$E = 2.05\times10^5 \text{ N/mm}^2$$

Then, $E = 2G\left(1+\dfrac{1}{m}\right)$

Substituting the known values

$$2.05\times10^5 = 2\times G\,(1+0.3)$$

Rearranging

Rigidity modulus $G = \dfrac{2.05\times10^5}{2\times1.3} = 78{,}846.15 \text{ N/mm}^2$

Again, $E = 3K\left(1-\dfrac{2}{\text{m}}\right)$

Substituting the known values

$$2.05\times10^5 = 3K\left[1-2(0.3)\right]$$

Rearranging Bulk modulus, $K = \dfrac{2.05}{3\times0.4} = 1.71\times10^5 \text{ N/mm}^2$

SOLVED PROBLEM 1.21

For a given material, Young's modulus is 1×10^5 N/mm^2 and modulus of rigidity is 4×10^4 N/mm^2. Find the bulk modulus and lateral contraction of a round bar of 50 mm diameter and 2.5 m length when stretched by 2.5 mm (Anna Univ., June 2007, CE).

Solution:

Given data: $E = 1\times10^5$ N/mm^2, $G = 4\times10^4$ N/mm^2, diameter of bar = 50 mm and length of bar = 2.5 m.

$$G = \frac{mE}{2(m+1)},$$

Then $4\times10^4 = \dfrac{m(1\times10^5)}{2m+2}$

i.e., $8m\times10^4 - m\times10^5 = 8\times10^4$

Rearranging,

$$m = \frac{8\times10^4}{0.2\times10^5} = 4$$

Bulk modulus, $$K = \frac{mE}{3(m-2)} = \frac{4\times1\times10^5}{3(4-2)} = 0.67\times10^5 \text{ N/mm}^2$$

Longitudinal strain $$= \frac{2.5}{2.5\times1,000} = 0.001$$

Lateral strain = (Poisson's ratio) × (Longitudinal strain)

$$= \frac{1}{4}\times0.001 = 0.00025$$

Lateral contraction in the diameter of rod = (Lateral strain) × (Original diameter)

$$= 0.00025 \times 50$$

$$= 0.0125 \text{ cm.}$$

SOLVED PROBLEM 1.22

A compound tube consists of a steel tube of 140 mm internal diameter and 5 mm thickness and an outer brass tube of 150 mm internal diameter and 5 mm thickness. The two tubes are of the same length. Compound tube carries an axial load of 600 kN. Find the stresses carried by each tube and the amount of shortening. Length of the tube is 120 mm, $E_s = 2\times10^5$ N/mm^2 and $E_b = 1\times10^5$ N/mm^2 (Anna Univ., May 2009, ME).

Solution:

Given data: Internal diameter of steel tube = 140 mm, internal diameter of brass tube = 150 mm, thickness = 5 mm, axial load = 600 kN, length of tubes, l = 120 mm, $E_s = 2\times10^5$ N/mm^2 and $E_b = 1 \times 10^5$ N/mm^2.

Internal diameter of steel tube = 140 mm

Outer diameter of steel tube = 140 + 5 + 5 = 150 mm

Area of steel tube, $$A_s = \frac{\pi}{4}(150^2 - 140^2) = 2,277.65 \text{ mm}^2$$

Internal diameter of brass tube = 150 mm

External diameter of brass tube = 150 + 5 + 5 = 160 mm.

Area of brass tube, $$A_b = \frac{\pi}{4}(160^2 - 150^2) = 2,434.73 \text{ mm}^2$$

Axial load carried by compound tube = 600 kN = 600 × 10^3 N.

Strain in steel tube = Strain in brass tube

i.e., $$\frac{p_s}{E_s} = \frac{p_b}{E_b}$$

$$\therefore p_s = \frac{E_s}{E_b}\times p_b = \frac{2\times10^5}{1\times10^5} p_b = 2p_b$$

$$\text{Axial load} = (\text{Load on steel}) + (\text{Load on brass})$$

$$p_s A_s + p_b A_b = 600{,}000$$

$$2 \times p_b \times 2{,}277.50 + p_b \times 2{,}434.73 = 600{,}000$$

$$\therefore \quad p_b = 85.84 \text{ N/mm}^2$$

and $\qquad p_s = 2 \times 85.84 = 171.71 \text{ N/mm}^2$

Decrease in length of the compound tube = Decrease in length of either of the tubes

= Decrease in the length of the steel tube

$$= \frac{p_s}{E_s} \times l$$

$$= \frac{171.71}{2 \times 10^5} \times 120$$

$$= 0.103 \text{ mm.}$$

SOLVED PROBLEM 1.23

A steel tube of 20 mm internal diameter and 30 mm external diameter encases a copper rod of 15 mm diameter to which it is rigidly joined at each end. If the temperature of the assembly is raised by 80°C, calculate the stresses produced in the tube. $E_s = 2 \times 10^5$ N/mm², $E_c = 1 \times 10^5$ N/mm², Coefficient of linear expansion of steel and copper are 11×10^{-6} per °C and 18×10^{-4} per °C, respectively (Anna Univ., May 2009, ME).

Solution:

Given data: Internal diameter of steel tube = 20 mm, external diameter of steel tube = 30 mm, diameter of copper rod = 15 mm and raised temperature = 80°C.

$$E_s = 2 \times 10^5 \text{ N/mm}^2, \; E_c = 1 \times 10^5 \text{ N/mm}^2$$

$$\alpha_s = 11 \times 10^{-6} \text{ per °C and } \alpha_c = 18 \times 10^{-4} \text{ per °C.}$$

Area of copper rod, $\qquad A_c = \frac{\pi}{4} \times 15^2 = 176.71 \text{ mm}^2$

Area of steel tube, $\qquad A_s = \frac{\pi}{4} \times (30^2 - 20^2) = 392.7 \text{ mm}^2$

As α_c is more than that of the steel, the copper rod would expand more than the steel tube. Further as the two are joined, copper will be prevented from expanding its full amount and will be put under compression and the steel under tension.

For equilibrium,

$$\text{Compressive load on copper} = \text{Tensile load on steel}$$

$$p_c A_c = p_s A_s$$

$$p_c = \frac{392.7}{176.62} p_s = 2.22 p_s$$

Actual expansion of copper = Actual expansion of steel

= (Free expansion of copper) – (Contraction due to compressive stress)

$$= \alpha_c Tl - \frac{p_c}{E_c} l$$

Similarly,

Actual expansion of steel = (Free expansion of steel) + (Expansion due to tensile stress)

$$= \alpha_s Tl - \frac{p_s}{E_s} l$$

Equating

$$\alpha_c Tl - \frac{p_c}{E_c} l = \alpha_s Tl + \frac{p_s}{E_s} l$$

i.e.,

$$\alpha_c T - \frac{p_c}{E_c} l = \alpha_s T + \frac{p_s}{E_s}$$

Substituting,

$$18 \times 10^{-6} \times 80 - \frac{2.22 p_s}{1 \times 10^5} = 11 \times 10^{-6} \times 80 + \frac{p_s}{2 \times 10^5}$$

Solving, we get

$$p_s = 20.588 \text{ N/mm}^2$$

and

$$p_c = 2.22 \times p_s \text{ N/mm}^2$$

$$= 2.22 \times 20.588 \text{ N/mm}^2$$

$$= 45.71 \text{ N/mm}^2$$

SOLVED PROBLEM 1.24

A steel rod of 25 mm diameter is placed inside a copper tube of 30 mm internal diameter and 5 mm thickness and the ends are rigidly connected. The assembly is subjected to a compressive load of 250 kN. Determine the stresses induced in the steel rod and copper tube. Take the modulus of elasticity of steel and copper as 200 GPa and 80 GPa, respectively (Anna Univ., Nov. 2006, ME).

Solution:

Given data: Diameter of steel rod = 25 mm, internal diameter of copper tube = 30 mm, thickness = 5 mm, total load = 250 kN,

$$E_{\text{steel}} = 200 \text{ GPa}, \ E_{\text{copper}} = 80 \text{ GPa}.$$

Steel

Area of steel rod,

$$A_s = \frac{\pi \times 25^2}{4} = 490.87 \text{ mm}^2$$

Copper tube

External diameter = (Internal diameter) + 2 × (Thickness)

$$= 30 + 2 \times 5 = 40 \text{ mm}$$

Area of copper tube, $A_c = \dfrac{\pi \times (40^2 - 30^2)}{4}$

$= 549.5 \text{ mm}^2$

Total load = (Load on steel rod) + (Load on copper tube).

i.e., $P = P_s + P_c$

As the ends are rigidly connected, strain is same in both, then

$$\frac{P_s}{E_s} = \frac{P_c}{E_c}$$

$$\therefore \quad P_s = \frac{E_s}{E_c} P_c$$

$$P_s = \frac{200 \times 10^3}{80 \times 10^3} = P_c$$

$$P_s = 2.5\, P_c$$

Substituting load in terms of stress, then

$$P = p_s A_s + p_c A_c$$

i.e., $250{,}000 = 2.5 \times p_c \times 490.2 + p_c \times 549.5$

i.e., $250{,}000 = 1{,}775\, p_c$

i.e., $p_c = 140.84 \text{ N/mm}^2$

$\therefore$ Stress in copper, $p_c = 140.84 \text{ N/mm}^2$

Stress in steel, $p_s = 2.5 \times 140.84$

$= 352.11 \text{ N/mm}^2$

SOLVED PROBLEM 1.25

A rod is 3 m long at a temperature of 15°C. Find the expansion of the rod, when the temperature is raised to 95°C. If this expansion is prevented, find the stress in the material of the rod. Take $E = 1 \times 10^5$ N/mm² and $\alpha = 1.2 \times 10^{-5}$/°C (Anna Univ., 2008, ME).

Solution:

Given data: Length, $l = 3$ m, $E = 1 \times 10^5$ N/mm², Initial temperature = 15°C, raised temperature = 95°C and $\alpha = 1.2 \times 10^{-5}$/°C.

Increase in temperature = 95° – 15° = 80°C

Expansion of the rod, $\delta l = \alpha T l$

$= 1.2 \times 10^{-5} \times 80 \times 3 \times 1{,}000 = 2.88 \text{ mm}$

If expansion is prevented, then

Thermal stress, $p = \alpha T E$

$= 1.2 \times 10^{-5} \times 1 \times 10^5 \times 80$

$= 96 \text{ N/mm}^2$

SALIENT POINTS

- If the properties of a substance are assumed to exist as continuous distribution of mass, then this idealized subtance is called a continuum.
- A rigid solid body may be defined as a body in which the relative positions of any two arbitrary points is invariant under any condition.
- A deformable solid body may be defined as a body which may change its linear or angular dimensions or deform under certain conditions.
- A body can be isolated from the actions of all the outside systems by representing such actions as a set of forces which are denoted as external forces. External forces comprise body and surface forces.
- Body forces are distributed once the entire volume of the body and are expressed as force per unit volume.
- Surface forces are the forces which act through a contact surface as the boundary of the body and are expressed as force per unit area.
- Forces of interaction between molecules which are brought into action to resist deformation are called internal forces.
- Elasticity is the property of a material which is under stress and recovers its original shape after the removal of the external load.
- Plasticity is the property of a material by which a strained material retains the deformed position even after the removal of the external load which caused the deformation.
- Ductility is the property of a material by which it can be drawn into a wire by external forces.
- Brittleness is the property of a material by which it is not capable of undergoing a significant deformation due to the application of an external load but breaks or ruptures suddenly.
- Malleability is the property of a material by which it can be uniformly lengthened or widened by hammering or rolling without rupture.
- Strength is the properly of a material determined by the greatest stress that the material can withstand prior to failure.
- Hardness of a material is the ability of a material to resist penetration by a hard material or object.
- Toughness is the property of a material which enables the material to absorb energy without fracture.
- Stiffness is the property that enables a material to withstand high stress without large deformation.
- Stability is the overall property of a member made of a material to maintain the overall equilibrium preventing complete collapse.
- Force (P) per unit area (A) is called the intensity of stress (p) or simply as stress, i.e., $p = P/A$
- The ratio of change of dimension, due to an application of external load to the original dimension is called strain.
- Volumetric strain is defined as the change in volume of an elastic body due to external force in unit original volume.
- Lateral strain is the ratio of change in lateral dimension (due to external load) to the original lateral dimension.
- Poisson's ratio ($\mu = 1/m$) is the ratio of the lateral strain to the axial strain.

- Hooke's law is defined as the stress induced in a material that is proportional to the strain within the elastic limit.
- For a given field problem, the magnitude of the stress which may be considered as a safe stress is referred to as the allowable stress or working stress.
- Allowable stress is obtained by dividing the ultimate stress or the field stress of a material by a suitable factor called the factor of safety.
- The ratio of stress to strain within the elastic limit in a material is a constant which is defined as the Modulus of elasticity or Young's modulus (E).
- When a material is subjected to three direct stresses along the three mutually perpendicular directions with equal intensity, then the ratio of the direct stress to the volumetric strain is denoted as the Bulk modulus (K).
- Modulus of rigidity or Shear modulus (G) is the ratio of shear stress to shear strain when the material is under the elastic limit.
- Young's modulus (E) and Shear modulus (G) are related as $E = 2G\left(\frac{m+1}{m}\right)$
- Young's modulus (E) and the Bulk modulus (K) are related as $E = 3K\left(1-\frac{2}{m}\right)$
- Young's modulus (E), Shear modulus (G) and Bulk modulus (K) are related as $E = \left(\frac{9KG}{G+3K}\right)$

QUESTIONS

1. Draw a stress–strain diagram for a brittle material. Explain different stress levels.
2. What is lateral strain? How is it related with Poisson's ratio? What is the maximum value of Poisson's ratio?
3. What is a compound bar? How will you compute the elongation of a compound bar subjected to axial load?
4. What are elastic constants? Define the interconnected relations.
5. What are temperature stresses? Give the practical applications.
6. Results from a tensile test on a mild steel bar are given below:
   Diameter of bar = 2.5 cm, Gauge length = 5.0 cm, Load at limit of proportionality = 82,000 N, Extension of the bar at the limit of proportionality = 0.042 mm, Yield load = 86,000 N, Maximum load = 150,000 N, Length between gauge point after test = 5.65 cm and Diameter of neck after test = 2.02 cm.

   Calculate the modulus of elasticity, stress at the limit of proportionality, the field stress, ultimate tensile stress, percentage of elongation and percentage of contraction.
7. An M.S. rod of square cross-section 40 mm × 40 mm is subjected to an axial force of 350 kN. The modulus of elasticity and the Poisson's ratio of material are 200 kN/mm^2 and 0.32, respectively. Calculate the change in volume and the volumetric strain of the rod.
8. A steel bar of 8 mm square cross section and 1 m length is subjected to an axial compression by a load of 20 kN. There is no buckling. Compute the percentage change in the volume of the bar.
9. A bar of uniform cross section is subjected to an axial tensile load such that the normal strain in the axial is 10.20 mm per metre. If the Poisson's ratio of the material is 0.3, find the volumetric strain.

10. A hollow steel column of 200 mm external diameter carries an axial load of 2.0 MN. The ultimate stress is found to be 480 N/mm^2. Calculate the internal diameter of the column considering a factor of safety of 4.
11. A rod of diameter 35 mm and the length 3.5 is loaded by a tensile load of 40 kN. A bore of diameter 15 mm is made carefully at the centre of the rod. Estimate the length of the bore to be made so that total increase in extension is 30% under the same tensile load. $E_S = 2 \times 10^5$ m/mm^2.
12. Two pieces of material A and B have the same bulk modulus but the value of modulus of elasticity of B is 1% greater than that of A. Determine the shear modulus for the material B in terms of modulus of elasticity and shear modulus for the Material A.
13. A bar of 30 mm diameter is subjected to a pull of 60 kN. The measured extension of a gauge length of 200 mm is 0.09 mm and the change in diameter is 0.0039 mm. Calculate the Poissons ratio and elastic constants.
14. Two vertical rods of equal length are of steel and the other of copper are each rigidly fixed at the top and 60 mm apart. Diameters of rods are 3 cm and 3.5 cm, respectively. A cross bar fixed to the rods at the lower end carries a load of 5,000 N such that the cross bar remains horizontal even after loading. Find the stress in each rod and the position of the load on the bar. Take $E_s = 2\times10^5$ N/mm^2 and $E_{cu} = 1\times10^5$ N/mm^2.
15. A hollow steel cylinder has a length of 30 cm with external and internal diameters as 24 cm and 18 cm, respectively. The cylinder is filled with concrete and compressed between inside parallel plates by a load of 5,000 kg. Calculate the compressive stress in each material and the total shortening of the cylinder.
16. A circular mild steel bar tapers, uniformly from 25 mm diameter to 10 mm diameter over a length of 80 cm. Estimate the change in length of the bar when subjected to an axial pull of 5 N. Find the uniform diameter of the bar which can produce the same change in length for the same axial pull.
17. A flat steel plate of 1.5 m length with 1.5 m thickness uniformly tapers from 10 cm width to 5 cm width. Determine the extension of the plate when an axial pull of 6 tonnes acts on the plate. E of material is 200 kN/mm^2. What will be the percentage of error in the extension of the plate assuming an average constant width throughout its length?
18. A copper tube of 2.5 cm external diameter and 1.9 cm internal diameter encloses a steel rod of 1.5 cm diameter to which it is rigidly joined at each end. At the temperature of 20°C, there is no longitudinal stress. Determine the stress in the tube and the rod when the temperature is raised to 180°C. E = 21,000 N/mm^2 and E_{cu} = 100,000 N/mm^2. The coefficient of linear expansion for steel and copper are 11×10^{-6}/°C and 18×10^{-6}/°C, respectively.
19. Three vertical steel wires in the same plane of equal length are suspended from a horizontal support. The wires carry a load by means of rigid cross bar attached at their lower ends. The load is increased and the temperature changed so that the stress in each wire increases by 10 N/mm^2. Compute the change of temperature. E_s = 205 GN/mm^2 and $\alpha_s = 11\times10^{-6}$/°C.

2

Principal Stresses and Strains

LEARNING OBJECTIVES

2.1 Stress at a Point
2.2 Two-Dimensional State of Stresses
2.3 Stresses on an Oblique Plane
2.4 Principal Stresses and Strains
2.5 Mohr's Circle

2.1 STRESS AT A POINT

As discussed in Chapter 1, materials are regarded as continuum and stresses and strains are evaluated considering an infinitesimal element having the same properties of the entire mass. Thus, stress at a point in a material has to be viewed as a large point with representative materials of the whole mass.

In such a system tensile stress is considered as positive, compressive stress as negative and shear stress causing a clockwise torque about the centre of a free body as positive.

An incremental element is considered with the stresses acting on the planes to represent the stress conditions at a point in a material as shown in Figure 2.1.

Here, p_x, p_y and p_z are the normal stresses and q_{xy}, q_{yz} and q_{zx} are the shear stresses. Let a two-dimensional (2D) view of the element be considered as shown in Figure 2.2.

Since the element is in equilibrium, the moment on any axis of all the forces acting on the element must be zero. Taking moments of all forces about a line passing through A and parallel to z-axis and equating to zero, we have

$$\left(q_{xy} \cdot dy \cdot dz\right) dx = \left(q_{yx} \cdot dx \cdot dz\right) dy \tag{2.1}$$

$$q_{xy} = q_{yx}$$

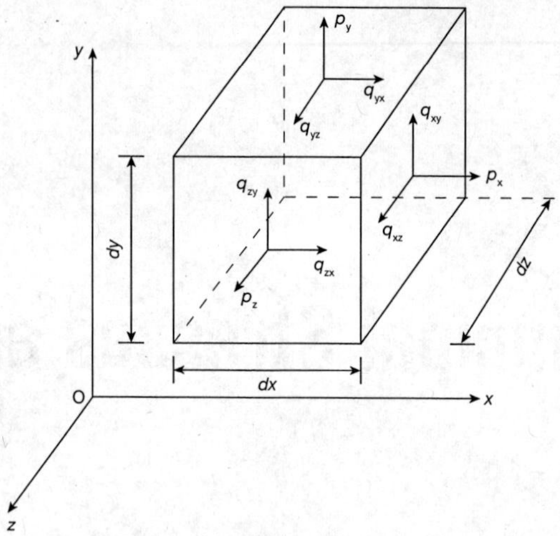

Fig. 2.1 Stresses acting as an element.

Similarly, it can be shown that

and

$$q_{xz} = q_x$$

$$q_{yz} = q_{zy}$$

It may be inferred that the shear stresses on planes at right angles are equal and are called *complementary shear stresses*, e.g., q_{xz} is a complementary shear to q_{zx}.

Thus, after satisfying the rotational moment equilibrium condition

$$q_{xy} = q_{yx} \tag{2.2}$$

$$q_{yz} = q_{zy} \tag{2.3}$$

$$q_{xz} = q_{zx} \tag{2.4}$$

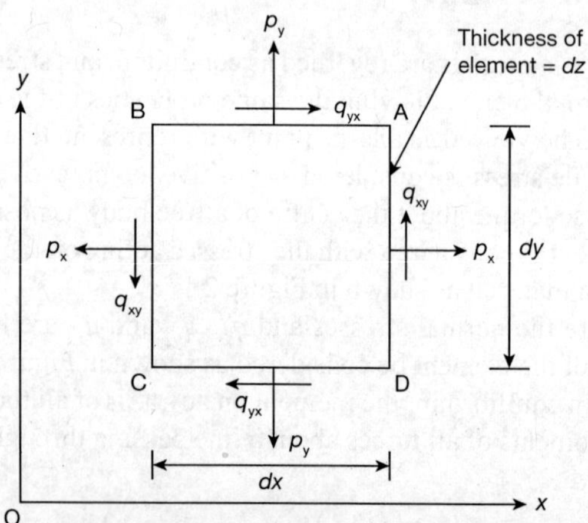

Fig. 2.2 Two-dimensional view of an element.

In a general 3D case, there are nine components of stresses out of which six are independent.

Let e_x, e_y and e_z be the strains in the directions x, y and z, respectively. Then, the stress–strain relationship may be written as

$$e_x = \frac{p_x}{E} - \frac{(p_y + p_z)}{mE} \tag{2.5}$$

$$e_y = \frac{p_y}{E} - \frac{(p_z + p_x)}{mE} \tag{2.6}$$

$$e_z = \frac{p_z}{E} - \frac{(p_x + p_y)}{mE} \tag{2.7}$$

2.2 TWO-DIMENSIONAL STATE OF STRESSES

It is convenient to presume one of the principal stresses (or stress components in a particular direction) or principal strains (or strain components in a particular direction) as zero and convert the problem to a 2D state. In many occasions, 2D problems are easy to solve.

Thus, in certain situations, the stress components in a particular direction may all be zero and only the components perpendicular to that direction may exist. This is referred to as case of *plane stress.*

For example, if $p_z = q_{xz} = q_{yz} = 0$ (i.e., all the stress components in the z-direction are zero) and the components p_x, p_y and q_{xy} exist, then it is a case of plane stress.

Thus, a state of plane stress may be defined as that condition in which there is no stress component acting on a pair of parallel faces. The plane stress conditions prevail in structural members and machine elements.

In certain situations, the strain components in a particular direction are zero (or ignoring the stress in that direction) and only the stress components perpendicular to that direction may exist. This is referred to as a case of *plane strain.*

For example, the strain $e_y = 0$, i.e., p_y is ignored. Many geotechnical engineering problems are plane strain problems.

Further discussion in the book is confined to plane stress problems.

2.3 STRESSES ON AN OBLIQUE PLANE

Let us consider a plane stress case in which p_z, q_{xz} and q_{yz} are zero. Other stresses acting on an element of unit thickness for this condition are shown in Figure 2.3. For simplicity, the complementary shear stress $q_{xy} = q_{yx}$ may be represented as q. Figure 2.3(a) represents a free body diagram of an element that is removed from a body in equilibrium.

Let us consider a free body diagram of the portion of the element (Figure 2.3(b)). On the arbitrarily chosen oblique plane AE, there will be normal stress p_n and shear or tangential stress p_t. As the parent element is in equilibrium, the triangular element also will be in equilibrium with the given set of stresses as shown in Figure 2.3(b). As the element is in equilibrium, the components of stresses at the triangular element must be zero.

Equating the forces perpendicular to the face AE to zero, we have

$$p_n \mathrm{AE} = p_x \mathrm{AD}\cos\theta + p_y \mathrm{ED}\sin\theta + q\mathrm{AD}\sin\theta + q\mathrm{ED}\cos\theta$$

i.e.,
$$p_n = p_x \frac{\mathrm{AD}}{\mathrm{AE}}\cos\theta + p_y \frac{\mathrm{ED}}{\mathrm{AE}}\sin\theta + q\frac{\mathrm{AD}}{\mathrm{AE}}\sin\theta + q\frac{\mathrm{ED}}{\mathrm{AE}}\cos\theta$$

or
$$p_n = \frac{(p_x + p_y)}{2} + \frac{(p_x - p_y)}{2}\cos 2\theta + q\sin\theta \tag{2.8}$$

Equating the forces parallel to the face AE to zero, we have

$$p_t \mathrm{AE} = p_x \mathrm{AD}\sin\theta - p_y \mathrm{ED}\cos\theta - q\mathrm{AD}\sin\theta + q\mathrm{ED}\sin\theta$$

or
$$p_t = p_x \frac{\mathrm{AD}}{\mathrm{AE}}\sin\theta - p_y \frac{\mathrm{ED}}{\mathrm{AE}}\cos\theta - q\frac{\mathrm{AD}}{\mathrm{AE}}\sin\theta + q\frac{\mathrm{ED}}{\mathrm{AE}}\sin\theta$$

i.e.,
$$p_t = \frac{(p_x - p_y)}{2}\sin 2\theta - q\cos 2\theta \tag{2.9}$$

The resultant stress p_r acting on the plane AD is given as

$$p_r = \sqrt{p_n^2 + p_t^2} \tag{2.10}$$

and the angel of inclination of resultant force

$$\phi = \tan^{-1}\left(\frac{p_t}{p_n}\right) \tag{2.11}$$

Equations (2.8)–(2.10) represent the normal stress, tangential stress or shear stress and resultant stress on the plane AE inclined at an angle θ to the vertical plane (or the plane on which p_x acts) for the most general case of stress system.

Special cases of the stress system and the corresponding normal and shear stresses on the oblique plane AE are discussed below.

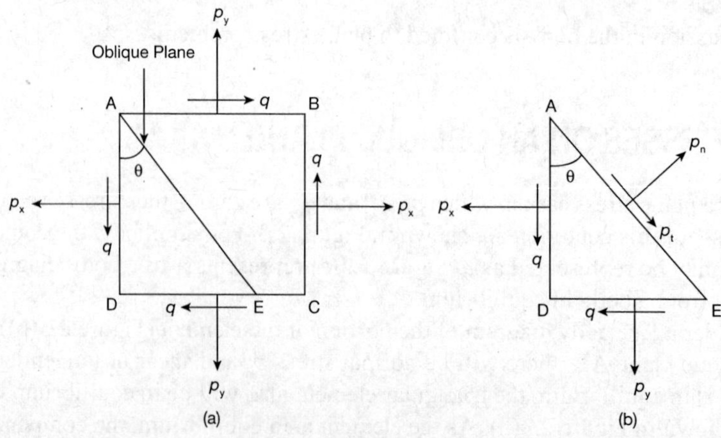

Fig. 2.3 Oblique plane.

2.3.1 Stresses Due to Uniaxial Stress

An element subjected to a tensile stress p is considered (Figure 2.4).

Substituting, $p_x = p$, $p_y = 0$ and $q = 0$ in the generalised Equations (2.8)–(2.10), we get

$$p_n = p\cos^2\theta \tag{2.12a}$$

$$p_t = \frac{p}{2}\sin 2\theta \tag{2.12b}$$

$$p_r = \sqrt{p_n^2 + p_t^2} = \sqrt{p^2\cos^4\theta + p^2\sin^2\theta\cos^2\theta}$$

or

$$p_r = p\cos\theta \tag{2.13}$$

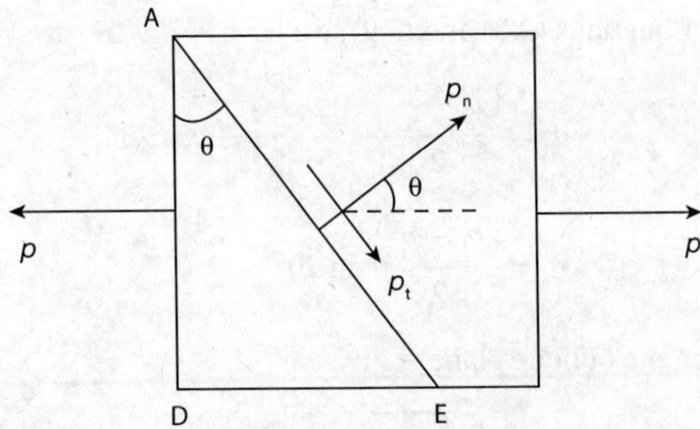

Fig. 2.4 Stresses due to uniaxial stress.

2.3.2 Stresses Due to Simple Shear

A state of simple shear is represented in Figure 2.5.

Here, substituting for $p_x = 0$ in the generalized Equations (2.10) and (2.11), we have

$$p_n = q\sin 2\theta \tag{2.14a}$$

and

$$p_t = q\cos 2\theta \tag{2.14b}$$

$$p_r = \sqrt{p_n^2 + p_t^2} = \sqrt{q^2\sin^2 2\theta + q^2\cos^2 2\theta}$$

i.e.,

$$p_r = q\sqrt{\sin^2 2\theta + \cos^2 2\theta} \tag{2.15}$$

The value of p_x will be maximum when $\theta = 45°$ and for this case $p_t = 0$.

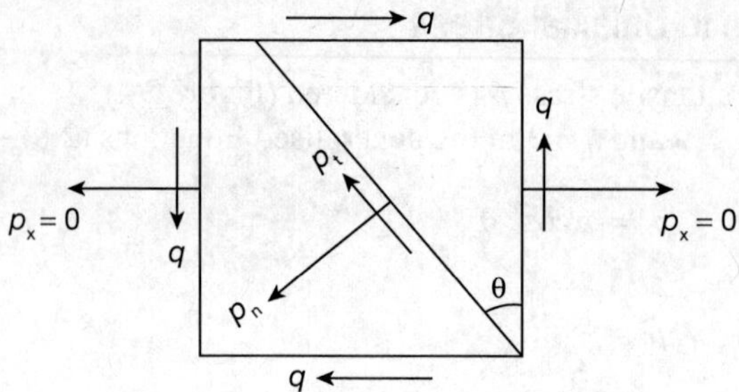

Fig. 2.5 Stresses due to simple shear.

2.3.3 Stresses Due to Two Tensile Stresses

Two direct tensile stresses acting on an element are considered (Figure 2.6) and there are no shear stresses acting on the planes.

Substituting $q = 0$ in Equations (2.8) and (2.9), we have

$$p_n = \frac{(p_x + p_y)}{2} + \frac{(p_x - p_y)}{2}\cos 2\theta \tag{2.16a}$$

and
$$p_t = \frac{(p_x - p_y)}{2}\sin 2\theta \tag{2.16b}$$

The resultant stress on the oblique plane is given as

$$p_r = \sqrt{p_n^2 - p_t^2} \tag{2.17}$$

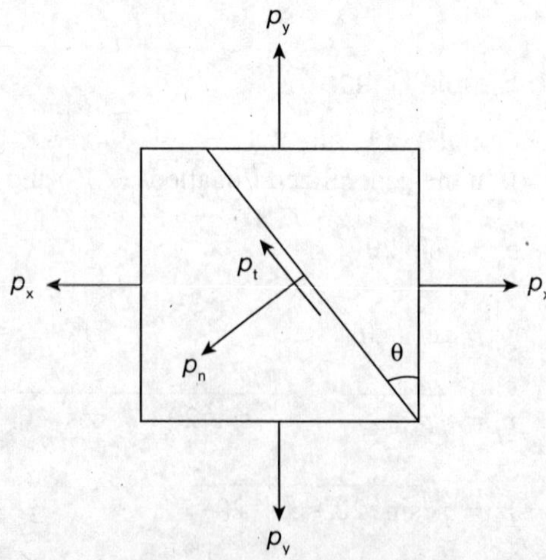

Fig. 2.6 Stresses due to two direct tensile stresses.

Maximum tangential stress occurs when sin 2θ = 1, i.e., θ = 45°, then

$$(p_t)_{max} = \frac{(p_x - p_y)}{2} \tag{2.18}$$

and the corresponding normal stress on the plane is

$$p_n = \frac{(p_x + p_y)}{2} \tag{2.19}$$

2.3.4 Stresses Due to Unlike Stresses

An element acted upon by a tensile stress p_x and a compressive stress p_y is considered (Figure 2.7) with no shear stresses acting on it.

Substituting for $p_x = p_x, p_y = -p_y$ in Equations (2.16a) and (2.16b) and $q = 0$, we have

$$p_n = \frac{(p_x - p_y)}{2} + \frac{(p_x + p_y)}{2}\cos 2\theta \tag{2.20}$$

and

$$p_t = \frac{(p_x + p_y)}{2}\sin 2\theta \tag{2.21}$$

$$p_r = \sqrt{p_n^2 + p_t^2}$$

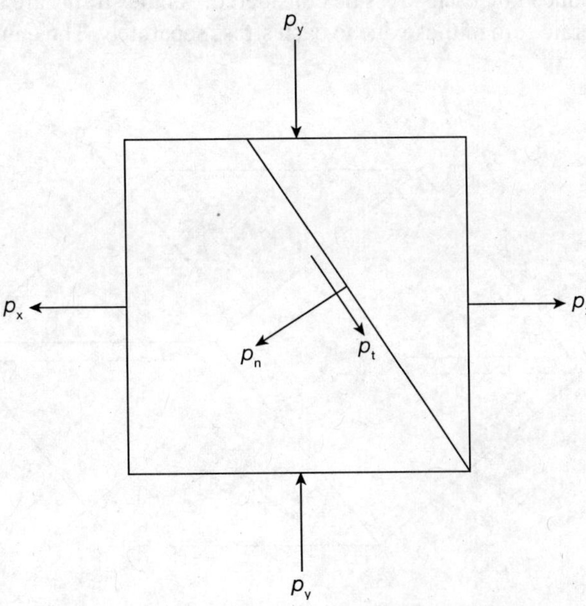

Fig. 2.7 Stresses due to unlike direct stresses.

2.4 PRINCIPAL STRESSES AND STRAINS

2.4.1 Principal Stresses

Let us consider all the planes passing through a point and locate such a plane on which there are no shear stresses. There are atleast three such planes at a point on which there are no shear stresses. The additional characteristics of these is that they are mutually perpendicular to each other (Figure 2.8).

On these planes, there are no shear stresses acting but only normal stresses. These normal stresses are called as principal stresses.

Thus, principal stress may be defined as a normal stress acting on a plane wherein there are no shear stresses. There are three principal stresses which are named as major principal stress, σ_1 (the largest stress), minor principal stress, σ_2 (the smallest stress) and intermediate principal stress, σ_3 (between the largest and the smallest stresses). Specifications of the principal stresses and their directions provide a convenient way of describing the state of stress at a point.

2.4.2 Principal Strains

Let σ_1, σ_2 and σ_3 be the three like principal stresses and ε_1, ε_2 and ε_3 be the principal strains in the respective directions.

Figure 2.9(a) represents the principal stresses acting on the mutually perpendicular planes.

Figure 2.9(b) shows a 2D stress system with σ_3 absent. This is called a uniaxial stress system which is applied in the analysis and design of thin pressure vessels, beams, shafts and many other structural components.

Let the strain produced by each stress be considered. As the strains are small, the resultant strain is given by the algebraic sum of those due to each stress separately. Thus, in a 2D condition, the principal strain

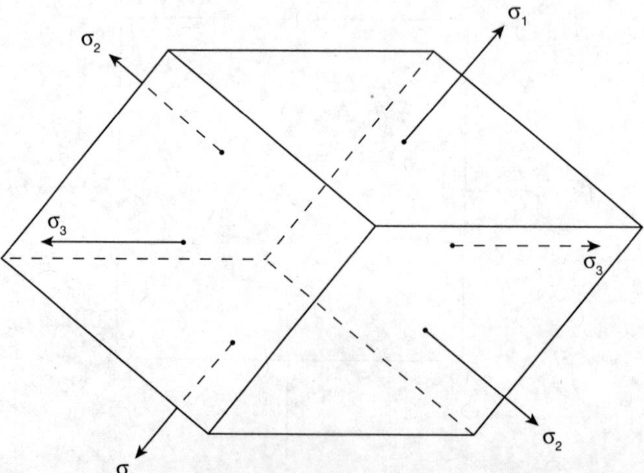

Fig. 2.8 Principal stresses on orthogonal planes.

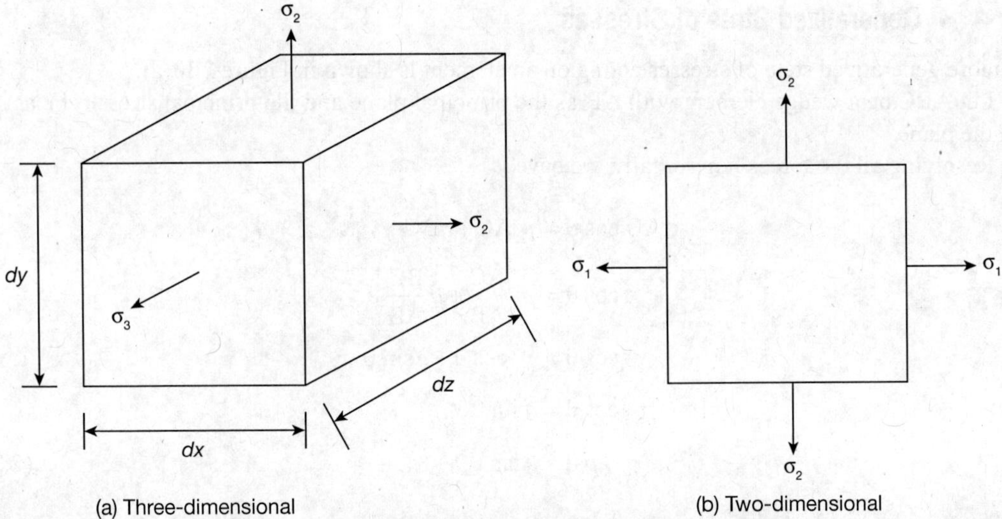

(a) Three-dimensional

(b) Two-dimensional

Fig. 2.9 Two- and three-dimensional principal stresses.

$$\varepsilon_1 = (\text{Strain due to } \sigma_1 \text{ in } \sigma_1\text{-direction}) + (\text{Strain due to } \sigma_2 \text{ in } \sigma_1\text{-direction})$$

$$= \frac{\sigma_1}{E} - \mu \frac{\sigma_2}{E}$$

i.e.,

$$\varepsilon_1 = \frac{\sigma_1}{E} - \frac{\sigma_2}{mE}$$

Similarly, the principal strain in the other direction is

$$\varepsilon_2 = \frac{\sigma_2}{E} - \frac{\sigma_1}{mE}$$

Thus, in a 3D case, the principal strain

$$\varepsilon_1 = (\text{Strain due to } \sigma_1 \text{ in } \sigma_1\text{-direction}) + (\text{Strain due to } \sigma_2 \text{ in } \sigma_1\text{-direction}) + (\text{Strain due to } \sigma_3 \text{ in } \sigma_1\text{-direction})$$

∴

$$\varepsilon_1 = \frac{\sigma_1}{E} - \frac{\sigma_2}{mE} - \frac{\sigma_3}{mE}$$

i.e.,

$$\varepsilon_1 = \frac{\sigma_1}{E} - \frac{(\sigma_2 + \sigma_3)}{mE} \tag{2.22}$$

Similarly

$$\varepsilon_2 = \frac{\sigma_2}{E} - \frac{(\sigma_3 + \sigma_1)}{mE} \tag{2.23}$$

and

$$\varepsilon_3 = \frac{\sigma_3}{E} - \frac{(\sigma_1 + \sigma_2)}{mE} \tag{2.24}$$

In the case of compressive principal stress, it has to be taken as negative.

2.4.3 Generalized State of Stresses

A more generalized state of stresses acting on an element is shown in Figure 2.10(a).

Let ABC be a wedge element with AB as the principal plane and the principal stress σ be acting on the plane.

Resolving all the forces horizontally, we have

$$\sigma AB \cos\theta = p_x AC + qBC$$

or $$\sigma\cos\theta = p_x\frac{AC}{AB} + q\frac{BC}{AB}$$

$$\sigma\cos\theta = p_x\cos\theta + q\sin\theta$$

or $$(\sigma - p_x)\cos\theta = q\sin\theta$$

or $$(\sigma - p_x) = q\tan\theta \tag{2.25}$$

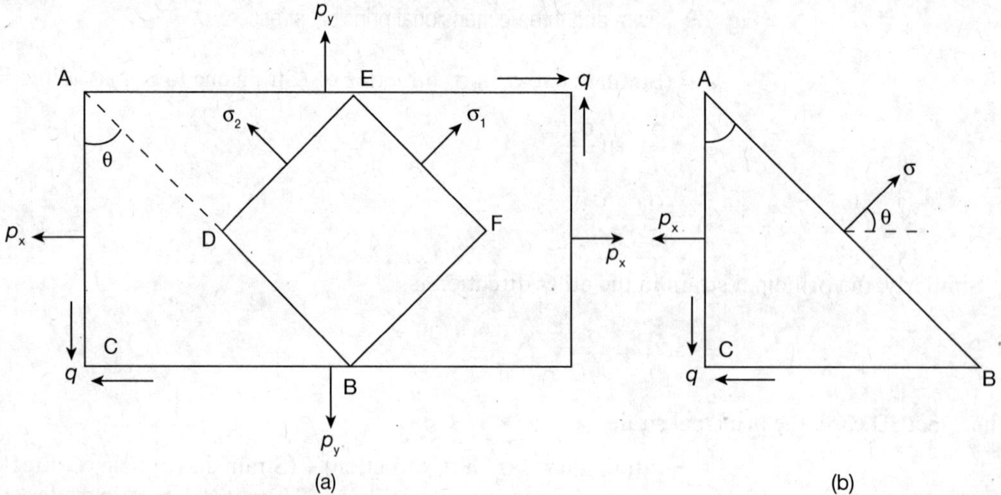

Fig. 2.10 Generalized state of stress.

Resolving all the forces vertically, we have

$$\sigma AB \sin\theta = p_y BC + qAC$$

or $$\sigma\sin\theta = p_y\frac{BC}{AB} + q\frac{AC}{AB}$$

or $$(\sigma - p_y)\sin\theta = q\cos\theta$$

or $$(\sigma - p_v) = q\cot\theta \tag{2.26}$$

Subtracting Equation (2.25) from Equation (2.26), we have

$$\left(p_x - p_y\right) = q\left(\cot\theta - \tan\theta\right)$$

$$= \frac{2q}{\tan 2\theta}$$

i.e.,
$$\tan 2\theta = \frac{2q}{\left(p_x - p_y\right)} \tag{2.27}$$

Equation (2.27) gives the position of principal planes.

Multiplying Equations (2.25) and (2.26), we have

$$\sigma^2 - \sigma(p_x + p_y) + (p_x p_y - q^2) = 0 \tag{2.28}$$

Solving Equation (2.28), we get the major principal stress, σ_1 and minor principal stress, σ_2 as

$$\sigma = \frac{(p_x + p_y)}{2} \pm \sqrt{\left(\frac{p_x + p_y}{2}\right)^2 + (q^2 - p_x p_y)}$$

i.e.,

$$\sigma = \frac{(p_x + p_y)}{2} \pm \sqrt{\left(\frac{p_x - p_y}{2}\right)^2 + q^2}$$

∴ Major principal stress,
$$\sigma_1 = \frac{(p_x + p_y)}{2} + \sqrt{\left(\frac{p_x - p_y}{2}\right)^2 + q^2} \tag{2.29}$$

Minor principal stress,
$$\sigma_2 = \frac{(p_x + p_y)}{2} - \sqrt{\left(\frac{p_x - p_y}{2}\right)^2 + q^2} \tag{2.30}$$

Maximum value of shear stress,
$$q_{max} = \frac{1}{2}\sqrt{(p_x - p_y)^2 + q^2} \tag{2.31}$$

The planes on which there is maximum shear stress will be the planes inclined at 45° to the principal stresses.

If the stress p_y is a compressive stress, then the sign for p_y is taken as $-p_y$ in Equations (2.29), (2.27) and (2.30), then

$$\sigma_1 = \frac{\left(p_x - p_y\right)}{2} + \sqrt{\left(\frac{p_x + p_y}{2}\right)^2 + q^2} \tag{2.32}$$

$$\sigma_2 = \frac{\left(p_x - p_y\right)}{2} - \sqrt{\left(\frac{p_x + p_y}{2}\right)^2 + q^2} \tag{2.33}$$

2.5 MOHR'S CIRCLE

If a plot is made with normal stresses used as abscissas and shearing stresses as ordinates, and if points are plotted to represent stress coordinates for all possible values of θ, it is found that the locus of these points is a circle, as shown in Figure 2.11.

This circle has its centre on the *x*-axis, and cuts it at abscissas of σ_2 and σ_1. Any point on the circle represents the coordinates of stress on some plane. Thus, point C in Figure 2.11(a) is for the θ value as given in Figure 2.11(b). This circle is the graphical representation of the state of stress in a lucid form which is known as Mohr's circle after Mohr (1882).

Mohr's diagram is an excellent visualization of the orientations of various planes. If through the coordinates of p_n and p_t (point G in Figure 2.11(b)) on the Mohr circle, a line is drawn parallel to the plane on which these stresses act, this line intersects the Mohr's circle at an unique point. If parallels are drawn from E(σ_1,0) and F(σ_2,0) to the respective planes, these planes pass through the same unique point.

This point is referred to as the *origin of planes* or *pole*, O_p. Thus, any line drawn through O_p, parallel to any arbitrarily chosen plane, intersects the circle at a point, the coordinates of which are stress components (i.e., the shear and normal stresses) on that plane.

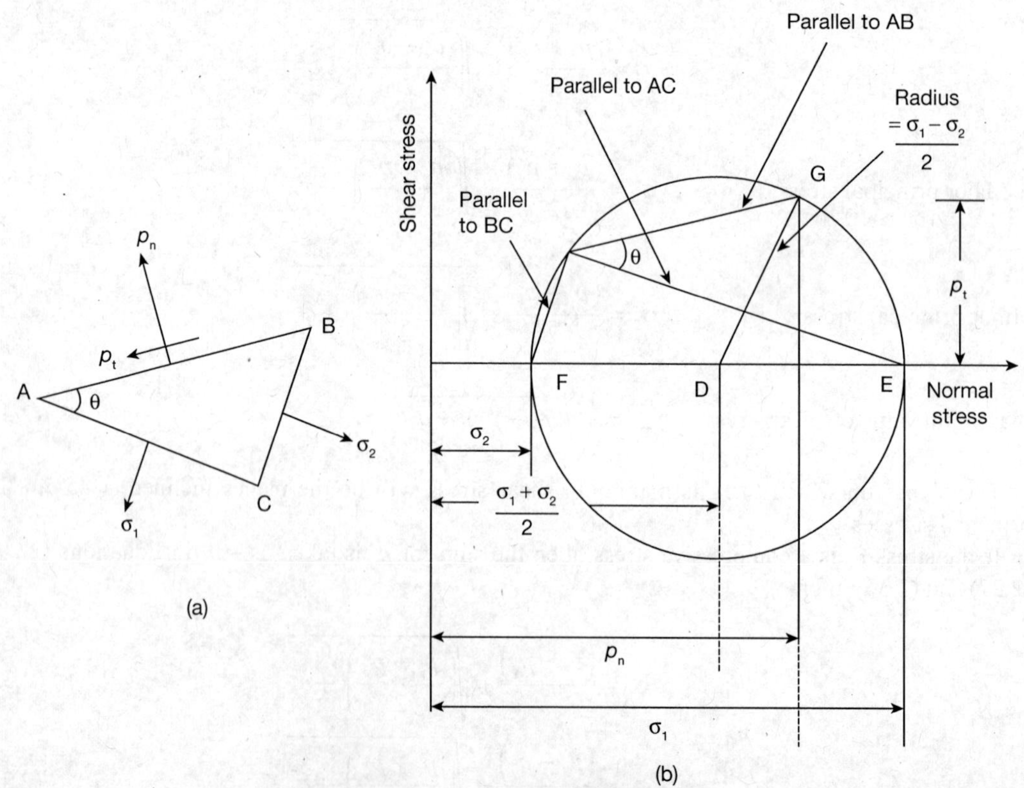

Fig. 2.11 Mohr's stress circle.

Mohr's circles for various states of stress conditions are discussed below.

2.5.1 Mohr's Circle for Generalized State of Stresses

Let a generalized case of a material subjected to tensile direct stresses p_x and p_y and shear stress, q (Figure 2.12(a)) be considered.

Choose a suitable scale such that the scale of the stresses p_x and p_y are denoted as OA and OB respectively (Figure 2.12). As the shear stress is making an anticlockwise torque q is scaled as AD downwards and BE upwards. Further, DE is joined to cut the x-axis at C. With C as centre and CE or CD as radius, a circle is drawn. The circle cuts the x-axis at F and G.

To determine the stresses, p_x and p_t, acting on the plane θ inclined to the vertical plane, CH is drawn at an angle 2θ with CD in the anticlockwise direction. In addition, HI is drawn perpendicular to the axis. Now, HI and OI represent p_t and p_n, respectively.

1. Proof of Construction

Let r be the radius of the stress circle, then

$$r = \sqrt{CA^2 + AD^2} = \sqrt{\left(\frac{p_x - p_y}{2}\right)^2 + q^2}$$

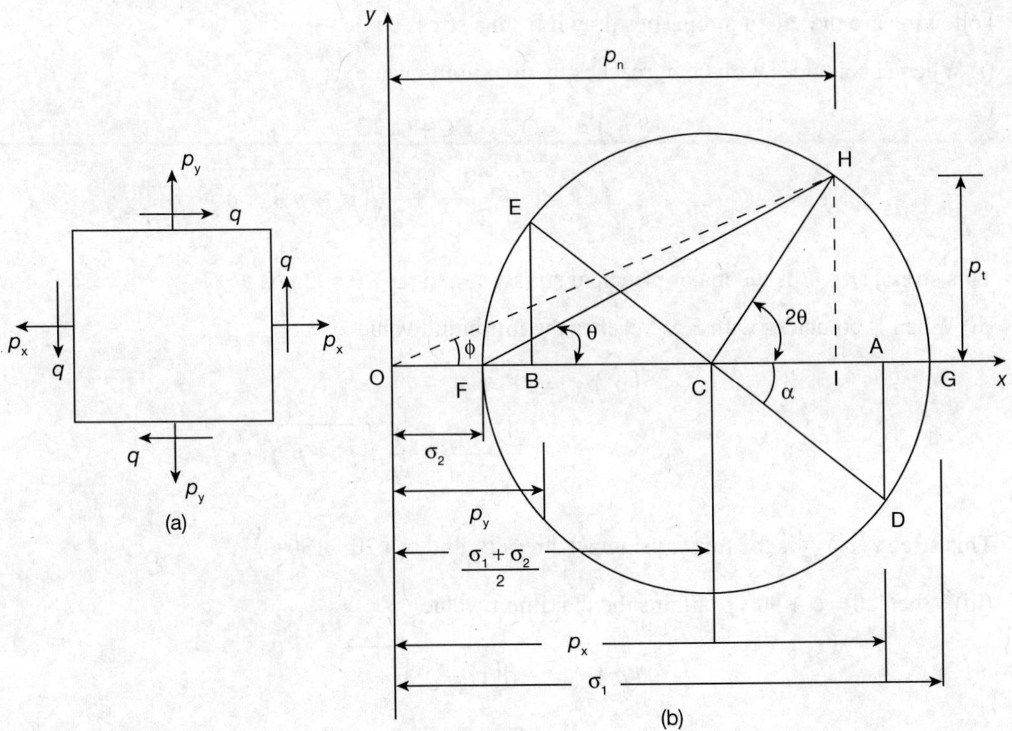

Fig. 2.12 Mohr's circle for generalized state of stress.

Now, $$r\cos\alpha = \text{CA} = \frac{p_x - p_y}{2}$$

$$r\sin\alpha = \text{AD} = q$$

i.e., $$\text{OI} = \text{OC} + \text{CI}$$

$$= \frac{p_x + p_y}{2} + r\cos(2\theta - \alpha)$$

$$= \frac{p_x + p_y}{2} + r\cos\alpha\cos 2\theta + r\sin\alpha\sin 2\theta$$

$$p_n = \frac{p_x + p_y}{2} + \frac{p_x - p_y}{2}\cos 2\theta + q\sin 2\theta$$

Similarly,

$$\text{HI} = r\sin(2\theta - \alpha)$$

$$= r\cos\alpha\sin 2\theta - r\sin\alpha\cos 2\theta$$

i.e., $$p_t = \frac{p_x - p_y}{2}\sin 2\theta - q\cos 2\theta$$

2. Properties

Following are the salient properties drawn from a stress circle.

(i) When H coincides with G, p_n reaches the maximum value,

i.e., $$(p_n)_{max} = \text{OG} = \text{OC} + \text{CG}$$

$$= \left(\frac{p_x - p_y}{2}\right) + \frac{1}{2}\sqrt{(p_x - p_y)^2 + q^2}$$

This stress, $(p_x)_{max}$ is the major principal stress, σ_1 and $\tan 2\theta = \tan\alpha$.

(ii) When H coincides with F, p_n reaches the minimum value

i.e., $$(p_n)_{min} = \text{OF} = \text{OC} - \text{CF}$$

$$= \left(\frac{p_x + p_y}{2}\right) - \frac{1}{2}\sqrt{(p_x + p_y)^2 + q^2}$$

This stress, $(p_n)_{min}$ is the minor principal stress σ_2 and $\tan 2\theta = 180 + \frac{\alpha}{2}$.

(iii) When $2\theta = \alpha + 90°$, p_t attains the maximum value

i.e., $$(p_t)_{max} = \frac{1}{2}\sqrt{(p_x - p_y)^2 + q^2}$$

$$= \left(\frac{\sigma_1 - \sigma_2}{2}\right)$$

and when $$2\theta = \alpha + 270°$$

$$(p_t)_{max} = -\frac{1}{2}\sqrt{(p_x - p_y)^2 + q^2}$$

$$= -\left(\frac{\sigma_1 - \sigma_2}{2}\right)$$

(iv) Angle ϕ is the angle of inclination of the resultant stress.

2.5.2 Mohr's Circle for Uniaxial Stress

An element acted upon by uniaxial stress is shown in Figure 2.13.

Here, $p_y = 0$ and $q = 0$. Mohr's circle is drawn with radius $= \frac{p_x}{2}$ and C as centre. Angle 2θ in set from C, to establish Point B. Then, BD and OD represent the shear and normal stresses and ϕ represents the angle of inclination of the resultant stress.

Here, the major principal stress is σ_1 and the minor principal stress σ_2 is zero.

2.5.3 Mohr's Circle for Simple Shear

An element acted upon by only shear stresses is shown in Figure 2.14(a).

In this case, $p_x = 0$, $p_y = 0$ and only the shear stress q exists. Since the stresses p_x and p_y are zero, the radius of Mohr's circle is q.

With O as centre and radius OA = q, a circle is drawn which cuts the x-axis at C and D (Figure 2.14(b)).

Then, the major principal stress, σ_1, is OD and the minor principal stress, σ_2, is OC. In this case, both σ_1 and σ_2 are of equal value.

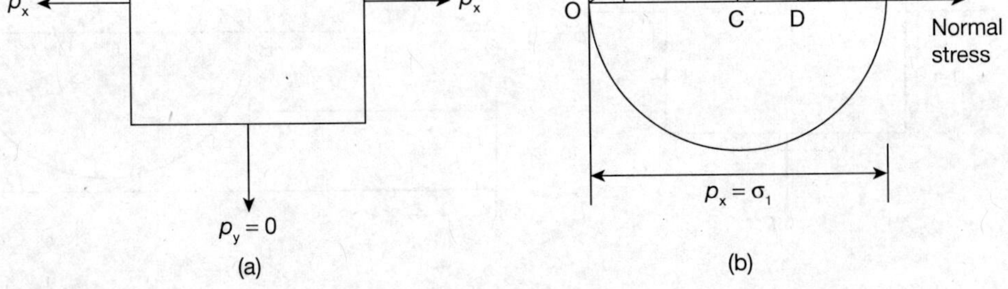

Fig. 2.13 Mohr circle for uniaxial tensile stress.

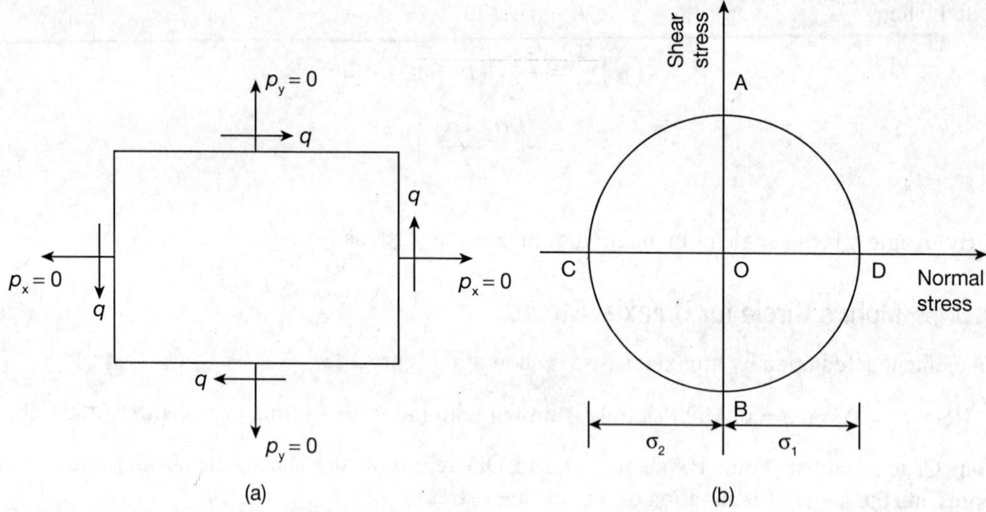

Fig. 2.14 Mohr circle for simple shear.

2.5.4 Mohr's Circle for Two Tensile Stresses

An element subjected to two tensile stresses is now considered (Figure 2.15(a)).

From Point O, Point A is fixed with a magnitude of p_x and OB with a magnitude of p_y. A circle is drawn with BA as diameter and C as centre. From C, a line is drawn with an angle of 2θ with CA which cuts on the Mohr's circle at E. A perpendicular is drawn from E and OE is joined.

Now, OD is the normal stress, p_n, and OE is the tangential stress, p_t. Further, OE is the resultant inclined at ϕ to the horizontal. Here, $p_x = \sigma_1$ and $p_y = \sigma_2$.

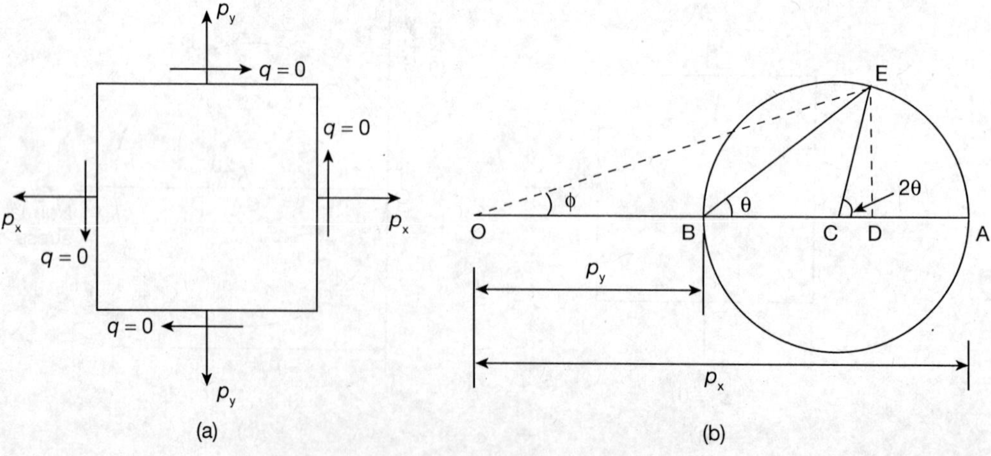

Fig. 2.15 Mohr's circle for two tensile stresses.

2.5.5 Mohr's Circle for Unlike Direct Stresses

An element of a material subjected to two unlike forces is considered (Figure 2.16(a)).

In this case, p_x is tensile and p_y is compressive. Any point B is chosen as the x-axis and BC and BA are scaled off as $BC = p_x$ and $p_y = BA$, as shown in Figure 2.16(b). The centre of the Mohr's circle O is fixed with $OA = \frac{p_n + p_y}{2}$ and the circle is drawn. Through O, a line OD is drawn with an angle of 2θ with OC. A perpendicular is dropped at DE.

Further, DE represents the tensile stress, AE the normal stress and BD the resultant stress, which is inclined at ϕ to the horizontal.

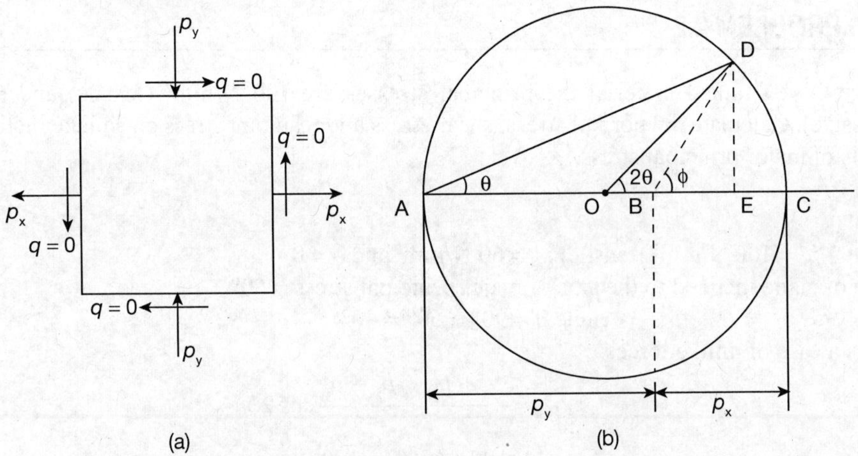

Fig. 2.16 Mohr's circle for unlike direct stresses.

SOLVED PROBLEM 2.1

The principal tensile stresses at a point across two mutually perpendicular planes are 120 N/mm² and 60 N/mm². Determine the normal, tangential and resultant stresses on a plane inclined at 40° to the axis of the minor principal stress.

Solution:

Given data: $p_x = 120$ N/mm², $p_y = 60$ N/mm², $\theta = 40°$ and $q = 0$.

It is the case of two tensile stresses acting on an element (Figure 2.16(a)).

Normal stress,

$$p_n = \left(\frac{p_x + p_y}{2}\right) + \left(\frac{p_x - p_y}{2}\right)\cos 2\theta$$

$$p_n = \left(\frac{120 + 60}{2}\right) + \left(\frac{120 - 60}{2}\right)\cos(2 \times 40°)$$

i.e.,

$$p_n = 95.20 \text{ N/mm}^2$$

Tangential stress, $p_t = \dfrac{p_x - p_y}{2} \sin 2\theta$

$$p_t = \frac{120-60}{2} \sin (2\times 40°)$$

$$p_t = 29.52 \text{ N/mm}^2$$

i.e., Resultant stress, $p_r = \sqrt{p_n^2 + p_t^2}$

$$= \sqrt{(95.20)^2 + (29.52)^2}$$

$$= 99.67 \text{ N/mm}^2$$

SOLVED PROBLEM 2.2

At a point in a strained material the principal stresses are 100 N/mm² (tensile) and 60 N/mm² (compressive). Calculate the normal stress, shear stress and resultant stress on a plane inclined at 50° to the axis of major principal stress.

Solution:

Given data: $p_x = 100$ N/mm² (tensile) $p_y = 60$ N/mm² and $q = 0$.

Angle of plane inclined to the axis of major principal stress = 50°

i.e., Angle $\theta = 90° - 50° = 40°$

This is a case of unlike forces.

$$p_n = \frac{p_x + p_y}{2} + \frac{p_x - p_y}{2} \cos 2\theta$$

$$= \frac{100+(-60)}{2} + \frac{100-(-60)}{2} \cos (2\times 40°)$$

$$= 33.89 \text{ N/mm}^2$$

$$p_t = \frac{p_x - p_y}{2} \sin 2\theta = \frac{100-(-60)}{2} \sin (2\times 40°)$$

$$= 85.76 \text{ N/mm}^2$$

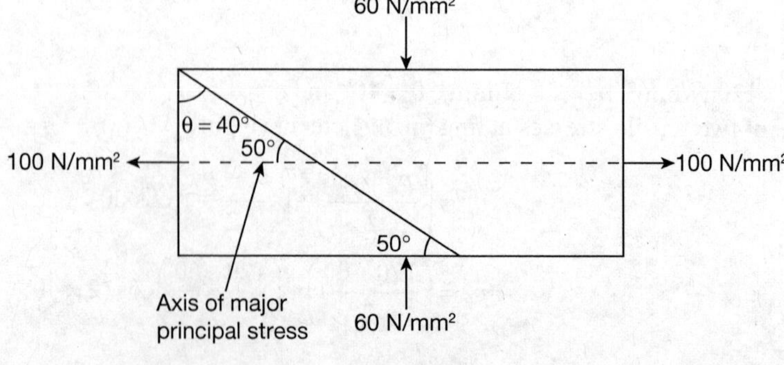

Fig. 2.17

SOLVED PROBLEM 2.3

A rectangular block of a material is subjected to tensile stress of 110 N/mm² on the plane and a compressive stress of 45 N/mm² on the plane at right angles to the former. Each of the above stresses is accompanied by a shear stress of 20 N/mm² which tends to rotate the block anticlockwise. Calculate the principal stresses and planes.

Solution:

Given data: $p_x = 110$ N/mm² (tensile), $q = 20$ N/mm² and $p_y = 45$ N/mm² (compressive).

We know,
$$\sigma_1 = \left(\frac{p_x - p_y}{2}\right) + \sqrt{\left(\frac{p_x + p_y}{2}\right)^2 + q^2}$$
$$= \left(\frac{110-45}{2}\right) + \sqrt{\left(\frac{110+45}{2}\right)^2 + 20^2}$$
$$= 32.5 + 80.039$$
i.e.,
$$\sigma_1 = 112.53 \text{ N/mm}^2 \text{ (tensile)}$$

Further,
$$\sigma_2 = \left(\frac{p_x - p_y}{2}\right) - \sqrt{\left(\frac{p_x + p_y}{2}\right)^2 + q^2}$$
$$= 32.5 - 80.039$$
i.e.,
$$\sigma_2 = -47.53 \text{ N/mm}^2 \text{ (compressive)}$$

SOLVED PROBLEM 2.4

At a point within a body subjected to two mutually perpendicular directions, the stresses are 100 N/mm² tensile and 50 N/mm² tensile. Each of the above stresses is accompanied by a shear stress of 70 N/mm². Determine the normal stress, shear stress and resultant stress on an oblique plane inclined at an angle of 45° with the major principal plane.

Solution:

Given data: $p_x = 100$ N/mm², $p_y = 50$ N/mm², $q = 70$ N/mm² and $\theta = 45°$.

Normal stress,
$$p_n = \left(\frac{p_x + p_y}{2}\right) + \left(\frac{p_x - p_y}{2}\right)\cos 2\theta + q \sin 2\theta$$
$$= \left(\frac{100+50}{2}\right) + \left(\frac{100-50}{2}\right)\cos(2\times 45) + 70 \times \sin(2\times 45)$$
$$= 75 + 0 + 70$$
$$= 145 \text{ N/mm}^2$$

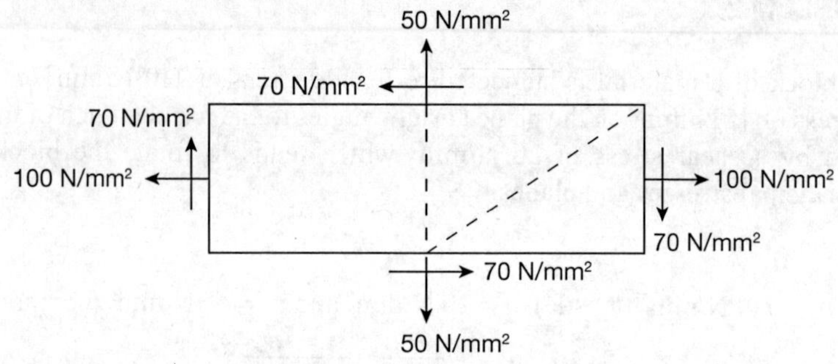

Fig. 2.18

Shear stress,
$$p_t = \left(\frac{p_x - p_y}{2}\right)\sin 2\theta - q\cos 2\theta$$
$$= \left(\frac{100-50}{2}\right)\sin(2\times 45) - 70\cos(2\times 45)$$
$$= 25\ \text{N/mm}^2$$

Resultant stress,
$$p_r = \sqrt{p_n^2 + p_t^2}$$
$$= \sqrt{145^2 + 25^2}$$
$$= 147.13\ \text{N/mm}^2$$

SOLVED PROBLEM 2.5

An elastic material at a point is subjected to a tensile stress of 90 N/mm² and a shear stress of 60 N/mm². Determine the maximum and minimum intensity of direct stress.

Solution:

Given data: $p_x = 90$ mN/m² (tensile), $p_y = 0$ and $q = 60$ mN/m² (shear).

We know the maximum intensity of direct stress when $p_y = 0$ and $p_x = 0$ and $\sigma_1 = (p_n)_{max}$ is

$$(p_n)_{max} = \frac{p}{2} + \sqrt{\left(\frac{p}{2}\right)^2 + q^2}$$
$$= \frac{90}{2} + \sqrt{\left(\frac{90}{2}\right)^2 + 60^2}$$
$$= 45 + 75$$
$$= 120\ \text{mN/m}^2\ (\text{tensile})$$

Similarly, when $p_y = 0, p_x = p$ and $\sigma_2 = (p_n)_{min}$ the minimum intensity of direct stress is

$$(p_x)_{min} = \frac{p}{2} - \sqrt{\left(\frac{p}{2}\right)^2 + q^2}$$

$$= \frac{90}{2} - \sqrt{\left(\frac{90}{2}\right)^2 + 60^2}$$

$$= 45 - 75$$

$$= 30 \text{ mN/m}^2 \text{ (compressive)}$$

SOLVED PROBLEM 2.6

At a certain point in an elastic body direct stresses of 150 N/mm² tensile and 100 N/mm² compressive stress act on perpendicular planes. Shear stress q also acts on these mutually perpendicular planes. Compute the magnitude of shear stress q. The greatest principal stress at a point is 180 N/mm². Also compute the maximum shearing stress at the point.

Solution:

Given data: $p_x = 150$ N/mm² (tensile), $p_y = 100$ N/mm² (compressive) and $\sigma_1 = 180$ N/mm².

We know,

$$\sigma_1 = \left(\frac{p_x + p_y}{2}\right) + \sqrt{\left(\frac{p_x + p_y}{2}\right)^2 + q^2}$$

i.e.,

$$180 = \left(\frac{150 - 100}{2}\right) + \sqrt{\left(\frac{150 - (-100)}{2}\right)^2 + q^2}$$

$$180 = 25 + \sqrt{(125)^2 + q^2}$$

$$125^2 + q^2 = 155^2$$

Solving,

$$q = 91.65 \text{ N/mm}^2$$

Maximum shear stress at the point is

$$(p_t)_{max} = \frac{1}{2}\sqrt{(p_x - p_y)^2 + q^2}$$

$$= \frac{1}{2}\sqrt{\left(\frac{150 + 100}{2}\right)^2 + 91.65^2}$$

$$= 77.49 \text{ N/mm}^2$$

SOLVED PROBLEM 2.7

A resultant tensile stress of 70 N/mm^2 inclined at 20° to the normal of that plane is acting at a point in a material. Another tensile direct stress of 40 N/mm^2 acts on a plane at right angles to the previous one. Find (i) the resultant stress on the second plane, (ii) the principal planes and stresses and (iii) the plane of maximum shear and intensity.

Solution:

Given data: Resultant tensile stress = 70 N/mm^2, $\alpha = 20°$ and direct stress = 40 N/mm^2.

Resolving the resultant stress, we get the normal and shear stresses on the horizontal plane.

Normal stress on the horizontal plane, $p_x = 70 \cos 20° = 65.78$ N/mm^2

Shear stress on the horizontal plane, $q = 70 \sin 20° = 23.94$ N/mm^2.

Shear stress is a complimentary one, hence the shear stress on the vertical plane is 23.94 N/mm^2.

$\therefore$ Resultant stress on the second (vertical) plane $= \sqrt{40^2 + 23.94^2}$

$= 46.61$ N/mm^2

The modified stress condition is shown in Figure 2.19.

Then

$$\sigma_1 = \frac{p_x + p_y}{2} + \sqrt{\left(\frac{p_x - p_y}{2}\right)^2 + q^2}$$

$$= \left(\frac{65.78 + 40}{2}\right) + \sqrt{\left(\frac{65.78 - 40}{2}\right)^2 + 23.94^2}$$

$$= 52.89 + 58.05$$

$$= 110.95 \text{ N/mm}^2 \text{ (tensile)}$$

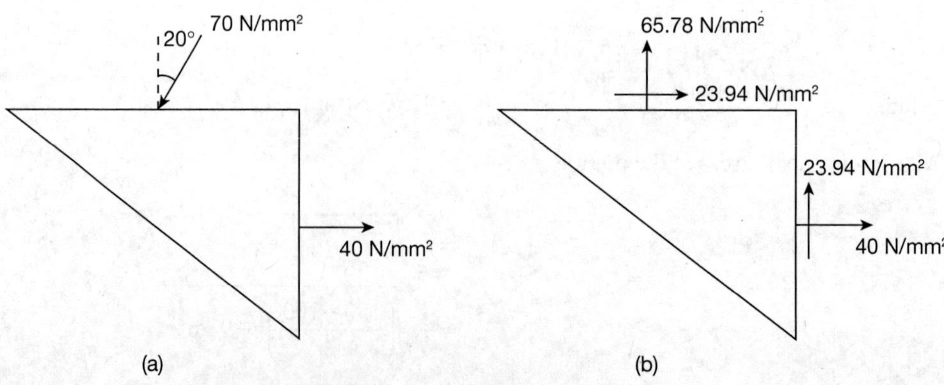

Fig. 2.19

Similarly,
$$\sigma_2 = \left(\frac{p_x - p_y}{2}\right) - \sqrt{\left(\frac{p_x - p_y}{2}\right)^2 + q^2}$$
$$= \left(\frac{65.78 - 40}{2}\right) - \sqrt{\left(\frac{65.78 + 40}{2}\right)^2 + 23.94^2}$$
$$= 45.2 \text{ N/mm}^2 \text{ (tensile)}$$
$$\tan 2\theta = \frac{2q}{p_x - p_y} = \frac{2 \times 23.94}{65.78 - 40} = 1.857$$
$$\therefore \quad 2\theta = 61.70°$$
$$\therefore \quad \theta = 30.85° = 30°51'$$

Inclination of planes, $\theta_1 = 30°51'$ and $\theta_2 = 120°51'$
$$(p_t)_{max} = \frac{\sigma_1 - \sigma_2}{2} = \frac{110.95 - 45.2}{2}$$
$$= 32.875 \text{ N/mm}^2$$

SOLVED PROBLEM 2.8

Determine the direction of principal planes, normal stress, and tangential stress of the strained material as shown in Figure 2.20.

Solution:
Given data: $p_x = 50$ N/mm² (tensile), $p_y = 30$ N/mm² (compressive) and $q = 20$ N/mm².

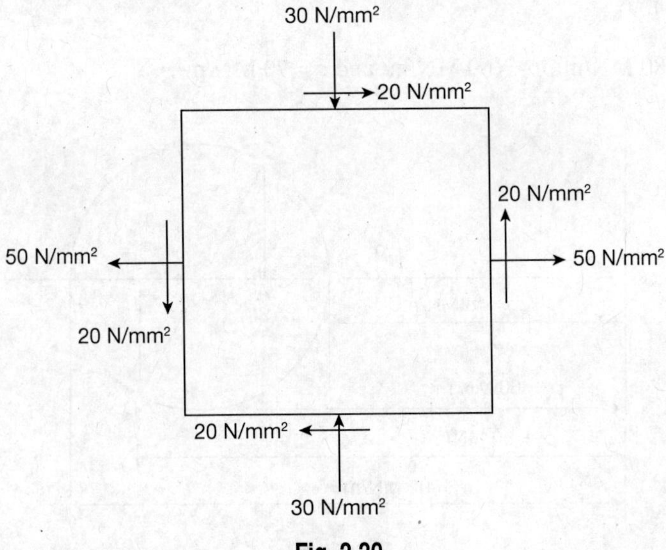

Fig. 2.20

$$\tan 2\theta = \frac{2q}{p_x - p_y} = \frac{2\times 20}{50-(-30)} = 0.5$$

$$\therefore \quad 2\theta = 26.56°$$

$$\therefore \quad \theta = 13.28° = 13°16.8'$$

Inclination of principal plane, $\theta_1 = 13°16.8'$ and $103°16.8'$.

Normal stress,
$$p_n = \left(\frac{p_x + p_y}{2}\right) + \left(\frac{p_x - p_y}{2}\right)\cos 2\theta + q\sin 2\theta$$

$$= \left(\frac{50-30}{2}\right) + \left(\frac{50+30}{2}\right)\cos(2\times 13.28) + 20\sin(2\times 13.28)$$

$$= 54.72 \text{ N/mm}^2$$

Tangential stress,
$$p_t = \left(\frac{p_x - p_y}{2}\right)\sin 2\theta - q\cos 2\theta$$

$$= \left(\frac{50-(-20)}{2}\right)\sin(2\times 13.28) - 20\cos(2\times 13.28)$$

$$= -2.24 \text{ N/mm}^2$$

SOLVED PROBLEM 2.9

At a certain point the stresses in a strained material acting on two planes at right angles to each other are 80 N/mm² and 60 N/mm², both tensile. They are accompanied by a shear stress of 20 N/mm². Find graphically the location of principal planes and evaluate the principal stresses.

Solution:

Given data: $p_x = 80$ MN/m², $p_y = 60$ MN/m² and $q = 70$ MN/m².

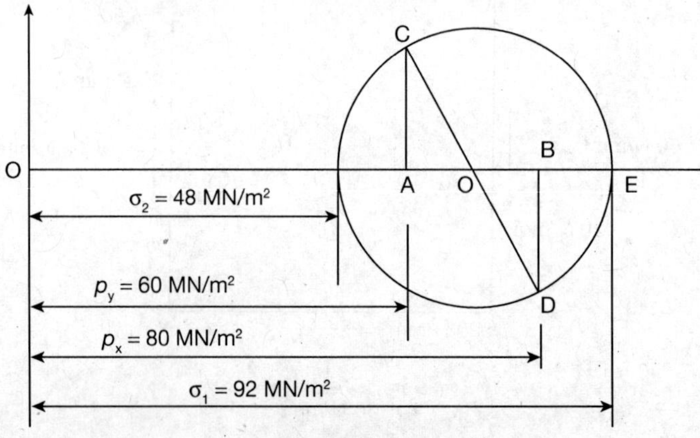

Fig. 2.21

A scale of 1 cm = 10 MN/m^2 is taken

The construction is made as explained in Section 2.5.1 with a stress scale of 1 cm = 10 MN/m^2

The major principal stress, $\sigma_1 = 92$ MN/m^2

The minor principal stress, $\sigma_2 = 48$ MN/m^2.

Further, OE and OA are the major and minor principal planes.

SOLVED UNIVERSITY QUESTIONS

SOLVED PROBLEM 2.10

At a certain point in a strained material the direct stresses on the two planes at right angles to each other are 40 N/mm^2 (along the horizontal direction) and 20 N/mm^2 (along the vertical direction). They are accompanied by a shear stress of 20 N/mm^2. Compute the principal stresses and any one of the principal planes (Anna Univ., June 2005, CE).

Solution:

Given data: $p_x = 40$ N/mm^2, $p_y = 20$ N/mm^2 and $q = 20$ N/mm^2.

$$\sigma_1 = \left(\frac{p_x + p_y}{2}\right) + \sqrt{\left(\frac{p_x - p_y}{2}\right)^2 + q^2}$$

$$= \left(\frac{40+20}{2}\right) + \sqrt{\left(\frac{40-20}{2}\right)^2 + 20^2}$$

$$= 30 + 20.25$$

$$= 52.25 \text{ N/mm}^2 \text{ (tensile)}$$

$$\sigma_2 = \left(\frac{p_x + p_y}{2}\right) - \sqrt{\left(\frac{p_x - p_y}{2}\right)^2 + q^2}$$

$$= 30 - 22.36$$

$$= 7.64 \text{ N/mm}^2 \text{ (tensile)}$$

$$\tan 2\theta = \frac{2q}{p_x - p_y} = \frac{2 \times 20}{40 - 20} = 2$$

$$\therefore \theta = 31°43'$$

SOLVED PROBLEM 2.11

At a point in a body, there are two mutually perpendicular stresses of 80 N/mm^2 and 40 N/mm^2, of tensile nature. Each stress is accompanied by a shear stress of 60 N/mm^2. Determine the normal, shear and resultant stress in an oblique plane at an angle of 45° with the axis of the major principal stress (Anna Univ., June 2009, ME).

Solution:

Given data: $p_x = 80$ N/mm², $p_y = 40$ N/mm², $q = 60$ N/mm² and $\theta = 45°$.

Normal stress
$$p_n = \left(\frac{p_x + p_y}{2}\right) + \left(\frac{p_x - p_y}{2}\right)\cos 2\theta + q \sin 2\theta$$

$$= \left(\frac{80+40}{2}\right) + \left(\frac{80-40}{2}\right)\cos(2\times 45°) + 60\sin(2\times 45°)$$

$$= 120 \text{ N/mm}^2$$

Shear stress,
$$p_t = \left(\frac{p_x - p_y}{2}\right)\sin 2\theta - q\cos 2\theta$$

$$= \left(\frac{80-40}{2}\right)\sin(2\times 45°) - 60\times\cos(2\times 45°)$$

$$= 20 \text{ N/mm}^2$$

Resultant stress,
$$p_r = \sqrt{p_n^2 + p_t^2}$$

$$= \sqrt{(120)^2 + (20)^2}$$

$$= 121.65 \text{ N/mm}^2$$

SOLVED PROBLEM 2.12

At a point in a strained material, there is a horizontal tensile stress of 100 N/mm² and an unknown vertical stress. There is also a shear stress of 30 N/mm² on this plane inclined at 30° to the vertical, and the normal stress is found to be 90 N/mm² tensile. Find the unknown vertical stress and also principal stresses and maximum shear stress (Anna Univ., Nov. 2008, ME).

Solution:

Given data: $p_x = 100$ N/mm² (tensile), $q = 30$ N/mm², inclination $\theta = 30°$ and $p_n = 90$ N/mm². Here, the unknown vertical stress, p_y, is to be found.

Normal stress,
$$p_x = \left(\frac{p_x + p_y}{2}\right) + \left(\frac{p_x - p_y}{2}\right)\cos 2\theta + q\sin 2\theta$$

$$= \left(\frac{100 + p_y}{2}\right) + \left(\frac{100 - p_y}{2}\right)\cos(2\times 30) + 30\sin(2\times 30)$$

Solving,
$$p_y = -3.92 \text{ MPa (compressive)}$$

Tangential stress, $p_t = \left(\dfrac{p_x - p_y}{2}\right)\sin 2\theta - q\cos 2\theta$

$$= \left(\frac{100-(-3.92)}{2}\right)\times \sin(2\times 30°) - 30\cos(2\times 30°)$$

$$= 47.09 \text{ MPa}$$

Principal stress, $\dfrac{\sigma_1}{\sigma_2} = \left(\dfrac{p_x + p_y}{2}\right) \pm \sqrt{\left(\dfrac{p_x - p_y}{2}\right)^2 + q^2}$

$$= \frac{100-3.92}{2} + \sqrt{\left(\frac{100+3.92}{2}\right)^2 + 30^2}$$

$$= 108.03$$

Solving, $\sigma_1 = 106$ MPa

i.e., $\sigma_2 = 49.4$ MPa

SALIENT POINTS

- Shear stresses acting on planes at right angles are equal and are called complementary shear stresses.
- State of a plane stress is defined as that condition in which there is no stress component acting on a pair of parallel faces.
- State of a plane strain is defined as that condition in which there is no strain component acting in a particular direction or ignoring the stress in that direction.
- Principal stress may be defined as a normal stress acting on a plane wherein there are no shear stresses. These principal stresses are named as major principal stress, σ_1, minor principal stress, σ_2 and intermediate principal stress, σ_3.
- Planes on which the principal stresses take place are called principal planes wherein there is no influence of shear stress.
- Maximum shear stresses occur on the planes inclined at 45° to those principal stresses.
- Principal strains are the strains which occur along the respective principal strain directions.
- Mohr's diagram is an excellent visualization of the orientations of various planes.
- Origin of planes or poles, O_p is an unique point on the Mohr's circle, from which a line drawn parallel to any arbitrarily chosen plane intersects the Mohr's circle at a point, the coordinates of which are stress components (i.e., the normal and shear stresses) at that plane.

QUESTIONS

1. What are the principal stresses and principal planes? Derive the expression for maximum shear stress in a 2D stress system.
2. Show that the sum of the normal components of the stresses on any two planes at right angles is constant in a material subjected to a 2D stress system.

3. Derive an expression for the stress on an oblique section of an elastic rectangular body where it is subjected to a direct stress in one plane only.
4. Explain the significance of Mohr's circle and its uses.
5. Using Mohr's circle, derive expressions for normal and tangential stresses on a diagonal plane of a piece of material in pure shear.
6. The direct tensile stress acting at a point in two mutually perpendicular planes are 500 N/mm^2 and 200 N/mm^2. The complementary stress acting on these planes is 250 N/mm^2. Determine the normal and shear stresses on the two planes, which are equally inclined to the planes carrying the direct stress mentioned above.
7. At a point in a strained material, the principal stresses are 150 N/mm^2 (tensile) and 60 N/mm^2 (compressive). Determine the resultant stress in magnitude and direction on a plane inclined at 45° to the axis of the major principal stress. What is the maximum intensity of shear in the material at that point?
8. At a certain point in an elastic material, on planes at right angles, direct stress 150 N/mm^2 (tension) and 50 N/mm^2 (compression) act. The greater principal stress is limited to 180 N/mm^2. Estimate the shearing stress that may be applied to the given planes and find the maximum shearing stress at that point.
9. Stresses at a point in a material under plane stress condition are $p_x = 90$ MN/m^2 and $q = 70$ MN/m^2. Find the maximum allowable stress p_y such that the major principal stress is not to exceed 150 MN/m^2.
10. At a certain point in a material under stress, the intensity of the resultant stress on a vertical plane is 1,000 kg/cm^2 inclined at 30° to the normal to that plane and the stress on a horizontal plane has a normal tensile component of intensity 600 kgf/cm^2. Find the magnitude and direction of the resultant stress on the horizontal plane and the principal stresses.
11. Normal stresses acting on two mutually perpendicular planes are 150 kN/m^2 and 60 kN/m^2 and the shear stress on each plane is 110 kN/m^2 and 60 kN/m^2, respectively. The shear stress on each plane is 110 kN/m^2. Draw Mohr's circle and find:
    (i) the principal stresses and planes and
    (ii) the shear and normal stress acting on a plane making an angle of 70° with the major principal stress.
12. In a strained material, the stress $p_x = 80$ MN/m^2 and $p_y = 40$ MN/m^2. Construct a Mohr's circle and find the angle of the plane for which the ratio of normal stress to shearing stress is a maximum.

3

Strain Energy

LEARNING OBJECTIVES

3.1 DEFINITIONS

External forces acting on a body do certain work causing deformation at their points of application. This means that there is a certain work done to strain the body from its undeformed state. As per the first law of thermodynamics, the work done on a body by the external forces in a certain time interval is equal to the change in the internal energy.

Strain energy is the energy which is stored in the body due to the effect of straining. Perfect elastic bodies show the same amount of work done during the loading and unloading process. Thus, all the work done during the elastic deformation is stored and is recovered at the release of loads.

Elastic strain energy is the recoverable stored energy in an elastic body.

Resilience is the total energy stored in a body within the elastic limit. When the external forces are removed from a strained body, the body starts doing work. Keeping this in view, resilience is also defined as the capacity of a strained body for doing work on the removal of the forces which caused the strain.

Proof resilience is the maximum strain energy stored in a body. The strain energy is maximum when a body is strained up to its elastic limit. Thus, proof resilience can also be defined as the quantity of strain energy stored in a body when strained up to its elastic limit.

Modulus of resilience is the greatest amount of strain energy per unit volume that a material can absorb without exceeding the elastic limit. That is,

$$\text{Modulus of resilience} = \frac{\text{Proof resilience}}{\text{Volume of the body}}$$

3.2 STRAIN ENERGY DUE TO LOADING

Strain energy stored depends on the type of loading, viz., gradual loading, sudden loading and impact loading.

3.2.1 Strain Energy Due to Gradual Loading

A prismatic bar loaded gradually is shown in Figure 3.1. The load deflection diagram is shown in Figure 3.2 which will be a straight line during the elastic behaviour of the material. Thus, for a load of P_g the corresponding elongation is δl_g.

Now, if the load is increased by a load increment of dP_g the deformation increases by an amount $d\delta l_g$ from δl_g. Thus, the load P_g does positive work $P_g d\delta l_g$ which is shown in the diagram by hatched strip. This area is the energy stored in the bar in the form of potential energy or strain energy.

If the increment load dP_g is removed, the stored energy $P_g d\delta l_g$ is transformed back in to the work of raising the load through the distance $d\delta l_g$, i.e., the lower end of the bar moves up through the distance $d\delta_{st}$. This property of an elastic body to absorb and release energy with changes in loading has an important practical application in the design of structures and machine parts subjected to dynamic loading.

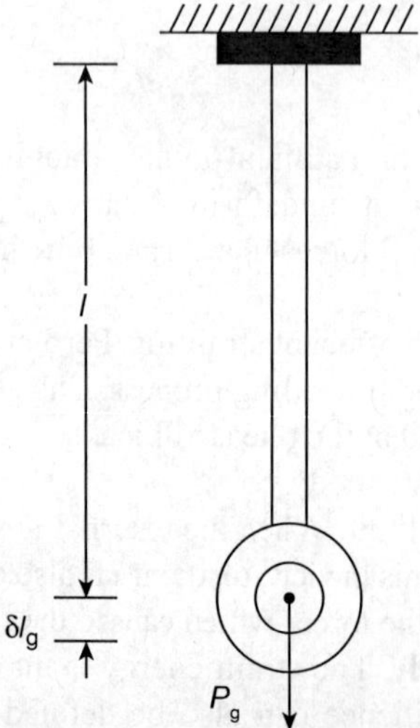

Fig. 3.1 Gradual loading.

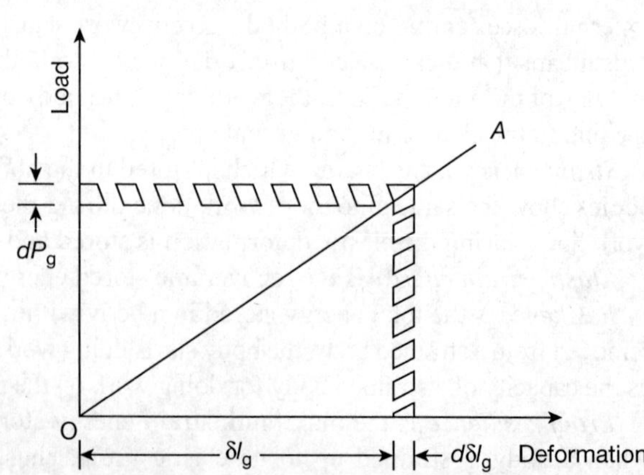

Fig. 3.2 Load-deformation diagram.

The total energy due to the load P is equal to the sum of all such elemental strips. This means that the strain energy U is given as (area of the diagram).

$$U = \frac{P_g \delta l_g}{2} \tag{3.1}$$

where δ_{st} is the final deformation.

$$U = \frac{\text{stress} \times \text{area} \times \delta}{2}$$

$$U = \frac{p_g A}{2} \times \delta l_g \tag{3.2a}$$

But

$$\delta l_g = \text{strength} \times \text{length}$$

$$= \frac{\text{stress}}{E} \times \text{length}$$

i.e.,

$$\delta l_g = \frac{p_g}{E} \times l$$

Substituting for δl_g in Equation (3.2a)

$$U = \frac{1}{2} p_g A \times \frac{p_g}{E} \times l$$

$$= \frac{p_g^2 A l}{2E} \tag{3.2b}$$

i.e.,

$$U = \frac{p_g^2 A l}{2E}$$

Substituting, $V = A \times l$, then

$$U = \frac{p_g^2 V}{2E} \tag{3.3}$$

Modulus of resilience

$$= \frac{\text{Strain energy}}{\text{Unit volume}}$$

$$= \frac{p_g^2}{2E} \tag{3.4}$$

If p_g is the stress at the elastic limit

Proof resilience

$$= \frac{p_{ge}^{\ 2}}{2E} \times V \tag{3.5}$$

This discussion for axial tension is also applicable for axial compression.

3.2.2 Strain Energy Due to Sudden Loading

In the case of sudden application of load, the load is constant throughout the process of the deformation of the body. If δl_S is the deformation due to the application of sudden load P_S, then

Work done by the load = $P_S \times \delta l_S$

Equating the strain energy stored in the body to the work done, we have

$$\frac{p_S^2}{2E} \times V = P_S \times \delta l_S$$

Substituting for δl_S

$$\frac{p_S^2}{2E} \times A \times l = P_S \times \frac{p_S}{E} \times l$$

$$\frac{p_S A}{2} = P_S$$

i.e.,

$$p_S = \frac{2P_S}{A} \tag{3.6a}$$

i.e.,

$$(p_S)_{max} = 2p_S \tag{3.6b}$$

Equation (3.6b) shows that the maximum stress induced due to suddenly applied load is twice the stress induced when the same load is applied gradually.

3.2.3 Strain Energy Due to Impact Loading

A vertical rod fixed at the top end and having a collar at lower end is considered (Figure 3.3). Let the load P_i be dropped from a height h, on the collar. Due to this impact load, there will be some extension in the rod.

Work done by the load = Load × (Distance moved) = $P_i(h + \delta l_i)$

Strain energy stored in the bar $= \dfrac{p_i^2}{2E} \times V$

Equating the work done to the strain energy stored, we have

$$P_i(h + \delta l_i) = \frac{p_i^2}{2E} \times A \times l \tag{3.7}$$

Substituting for $\delta l_i = \dfrac{p_i}{E} \times l$ and

Rearranging $\quad \dfrac{p_i^2}{2E} Al - P_i \dfrac{p_i}{E} l - P_i h = 0$

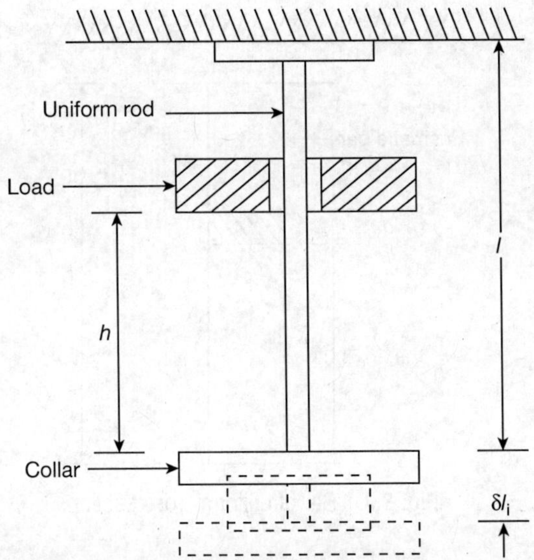

Fig. 3.3 Impact load arrangement.

Multiplying by $\dfrac{2E}{Al}$ on both the sides, we get

$$p_i^2 - P_i\frac{p_i}{E}l\times\frac{2E}{Al} - P_i h\times\frac{2E}{Al} = 0$$

$$p_i^2 - \left(2\frac{P_i}{A}\right)p_i - \frac{2P_i Eh}{Al} = 0$$

Solving, we have

$$p_i = \frac{\dfrac{2P_i}{A} \pm \sqrt{\left(\dfrac{2P_i}{A}\right)^2 + 4\times\dfrac{2P_i Eh}{Al}}}{2\times 1}$$

Neglecting negative root, we have

$$p_i = \frac{P_i}{A}\left(1+\sqrt{1+\frac{2AEh}{P_i l}}\right) \tag{3.8}$$

3.3 STRAIN ENERGY DUE TO SELF-WEIGHT

A bar of uniform cross section A and of length l in a vertically hung position is considered (Figure 3.4). Further, a strip of d_x at a distance x from the lower end is also considered.

The strip is acted upon by the weight of the bar of length, x.

Let ρ be the density of the material of the bar.

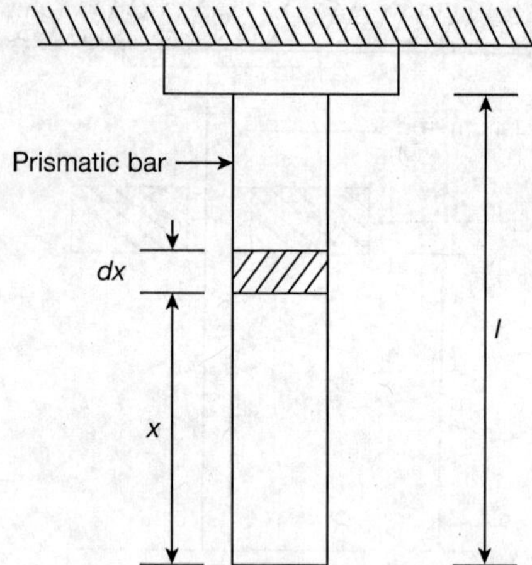

Fig. 3.4 Bar of uniform cross section.

Weight acting on the strip $= (A \times x) \times \delta = \delta A x$

Strain in the strip of thickness, $dx = \dfrac{\text{Elongation in } dx}{dx}$

$$= \frac{\delta x_s}{dx}$$

where δx_s is the elongation in dx

Stress in the strip, $p_{st} = \dfrac{W_x}{A} = \dfrac{\delta A x}{A} = \rho x$

$$E = \frac{\text{Stress}}{\text{Strain}} = \frac{p_S}{\delta x_S / dx} = \frac{\delta x}{\delta x_S / dx}$$

i.e., $\delta x_S = \dfrac{\rho x dx}{E}$

Strain energy stored in the strip, dU, is given as

$$dU = (\text{Average weight}) \times (\text{Elongation of } dx)$$

$$= \left(\frac{W_x}{2}\right) \delta x_s$$

$$= \frac{\rho A x}{2} \times \frac{\rho x dx}{E}$$

$$= \frac{1}{2E} \rho^2 A x^2 dx$$

Total energy stored over the length l due to self-weight

$$U = \int_0^l l dU = \frac{\rho^2 A}{2E} \int_0^l x^2 dx$$

After integrating and fixing the limits

$$U = \frac{A\rho^2 l^3}{6E} \tag{3.9}$$

3.4 STRAIN ENERGY DUE TO SHEAR FORCE

A rectangular block of length l, height h and breadth b is being fixed at the bottom face DC and subjected to a shear force P_f (Figure 3.5).

The face AB moves through a distance δl relative to the face DC as the shear force increased gradually from zero to its final value P_f. Let q be the shear stress produced and ϕ be the shear strain.

Shear stress, $$q = \frac{P_f}{l \times b}$$

Shear strain, $$\phi = \frac{BB_1}{BC}$$

$\therefore$ $$BB_1 = \delta l = \phi\, BC = h\theta$$

Then, Shear force, $P_f = qlb$

As the shear force is applied gradually, the average load is $\frac{P_f}{2}$.

$\therefore$ Work done by the gradually applied load = (Average load) × (Distance)

$$= \frac{P_f}{2}\delta l$$

$$= \frac{qlb}{2} h\theta$$

$$= \frac{1}{2} \times q \times \theta\,(l \times b \times h)$$

$$= \frac{1}{2} q\theta \text{ (Volume of block)}$$

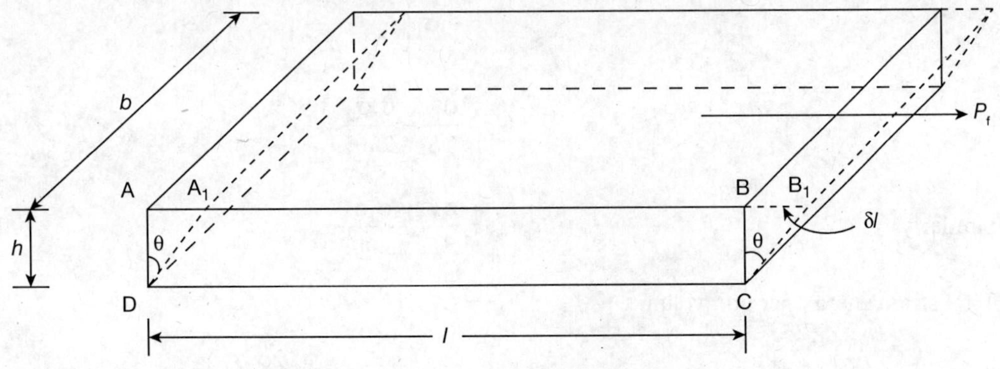

Fig. 3.5 Rectangular block under shear force.

Shear modulus, $G = \dfrac{\text{Shear stress}}{\text{Shear strain}}$

i.e., Shear strain, $\phi = \dfrac{q}{G}$

i.e., Work done $= \dfrac{q^2}{2G}V$ (3.10)

and V = volume of block and $\phi \approx \theta$

∴ Strain energy stored due to shear force $= \dfrac{q^2}{2G}V$ (3.11)

If q_p is the proof shear stress, then the strain energy per unit volume is equal to $(q_p^2/2G)$ which is known as modulus of shear resilience.

Strain energy due to flexural and torsional shear are discussed elsewhere.

3.5 STRAIN ENERGY IN TERMS OF PRINCIPAL STRESSES

Let the major and minor principal stresses in a strained material be σ_1 and σ_2 and of the same sign. Then, the principal strains are given as

$$\varepsilon_1 = \frac{\sigma_1}{E} - \frac{\sigma_2}{mE}$$

and

$$\varepsilon_2 = \frac{\sigma_2}{E} - \frac{\sigma_1}{mE}$$

Strain energy per unit volume due to σ_1, $U_1 = \dfrac{1}{2}(\text{Stress})(\text{Strain})$

$$= \frac{1}{2}\sigma_1\left(\frac{\sigma_1}{E} - \frac{\sigma_2}{mE}\right)$$

i.e.,

$$U_1 = \frac{\sigma_1^2}{2E} - \frac{\sigma_1\sigma_2}{2mE} \quad (3.12)$$

Similarly,

$$U_2 = \frac{\sigma_2^2}{2E} - \frac{\sigma_1\sigma_2}{2mE} \quad (3.13)$$

Total strain energy per unit volume $U = U_1 + U_2$

i.e.,

$$U = \frac{1}{2E}\left(\sigma_1^2 + \sigma_2^2 - \frac{2\sigma_1\sigma_2}{m}\right) \quad (3.14)$$

In the case of 3*D* principal stress (with same sign) condition, strains in three directions are

$$\varepsilon_1 = \frac{\sigma_1}{E} - \frac{\sigma_2 + \sigma_3}{mE}$$

$$\varepsilon_2 = \frac{\sigma_2}{E} - \frac{\sigma_3 + \sigma_1}{mE}$$

and

$$\varepsilon_3 = \frac{\sigma_3}{E} - \frac{\sigma_1 + \sigma_2}{mE}$$

Total strain energy per unit volume $U = \frac{1}{2}\sigma_1\varepsilon_1 + \frac{1}{2}\sigma_2\varepsilon_2 + \frac{1}{2}\sigma_3\varepsilon_3$

Substituting for ε_1, ε_2 and ε_3 and reducing, we have

$$U = \frac{1}{2E}\left(\sigma_1^2 + \sigma_2^2 + \sigma_3^2 - \frac{2\sigma_1\sigma_2 + 2\sigma_3\sigma_2 + 2\sigma_3\sigma_1}{m}\right) \quad (3.15)$$

SOLVED PROBLEM 3.1

A square steel bar of 4 cm side and 500 cm length is subjected to an axial pull of 60 kN. Compute the alterations in length and sides of the bar. Also, compute the amount of energy stored in the bar during the extension. The modulus of elasticity and Poisson's ratio of the material are 2×10^5 N/mm^2 and 0.30, respectively.

Solution:

Given data: Side of steel bar = 4 cm, length of bar = 500 cm, P = 60 kN, $E = 2 \times 10^5$ N/mm^2 and $\mu = 0.30$.

Area of cross section of the bar, $A = 4 \times 4 = 16$ cm^2

Volume of the bar, $V = 16 \times 500 = 8{,}000$ cm^3

Alteration in the length of the bar,

$$\delta l = \frac{Pl}{AE}$$

$$= \frac{60 \times 1{,}000 \times 500}{16 \times 100 \times 2 \times 10^5}$$

$$= 0.093$$

Linear strain, $e = \frac{\delta l}{l} = \frac{0.093}{500 \times 10} = 0.00001875$

Lateral strain = $e \times \mu = 0.00001875 \times 0.3 = 0.00000544$

Change in the side length $= 0.00000564 \times 40 = 0.000226$ mm

Energy stored in the bar, $U = \frac{p_g^2}{2E} \times V$

$$= \left(\frac{60 \times 1,000}{1,600}\right)^2 \times \frac{1,600 \times 5,000}{2 \times 2 \times 10^5}$$

$$= 28,125 \text{ N} \cdot \text{mm}$$

SOLVED PROBLEM 3.2

A wire rope of a crane has a cross-sectional area of 8 cm^2 and carries a load of 12 kN. The load was lowered at a uniform rate of 0.5 m/s. During the process of lowering, the wire got jammed suddenly and at that moment the unwound length of the rope is 8 m. Compute the stress induced in the wire rope due to the sudden stoppage. Take $E = 2 \times 10^5$ N/mm^2. The self-weight of the rope may be neglected.

Solution:

Given data: Cross-sectional area = 8 cm^2, P_g = 12 kN, rate of loading = 0.5 m/s and $E = 2 \times 10^5$ N/mm^2.

$$\text{Kinetic energy of the crane} = \frac{1}{2}(\text{Mass})(\text{Velocity})^2$$

$$= \frac{1}{2}\left(\frac{10}{8}\right)V^2$$

$$= \frac{1}{2}\left(\frac{12 \times 1,000}{9.81}\right)(0.5)^2$$

$$= 152.905 \text{ N} \cdot \text{m}$$

$$= 152,905 \text{ N} \cdot \text{mm}$$

When the wire rope got suddenly jammed, the entire kinetic energy of the crane is absorbed in the wire rope.

$$\text{Strain energy stored for gradual application of load,} \quad U = \frac{p_g^2}{2E}V$$

$$= p_g^2\left(\frac{8 \times 100 \times 8 \times 1,000}{2 \times 2 \times 10^5}\right)$$

$$= 16p_g^2$$

Now

$$\text{Energy stored in the wire rope} = \text{Kinetic energy of the rope}$$

$$16p_g^2 = 305,800$$

$$p_g = \sqrt{\frac{305,800}{16}}$$

$$= 138.24 \text{ N/mm}^2$$

Stress induced in the wire rope = 138.24 N/mm^2

SOLVED PROBLEM 3.3

A bar has a cross-sectional area of 650 mm^2 and a length of 1.2 m. Compute the proof resilience if the stress at the elastic limit is 150 N/mm^2. E of the material is 2×10^5 N/mm^2. Estimate the value of the applied load which may be suddenly applied without exceeding the elastic limit. Also, calculate the value of a gradually applied load which produces the same extension as that produced by the suddenly applied load above.

Solution:

Given data: Cross-sectional area = 650 m^2, length = 1.2 m, elastic limit = 150 N/mm^2 and $E = 2 \times 10^5$ N/mm^2.

Volume of the bar, $V = 650 \times 1.2 \times 1,000 = 780,000$ mm^3

Proof resilience $$= \frac{p_{ge}^2}{2E} \times V$$

$$= \frac{150^2}{2 \times 2 \times 10^5} 780,000$$

$$= 43,875 \text{ N} \cdot \text{m}$$

For suddenly applied load, $$p_s = \frac{2P_S}{A}$$

Here, the suddenly applied load should not induce stress exceeding the elastic limit.

i.e., $$p_S = \frac{(p_S)_{max}}{2}$$

$$P_S = p_S \times A$$

$$P_S = \frac{150 \times 650}{2} = 48,750 \text{ N}$$

$\therefore$ $$P_S = 48.75 \text{ kN}$$

Deformation produced due to suddenly applied load is twice as great as that of the load applied gradually.

Thus, the value of gradually applied load which can produce the same deformation is

$$P_g = 2P_S = 2 \times 48.75 = 97.50 \text{ kN}$$

SOLVED PROBLEM 3.4

A load of 200 N falls through a height of 25 mm on to a collar rigidly attached to the lower end of a vertical bar of 1,500 mm length and 160 mm² cross-sectional area. The upper end of the bar is fixed. Find the maximum (i) instantaneous stress and (ii) instantaneous elongation of the bar $E = 2 \times 10^5$ N/mm².

Solution:
Given data: Impact load, $P_i = 200$ N, $l = 1{,}500$ mm, $h = 25$ mm, $A = 160$ mm² and $E = 2 \times 10^5$ N/mm².

$$p_i = \frac{P_i}{A}\left(1+\sqrt{1+\frac{2AEh}{P_i l}}\right)$$

$$= \frac{200}{160}\left(1+\sqrt{1+\frac{2\times 160\times 2\times 10^5\times 25}{200\times 1{,}500}}\right)$$

Instantaneous stress = 92.55 N/mm²

$$E = \frac{\text{Stress}}{\text{Strain}} = \frac{p_i}{\frac{\delta l}{l}}$$

i.e.,

$$\delta l = \frac{p_i l}{E}$$

$$= \frac{92.55\times 1{,}500}{2\times 10^5}$$

$$= 0.694 \text{ mm}$$

SOLVED PROBLEM 3.5

A bar of 1.5 cm diameter gets stretched by 2.5 mm under a steady load of 10 kN. What stress would be produced in the same bar by a weight of 12 kN which falls vertically a distance of 5 cm on to a rigid circular collar attached into its end? The bar is initially unstressed. $E = 2 \times 10^5$ N/mm².

Solution:
Given data: Diameter of bar = 1.5 cm, $\delta = 2.5$, $P = 10$ kN, $h = 5$ cm and $E = 2 \times 10^5$ N/mm².

Area of the bar,

$$A = \frac{\pi}{4}\times (15)^2 = 176.71 \text{ mm}^2$$

Steady load condition

$$E = \frac{\text{Stress}}{\text{Strain}} = \frac{(\text{Steady load})/\text{Area}}{\delta l/l}$$

$$2.5\times 10^5 = \frac{10\times 1{,}000/176.71}{2.5/l}$$

Rearranging,

$$l = 11{,}044.375 \text{ cm}$$

Instantaneous falling head condition

$$p_i = \frac{P_i}{A}\left[1+\sqrt{1+\frac{2AEh}{P_i l}}\right]$$

$$= \frac{12\times 1{,}000}{176.71}\left[1+\sqrt{1+\frac{2\times 176.71\times 2\times 10^5\times 5\times 10}{12\times 1{,}000\times 8{,}835.5}}\right]$$

Solving

Stress produced = 465.8 N/mm^2

SOLVED PROBLEM 3.6

A vertical steel rod of 30 mm diameter checks the fall on its end of weight of 3.5 kN which drops through a distance of 5 mm before it strikes the rod. Find the shortest length of rod which will be at the impact if the stress is not to exceed 130 MN/m^2. E = 210 GN/m^2.

Solution:

Given data: Diameter of rod = 30 mm, h = 5 mm, impact stress = 130 MN/m^2 and E = 2,106 GN/m^2.

Let l be the shortest possible length of the rod.

Total distance through which the weight fall = (Height of drop) + (Elongation of bar)

$$= 0.005 + \delta l_i$$

$$= 0.005 + \left(\frac{(p_i)_m \times l}{E}\right)$$

where $(p_i)_m$ is the maximum stress allowed.

Then

Total distance $$= 0.005 + \frac{130\times 10^6 \times l}{210\times 10^9}$$

$$= 0.005 + 0.0006141$$

Also

Potential energy given by the weight = Strain energy stored in the rod

$$P_i(h+\delta l) = \frac{p_i^2 V}{2E}$$

$$P_i(h+\delta l) = \frac{\left(p_i^2\right)_m \times A \times l}{2E}$$

$$3.5\times 10^3\,(0.005+0.0006141) = \frac{\left(130\times 10^6\right)^2 \times \frac{\pi}{4}\times 0.03^2 \times l}{2\times 210\times 10^9}$$

Rearranging and solving for l, we have l = 0.6908 m

Shortest length of the rod = 0.6908 m

SOLVED PROBLEM 3.7

An unknown weight falls through a height of 12 mm on a collar rigidly attached to the lower end of a vertical bar 5,000 mm long and 800 mm^2 in cross section. If the maximum extension of the rod is to be 4 mm, what is the corresponding stress and magnitude of the unknown weight? $E = 2 \times 10^5$ N/mm^2.

Solution:

Given data: $h = 12$ mm, $l = 5{,}000$ mm, $A = 800$ mm^2; $\delta l_i = 4$ mm and $E = 2 \times 10^5$ N/mm^2.

$$\text{Stress} = E \times \text{strain}$$

$$= \frac{2\times10^5}{1}\times\frac{4}{5{,}000} = 160 \text{ N/mm}^2$$

$$\therefore \quad P_i(h+\delta l_i) = \frac{p_i^2}{2E}V$$

$$\therefore \quad P_i = \frac{p_i^2}{2E}\ \frac{V}{h+\delta l_i}$$

$$= \frac{160^2}{2\times2\times10^5}\times\frac{5{,}000\times800}{12+4}$$

$$= 16{,}000 \text{ N}$$

$$\text{Unknown weight} = 16{,}000 \text{ N}$$

SOLVED PROBLEM 3.8

A vertical rod of 5 m length with 35 mm diameter is fixed at the upper end and provided with a collar at the lower end. A weight of 1.5 kN is dropped at the collar. Compute the height of drop if the maximum instantaneous stress is not to exceed 150 N/mm^2. Also, find the instantaneous elongation. Take $E = 0.205 \times 10^6$ N/mm^2.

Solution:

Given data: l = 5 m, diameter of the rod = 35 mm, P_i = 1.5 kN, $(p_i)_{max}$ = 105 N/mm^2 and $E = 0.205 \times 10^6$ N/mm^2

Area of cross section, $A = \frac{\pi}{4}\times35^2 = 961.63$ mm^2.

Maximum instantaneous elongation $= \frac{(p_i)_{max}\, l}{E}$

$$= \frac{150\times5\times1{,}000}{0.205\times10^6} = 3.66 \text{ mm}$$

Work done by the falling weight $= P(h+\delta l)$

$$= 1.5\times1{,}000(h+3.66)$$

Strain energy stored in the bar $= \dfrac{(p_i)^2_{max} V}{2E}$

$$= \frac{150^2}{2 \times 0.205 \times 10^6}(961.63 \times 5 \times 1{,}000)$$

$$= 263{,}861.89 \text{ N} \cdot \text{mm}$$

Equating the work done to the strain energy stored $1{,}500(h + 3.66) = 263{,}861.89$

i.e., $h + 3.66 = 175.91$

Hence, the height of drop = 172.25 mm

SOLVED PROBLEM 3.9

A steel bar of 25 mm diameter is encased in a brass tube of 35 mm outer diameter and 25 mm inner diameter and the length of the composite bar is 1.2 m. The composite bar is fixed at a end and has an arrangement to arrest a falling weight (Figure 3.6). It is loaded suddenly to produce tension due to impact by a weight of 10 kN falling through a height of 4 mm. Compute the maximum stress in the steel and brass. Take $E_s = 2 \times 10^5$ N/mm² and $E_b = 1 \times 10^5$ N/mm². Assume that there is no energy loss during the impact.

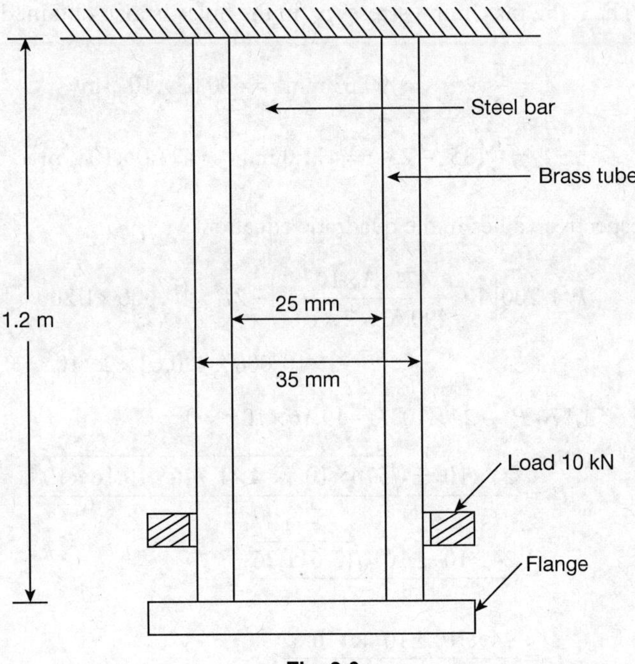

Fig. 3.6

Solution:

Given data: Diameter of the steel bar = 25 mm, outer diameter of the brass tube = 35 mm, inner diameter of the brass tube = 25 mm, l = 1.2 m, P_i = 10 kN, h = 4 mm, $E_s = 2\times10^5$ N/mm^2 and $E_b = 1\times10^5$ N/mm^2

Work done by the falling weight $= P_i(h+\delta l)$

This energy is completely stored in the composite bar.

Let P_s and P_b be the forces in steel and brass, respectively, which are applied gradually and without shock on the bar which will produce the same stress and deflection in the bar as the falling weight.

Strain energy stored in the composite bar will be $= \frac{1}{2}P_s\delta l + \frac{1}{2}P_b\delta l$

Equating the work done and the energy stored $W(h+\delta l) = \frac{\delta l}{2}(P_s + P_b)$

Let A_s and A_b be the area of cross section of steel bar and brass, respectively.

Then,

$$\delta l = \frac{P_s l}{A_s E_s} - \frac{P_b l}{A_b E_b}$$

Substituting for δl in the energy or workdone equation and rearranging the terms, we get a quadratic equation in P_s as

$$P_s^2\left(l + \frac{A_s E_b}{A_s E_a}\right) - 2P_s Wl - 2WhA_s E_s = 0$$

Solving for P_s and computing the maximum stress P_s in steel by dividing P_s by the area of cross section A_s. Then, the maximum stress p_b in the brass is also obtained.

$$A_s = \frac{\pi}{4}(25)^2 = 490.63 \text{ mm}^2 = 490.63\times10^{-6} \text{ m}^3$$

$$A_b = \frac{\pi}{4}(35^2 - 25^2) = 471.0 \text{ mm}^2 = 471.0\times10^{-6} \text{ m}^3$$

Substituting the respective values in the quadratic equation

$$P_s^2 1,200\left(1 + \frac{471\times1\times10^5}{490.63\times2\times10^3}\right) - 2P_s\times10,000\times1,200$$

$$-2\times10,000\times490.63\times2\times10^3 = 0$$

i.e., $$1,176\,P_s^2 - 24\times10^6 P_s - 19.16\times10^{11} = 0$$

$$P_s = \frac{24\times10^6 \pm \sqrt{570\times10^{12} + 4\times1,776\times19.16\times10^{11}}}{2\times1,776}$$

$$= \frac{24\times10^6 \pm 10^6\sqrt{13,611.26}}{3,552}$$

$$= \frac{24\times10^6 \pm 10^6\times116.67}{3,552}$$

Taking the positive value

$$P_s = \frac{24\times10^6 + 116.67\times10^6}{3,552} = 56.85 \text{ kN}$$

P_b is obtained from the expression

$$\frac{P_s l}{A_s E_s} = \frac{P_b l}{A_b E_b}$$

$$\therefore \quad P_b = P_s \frac{A_b E_b}{A_s E_s}$$

$$= 56.85 \times \frac{471\times1\times10^5}{490.63\times2\times10^5}$$

$$P_b = 27.29 \text{ kN}$$

$$\text{Maximum stress in steel} = \frac{P_s}{A_s} = \frac{56.85\times1,000}{490.63} = 115.87 \text{ N/mm}^2$$

$$\text{Maximum stress in brass} = \frac{P_b}{A_b} = \frac{27.29\times1,000}{471} = 57.94 \text{ N/mm}^2$$

SOLVED PROBLEM 3.10

The maximum stress produced by a pull in a bar of length 1 m is 150 MN/m^2. The area of cross section and length are shown in Figure 3.7. Calculate the strain energy stored in the bar if E = 200 GN/m^2.

Solution:

Given data: l = 1 m, stress in bar = 150 MN/m^2 and E = 200 GN/m^2.

As the axial pull for the bar is same, the stress will be maximum when the area is minimum.

Maximum stress in the 100 mm^2 portion, p_1 = 150 MN/m^2

Maximum stress in the 200 mm^2 portion, p_2

Then

$$P_1 A_1 = P_2 A_2$$

$$150\times1\times10^{-4} = P_2 \times 2\times10^{-4}$$

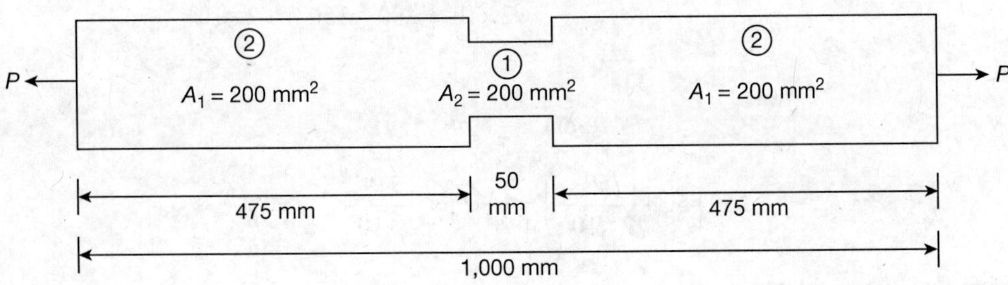

Fig. 3.7

Therefore, $P_2 = 75\ \text{MN/m}^2$
Strain energy stored in the bar

$$U = \frac{p_1^2 V_1}{2E} + \frac{p_2^2 V_2}{2E}$$

$$U = \frac{p_1^2 A_1 l_1}{2E} + \frac{p_2^2 A_2 l_2}{2E}$$

$$= \frac{\left(150\times10^6\right)^2 \times\left(1\times10^{-4}\right)\times 50/1,000}{2\times\left(200\times10^9\right)} + \frac{\left(75\times10^6\right)^2 \times\left(2\times10^{-4}\right)\times 475/1,000}{2\times\left(200\times10^9\right)}$$

$$= 0.28125 + 2.6718$$

$$= 2.953\ \text{N}\cdot\text{m}$$

SOLVED PROBLEM 3.11

A bar 1,200 mm long is 200 mm² in cross-sectional area for 600 mm of its length and 150 mm² for the remaining 600 mm. If a load of 150 N falls on the collar which is provided at one end of the rod, the other being fixed, from a height of 50 mm, find the maximum stress induced in the bar. Assume $E = 200\ \text{kN/mm}^2$.

Solution:

Given data: $l = 1,200$ mm, $A_1 = 200$ mm², $A_2 = 150$ mm², $l_1 = 600$ mm, $l_2 = 600$ mm, $P_i = 150$ N, $h = 50$ mm and $E = 200\ \text{kN/mm}^2$.
Let P_e be the equivalent gradually applied load which produces the same maximum stress and extension $A_1 = 200$ mm² as is caused by the falling load P_i.

Total extension,
$$\delta l_i = \left(\delta l_i\right)_1 + \left(\delta l_i\right)_2$$

$$= \frac{P_e l_1}{A_1 E} + \frac{P_e l_2}{A_2 E}$$

$$= \frac{P_e}{E} + \left[\frac{l_1}{A_1} + \frac{l_2}{A_2}\right]$$

$$= \frac{P_e}{2\times10^5}\left(\frac{600}{200} + \frac{600}{150}\right) = \frac{7P_e}{2\times10^5}$$

Also,

$$P(h+\delta l) = \frac{l}{2} P_e \delta l_i$$

$$150\left(50 + \frac{7P_e}{2\times10^5}\right) = \frac{1}{2}\times P_e \times \frac{7P_e}{2\times10^5}$$

Solving for P_e, we obtain
$P_e = 14,658.5$ N (Considering only positive sign for maximum stress)

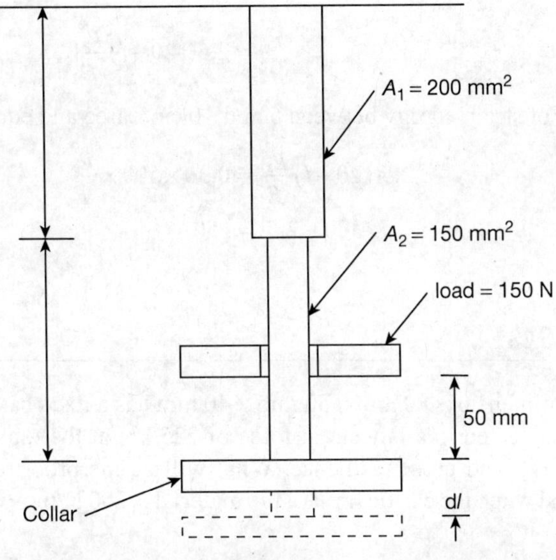

Fig. 3.8

Maximum stress in smaller section, $P = \dfrac{P_e}{A_e}$

$$= \frac{14,658.5}{150 \times 10^{-4}}$$

$$= 977.233 \times 10^3 \text{ N/mm}^2$$

SOLVED PROBLEM 3.12

A bar of 1.2 m length has 10 cm^2 area of cross section for a portion of its length and 5 cm^2 area of cross section for the remainder. The strain energy of the 5 cm^2 area portion of the bar is 30% of the bar of 10 cm^2 in cross section and 1 m length under the same maximum stress. Find the length of the portion of 10 cm^2 area of cross section.

Solution:

Given data: $l = 1.2$ m, $A_1 = 5 \text{ cm}^2$ and $A_2 = 10 \text{ cm}^2$.

Strain energy of 1.2 m length of 10 cm^2 area of cross section.

$$= \frac{p_s^2 V}{2E} = \frac{p_s^2}{E} \times \frac{10 \times 100}{2}$$

$$= 500 \times \frac{p_s^2}{E}$$

Let l be the length of portion with 10 cm^2 area of cross section.

Strain energy on the 5 cm^2 area length $= \dfrac{p_s^2}{E} \times \dfrac{5(120 - l)}{2}$

$$= 2.5(120 - l)\frac{p_s^2}{E}$$

Considering 30% ratio of strain energy between 5 and 10 cm^2 area and equating

$$2.5(120 - l)\frac{p_s^2}{E} = 0.30 \times 500 \times \frac{p_s^2}{E}$$

$$300 - 2.5l = 150$$

i.e.,

$$\therefore\ l = 60 \text{ cm}$$

SOLVED PROBLEM 3.13

A rectangular block of dimensions 80 mm × 50 mm × 40 mm has a fixed base of size 50 mm × 40 mm as in Figure 3.9. It is subjected to a tangential force of 225 kN at the top face. Calculate the shear strain, shear strain energy and proof resilience. What will be the proof resilience and modulus of resilience of the material when the elastic limit of the material is 250 N/mm^2 and modulus of resilience $G = 84 \times 10^3$ N/mm^2?

Solution:

Given data: $b = 40$ mm, $l = 40$ mm, $d = 80$ mm, tangential force = 225 kN, $G = 84 \times 10^3$ N/mm^2 and Elastic limit = 250 N/mm^2.

$$\text{Shear stress} = \frac{225 \times 10^3}{50 \times 40} = 112.5 \text{ N/mm}^2$$

$$\text{Shear strain} = \frac{\text{Shear stress}}{G}$$

$$= \frac{112.5}{84 \times 10^3}$$

$$= 1.339 \times 10^{-3}$$

$$\text{Shear strain energy} = \frac{q^2}{2G} \times \text{volume}$$

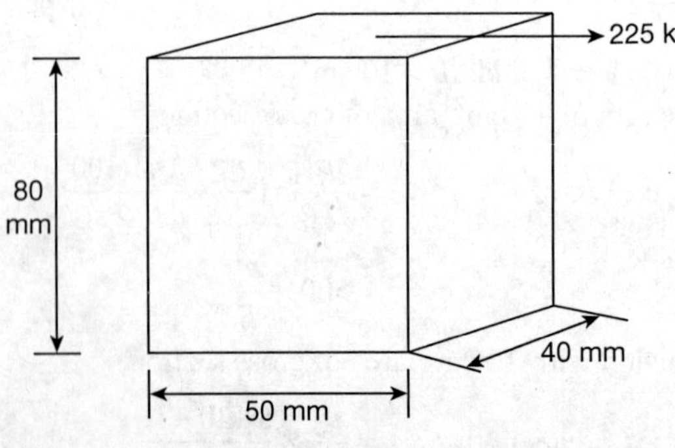

Fig. 3.9

$$= \frac{(112.5)^2}{2\times 84\times 10^3}\times(80\times 50\times 40)$$

$$= 12{,}053.57\ \text{N}\cdot\text{mm}$$

$$\text{Proof resilience} = \frac{250^2}{2\times 84\times 10^3}(80\times 50\times 40)$$

$$= 59{,}523.8\ \text{N/m}$$

$$\text{Modulus of resilience} = \text{Proof resilience per unit volume}$$

$$= \frac{250^2}{2\times 84\times 10^3}$$

$$= 0.372\ \text{N}\cdot\text{mm/mm}^2$$

SOLVED UNIVERSITY QUESTIONS

SOLVED PROBLEM 3.14

The maximum instantaneous extension produced by an unknown falling weight through a height of 40 mm in a vertical bar of length 3 m and cross-sectional area of 500 mm^2 is 2.1 mm. Determine the instantaneous stress induced in the vertical bar and the value of unknown weight. Take $E = 2\times 10^5$ N/mm^2 (Anna Univ., Nov. 2008, ME).

Solution:

Given data: $h = 40$ mm, $l = 3$ m, $A = 500$ mm^2, $\delta l = 2.1$ mm and $E = 2\times 10^5$ N/mm^2.

$$\text{Volume of bar} = 3{,}000\times 500 = 1{,}500{,}000\ \text{mm}^3$$

$$\text{Instantaneous stress} = E\times(\text{Instantaneous strain})$$

$$= E\times \delta l\times l$$

$$= 2\times 10^5\times\frac{2.1}{3{,}000} = 140\ \text{N/mm}^2$$

Equating the work done by the falling weight to the strain energy stored is

$$P(h+\delta l) = \frac{p_i^2}{2E}\times V$$

$$P(40+2.1) = \frac{(140)^2}{2\times 2\times 10^5}\times 1{,}500{,}000$$

Solving

Unknown weight, $P = 1{,}745.8$ N

SOLVED PROBLEM 3.15

Find the total strain energy stored in a steel bar of diameter 50 mm and length 300 mm when it is subject to an axial load of 150 kN. Take $E_s = 200\times10^3$ MPa (Anna Univ., 2006, ME).

Solution:

Given data: Diameter of the bar = 50 mm, $l = 300$ mm, $P = 150$ kN and $E_s = 200\times10^3$ MPa.

$$\text{Stress} = \frac{\text{Load}}{\text{Area}}$$

$$\text{Area} = \frac{\pi}{4}\times 50^2 = 1{,}962.5\ \text{mm}^2$$

$$= \frac{150\times1{,}000}{1{,}962.5} = 76.43\ \text{N/mm}^2$$

Strain energy,

$$U = \frac{p_s^2 \times A\times l}{2E}$$

$$= \frac{(76.43)^2\times1{,}962.5\times300}{2\times10^5}$$

$$= 17{,}196.05\ \text{N}\cdot\text{mm}$$

SALIENT POINTS

- Strain energy is the energy which is stored in the body due to the effect of straining.
- Elastic strain energy is the recoverable stored energy in the elastic body.
- Resilience is the total energy stored in a body within its elastic limit.
- Proof resilience is the maximum strain energy stored in a body.
- Modulus of resilience is the greatest amount of strain energy per unit volume that a material can absorb without exceeding the elastic limit.
- Strain energy stored depends on the type of loading, viz., gradual loading, sudden loading and impact loading.
- Strain energy due to gradual loading is $U = P_g^2 V/2E$
  where P_g = gradually applied load
  V = volume of the body
  E = Modulus of elasticity of the material
- Maximum stress induced due to suddenly applied load is twice the stress induced when the same load is applied gradually.
- Stress due to impact is given as $p_i = \dfrac{P_i}{A}\left(\dfrac{l+\sqrt{2AEh}}{P_i l}\right)$
  where P_i = impact load
  A = area of cross section
  E = Modulus of elasticity
  h = height of fall
  l = length

- Strain energy due to self-weight is $U = \dfrac{A\rho^2 l^3}{6E}$
where ρ = density of the material
- Strain energy due to shear force is $U = \dfrac{q^2}{2G}V$
where q = shear stress
V = volume of the material
G = shear modulus.
- Total strain energy per unit volume in terms of principal stresses is

$$U = \frac{l}{2E}\left(\sigma_1^2 + \sigma_2^2 + \sigma_3^2 - \frac{2\sigma_1\sigma_2 + 2\sigma_3\sigma_1 + 2\sigma_2\sigma_3}{m}\right)$$

where $\sigma_1, \sigma_2, \sigma_3$ = Major, minor and intermediate principal stresses
$\mu = 1/m$ = Poisson's ratio.

QUESTIONS

1. Define resilience, modulus of resilience and proof resilience.
2. Derive the expression for strain energy stored in an axially loaded bar.
3. Show the relationship between suddenly loaded and gradually loaded rod.
4. What is shear modulus? How strain energy stored in a body due to shear is calculated?
5. How the principal stresses are related to principal stresses and Poisson's ratio?
6. A rod is subjected to a gradually increasing load which causes a normal stress of 8 N/mm^2. If the same load is applied suddenly, what is the normal stress set-up?
7. The shear stress in a material is 60 N/mm^2. Calculate the strain energy per unit volume assuming C to be 8×10^4 N/mm^2.
8. An unknown weight falls 3 cm on to a collar attached to the lower end of a vertical bar 4 m long and 10 cm^2 in section. If the maximum instantaneous extension is found to be 4.2 mm, find the corresponding stress and the value of the unknown weight.
9. A bar of 1.5 cm diameter gets stretched by 2.5 mm under a gradually applied load of 10 kN. What stress would be produced in the same bar by a weight of 12 kN which falls vertically a distance of 5 cm on to a rigid collar attached at its ends? The bar is initially unstressed. Take $E = 2 \times 10^5$ N/mm^2.
10. A cage weighing 40 kN is fixed to a steel wire rope. The cage is lowered down in a mine shaft with a uniform velocity of 0.80 m/s. Compute the maximum stress produced in the steel rope when its supporting drum gets jammed suddenly. At the time of jamming, the free length of the rope is 12 m. The net cross-sectional area of the rope is 20 cm^2 and $E = 2 \times 10^5$ N/mm^2. Neglect the self-weight of the wire rope.
11. A bar is of 5 m length and is made of two parts, 3 m of its length has a cross-sectional area of 12 cm^2 and the remaining length has a cross-sectional area of 24 cm^2. An axial tensile load of 50 kN is gradually applied. Compute the total strain energy produced in the bar. Also, find the total strain energy produced in a uniform bar of the same length and having the same volume when under the same load. $E = 2 \times 10^5$ N/mm^2.

12. A vertical composite tie bar is rigidly fixed at its upper end and provision is made for a weight to freely slide and stop at the lower end. The composite bar consists of a brass rod of 25 mm diameter and 3 m length placed within an equally long steel tube of 25 mm internal diameter and 35 mm external diameter. The composite bar is suddenly loaded in tension by a weight of 55 kN falling through a height of 5 mm. Find the maximum stress in steel and brass.
13. Two bars A and B are each of 30 cm long and made of the same material. Bar A has a diameter of 2.5 cm with a length of 20 cm. Bar B has a diameter of 2.5 cm for a length of 20 cm and a diameter of 5 cm for the remaining length of 5 cm. If each bar is stressed up to an elastic limit of 150 N/mm^2, compute the ratio of energy stored by A and B at proof stress. $E = 0.205\times10^6$ N/mm^2.
14. At a point in a material, the shear stress is 55 N/mm^2. Compute the local strain energy stored per unit volume in the material due to shear stress. Take $C = 7.6\times10^4$ N/mm^2.

4

Shear Force and Bending Moment

LEARNING OBJECTIVES

4.1 BEAMS AND SUPPORTS

A structural member that is reasonably long compared to its lateral dimensions, when supported at one or more points, and subjected to forces cause bending of the member in an axial plane is called a beam.

For an efficient functioning of a beam, it should be supported at least at one point. Depending on the type of support, the load-bearing capacity of the beam varies.

There are three types of supports. They are:

1. Simply supported or roller support
2. Pinned support or hinged support
3. Fixed support or rigid support or encastred support.

4.1.1 Simply Supported or Roller Support

It is the simplest form of support. The reaction is normal to the plane of the roller or the support. A beam resting on a wall is a simply supported case (Figure 4.1).

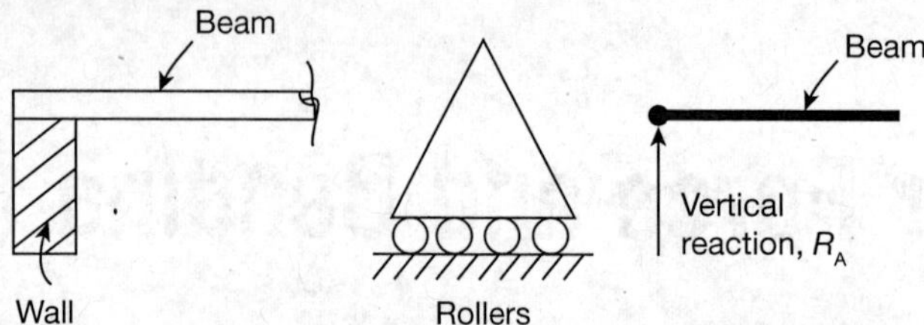

Fig. 4.1 Representation of simply supported and roller support.

4.1.2 Pinned Support or Hinged Support

It is one type of support in which the beam can rotate about the hinge or the pin. It can withstand force in any direction. The reaction is normal to the plane of the support and it has a horizontal reaction along the axis of the beam (Figure 4.2).

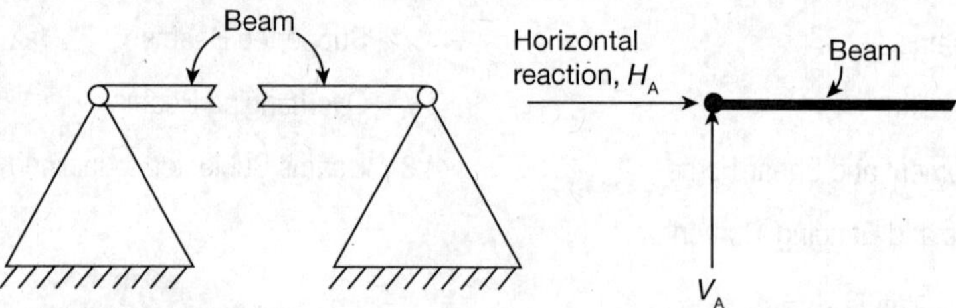

Fig. 4.2 Representation of pinned support and hinged support.

4.1.3 Fixed or rigid or encastred support

It is one type of support which prevents the beam not only from rotation but also from linear movement or translation. A beam embedded in a wall is an example of this type. In addition to the horizontal and vertical component of forces (reactions), a fixed moment is present at the support to prevent rotation of beam (Figure 4.3).

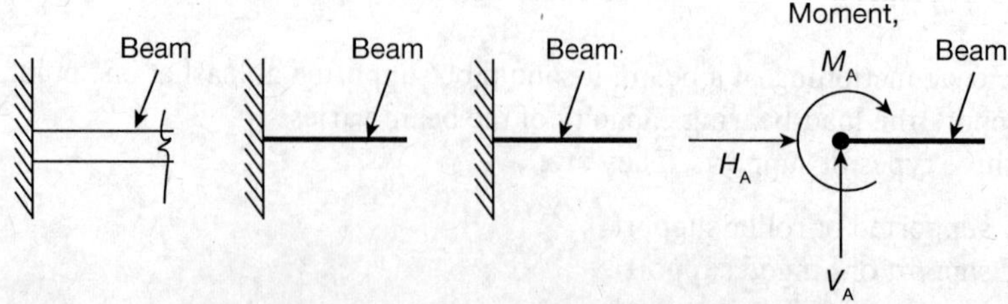

Fig. 4.3 Representation of fixed support.

4.2 TYPES OF BEAMS

Beams are classified according to their support conditions and not on any other parameter. There are six different methods of supporting a beam.

4.2.1 Simply Supported Beam

Simply supported beam is supported at two ends. This is the most common type of beam. In this case, both ends can be simply supported or one end simply supported and the other end be a hinged support or on a roller (Figure 4.4). Both supports cannot be unstable. Similarly, both ends cannot be hinged support since the beam will produce change in span which is not permitted. When the beam is horizontal, the reaction at the supports is vertical. Depending on the type of loading, the reaction may be upwards or downwards.

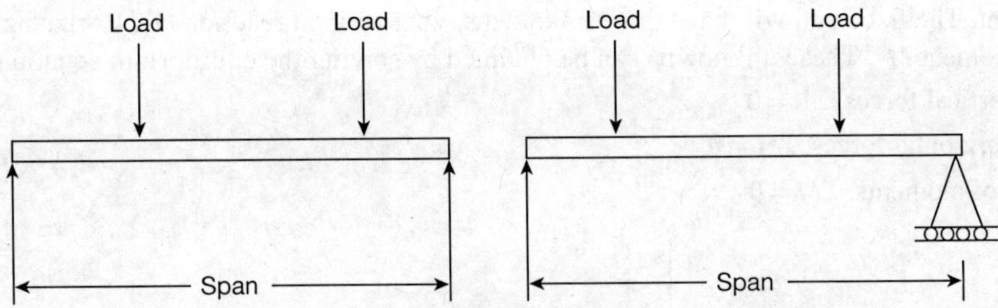

Fig. 4.4 Simply supported beams.

4.2.2 Cantilever Beam

Cantilever beam is fixed or built in at one end and the other end is free (Figure 4.5).

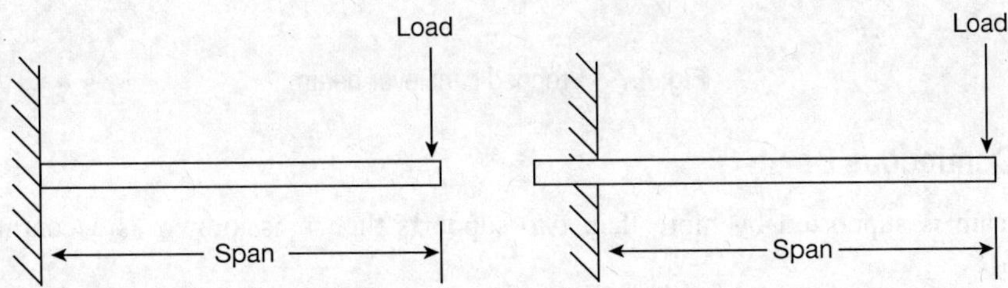

Fig. 4.5 Cantilever beams.

4.2.3 Overhanging Beam

When a beam extends beyond the support, then it is known as an overhanging beam. The overhang may be on one end, called single overhang, or on both ends, called double overhang (Figure 4.5).

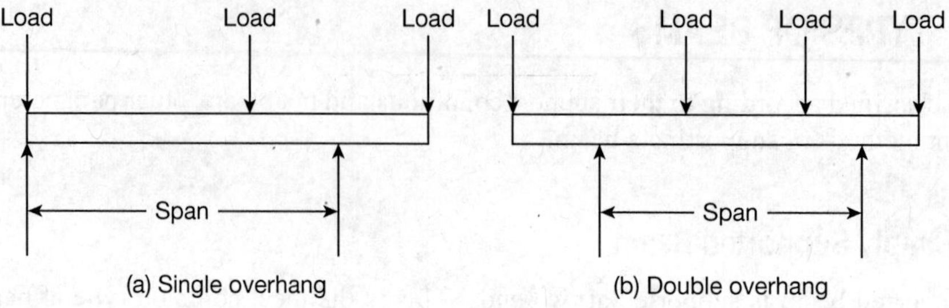

Fig. 4.6 Overhanging beams.

4.2.4 Propped Cantilever Beam

When a cantilever beam is supported at its free end, then it is called a propped cantilever beam (Figure 4.7). The introduction of one additional support at the free end induces one more reaction component. The fixed end will have three unknowns, viz., vertical reaction, V_A, Horizontal reaction, H_A and moment M_A. These unknowns can be obtained by solving the equilibrium equations, viz.

Sum of vertical forces, $\Sigma V = 0$

Sum of horizontal forces, $\Sigma H = 0$

and Sum of moments, $\Sigma M = 0$

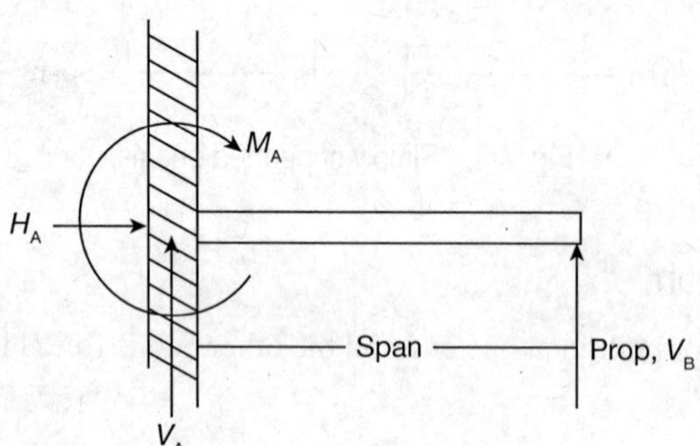

Fig. 4.7 Propped cantilever beam.

4.2.5 Continuous Beam

When a beam is supported by more than two supports then it is known as a continuous beam (Figure 4.8).

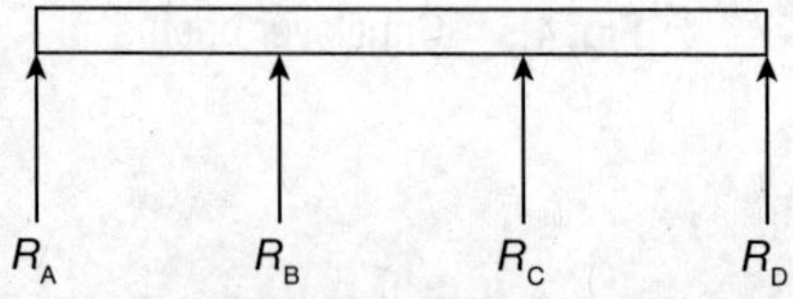

Fig. 4.8 Continuous beam.

4.2.6 Fixed beam or Restrained Beam

When both the ends of a beam is fixed or built in, then it is known as a fixed beam or a restrained beam (Figure 4.9). As there are two fixed ends there are four reactions (V_A, V_B, H_A and H_B) and two moments (M_A and M_B). Normally, the beams are horizontal in nature and there will not be any horizontal forces. In that case, $H_A = H_B = 0$. Hence, there are four unknowns and only two equilibrium equations.

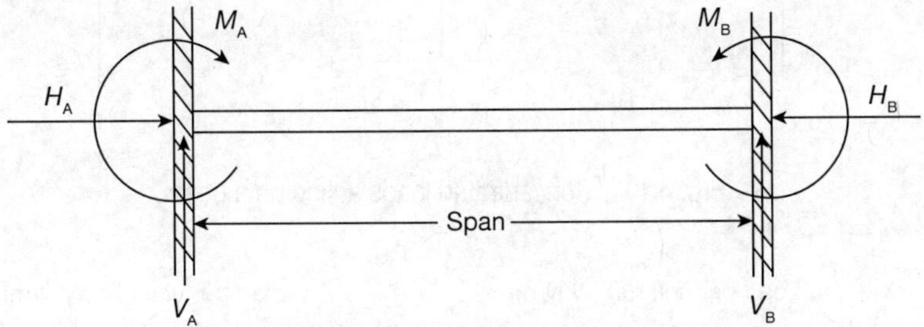

Fig. 4.9 Fixed beam.

In the first three types of beams, viz., simply supported, cantilever and overhanging beams the unknowns can be found by solving three equilibrium equations, viz., $\Sigma V_A = 0$, $\Sigma H_A = 0$ and $\Sigma M_A = 0$. Thus, these beams are called as *statically determinate beams.* In the case of the other three beams, viz., propped cantilever, continuous and fixed beams, the number of unknowns are more than three, hence they cannot be solved by using equilibrium equations. Other properties of the material like modulus of elasticity, E, and the second moment of area I (discussed elsewhere) are used to determine the unknowns. These beams are called *statistically indeterminate beams.*

4.3 TYPES OF LOADING

Beams are subjected to loading. There are four type of loads. They are discussed below.

4.3.1 Concentrated Load or Point Load

Concentrated or Point Load is a load acting on a small elemental area. In practice, a load cannot be assumed to act on a single point just like a contact made by a sharp needle. However, when a load is transferred through a roller or a sphere on to the beam the contact will be through a point. In all other cases, the load is presumed to act on a small restricted area of the beam. In general, the concentrated loads are vertical. In certain cases it can be inclined, horizontal and act below the beam (Figure 4.10).

4.3.2 Uniformly Distributed Load

If some magnitude of the load is distributed or spread over the length of a beam partly or fully, then it is known as uniformly distributed load (UDL). It is represented as unit load per unit length of run (Figure 4.11).

For the purpose of calculating moment, an UDL can be considered as a series of continuous point loads so closely placed such that their action cannot be separated from each other.

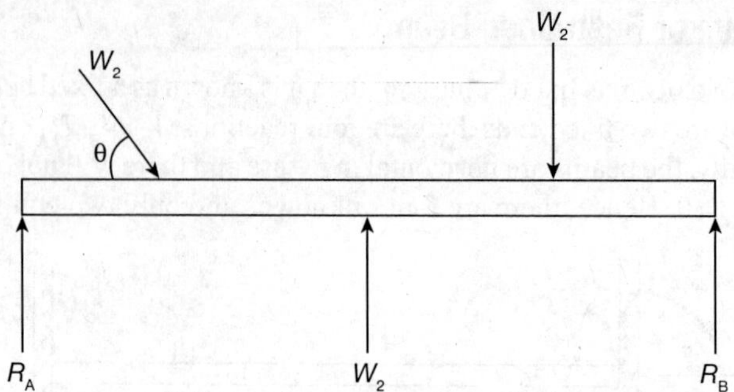

Fig. 4.10 Concentrated loads acting on a beam.

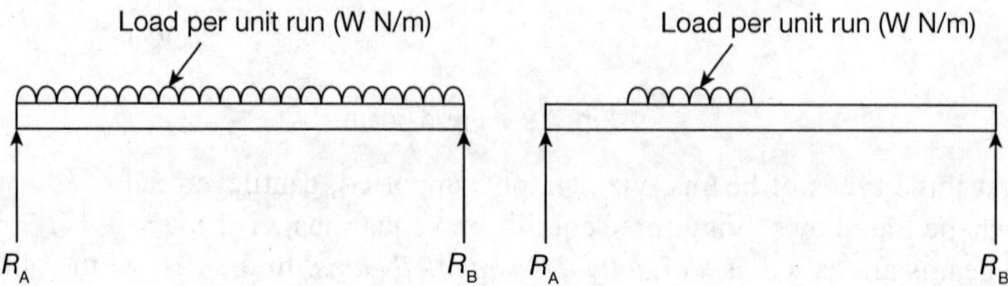

Fig. 4.11 Uniformly distributed load acting on beams.

4.3.3 Non-uniform Load

Non-uniform load is one in which the magnitude of loading varies along the length of the beam. Variations may be of different types, viz, triangular variation, trapezoidal variation, parabolic variation, etc. Variation can be of any shape (Figure 4.12).

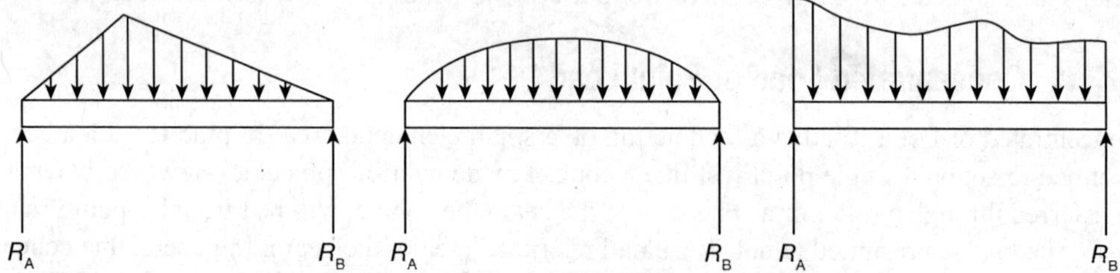

Fig. 4.12 Non-uniform load acting on beams.

4.3.4 Couple or Moment

A couple is defined as the action of two parallel, equal and opposite forces. The magnitude of the couple is given as the product of the force and the perpendicular distance between the two parallel forces. The unit of the couple will be in Nm (force × distance). Moments may be clockwise or anticlockwise. The point of application of a couple is an important factor to be considered (Figure 4.13).

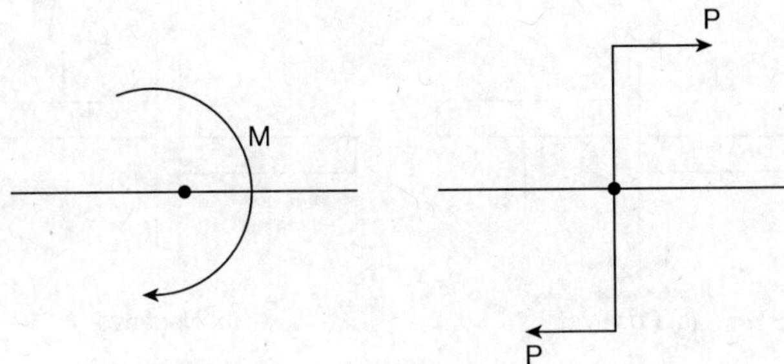

Fig. 4.13 Couple acting on a beam.

4.4 BENDING MOMENT AND SHEAR FORCE

Bending moment at a section is defined as the algebraic sum of the moments about the section of all the forces (including the reaction) acting on the beam, either to the left or to the right of the section.

Shear force at a section in a beam is defined as the algebraic sum of all the forces including the reactions acting normal to the axis of the beam either to the left or to the right of the section.

4.4.1 Sign Convention

1. Bending Moment

Bending moment is said to be positive (sagging) moment at a section when it is acting in an anticlockwise (ACW) direction to the right and negative (hogging) moment when acting in a clockwise (CW) direction (Figure 4.14).

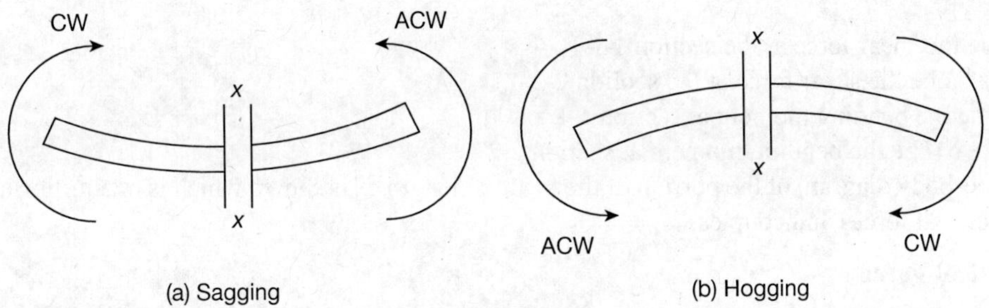

Fig. 4.14 Sign convention for bending moment.

2. Shearing Force

A shear force having an upward direction to the right-hand side of a section or downward to the left of the section is taken as *positive*. Similarly, a *negative* shearing force will be the one that has a downward direction to the right of the section or upward direction to the left of the section (Figure 4.15).

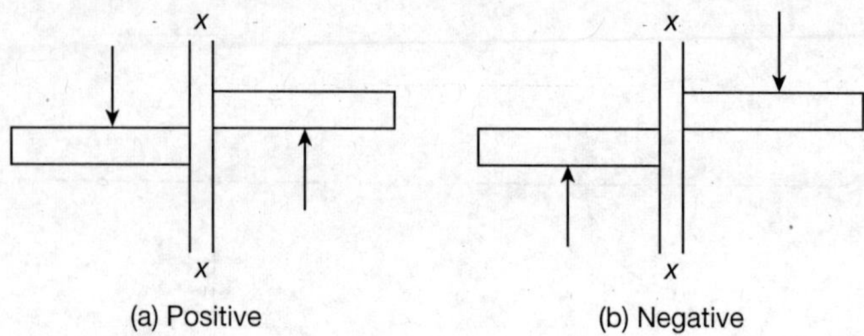

Fig. 4.15 Sign convention for shear force.

4.4.2 Relation Between Bending Moment and Shear Force

Figure 4.16 shows a beam carrying uniformly distributed load of w per unit length. The equilibrium of the portion of the beam between sections 1-1 and 2-2 is considered. This portion is at a distance of x from left support and its length is δx.

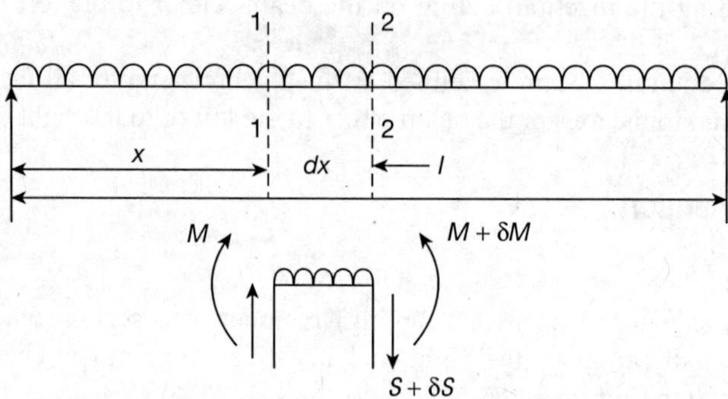

Fig. 4.16 Beam loaded with UDL.

Let S be the shear force at the section 1-1
Let $S + \delta S$ be the shear force at the section 2-2
Let M be the bending moment at section 1-1
Let $M + \delta M$ be the bending moment at section 2-2
The free-body diagram of the portion of the beam of length δx is shown which is in equilibrium under both vertical forces and couples.

1. Vertical Forces

$$(S + \delta S) - S = w.\delta x$$

i.e.,

$$\delta S = w.\delta x$$

or

$$\frac{\delta S}{\delta x} = w$$

or

$$\frac{dS}{dx} = w \qquad (4.1)$$

2. Couples

ie.,
$$M-(M+\delta M)=S.\delta x+w.\delta x\left(\frac{\delta x}{2}\right)$$

$$-\delta M=S.\delta x+\frac{w}{2}\delta x^2$$

Neglecting higher powers of small quantities

$$-\delta M=-S\delta x$$

or
$$S=\frac{\delta M}{\delta x} \tag{4.2}$$

or
$$S=\frac{dM}{dx}$$

Equation (4.1) shows that the rate of change of shear force is equal to the rate of loading, whereas Equation (4.2) shows that the rate of change of bending moment is equal to the shear force at the section.

4.4.3 Shear Force and Bending Moment Diagrams

Following are the steps to be followed for drawing shear force diagram (SFD) and bending moment diagram (BMD).

(i) Any section *x-x* at a distance *x* from a fixed point is considered. In the case of a simply supported beam, the section may be considered from any one support. In the case of cantilever, it has to be considered from the free end.

(ii) Shear force equation and Bending moment equation are written considering section *x-x* in terms of *x*, i.e., S_x and M_x respectively.

(iii) Various values for *x* are allotted and the shear forces and bending moments are calculated at various points of the beam.

(iv) These values are plotted as ordinates to some suitable scale at the respective points as abscissa. While plotting the ordinates, positive values are plotted above the line and negative values below the line.

(v) These points are joined by

(a) straight line if S_x or M_x is a function of first degree (i.e., x)

(b) by a smooth curve if S_x or M_x is a function of second degree (i.e., x^2) and above.

4.5 SHEAR FORCE AND BENDING MOMENT OF CANTILEVER BEAMS

In a cantilever beam, shear force at any section is equal to the sum of the loads between the sections and the free end. Bending moment at a given section is equal to the sum of the moments about the section of all the loads between the section and the free end of the cantilever.

4.5.1 Cantilever Beam Subjected to a Concentrated Load at the Free End

A cantilever beam AB carrying a concentrated load W at the free and B (Figure 4.17) is considered. Any section x-x at a distance x from the free end is taken.

1. Shear Force

Shear force at section x-x is

$$S_x = -W$$

Negative sign is assigned as the force is downward to the right of the section.

Here, the shear force is independent of the distance x:

$$\therefore \qquad S_A = S_B = -W \tag{4.3}$$

2. Bending Moment

Bending moment at the section x-x is

$$M_x = -W \times x \tag{4.4}$$

Negative sign is assigned as the moment is clockwise to the right of the section.

At B, $x = 0$ then $\qquad M_B = 0$

At A, $x = l$ then $\qquad M_A = -Wl$

Shear force and bending moment diagrams are drawn below a reference line since they have a negative value.

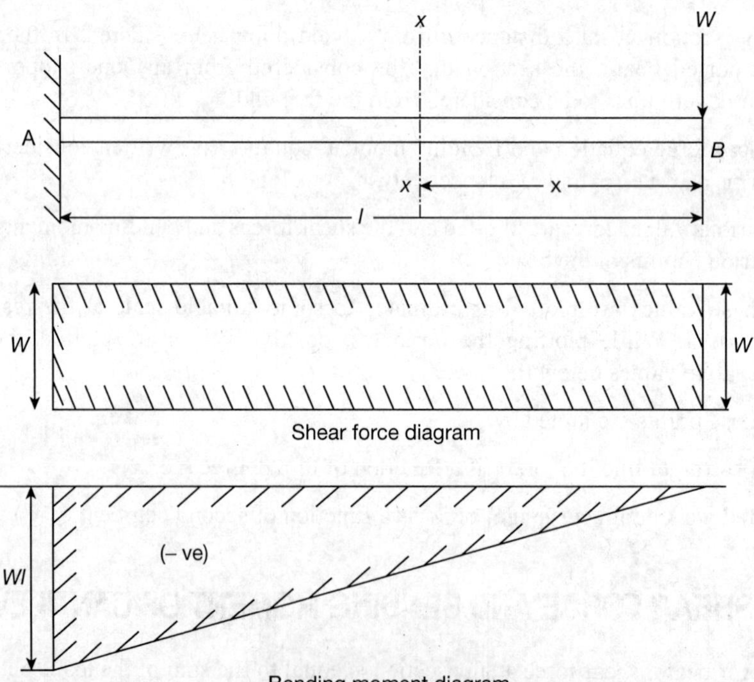

Fig. 4.17 Cantilever beam with point load at free end.

4.5.2 Cantilever Beam Subjected to Several Concentrated Loads

A cantilever beam *AB* acted upon by several concentrated loads of W_1, W_2 and W_3 at B, C and D at distances l, l_1 and l_2 from the fixed end is considered (Figure 4.18).

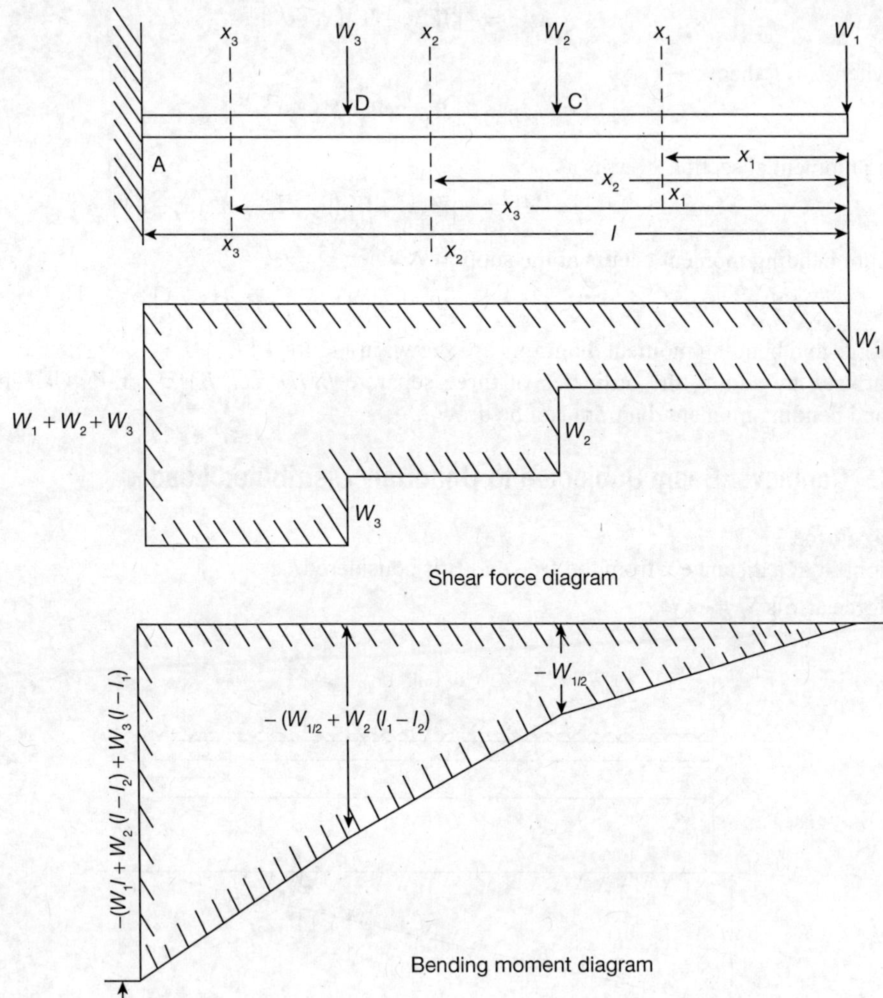

Fig. 4.18 Cantilever beam loaded with several concentrated loads.

1. Shear Force

Shear force at section x_1–x_1 is, $S_{x1} = -W_1$

Shear force at section x_2–x_2 is $S_{x2} = -(W_1 + W_2)$

Shear force at section x_3–x_3 is $S_{x2} = -(W_1 + W_2 + W_3)$

Alternatively, shear force can be drawn for individual load and then combined together.

2. Bending Moment

Bending moment at section x_1–x_1 is $M_{x1} = -W_1 x_1$

At C, when $x_1 = l_2$, $M_C = W_1 l_2$

Bending moment at section x_2–x_2 is

$$M_{x2} = -\left[W_1 x_2 + W_2\left(x_2 - l_2\right)\right]$$

At D, when $x_2 = l$, then

$$M_D = -\left[W_1 l_1 + W_2\left(l_1 - l_2\right)\right]$$

Bending moment at section x_3–x_3 is

$$M_{x3} = -\left[W_1 x_3 + W_2\left(x_3 - l_2\right) + W_3\left(x_3 - l_1\right)\right]$$

Maximum bending moment occurs at the support A when $x_3 = l$.

Then $$M_{max} = -\left[W_1 l + W_2\left(l - l_2\right) + W_3\left(l - l_1\right)\right]$$

Shear force and bending moment diagrams are shown in Figure 4.18.

Alternatively, by adding the ordinates of three separate *BMD*, i.e., *BMD* for W_1, W_2 and W_3, the combined bending moment diagram can be drawn.

4.5.3 Cantilever Beam Subjected to Uniformly Distributed Load

1. Shear Force

A section *x-x* at a distance *x* from the free end B is considered.

Shear force at *x* is $S_x = -wx$

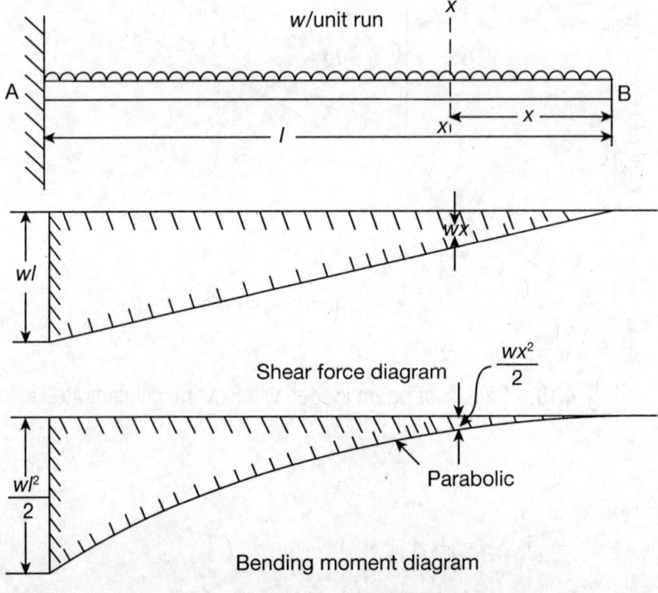

Fig. 4.19 Cantilever beam loaded with UDL.

At $x = 0$, $S_B = 0$
At $x = l$, $S_A = -wl$

2. Bending Moment

Bending moment at x is $M_x = -wx \times \frac{x}{2} = \frac{wx^2}{2}$

At $x = 0$, $M_B = 0$

At $x = l$, $M_A = -\frac{wl^2}{2}$

Shear force and bending moment diagrams are shown in Figure 4.19.

4.5.4 Cantilever Beam Carrying an UDL on Entire Span and a Point Load at the Free End

1. Shear force

A section $x - x$ at a distance x from the free end is considered.
Shear force at section x-x is $S_x = -(W_1 + wx)$
At B, when $x = 0$ $S_B = -W_1$
At A, when $x = l$ $S_A = -W_1 + wl$
Shear force diagram is shown in Figure 4.20.
Alternatively, the shear force diagram for the point load W_1 and the shear force diagram for the UDL, w_1 per unit run can be separately drawn and combined together to get the final SFD.

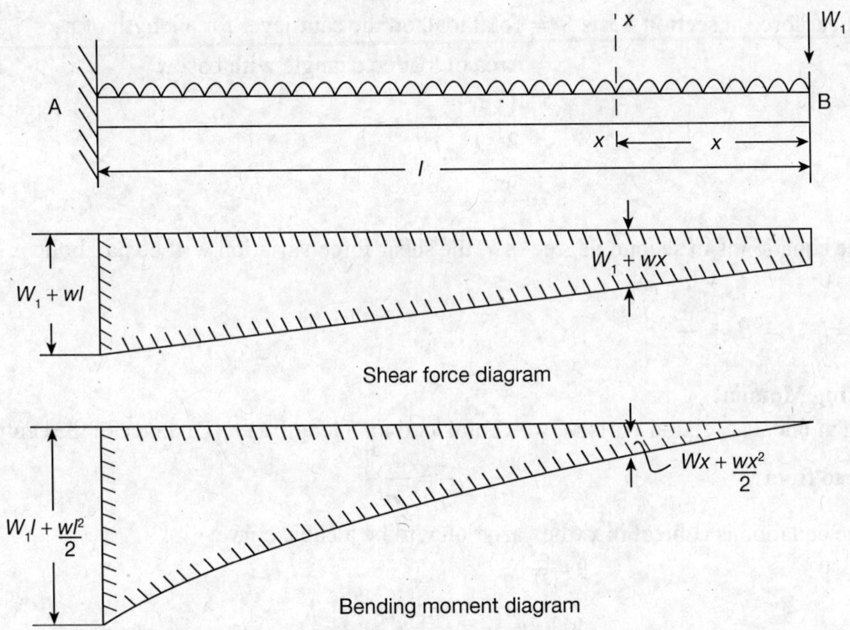

Fig. 4.20 Cantiliver beam loaded with UDL and point load.

2. Bending Moment

Bending moment at section *x-x*

$$M_x = \left[W_1 x + w \cdot x \cdot \frac{x}{2}\right]$$

$$= \left[W_1 x + \frac{wx^2}{2}\right]$$

At B, when $x = 0$ $\qquad M_B = 0$

At A, when $x = l$ $\qquad M_A = -\left[W_1 l + \frac{wl^2}{2}\right]$

The bending moment diagram will be a parabolic curve varying from zero at B to $\left(W_1 l + \frac{wl^2}{2}\right)$ at A. Shear force and bending moment diagrams are shown in Figure 4.20.

Alternatively, the bending moment diagram for the point load W_1 and the bending moment for the UDL, *w*/unit run, can be separately drawn and combined together to get the final BMD.

4.5.5 Cantilever Beam Carrying Triangular Varying Load

A cantilever beam AB of length *l*, fixed at A and free at B subjected to a triangular loading is considered (Figure 4.21). A section *x-x* at a distance *x* from the free end B is taken.

The rate of loading at section *x-x* should be found first. Loading at B is zero and at A, it is *w*/unit run. Hence, the rate of loading at *x* will be $\frac{w}{l} \times x$ per unit run.

1. Shear Force

Shear force at section *x-x* is S_x = Total load on the cantilever for a length of *x*

= Area of loaded triangle with base *x*

$$= \frac{1}{2}\left(\frac{w}{l} \times x\right) x$$

$$= \frac{wx^2}{2l}$$

Since the equation is in second degree of *x*, the shear force variation will be parabolic.

At B, $x = 0$ $\qquad S_B = 0$

At A, $x = l$ $\qquad S_A = \frac{wl}{2}$

2. Bending Moment

Bending moment at section *x-x* is $M_x = -$(Total load for a length *x*) × (Distance of the centre of gravity of the load from *x*) $\quad = \frac{wx^2}{2l} \times \frac{x}{3} = \frac{wx^3}{bl}$

Since the equation is cubical in *x*, the variation will be a cubic curve.

At B, $x = 0$ $\qquad M_B = 0$

At A, $x = l$ $\qquad M_A = \frac{wl^3}{6l} = \frac{wl^2}{6}$.

Shear force and Bending moment diagrams are shown in Figure 4.21.

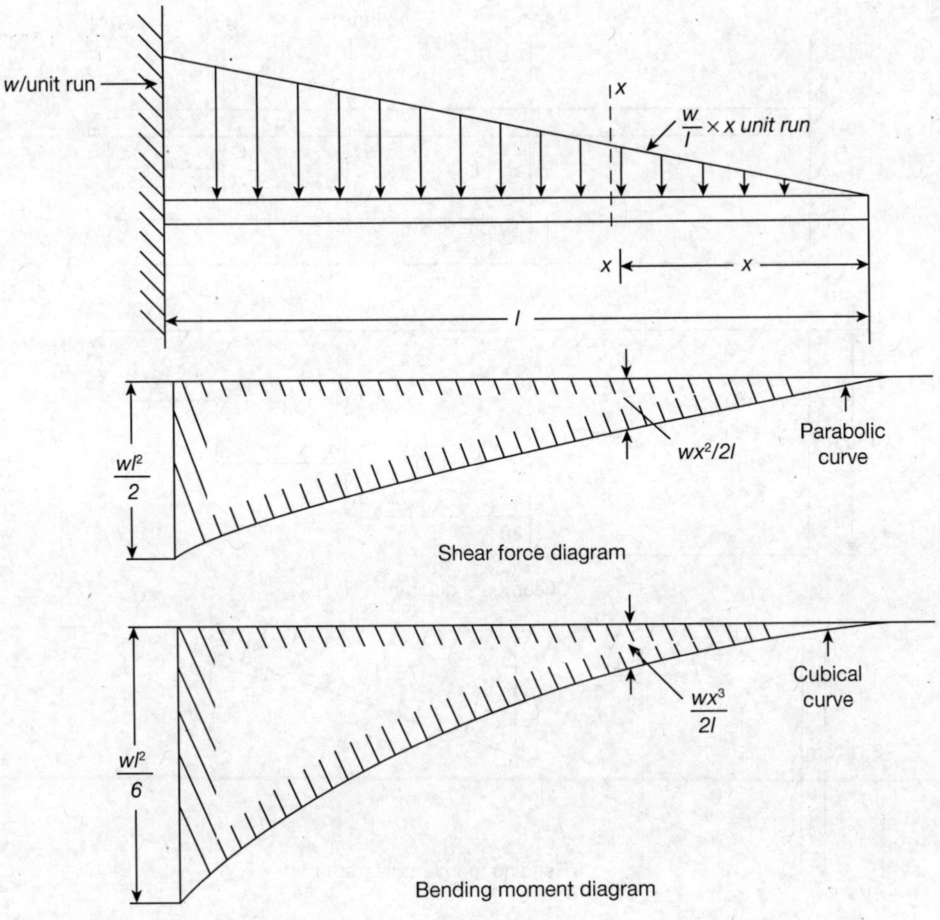

Fig. 4.21 Cantilever beam carrying triangular load.

SOLVED PROBLEM 4.1

A cantilever beam of 3 m long carries four concentrated loads of 50 N, 40 N, 30 N and 20 N at distances 0.0 m, 0.5 m, 1.1 m and 1.8 m, respectively, from the free end. Calculate the maximum bending moment and the shear force below the loads and draw the shear force and bending moment diagrams to a scale.

Given data: Length of beam 3.0 m, loads = 50 N, 40 N, 30 N and 20 N, spacing from end = 0.0 m, 0.5 m, 1.1 m and 1.8 m.

Let AB be the cantilever beam of length 3 m carrying points loads at B, C, D and E as shown in Figure 4.22.

1. Shear Force

Shear force below point B, S_B = 50 N (negative)

Shear force below point C, S_c = (50 + 40) N (negative)

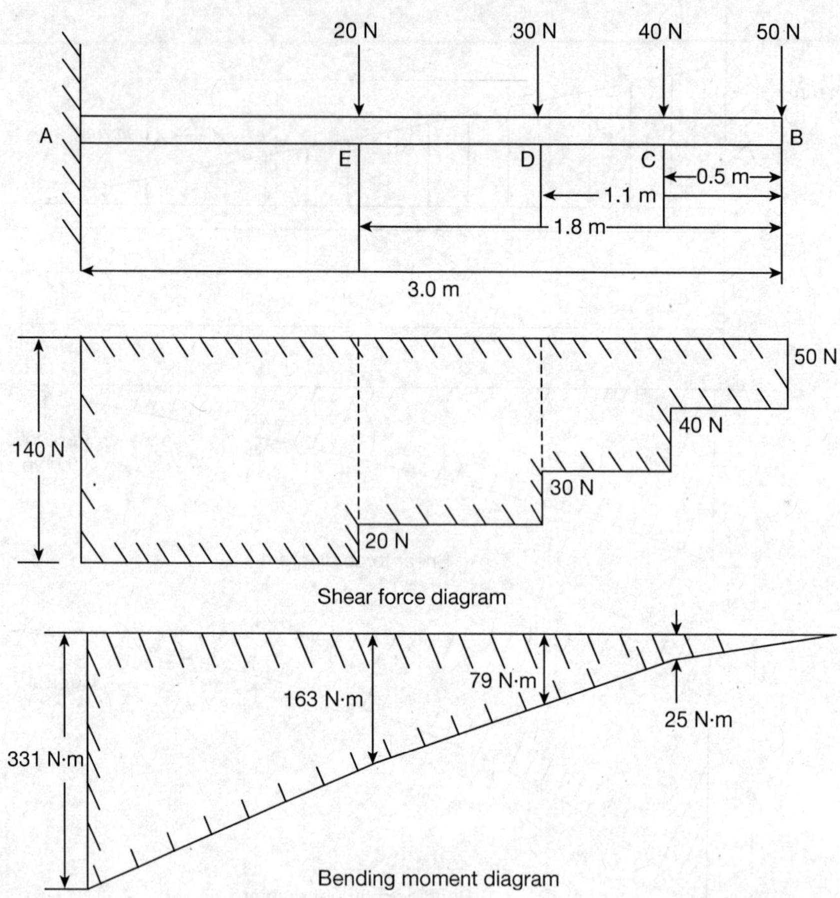

Fig. 4.22

Shear force below point D, $S_D = (50 + 40 + 30)$ N
$= 120$ N (negative)

Shear force below point A, $S_A = (50 + 40 + 30 + 20)$ N
$= 140$ N (negative)

Since all the loads are downward to the right of assumed section, it is marked as negative.

2. Bending moment

Bending moment at B, $M_B = 0$

Bending moment at C, $M_C = 50 \times 0.5 = 25 \text{ N}\cdot\text{m}$ (negative)

Bending moment at D, $M_D = (50 \times 1.1) + 40\,(1.1 - 0.50) = (55 + 24) \text{ N}\cdot\text{m}$
$= 79 \text{ N}\cdot\text{m}$ (negative)

Bending moment at E $M_E = 50 \times 1.8 + 40(1.8 - 0.5) + 30(1.8 - 1.1) = 163 \text{ N}\cdot\text{m}$ (negative).

$M_A = 50 \times 3.0 + 40(3.0 - 0.5) + 30(3.0 - 1.1) + 20(3.0 - 1.8) = 331\ \text{N} \cdot \text{m}$ (negative)

Since all the moments are clockwise to the right of the assured section, it is marked as negative. Shear force and bending moment diagrams are shown in Figure 4.22.

SOLVED PROBLEM 4.2

A cantilever beam 2 m long carries concentrated loads of 20 kN at the free end and 30 kN at a distance of 0.5 m from the fixed end, respectively. It also carries an UDL of intensity 20 kN/m run for a length of 1 m at a distance of 0.75 m from the fixed end. Calculate the maximum shear force and the bending moment and draw the shear force and bending moment diagrams.

Solution:

Given data: Point loads = 30 kN and 20 kN, span = 2 m and UDL = 20 kN/m run over a length of 1.0 m.

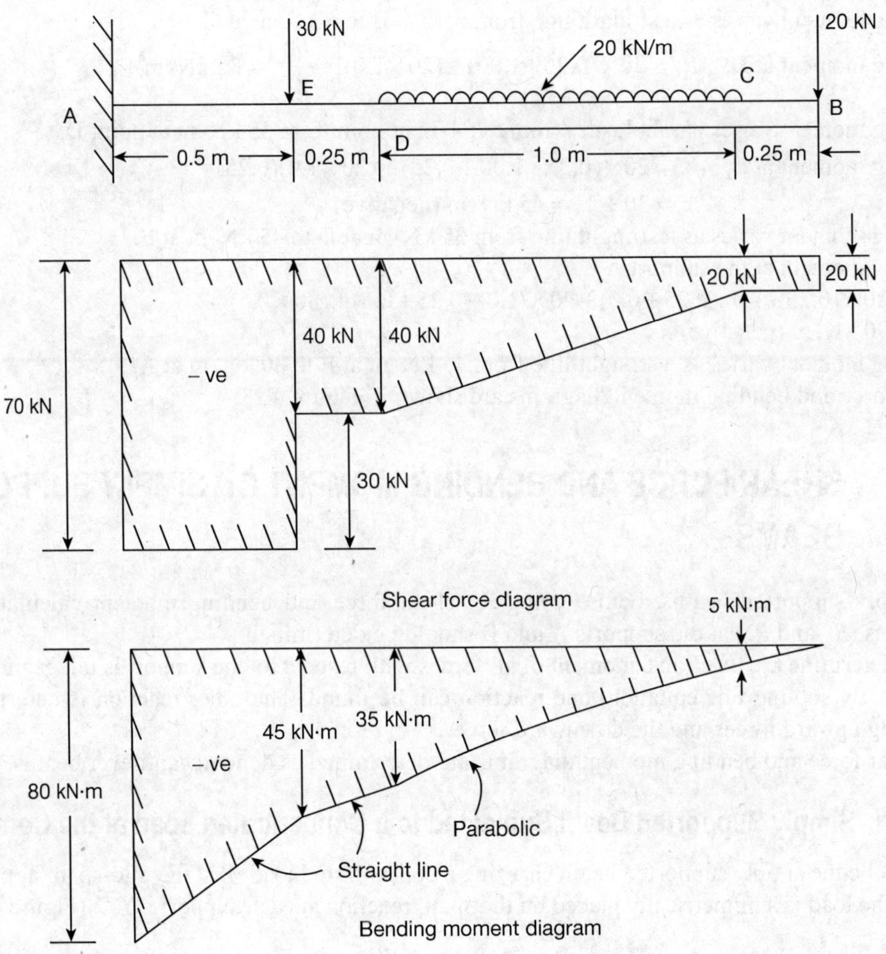

Fig. 4.23

1. Shear Force

Shear force below point B = 20 kN (negative)
Shear force between B and C = 20 kN (negative)
Shear force below point C = 20 kN (negative)
Shear force below point D = $-[20+(20\times 10)] = 40$ kN (negative)
Shear force from point C varies as a straight line from 20 kN to 40 kN (negative)
Shear force from D to E = 40 kN (negative)
Shear force from E to A = 70 kN (negative)

2. Bending Moment

Bending moment at the free end B is zero, i.e., $M_B = 0$.
Bending moment at C, $M_C = 20\times 0.25 = 5\ \text{kN}\cdot\text{m}$ (negative)
Bending moment varies as a straight line from zero at B to 5 kN·m at C.

Bending moment at D, $M_D = 20\times(0.25+1.00)+(20\times 1.0)\left(\dfrac{1.00}{2}\right) = 35\ \text{kN}\cdot\text{m}$

Bending moment varies parabolically from 5 kN·m at point C to 35 kN·m at point D.
Bending moment at E, $M_E = 20\times(0.25+1.00+0.25)+(20\times 1.0)(0.25)$
$= 30 + 15 = 45\ \text{kN}\cdot\text{m}$ (negative)
Bending moment varies as a straight line from 35 kN·m at D to 45 kN·m at E.
Bending moment at the support A

$$M_A = 20\times(0.25+1.0+0.25+0.5)+20\times(1.0+0.25+0.5)+(30\times 0.5)$$
$$= 90\ \text{kN}\cdot\text{m (negative)}$$

Bending moment varies as a straight line from 45 kN·m at E to 80 kN·m at A.
Shear force and bending moment diagrams are shown in Figure 4.23.

4.6 SHEAR FORCE AND BENDING MOMENT ON SIMPLY SUPPORTED BEAMS

In simply supported beams, before going for shear force and bending moment calculations, the reactions (R_A and R_B) at the supports A and B should be determined.

To determine the reactions, moment of all forces with respect to one support is taken and equated to zero. By solving this equation, one reaction can be found. The other reaction is determined by equating upward forces and the downward forces.

Shear force and bending moment diagrams are determined as done in cantilever beams.

4.6.1 Simply Supported Beam Subjected to a Concentrated Load at the Centre

Let AB be the simply supported beam carrying a concentrated load W at the mid-span (at point C).

As the load is symmetrically placed on the span, reaction at each support is $R/2$ (Figure 4.24).

$$\therefore \qquad R_A = R_B = \frac{W}{2}$$

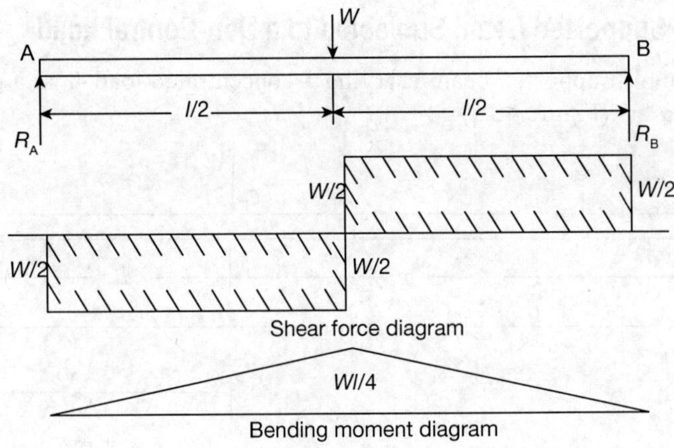

Fig. 4.24 Simply supported beam with central point load.

1. Shear Force

For a section between B and C,

$$\text{Shear force} = \frac{W}{2}$$

For a section between C and A,

$$\text{Shear force} = \frac{W}{2} - W = -\frac{W}{2}$$

2. Bending Moment

Bending moment at a distance x from B in BC is

$$M_x = \frac{Wx}{2}$$

At B, $x = 0$ $\qquad M_B = 0$

and at C, $x = \frac{l}{2}$ $\qquad M_C = \frac{Wl}{4}$

Bending moment at a distance x from B in CA item

$$M_x = \frac{W(l-x)}{2}$$

At C, $x = \frac{l}{2}$ $\qquad M_C = \frac{Wl}{4}$

At A, $x = l$ $\qquad M_A = 0$

It is to be noted that the maximum bending moment occurs at the point where shear force changes the sign from positive to negative or vice versa.

Also, the bending moment at supports in the case of simply supported beams is always zero.

4.6.2 Simply Supported Beam Subjected to a Non-Central Load

Let AB be the simply supported beam carrying a concentrated load W at a distance a from the support A where $a > b$ (Figure 4.25).

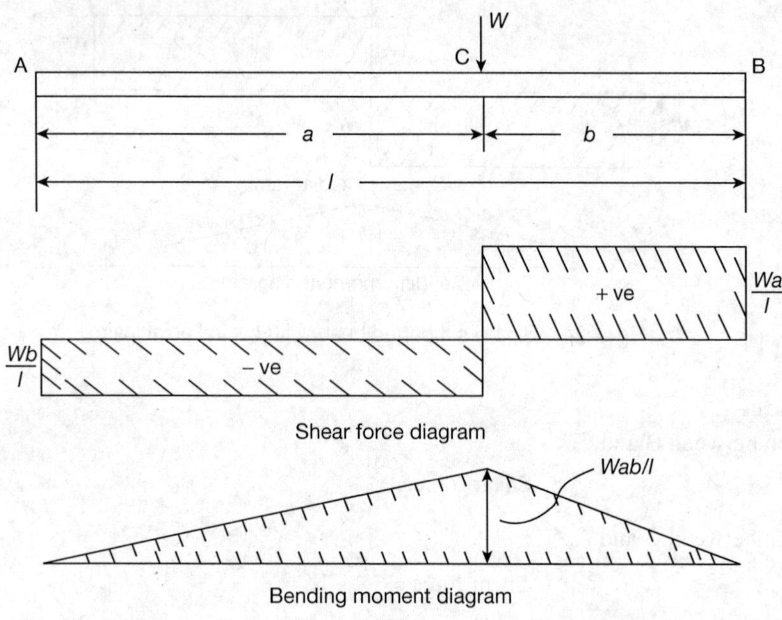

Fig. 4.25 Simply supported beam with non-central load.

Shear force and bending moment diagrams are shown in Figure 4.25.

To find the support reactions R_A and R_B moments of all forces are taken with respect to support A. Then

$$R_B \times l = Wa$$

$$R_B = \frac{Wa}{l}$$

But

$$R_A + R_B = W$$

i.e.,

$$R_A = W - \frac{Wa}{l} = \frac{W(l-a)}{l}$$

$$R_A = \frac{Wb}{l}$$

1. Shear Force

Shear force from B to C is $\frac{Wa}{l}$ (positive)

Shear force from C to A is $\frac{Wb}{l}$ (negative)

Shear force at A is $\frac{Wb}{l}$ (negative)

2. Bending Moment

Bending moment at A and B = 0

Bending moment at C,
$$M_C = \frac{Wa}{l} \times b$$
$$M_C = \frac{Wab}{l}$$

Shear force diagram and bending moment diagram are shown in Figure 4.25.

4.6.3 Simply Supported Beam Subjected to More than One Concentrated Load

Let AB be the simply supported beam with concentrated load W acting at C, D and E at equal distance (Figure 4.26).

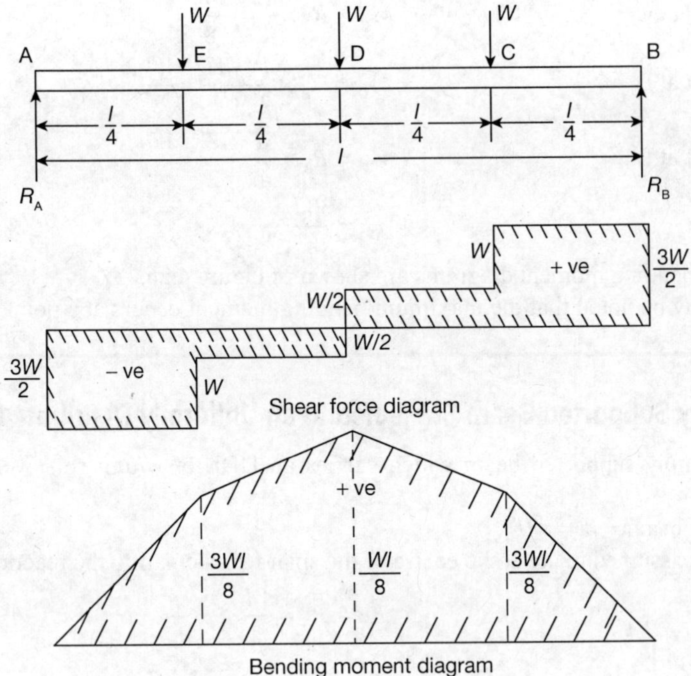

Fig. 4.26 Simply supported beam with several point loads.

Reactions R_A and R_B are equal to $\frac{3W}{2}$.

1. Shear Force

Shear force at B is $\frac{3W}{2}$ (positive)

Shear force from B to C is $\frac{3W}{2}$ (positive)

Shear force from C to D is $\left(\frac{3W}{2} - W\right) = \frac{W}{2}$ (positive)

Shear force from D to E is $\left(\frac{3W}{2} - 2W\right) = \frac{W}{2}$ (negative)

Shear force from E to A is $\left(\frac{3W}{2} - 3W\right) = \frac{3W}{2}$ (negative)

Shear force at A is $\frac{3W}{2}$ (negative)

2. Bending Moment

As the beam is simply supported, bending moment at the supports is zero, i.e., $M_A = M_B = 0$.

Bending moment at C, $$M_C = R_B \times \frac{l}{4} = \frac{3Wl}{8}$$

Bending moment at D, $$M_D = R_B \times \frac{l}{2} - \frac{Wl}{4} = \frac{Wl}{2}$$

Bending moment at E, $$M_E = R_B \times \frac{3l}{4} - \frac{Wl}{2} - \frac{Wl}{4}$$

$$M_E = \frac{3Wl}{8}.$$

Shear force and bending moment diagrams are shown in Figure 4.26.

Once again it may be noted that the maximum bending moment occurs at a point where shear force changes sign.

4.6.4 Simply Supported Beam Subjected to an Uniformly Distributed Load

Let AB be a simply supported beam which carries an UDL of *w*/unit run over its entire length (Figure 4.27).

Total load on the beam = $W = wl$.

This total load is assumed to act at the centre of the span for calculating the reactions.

$\therefore$ $$R_A = R_B = \frac{wl}{2}$$

1. Shear Force

A section *x-x* at a distance *x* from B is considered.

Shear force, $$S_x = \frac{wl}{2} - wx$$

At B, $x = 0$ $$S_B = \frac{wl}{2}$$

At A, $x = 1$ $$S_A = \frac{wl}{2} - wl = \frac{wl}{2}$$

Since the shear force equation $S_x = \frac{wl}{2} - wx$ is a single degree equation, the variation of the shear force is a straight line.

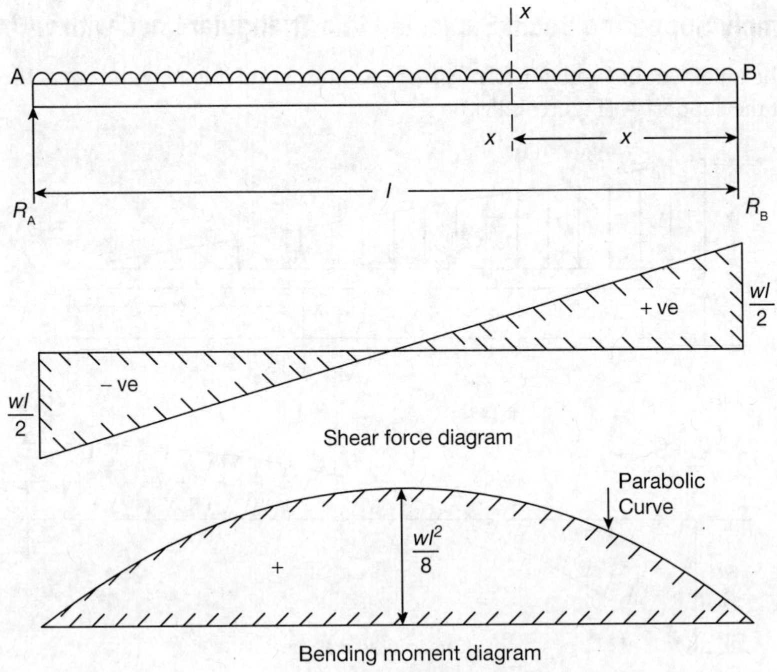

Fig. 4.27 Simply supported beam with UDL.

The shear force is $+\frac{wl}{2}$ at support B and $-\frac{wl}{2}$ at support A. The sign changes at mid-span. Hence, the bending moment will be maximum at this point.

2. Bending Moment

Bending moment with respect to section *x-x*, $M_x = \frac{wl}{2}x - \frac{w \times x}{2}$

Since the equation M_x is a second-degree equation in *x*, the variation of the value will be a parabolic curve.

At B, $x = 0$ $M_B = 0$

At A, $x = l$ $M_A = 0$

Maximum bending moment occurs at $x = \frac{l}{2}$ where the shear force changes sign.

$$M_{max} = \frac{wl}{2} \times \frac{l}{2} - \frac{w}{2}\left(\frac{l}{2}\right)^2$$

$$M_{max} = \frac{wl^2}{8}$$

That is the maximum bending moment is $\frac{wl^2}{8}$ occurs at the middle of the span.

Shear force and bending moment diagrams are shown in Figure 4.27.

4.6.5 Simply Supported Beam Subjected to a Triangular Load with w/Unit Run

Let AB be a beam of length l with uniformly increasing triangular load with zero at the support B and w/unit run at the support A (Figure 4.28).

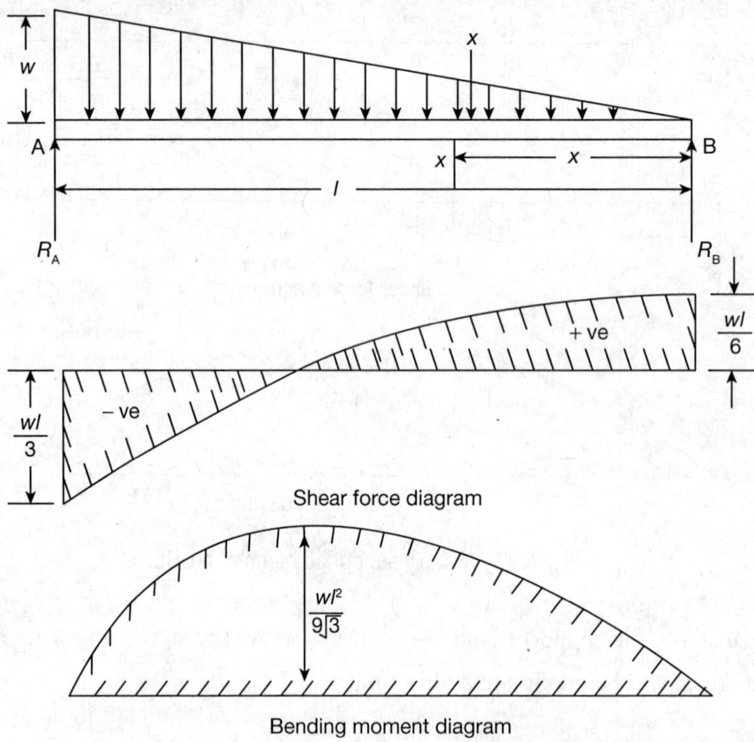

Fig. 4.28 Simply supported beam with triangular load.

Total load = (Area of the triangle) = $\frac{wl}{2}$

Taking moment about A

$$R_B \times l = \left(\frac{1}{2} \times w \times l\right) \times \frac{l}{3}$$

i.e.,

$$R_B = \frac{wl}{6}$$

$$R_A + R_B = \frac{wl}{2}$$

$\therefore$

$$R_A = \frac{wl}{2} - \frac{wl}{6} = \frac{wl}{3}$$

1. Shear Force

Any section x-x at a distance x from support B is considered.
Intensity of loading at this section is $w \times l$.
By similar triangles

$$\frac{w}{l} = \frac{w_x}{x}$$

$$\therefore \quad w_x = \frac{wx}{l}$$

Shear force at section x-x,

$$S_x = R_B - (\text{Load on portion right of section } x\text{-}x) = \frac{wl}{6} - \frac{1}{2} x \times \frac{wx}{l} = \frac{wl}{6} - \frac{wx^2}{2l}.$$

Shear force equation is in second degree, hence the variation is parabolic

At A, $x = l$
$$S_A = \frac{wl}{6} - \frac{wl^2}{2l} = \frac{wl}{3}$$

At B, $x = 0$
$$S_B = \frac{wl}{6}$$

Shear force changes sign at $S_x = 0$

$$\therefore \quad \frac{wl}{6} - \frac{wx^2}{2l} = 0$$

$$\therefore \quad x^2 = \frac{l^2}{3}$$

i.e.,
$$x = \frac{l}{\sqrt{3}}$$

i.e., Shear force changes sign at a distance of $\frac{l}{\sqrt{3}}$ from the support B.

At this point, the bending moment will be maximum.

2. Bending Moment

Bending moment at distance x from support B is

$$M_x = R_B \times x - \left(\frac{1}{2} \times x \times \frac{wx}{l}\right) \times \frac{x}{3}$$

$$= \frac{wlx}{6} - \frac{wx^2}{2l} \times \frac{x}{3} = \frac{wlx}{6} - \frac{wx^3}{6l}$$

The above equation is cubical and hence the curve will be of cubical variation.

At A, $x = l$
$$M_A = \frac{wl^2}{6} - \frac{wl^3}{6l} = 0$$

At B, $x = 0$
$$M_B = 0$$

At C, $x = \frac{l}{\sqrt{3}}$ $\quad M_C = \frac{wl}{\sqrt{3}} \times \frac{l}{6} - \frac{wl^3}{6 \times 3 \times \sqrt{3}l} = \frac{wl^2}{9\sqrt{3}}$

Maximum bending moment is $\frac{wl^2}{9\sqrt{3}}$.

Shear force and bending moment diagrams are shown in Figure 4.28.

4.6.6 Simply Supported Beam Subjected to Symmetrical Triangular Load

Let AB be a simply supported beam of length l subjected to symmetrical triangular load with zero load at support and w/unit run at the middle of the span (Figure 4.29).

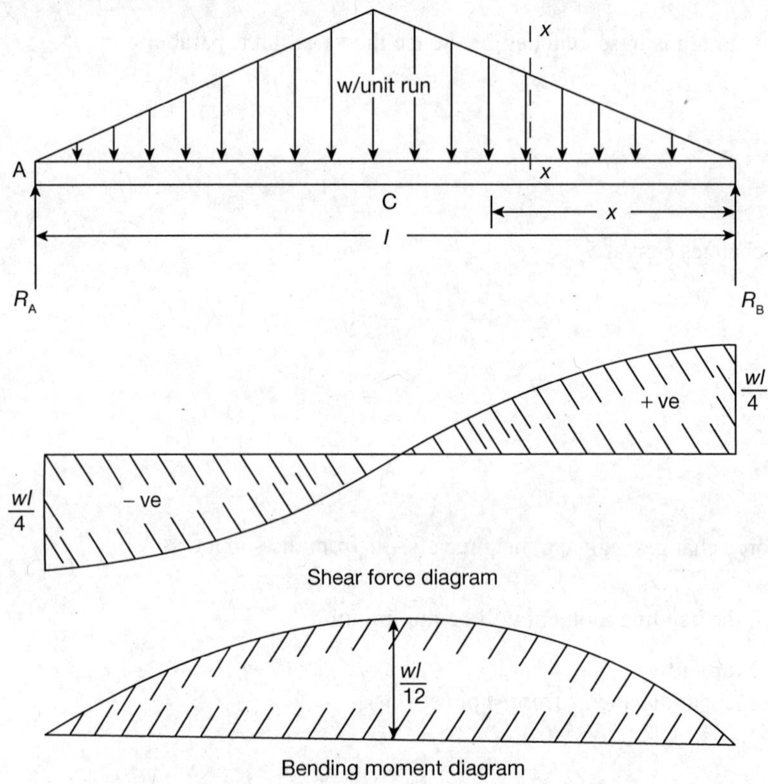

Fig. 4.29 Simply supported beam with symmetrical triangular load.

$$\text{Total load on the beam} = \frac{1}{2} \times l \times w = \frac{wl}{2}$$

Because of symmetrical load, the reaction R_A and R_B are equal.

$$\therefore \quad R_A = R_B = \frac{wl}{4}$$

A section *x-x* at a distance *x* from the support B is considered.

$$\text{Intensity of load at section } x\text{-}x \text{ is } w_x = \frac{wx}{\frac{l}{2}} = \frac{2wx}{l}$$

$$\text{Load over the distance } x = \frac{1}{2} \times x \times \frac{2wx}{l} = \frac{wx^2}{l}$$

1. Shear Force

Shear force at section *x-x* is $S_x = R_B - (\text{load on } x) = \frac{wl}{4} - \frac{wx^2}{l}$

The above equation shows that the variation is parabolic.

At B, $x = 0$, $S_B = \frac{wl}{4}$

At C, $x = l/2$, $S_C = \frac{wl}{4} - \frac{wl^2}{l \times 4} = 0$

Because of symmetry, the shear force at *A* will be $-\frac{wl}{4}$

2. Bending Moment

Bending moment at section *x-x* is $M_x = R_B x - (\text{Load on } x) \times \frac{x}{3}$

$$= \frac{wl}{4}x - \left(\frac{1}{2} \times x \times \frac{2wx}{l}\right)\frac{x}{3}$$

$$= \frac{wlx}{4} - \frac{wx^3}{3l}$$

This is a third-degree equation and can be a curve.

At B, $x = 0$, $M_B = 0$

At C, $x = l/2$, $M_C = \frac{wl^2}{8} - \frac{wl^3}{24l} = \frac{wl^2}{12}$

At A, $x = 0$, $M_A = 0$

Maximum bending moment is $\frac{wl^2}{12}$.

Shear force and bending moment diagrams are shown in Figure 4.29.

SOLVED PROBLEM 4.3

Draw the shear force and bending moment diagrams for the simply supported beam shown in Figure 4.30.

Solution:

Given data: Load, UDL = 4 kN/m over 5 m length and span = 8 m.

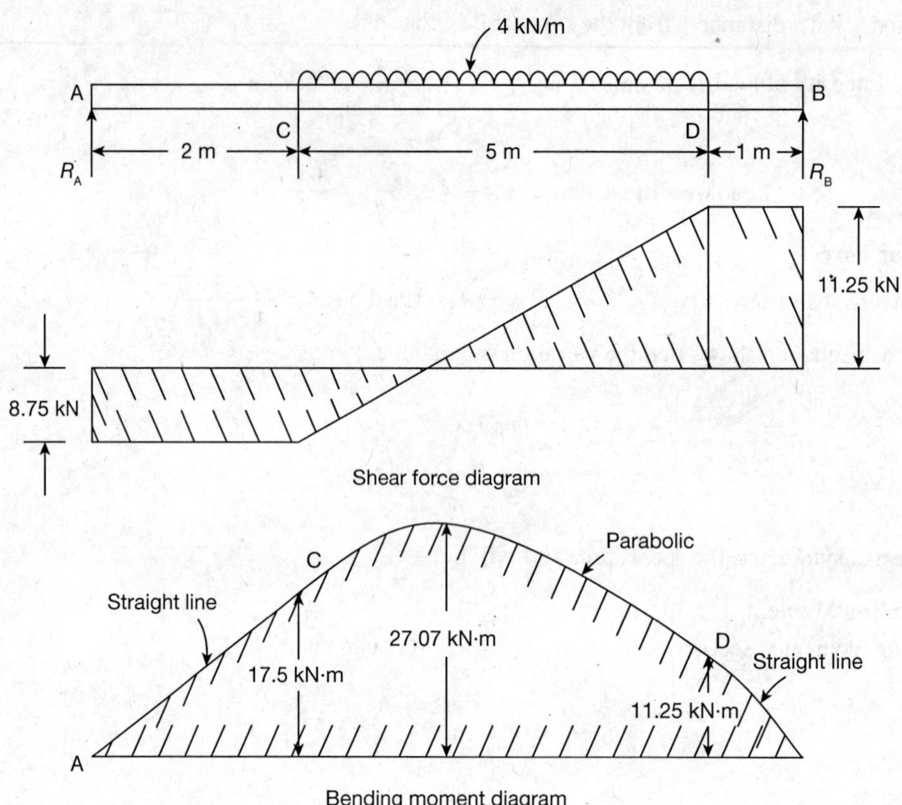

Fig. 4.30

Taking moment of all the forces about A

$$R_B \times 8 = 4 \times 5(2 + 2.5)$$

$\therefore$ $$R_B = 11.25 \text{ kN}$$

$$\text{Total load} = 4 \times 5 = 20 \text{ kN}$$

$\therefore$ $$R_A = 20 - 11.25 = 8.75 \text{ kN}$$

1. Shear Force

A section *x-x* at a distance *x* from B is considered.

Shear force at section *x-x*,

$$S_x = 11.25 - w(x-1) = 11.25 - 4(x-1)$$

i.e., $$S_x = +15.25 - 4x$$

At D, $x = 1$ m, $$S_D = 15.25 - 4 \times 1 = 11.25 \text{ kN}$$

At C, $x = 6$ m, $$S_C = 15.25 - 4 \times 6 = -8.75 \text{ kN}$$

Variation of shear force between D and C will be an inclined straight line.

Shear force changes signs at a distance x, when S_x is equal to zero.

i.e, $$S_x = 15.25 - 4x = 0$$

Solving $$x = \frac{15.25}{4} = 3.8125 \text{ m}$$

Shear force at A, $$S_A = 8.75 \text{ kN}.$$

2. Bending Moment

As the beam is simply supported, the bending moment is zero at the supports, i.e., $M_A = M_B = 0$.

Bending moment at D, $$S_D = 11.25 \times 1.0 = 11.25 \text{ kN} \cdot \text{m}.$$

Bending moment at section x-x is $$M_x = 11.25 \times x - w(x-1)\left(\frac{x-1}{2}\right)$$

$$= 11.25x - \frac{4}{2}\left(x^2 - 2x + 1\right) = 13.25x - 2.$$

At D, $x = 1$ m $$M_D = 13.25 \times 1 - 2 = 11.25 \text{ kN} \cdot \text{m}$$

At C, $x = 6$ m $$M_C = 13.25 \times 6 - 2 = 17.50 \text{ kN} \cdot \text{m}$$

At $x = 3.8125$ m, the shear force changes sign. At this point, the bending moment will be maximum.

$$M_{max} = 13.25 \times 3.8125 - 2 = 48.515$$

Shear force and bending moment diagrams are shown in Figure 4.30.

SOLVED PROBLEM 4.4

Draw the shear force and bending moment diagrams for the simply supported beam as shown in Figure 4.31.

Taking moment of all the forces about A

$$R_B \times 6 = (2 \times 1.5) \times (4.5 + 0.75) + (8 \times 4.5) + (5 \times 3) + (4 \times 1.5) + (3 \times 3)(1.5)$$

$$= 86.25 \text{ kN}$$

$\therefore$ $$R_B = 86.25/6 = 14.375 \text{ kN}$$

$$\text{Total load} = (3 \times 3) + 4 + 5 + 8 + 2 \times 1.5 = 29 \text{ kN}$$

$\therefore$ $$R_A = 29 - 14.375 \text{ kN}$$

$$= 14.625 \text{ kN}$$

1. Shear Force

Shear force at B, $$S_B = 14.375 \text{ kN}$$

Shear force at E, $$S_E = 14.375 - (2 \times 1.5)$$

$$= 11.375 \text{ kN (top value)}$$

Due to UDL the variation from B to E will be an inclined line.

Further, there is a downward load of 8 kN at E, then

$$S_E = 11.375 - 8$$

$$= 3.375 \text{ kN (bottom value)}$$

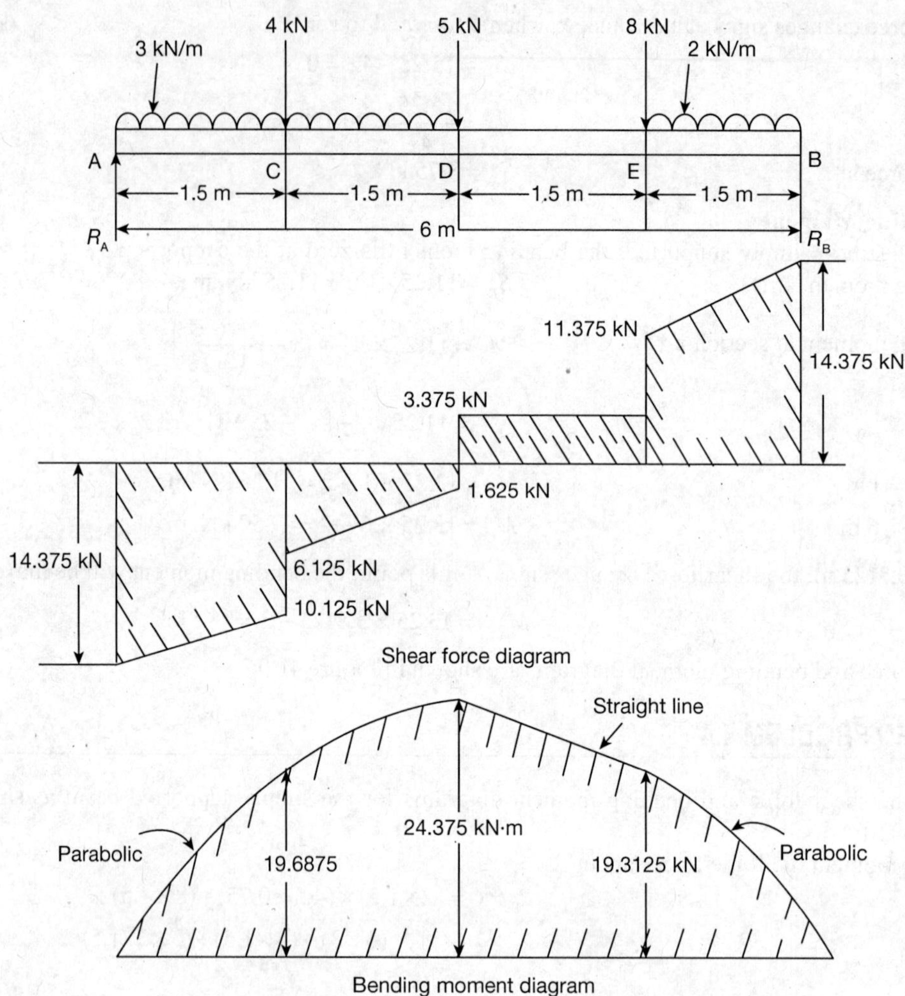

Fig. 4.31

Shear force at D, $S_D = 3.375$ kN (top value)

Since there is no load acting between E and D, the shear force variation is a horizontal straight line.

There is a downward load at D, then $S_D = 3.375 - 5 = -1.625$ (bottom value)

Shear force at C, $S_C = -1.625 - (3 \times 1.5) = -6.125$ kN (top value)

The variation from D to C is an inclined straight line from −1.625 to −6.125 kN.

There is a downward point load of 4 kN at C, then $S_C = -6.125 - 4 = 10.125$ kN (bottom value)

Shear force at A, $S_A = -10.125 - (3 \times 1.5) = 14.625$ kN

2. Bending Moment

Bending moment at supports A and B are zero, i.e., $M_A = M_B = 0$.

Bending moment at E, $M_E = (14.375 \times 1.5) - (2 + 1.5 \times 0.75) = 19.3125$ kN·m

Variation of bending moment from E to B is a parabolic curve.

Bending moment at D, $M_D = (14.375 \times 3) - (2 \times 1.5 \times 2.25) - (8 \times 1.5) = 24.375$ kN·m

Variation of bending moment from E to D is a straight line.

Bending moment at C, $M_C = (14.375 \times 4.5) - (2 \times 1.5 \times 3.25) - (8 \times 3.0) - (5 \times 1.5) - (3 \times 0.75)$
$= 19.6875$ kN·m.

Variation of bending moment from D to A is a parabolic curve.

Bending moment at A, $M_A = (14.375 \times 6) - (2 \times 1.5 \times 5.25) - (8 \times 4.5) - (5 \times 3) - (3 \times 3 \times 1.5) = 0.$

Shear force and Bending moment diagrams are shown in Figure 4.31.

SOLVED PROBLEM 4.5

A simply supported beam of length 5 m is subjected to an uniformly increasing load of 400 N/m from one end to 800 N/m to the other end. Draw the SFD and BMD indicating the important values.

Solution:

Given data: Varying UDL = 400 N/m to 800 N/m over 5 m span.

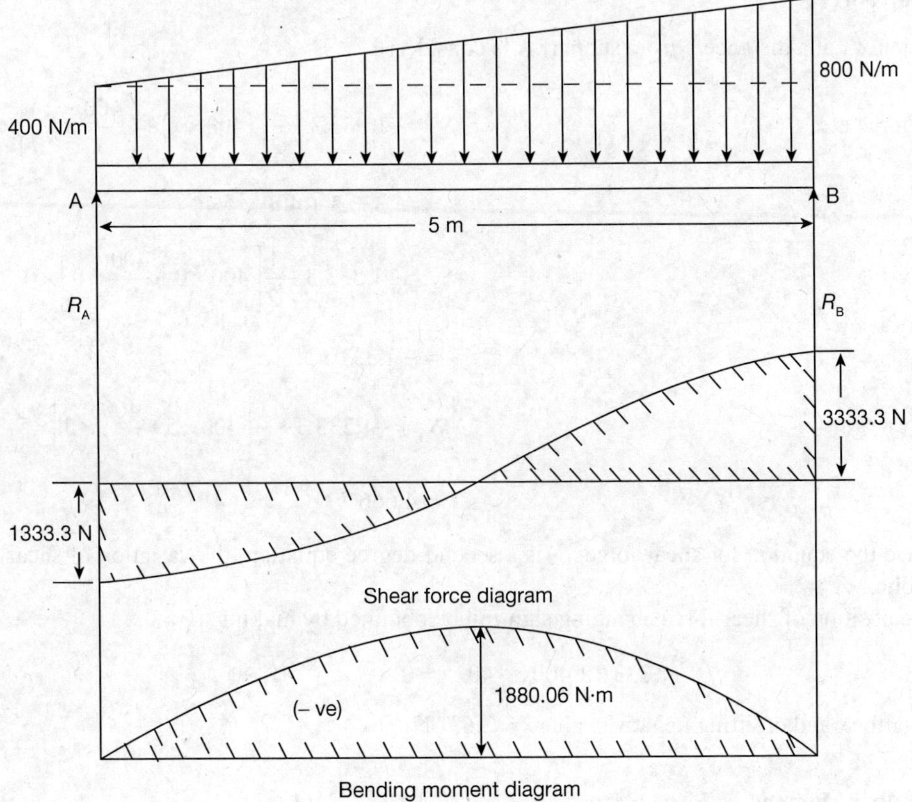

Fig. 4.32

Uniformly increasing load on the beam may be split into two types of loading as

(i) a uniformly distributed load of 400 N/m over the entire span length and

(ii) a triangular load of zero intensity at A to 400 N/m intensity at B.

Taking moment of all the forces about the support A.

$$R_B \times 5 = \left(400 \times 5 \times \frac{5}{2}\right) + \left(\frac{1}{2} \times 5 \times 400 \times \frac{2}{3} \times 5\right)$$

$$\therefore \quad R_B = \frac{8,333.3}{5} = 1,666.7 \text{ N}$$

$$\text{Total load} = (400 \times 5) + \left(\frac{1}{2} \times 5 \times 400\right) = 3,000 \text{ N}$$

$$\therefore \quad R_A = 3,000 - R_B = 3,000 - 1,666.7 = 1,333.3 \text{ N}$$

1. Shear Force

A section *x-x* at a distance *x* from support A is considered.

Shear force at *x*,
$$S_x = -1,333.3 + \frac{1}{2}\left[(400 \times x) + \left(\frac{400}{5} \times x\right)\right] \times x$$

$$= -1,333.3 + 400\,x + 40\,x^2$$

At A, $x = 0$
$$S_A = -1,333.3 + \frac{1}{2}\left[400 \times 0 + \frac{400}{5} \times 0\right] \times 0$$

$$= -\,1,333.3 \text{ N}$$

At B, $x = 5$ m
$$S_B = -1,333.3 + \frac{1}{2}\left[400 \times 5 + \frac{400}{5} \times 5\right] \times 5$$

$$= 4,666.7 \text{ N}$$

Since the equation for shear force, S_x is a second-degree equation, the variation of shear force is parabolic.

The position of shear force changing sign can be obtained by making, $S_x = 0$.

i.e., $-1,333.3 + 400x + 40x^2 = 0$

Solving for *x* and omitting negative value $x = 2.637$ N

2. Bending Moment

Bending moment at *x*,

$$M_x = R_A \times x\left(w \times x \times \frac{x}{2}\right) - \frac{1}{2}\left(\frac{wx}{l} \times x \times \frac{x}{3}\right)$$

$$= 1{,}333.3x - \left(400 \times x \times \frac{x}{2}\right) - \left[\frac{1}{2}\left(\frac{400 \times x}{5} \times x \times \frac{x}{3}\right)\right]$$

$$= 1{,}333.3x - \frac{400x^2}{2} - \frac{40x^3}{3}$$

$$M_{max} = 1{,}333.3 \times 2.637 - \frac{400 \times (2.637)^2}{2} - \frac{40}{3}(2.637)^3$$

$$= 1{,}880.06\ \text{N} \cdot \text{m}$$

Bending moment at A, when $x = 0$; $M_A = 0$

Bending moment at B, when $x = 5$ m;

$$M_B = 1{,}333.3 \times 5 - \left(400 \times 5 \times \frac{5}{2}\right) - \left(\frac{1}{2} \times \frac{400 \times 5}{5} \times 5 \times \frac{5}{3}\right)$$

$$M_B = 0$$

Shear force and bending moment diagrams are shown in Figure 4.32.

SOLVED PROBLEM 4.6

A simply supported beam of length 10 m is subjected to a clockwise couple of 1,500 N·m at a distance of 4 m from the right-hand support. Draw the shear force and bending moment diagrams.

Solution:

Given data: A clockwise couple of 1,500 N·m and span = 10 m.

Let AB be the simply supported beam subjected to a clockwise couple at C from the right-hand support.

Taking moment of the forces about the support B, we have

$$R_A \times 10 + 1{,}500 = 0$$

$$\therefore \quad R_A = -150\ \text{N}$$

Hence, the reaction at A will be downward at the support A.

Taking moment of the forces about B

$$R_B \times 10 - 1{,}500 = 0$$

$$\therefore \quad R_B = 150\ \text{N}$$

Hence, the reaction at B will be upwards at the support B.

1. Shear Force

The effect of moment will not be there for drawing the shear force diagram. Starting from the right-hand support, $S_{B-A} = 150$ N (positive).

2. Bending Moment

Bending moment at B = 0, i.e., $M_B = 0$

Bending moment at C = 150 × 4 = 600 N·m (positive)

At C, there is a clockwise moment of 1,500 N·m.

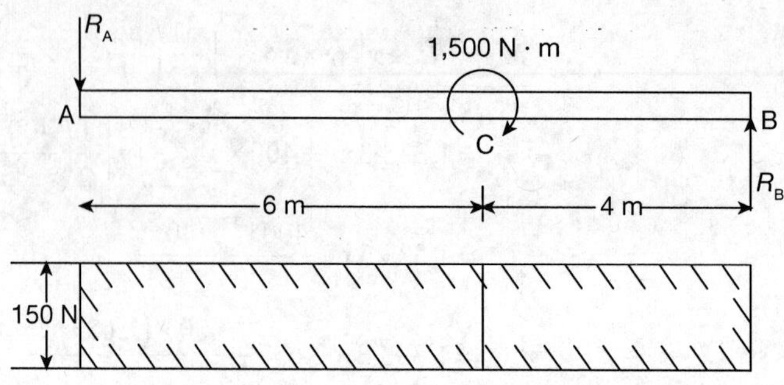

Shear force diagram

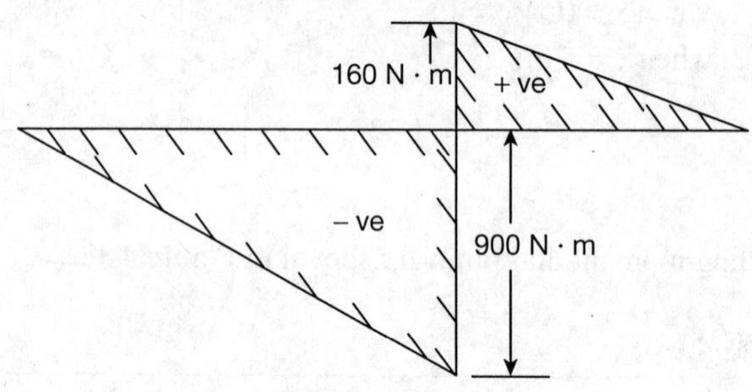

Bending moment diagram

Fig. 4.33

Bending moment at C = $150 \times 6 = 900$ N·m (negative)
At A, the bending moment will be zero.
i.e., $M_A = 150 \times 10 - 1{,}500 = 0.$
Shear force and Bending moment diagrams are shown in Figure 4.33.

4.7 OVERHANGING BEAMS

If the end portion of a beam extends beyond the support, then the beam is known as overhanging beam. Overhanging may be on one of the supports or on both the sides.

In the case of an overhanging beam, the bending moment is positive between the supports, whereas the bending moment is negative for the overhanging portion.

At the same point on the beam, the bending moment is zero or changes sign from positive to negative value or vice versa. This point where the bending moment changes sign is called as the *point of contraflexure* or *point of inflexion*.

4.8 BEAMS SUBJECTED TO INCLINED LOADS

So far, the discussion has been with reference to loads which are acting vertically upwards or downwards. However, in practice, there are many situations where the beams are subjected to inclined loads and to transverse loads. In such cases, ordinary supports cannot create an equilibrium condition because horizontal supports cannot be balanced. Hence, one of the supports is provided as a hinged support and the other as a roller support. Hinged support takes care of the horizontal unbalanced force and the roller support the vertical forces.

All inclined forces are resolved into two mutually perpendicular components one perpendicular to the axis of the beam and the other parallel to the axis.

Perpendicular or vertical components of forces cause shear force and bending moment, whereas parallel component of force cause an axial thrust on the beam. When the thrust acts away from the beam then it creates tension and when it acts towards the beam it causes compression.

In addition to shear force and bending moment diagrams, thrust diagram is also to be drawn. Thrust causing tension is plotted above the horizontal line and thrust causing compression is plotted below the horizontal line.

SOLVED PROBLEM 4.7

A beam of 10 m long has an overhang of 2 m at one end and is subjected to UDL of 50 N/m length over its entire length. Draw the SFD and BMD. Determine the point of contraflexure.

Solution:

Given data: load, UDL = 50 N/m and span = 8 m.

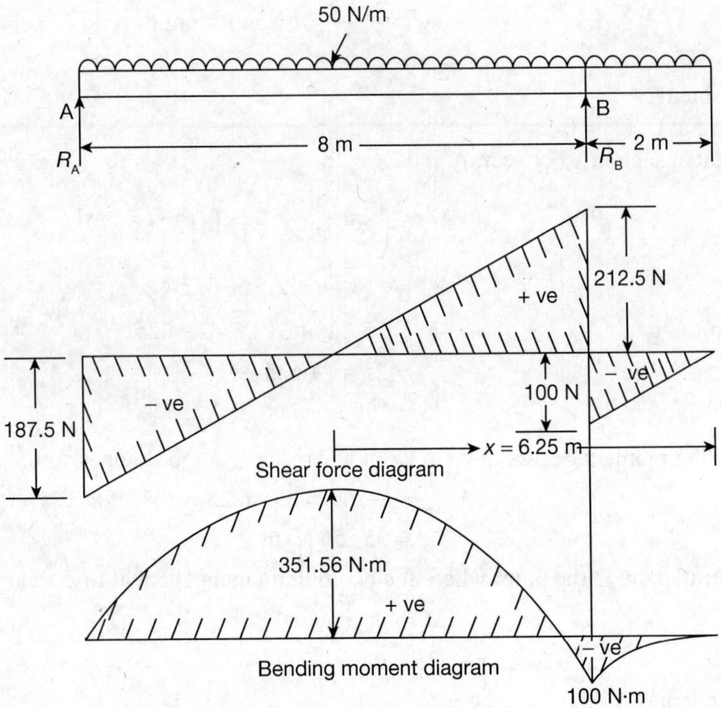

Fig. 4.34

Taking moment of all forces with respect to A

$$R_B \times 8 = 50 \times 10 \times \frac{10}{2}$$

$$\therefore \quad R_B = 312.5 \text{ N}$$

$$\text{Total load} = 50 \times 10 = 500 \text{ kN. } R_A = 500 - R_B = 500 - 312.5 = 187.5 \text{ N}$$

1. Shear Force

Shear force at C, $S_C = 0$

Shear force at B, $S_B = -50 \times 2 = -100$ N (bottom point)

Shear force at B, $S_B = -100 + 312.5 = 212.5$ (upper point)

For shear force between A and B, a section *x-x* at a distance *x* from point C is considered

$$S_x = -50\,x + 312.5$$

At $x = 2$ m $\quad S_B = (-50 \times 2) + 312.5 = 12.5$ N

At $x = 10$ m $\quad S_A = (-50 \times 10) + 312.5 = 187.5$ N

At A, there is an upward support reaction of 187.5 N.

Shear force changes sign between A and B.

Then, $\quad -50\,x + 312.5 = 0$

$$\therefore \quad x = \frac{312.5}{50} = 6.25 \text{ m from the end C.}$$

2. Bending Moment

Bending moment at C, $M_C = 0$

Bending moment at section *x* between A and B.

$$M_x = \left(-50 \times x \times \frac{x}{2}\right) + [312.5(x-2)]$$

i.e., $\quad M_x = 25x^2 + 312.5x - 625$

At B, $x = 2$ m

$$M_B = 25 \times 2^2 + 312.5 \times 2 - 625$$
$$= 25 \times 4 + 625 - 625$$
$$= 100 \text{ N} \cdot \text{m}$$

Maximum bending moment occurs at $x = 6.25$ m where shear force changes sign.

i.e, $\quad M_{max} = (-25 \times 6.25 \times 6.25) + (312.5 \times 6.25) - 625$

$$= 351.56 \text{ N} \cdot \text{m}$$

Point of contraflexure is the point where the bending moment is equal to zero and changes sign.

i.e., $\quad M_x = 0$

i.e., $\quad -2x^2 + 312.5x - 625 = 0$

Solving $x = 10$ m and 2.5 m

Points of contraflexure are at the support A and at 2.5 m from the point C.

Shear force and bending moment diagrams are shown in Figure 4.34.

SOLVED PROBLEM 4.8

Draw the SFD and BMD for the beam shown in Figure 4.35. Determine the points of contraflexure.

Solution:

Given data: Load, UDL = 100 N/m for 9 m length and point load = 100 N.

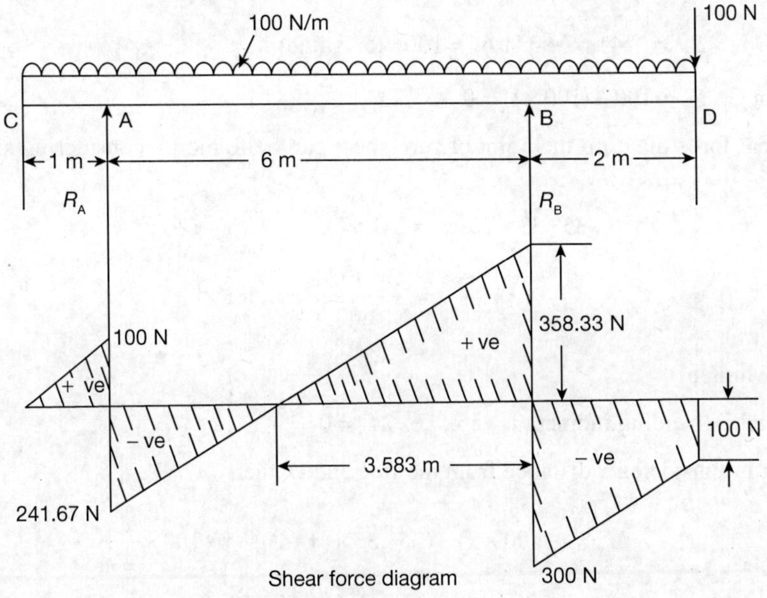

Shear force diagram

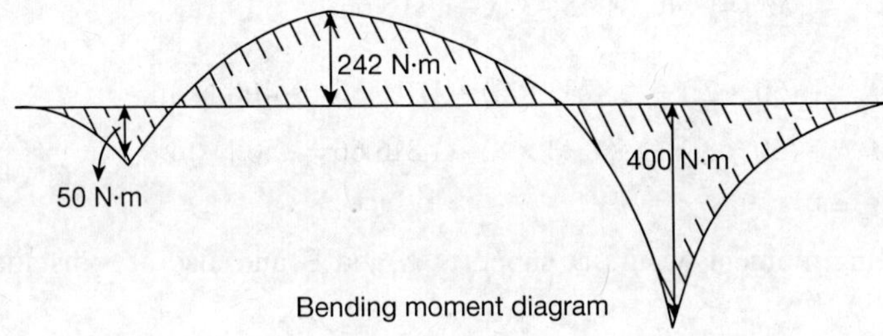

Bending moment diagram

Fig. 4.35

Taking moment of all the forces with respect to A

$$R_B \times 6 = 100 \times 8 + (100 \times 9 \times 3.5) = 3,950 \text{ N}$$

$$R_B = \frac{3,950}{6} = 658.33 \text{ N}$$

$$\text{Total load} = 100 \times 9 + 100 = 1,000 \text{ N}$$

$$\therefore \quad R_A = 1,000 - 658.33 = 341.67 \text{ N}$$

1. Shear Force

Shear force at D, $S_D = 100$ N

Shear force at B, $S_B = 100 - 2(100) = 300$ N (bottom value)

$$S_B = -300 + 658.33 = 358.33 \text{ (top value)}$$

Shear force at A, $S_A = 358.33 - (6 \times 100) = -241.67$ N (bottom value)

$$S_A = -241.67 + 341.67 = 100 \text{ (top value)}$$

Shear force at C, $S_A = 100 - (100 \times 1) = 0$

From the shear force diagram the point of zero shear can be found by considering similar triangles principle.

$$358.33 \times (6 - x) = 241.67\, x$$

$$\therefore \quad x = \frac{2{,}149.98}{600} = 3.583 \text{ m}$$

2. Bending Moment

At the free end D, bending moment is zero, i.e., $M_D = 0$

A section *x-x* is considered *x* distance from the free end D, then

$$M_x = (-100 \times x) + (658.33 \times (x - 2)) + \left(-100 \times \frac{x}{2}\right)$$

$$= -100x + 658.33x - 1{,}316.66 - 50x^2$$

$$M_x = -50x^2 + 558.33x - 1{,}316.66$$

At B, $x = 2$ m, $M_B = -(50 \times 2^2) + (558.33 \times 2) - 1{,}316.66 = -400$ N·m.

At A, $x = 8$ m, $M_A = -(50 \times 8^2) + (558.33 \times 8) - 1{,}316.66 = -50$ N·m.

At C, free end, $M_C = 0$.

Maximum bending moment occurs at supports A and B and also at a distance 3.583 m from supports B towards A.

Distance x from free end D is, $x = 3.583 + 2 = 5.583$ m.

Maximum bending moment at $x = 5.583$ m is

$$M_{max} = -(50 \times 5.583^2) + (558.33 \times 5.583) - 1{,}316.66 = 242 \text{ N}\cdot\text{m}$$

Points of contraflexure may be found by solving the equation $-50x^2 + 558.3x - 1{,}316.66 = 0$

$\therefore$ Solving for x we have $x = 7.79$ m or 3.38 m.

i.e., Point of contraflexure occurs at two points, i.e., 7.79 m and 3.38 m from free end D and 1.38 m and 5.79 m from B.

Shear force and bending moment diagrams are shown in Figure 4.35.

SOLVED PROBLEM 4.9

Draw the shear force, bending moment and thrust diagrams for the loading shown in Figure 4.36. Support A is hinged and support B is on rollers.

Solution:

Given data: Inclined load 350 N, 30° to horizontal inclined load 45°, point vertical load = 250 N and span = 10 m.

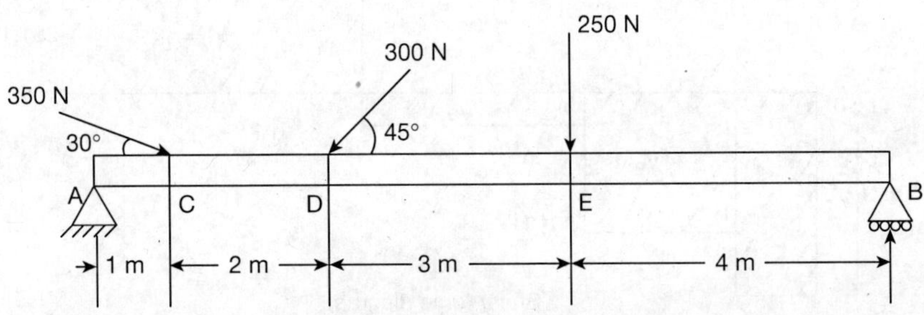

Fig. 4.36

The forces are resolved vertically and horizontally.
Vertical component of force 350 N = 350 sin 30° = 175 N
Horizontal component of force 350 N = 350 cos 30° = 303.1 N
Vertical component of force 300 N = 300 sin 45° = 212.13 N
Horizontal component of force 300 N = 300 cos 45° = 212.13 N.
The component of the forces are shown in Figure 4.37.
Taking moment of all vertical forces about A

$$R_B \times 10 = (250 \times 6) + (212.13 \times 3) + (175 \times 1)$$

$$\therefore \quad R_B = \frac{2,311.39}{10} = 213.13 \text{ N}$$

$$\text{Total load} = (250 + 212.13 + 175) = 637.13 \text{ N}$$

$$\therefore \quad R_A = 637.13 - R_B = 637.13 - 231.13$$

$$= 406 \text{ N}$$

For calculation of shear force and bending moment, only the vertical load components are considered.

1. Shear Force

Shear force at B, $S_B = 231.3$ N

Shear force at E, $S_E = 231.13 - 250 = -18.87$ N (bottom value)

Shear force at D, $S_D = -18.87 - 212.13 = -231$ N (bottom value)

Shear force at C, $S_C = -231 - 175 = -406$ N (bottom value)

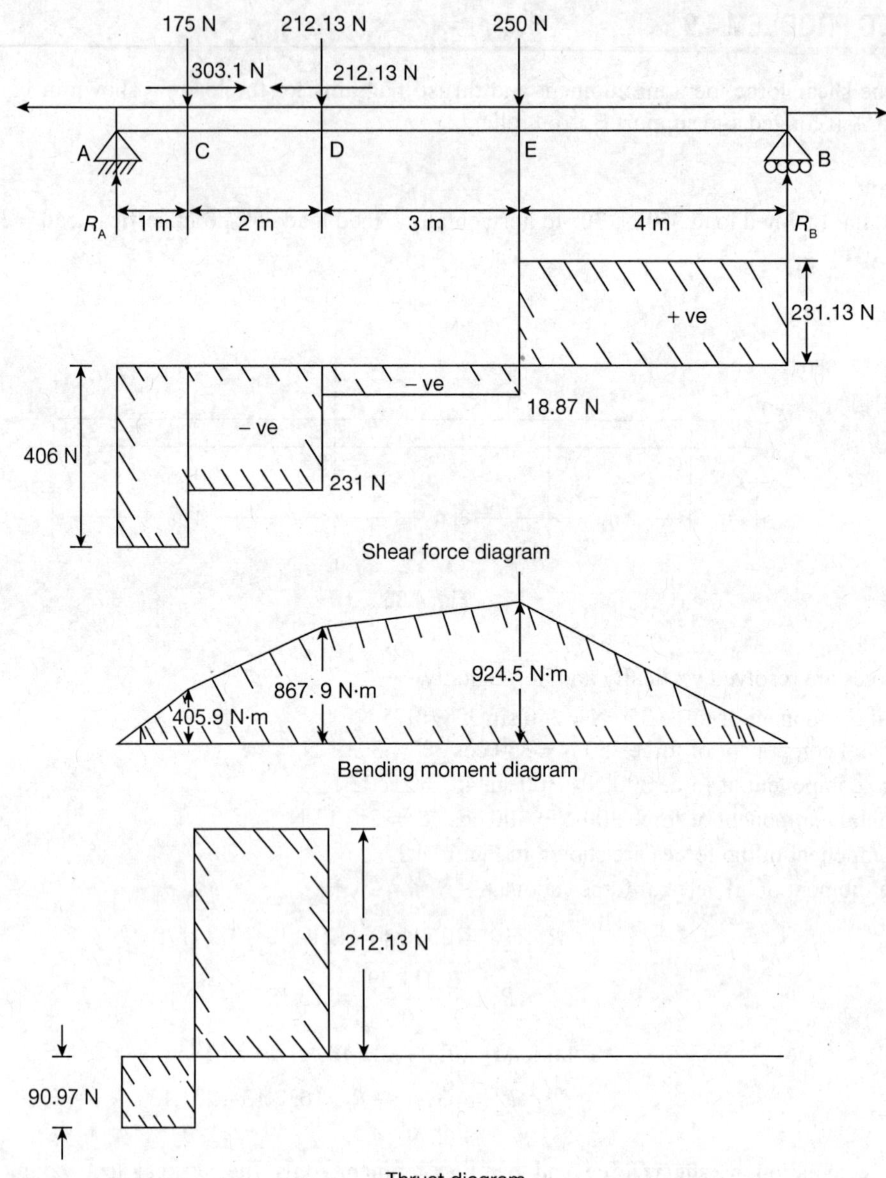

Fig. 4.37

2. Bending Moments

Bending moment at B, $M_B = 0$

Bending moment at E, $M_E = 231.13 \times 4 = 924.53 \text{ N} \cdot \text{m}$

Bending moment at D, $M_D = 231.13 \times 7 - 250 \times 3 = 807.91 \text{ N} \cdot \text{m}$

Bending moment at C, $M_C = 231.13 \times 9 - 250 \times 5 - 212.13 \times 2 = 405.91\ \text{N}\cdot\text{m}$

Bending moment at A, $M_A = 231.13 \times 10 - 250 \times 6 - 212.13 \times 3 - 175 \times 1 = 0$

Bending moment variation is a straight line from point to point with maximum value at E where the shear force changes sign.

3. Thrust

No thrust is acting between EB and DE. Portion CD is subjected to a compression of 213.13 N.

Portion AC is subjected to a tension (303.1 – 212.13) = 90.97 N

The hinged end A will balance the horizontal force and the reaction at A will be included. The resultant force is obtained as, $R_A = \sqrt{406^2 + 90.97^2}$

i.e., $R_A = 416.07$ N

Inclination, $\alpha = \tan^{-1}\left(\dfrac{90.97}{400}\right) = 12°37'46''$

Shear force and bending moment and thrust diagrams are shown in Figure 4.37.

SOLVED UNIVERSITY QUESTIONS

SOLVED PROBLEM 4.10

A cantilever beam of span 4 m carries point loads of 1 kN, 2 kN and 4 kN at 1 m, 2 m and 4 m, respectively from the fixed end. Draw the shear force and bending moment diagrams (Anna Univ., June 2005, ME).

Solution:

Given data: Span, $l = 4$ m, loads = 1 kN, 2 kN and 4 kN, spacing, 1 m, 2 m and 4 m.

1. Shear Force

Considering from the free end,

Shear force at D = +4 kN

Shear force at C = +4 + 2 = +6 kN

Shear force at B = +4 + 2 + 1 = +7 kN

Shear force at A = +4 + 2 + 1 = +7 kN

2. Bending Moment

Bending moment at D = – 4 × 0 = 0

Bending moment at C = (–4 × 2) = –8 kN·m

Bending moment at B = (–4 × 3) + (–2 × –1) = –12–2 = –14 kN·m

Bending moment at A = (–4 × 4) + (–2 × 2) + (–1 × 1) = –16–4–1 = –21 kN·m

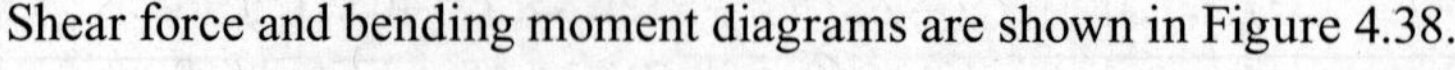

Shear force and bending moment diagrams are shown in Figure 4.38.

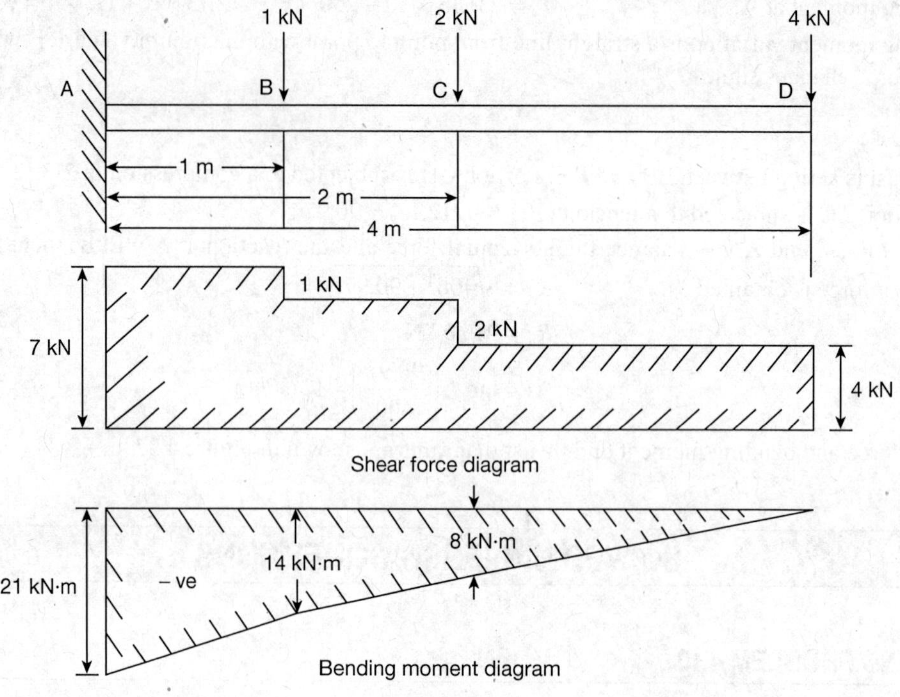

Fig. 4.38

SOLVED PROBLEM 4.11

A simply supported beam of span 6 m is carrying an UDL 2 kN/m over the entire span. Calculate the magnitude of shear force and bending moment at every section of 2 m from the left support. Also, draw SFD and BMD (Anna Univ., June 2005, ME).

Solution:

Given data: Load = UDL 2 kN/m and span = 6 m.
Taking moment of the UDL about support A

$$R_B \times 6 = 2 \times 6 \times 6/2$$

$$\therefore \quad R_B = 6 \text{ kN}$$

$$R_A + R_B = 2 \times 6 = 12 \text{ kN}$$

$$\therefore \quad R_A = 12 - R_B = 12 - 6 = 6 \text{ kN}$$

1. Shear Force

Shear force at B = $-R_B = -6$ kN

Shear force at D = $-R_B + (2+2) = -6 + 4 = -2$ kN

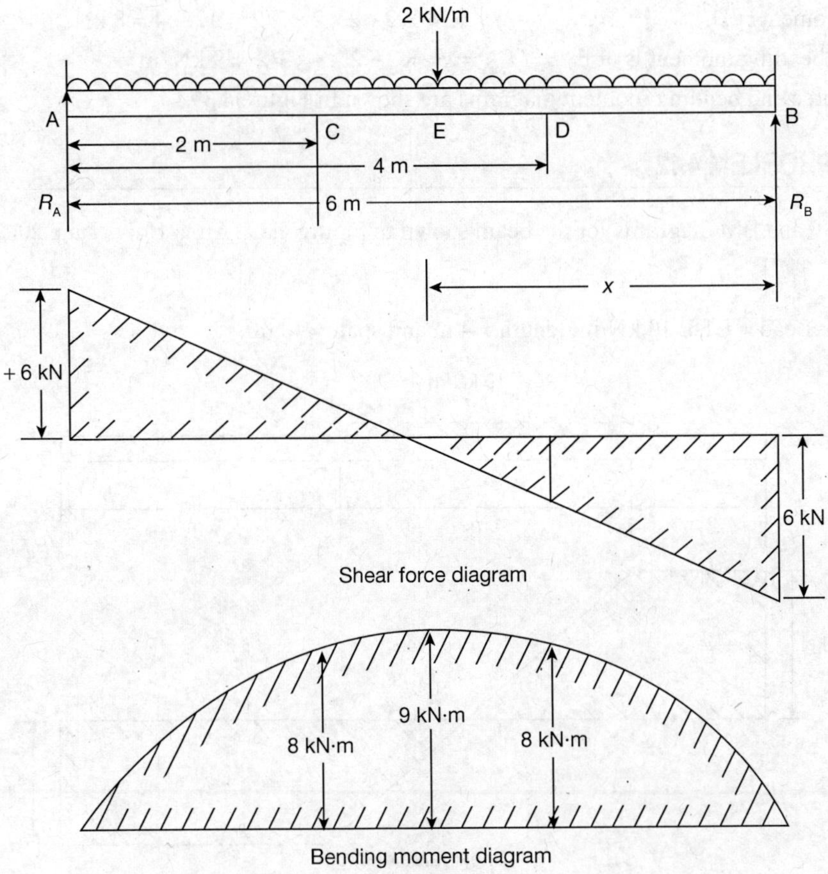

Fig. 4.39

Shear force at C $= -R_B + (2 \times 4) = -6 + (2 \times 4) = +2$ kN

Shear force at A $= +R_A = +6$ kN.

Maximum bending moment occurs when the shear force is zero. Let x be the distance from right end.

Shear force at E $= -R_B + (2 \times x) = 0$

Solving $x = 3$ m, i.e., the centre of the span.

2. Bending Moment

Bending moment at the supports A and B is zero.

i.e., $M_A = M_B = 0$

Bending moment at C, $M_C = R_B \times 4 - 2 \times 4 \times 4/2 = 24 - 16 = 8$ kN·m

Bending moment at D, $M_D = R_B \times 2 - 2 \times 2 \times 2/2 = 12 - 4 = 8$ kN·m

Maximum bending moment is at E, $M_{max} = R_B \times 3 - 2 \times 3 \times 3/2 = 9$ kN·m

Shear force and bending moment diagrams are shown in Figure 4.39.

SOLVED PROBLEM 4.12

Draw the SF and BM diagrams for the beam shown in Figure 4.39 (Anna Univ., June 2007, ME).

Solution:

Given data: Load = UDL 10 kN/m, length = 4 m and span = 10 m.

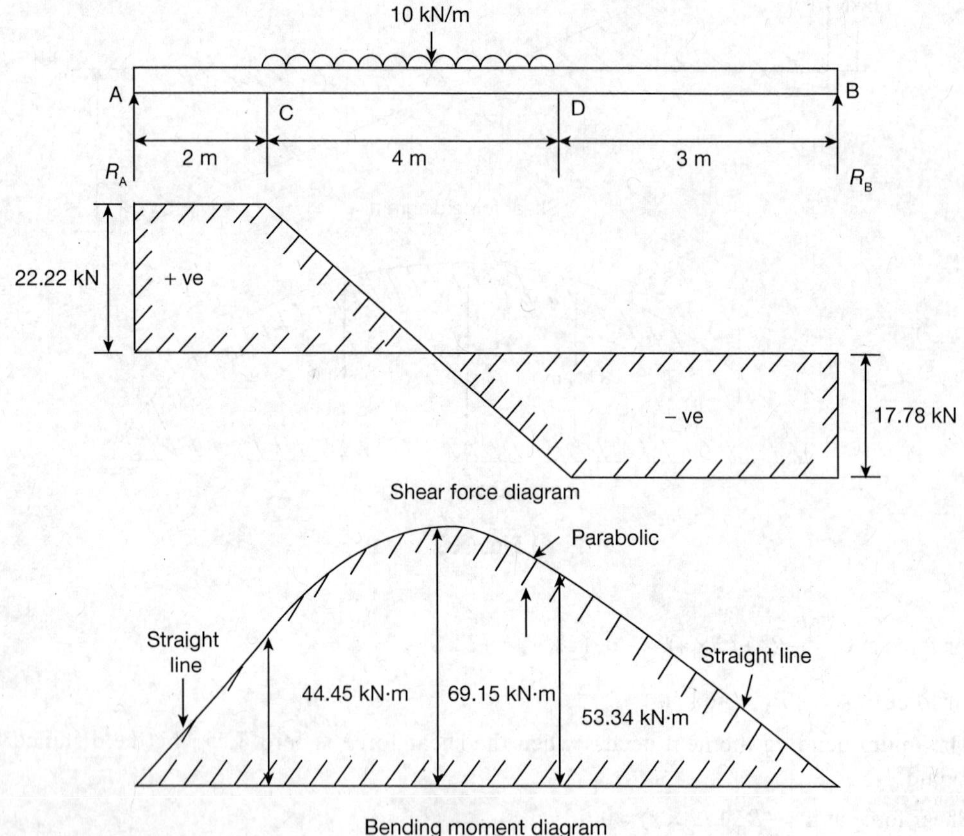

Fig. 4.40

Taking moment of the forces about A

$$R_B \times 9 = 10 \times 4\left(\frac{4}{2} + 2\right)$$

$$\therefore \quad R_B = 17.78 \text{ kN}.$$

$$R_A = \text{(Total load on beam)} - R_B$$
$$= 10 \times 4 - 17.78$$
$$= 22.22 \text{ kN}$$

1. Shear Force

Shear force at A, $R_A = 22.22$ kN

Shear force at B, $R_B = -17.78$ kN

Shear force at any section between C′ and D at a distance x from A is given by $S_x = 22.22 - 10(x - 2)$

At C, $x = 2$ $\quad S_C = 22.22 - 10(2 - 2) = 22.22$ kN

At D, $x = 6$ $\quad S_D = 22.22 - 10(6 - 2) = -17.78$ kN

Shear force between C and D varies as a straight line. That is somewhere between C and D, the shear force is zero. Let the shear force be zero at x metre form A, i.e., $S_x = 0$.

$$\therefore \quad 0 = 22.22 - 10(x - 2)$$

Solving $x = 4.78$ m from B.

2. Bending Moment

Bending moment at A and B are zero, i.e., $M_A = M_B = 0$

Bending moment at D is $\quad R_B \times 3 = 17.78 \times 3 = 53.34 \text{ kN}\cdot\text{m}$

Bending moment at C is $\quad R_B \times 7 - 10 \times 4 \times 4/2 = 44.45 \text{ kN}\cdot\text{m}$

Maximum bending moment $\quad M_{max} = R_B \times x - 10 \times (x-3) \times \left(\dfrac{x-3}{2}\right)$

$$= 17.78 \times 4.78 - \frac{10}{2}(4.78 - 3)^2$$

$$= 69.15 \text{ kN}\cdot\text{m}$$

Shear force and bending moment diagrams are shown in Figure 4.40.

SOLVED PROBLEM 4.13

For the simply supported beam loaded as shown in Figure 4.41, draw the shear force and bending moment diagrams. Also, obtain the maximum bending moment (Anna Univ., Apr. 2007, ME).

Solution:

Given data: Load = UDL 15 kN/m for 3 m, point load = 20 kN and span = 7 m.

Taking moments of all the forces about A

$$R_B \times 7 - 20 \times 5.5 - (15 \times 3)(1.5 + 1) = 0$$

Solving $\quad R_B = 31.8$ kN

$$R_A + R_B = 20 + 15 \times 3 = 65 \text{ kN}$$

$$\therefore \quad R_A = 65 - 31.8 = 33.2 \text{ kN.}$$

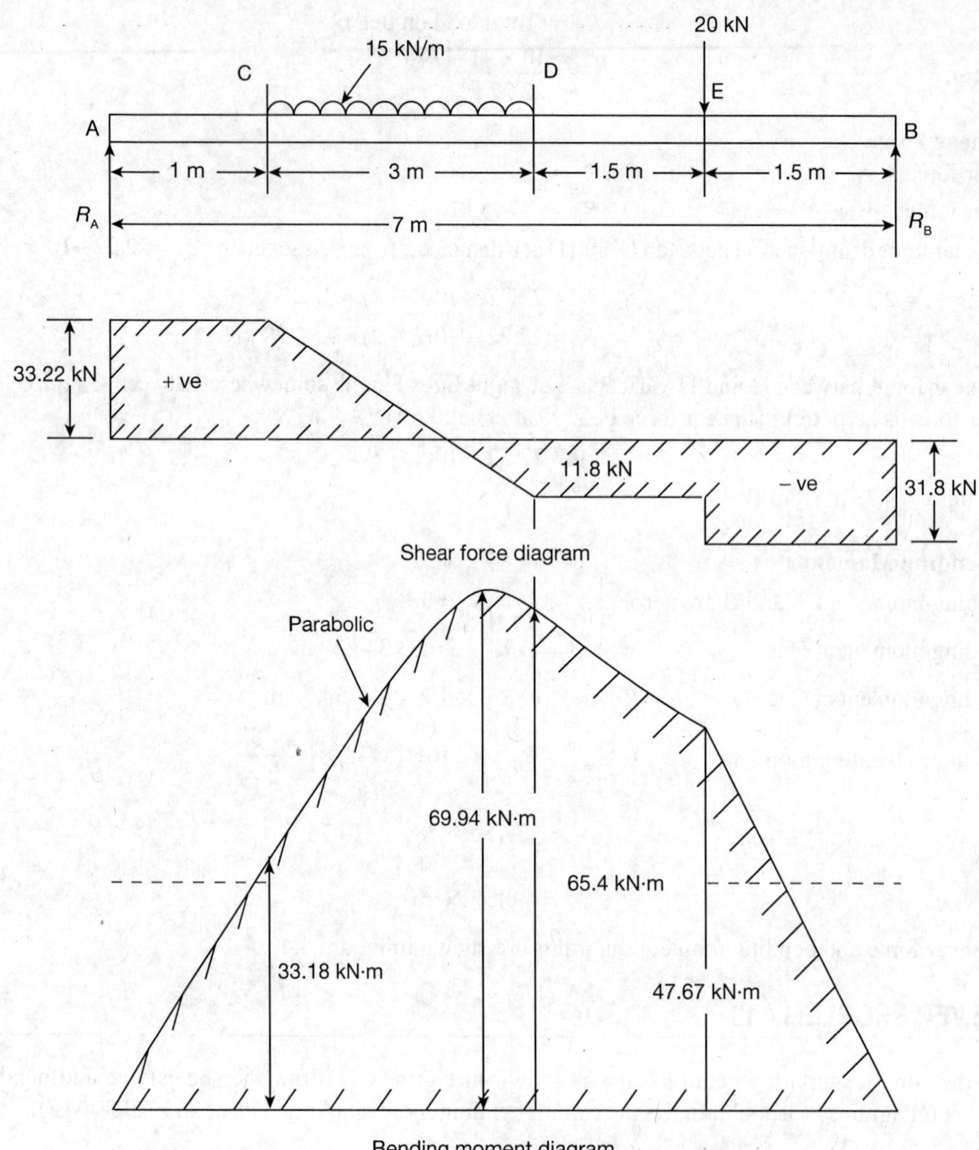

Fig. 4.41

1. Shear Force

Shear force at A = + R_A = 33.2 kN

Shear force between A and C = 33.2 kN

Shear force at D = R_A – 15 × 3 = 33.2 – 45 = 11.8 kN

Shear force at E = R_A – 15 × 3 – 20 = –31.8 kN.

2. Bending moment

Bending moments at A and B, $M_A = M_B = 0$

Bending moment at C = $31.8 \times 6 - 20 \times 4.5 - 15 \times 3 \times 3/2 = 33.18$ kN·m

Bending moment at D = $31.8 \times 3 - 20 \times 1.5 = 65.4$ kN·m

Bending moment at E = $31.78 \times 1.5 = 47.67$ kN·m

Maximum bending moment occurs when shear force changes sign. Let x be the point from A.

Then, the shear force equation is
$$S_x = R_A - 15(x-1)$$
$$0 = 33.2 - 15x + 15$$
$$x = 3.2 \text{ m}$$

Maximum bending moment is
$$M_{max} = R_A x - 15(x-1)\left(\frac{(x-1)}{2}\right)$$
$$= R_A \times 3.2 - 15(3.2-1)\left(\frac{3.2-1}{2}\right)$$
$$= 69.94 \text{ kN}\cdot\text{m}$$

Shear force and bending moment diagrams are shown in Figure 4.41.

SOLVED PROBLEM 4.14

A beam of length 10 m is simply supported at its ends, carries two concentrated loads of 5 kN each at a distance of 3 m and 7 m from the left support and also a uniformly distributed load of 1 kN/m between the point loads. Draw the shear force and bending moment diagrams. Calculate the maximum bending moment (Anna Univ., June 2006, ME).

Solution:

Given data: Span = 10 m, point load = 5 kN and UDC = 1 kN/m over 4 m length.

Taking of moment of all the forces about A
$$R_B \times 10 - 5\times 7 - (4\times 1)\times\left(\frac{4}{2}+3\right) - 5\times 3 = 0$$

Solving,
$$R_B = 7 \text{ kN}$$
$$R_A + R_B = 5 + 1\times 4 + 5 = 14 \text{ kN}$$

∴
$$R_A = 7 \text{ kN}$$

1. Shear Force

Shear force at A, $R_A = 7$ kN

Shear force at C, $R_A - 5 = 7 - 5 = 2$ kN

Shear force at D, $R_A - 5 - 1\times 4 = -2$ kN

Shear force at B, $-R_B = -7$ kN

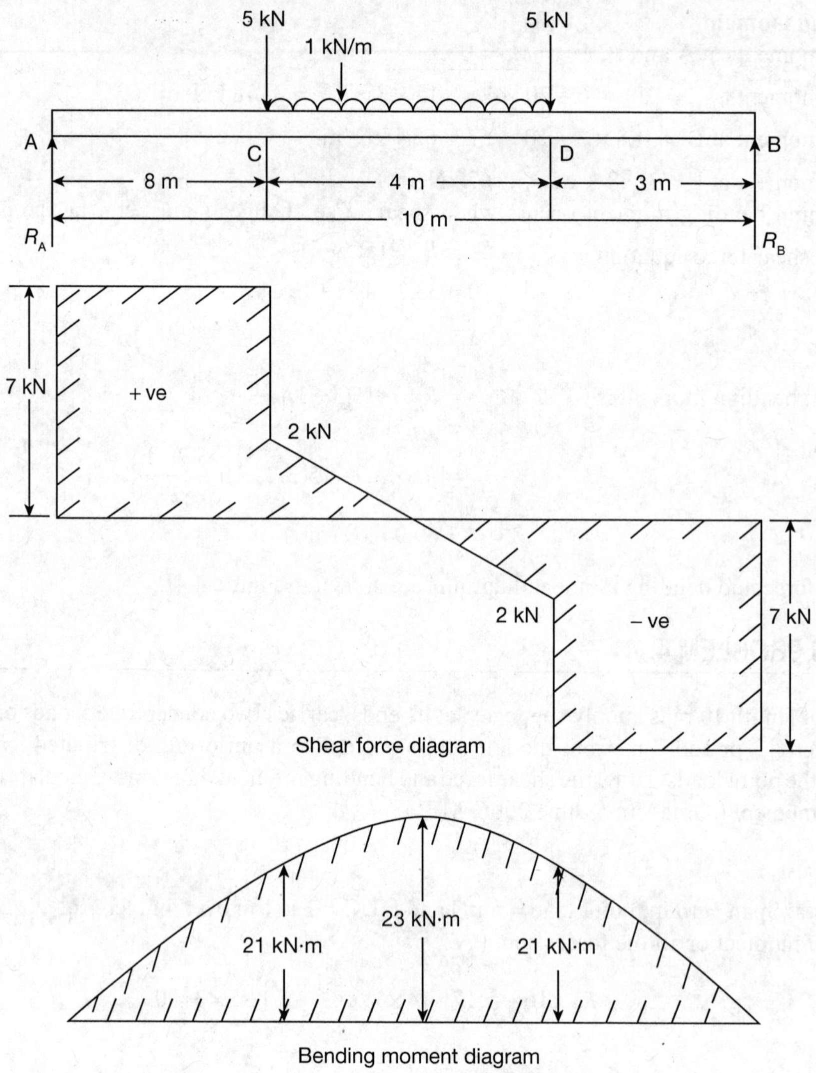

Fig. 4.42

2. Bending Moment

Bending moments at A and B are zero, i.e., $M_A = M_B = 0$

Bending moment at C, $M_C = R_A \times 3 = 7 \times 3 = 21 \text{ kN} \cdot \text{m}$

Bending moment at D, $M_D = R_B \times 3 = 7 \times 3 = 21 \text{ kN} \cdot \text{m}$

Maximum bending moment occurs at a distance x from point A where the shear force changes sign.

i.e., $S_x = 7-5-1(x-3)x = 0$

i.e., $x = 5\text{ m}$

i.e., $M_{max} = R_A \times 5 - 5\times 2 - 1\times 2\times \frac{2}{2}$

$= 7\times 5 - 10 - 2$

$= 23\text{ kN}\cdot\text{m}$

Shear force and bending moment diagrams are shown in Figure 4.42.

SOLVED PROBLEM 4.15

Draw the shear force and bending moment diagrams for the over hanged bear shown in Figure 4.9. Find the maximum sagging bending moment and also the point of contraflexure (Anna Univ., June 2006, ME).

Solution:

Given data: Span = 4 m, overhang = 2 m, point load = 2 kN and UDL = 2 kN/m over 6 m length. Taking moment of all the forces from the right end of the beam.

$$(2\times 6) + \left(2\times 6\times \frac{6}{2}\right) - R_B \times 4 = 0$$

Solving $R_B = 12\text{ kN}$

$R_A + R_B = \text{Total load} = (2\times 6) + 2 = 14\text{ kN}$

∴ $R_A = 14 - 12 = 2\text{ kN}$

1. Shear Force

Shear force at C = + 2 kN

Shear force at B = + 2 + (2 × 2) – R_B = 2 + 4 – 12 = –6 kN

Shear force at A = + R_A = +2 kN.

2. Bending Moment

Bending moment at A, $M_A = 0$

Bending moment at B $= (-2\times 2) - (2\times 2) + R_B \times 0 = -8\text{ kN}\cdot\text{m}$

Maximum bending moment occurs where the shear force changes sign.

At D, shear force = +2 + (2 × x) – R_B = 0

i.e., $2 + (2x) - 12 = 0$

∴ $x = 5$

Maximum bending moment at D $= (-2\times x) - \left(2\times x\times \frac{x}{2}\right) + R_B \times (x-2)$

$= (-2\times 5) - \left(2\times 5\times \frac{5}{2}\right) + 12\times (5-2)$

$= -10 - 25 + 36 = 1\text{ kN}\cdot\text{m}$

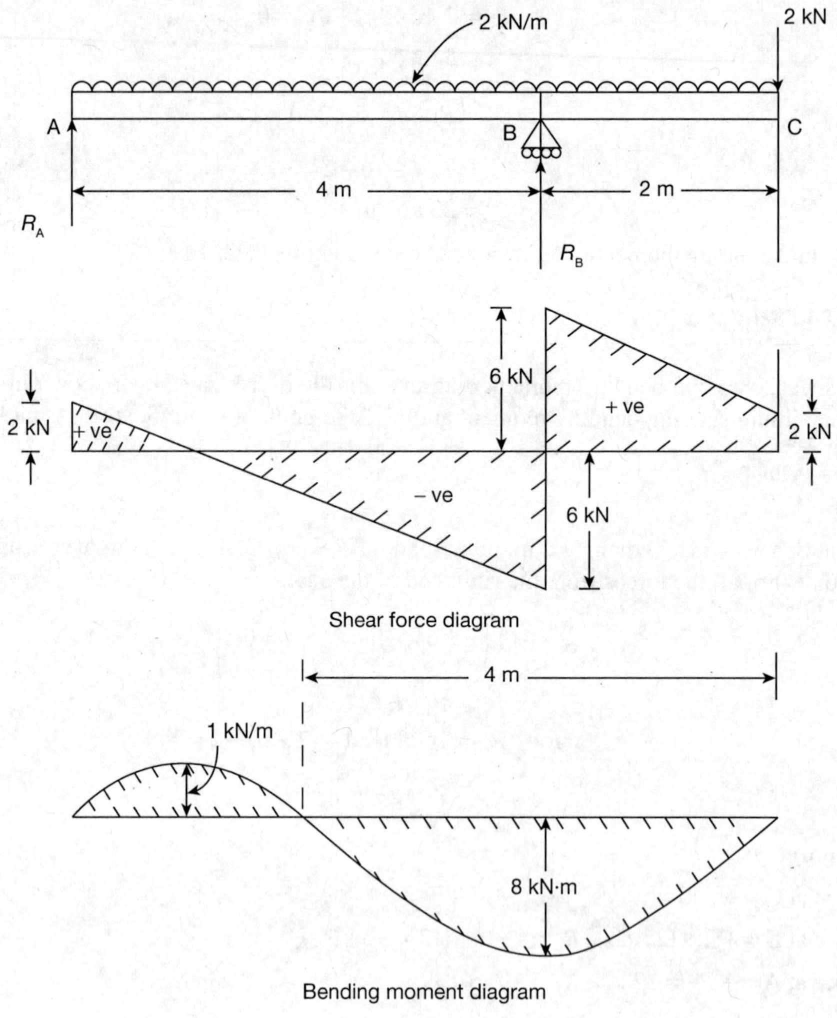

Fig. 4.43

Point of contraflexure occurs where the bending moment is zero. Let x' be the distance from right end.

Then, $(-2\times x')-\left(2\times x'\times\frac{x'}{2}\right)+R_B(x'-2)=0$

Solving, $x'=4$ m.

∴ The point of contraflexure occurs at 4 m from the right end.

The maximum sagging bending moment is at B and equals to $-8\text{ kN}\cdot\text{m}$.

Shear force and bending moment diagrams are shown in Figure 4.43.

SOLVED PROBLEM 4.16

Draw the shear force and bending moment diagrams of the beam loaded as shown in Figure 4.43. Also find the point of contraflexure if any (Anna Univ., Nov. 2008, ME).

Solution:

Given data: Point load = 25 kN, span = 4 m, overhang = 1 m, UDL = 10 kN/m for 4 m and UDL = 5 kN/m.

Taking moment of all the forces about A

$$R_B \times 6 - 10 \times 4 \times \frac{4}{2} - 5 \times 2 \times \left(\frac{2}{2} + 4\right) - 2.5 \times (4+2) = 0$$

Solving $R_B = 70$ kN

$$\text{Total load} = 10 \times 4 + 5 \times 2 + 25 = 75 \text{ kN}$$

$$R_A = 75 - R_B = 75 - 70$$

$$= 5 \text{ kN}$$

1. Shear Force

Shear force at A = 5 kN

Shear force at B (left) = 5 – 10 × 4 = –35 kN

Shear force at B (right) = 5 – 10 × 4 + 70 = 35 kN

Shear force at C = 25 kN

2. Bending Moment

Bending moment at A is zero

Bending moment at B $= R_A \times 4 - 10 \times 4 \times \frac{4}{2}$

$$= -60 \text{ kN} \cdot \text{m}$$

Bending moment at C = 0

Bending moment is maximum where shear force changes sign, i.e., $0 = 5 - 10 \times x$

i.e., $x = 0.5$ m from A

$\therefore$ Maximum bending moment $= R_A \times 0.5 - 10 \times 0.5 \times \frac{0.5}{2}$

i.e., $M_{max} = 1.25 \text{ kN} \cdot \text{m}$

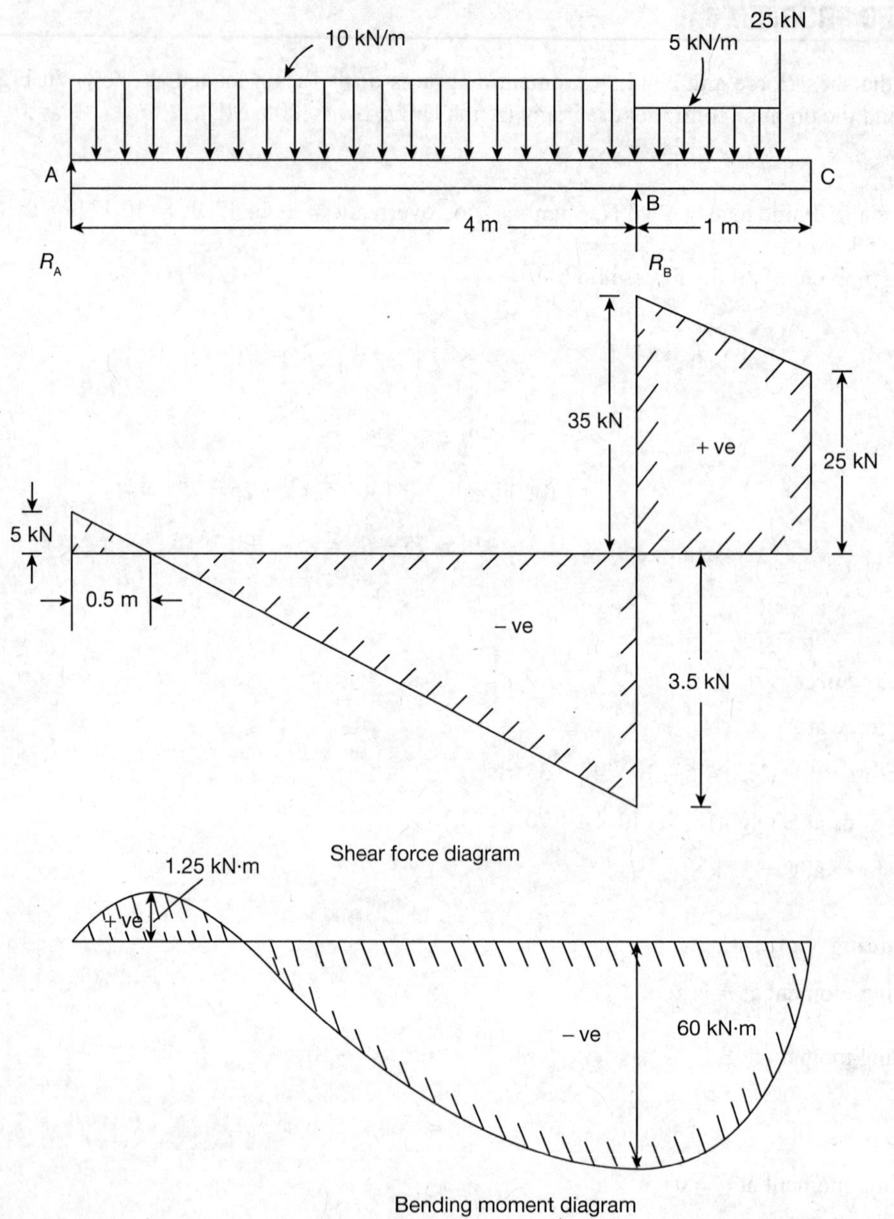

Fig. 4.44

SOLVED PROBLEM 4.17

Draw the shear force and bending moment diagrams for the beam shown in Figure 4.45. Determine the point of contraflexure (Anna Univ., Nov. 2006, ME).

Solution:

Given data: Point loads = 1,600 N, 4,000 N and 2,000 N and span = 8 m.

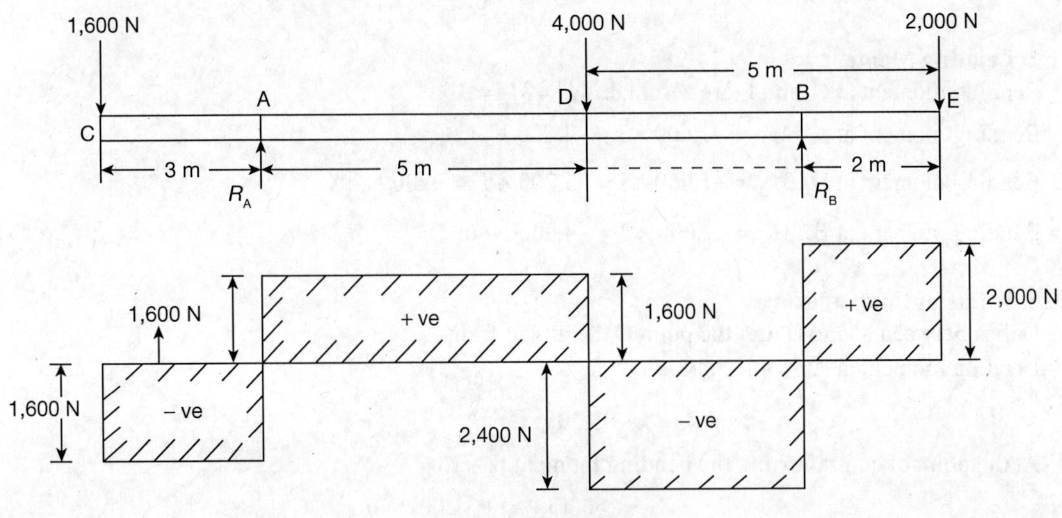

Shear force diagram

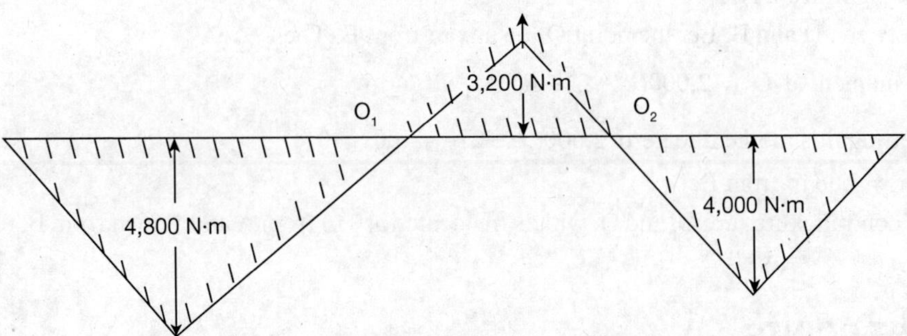

Bending moment diagram

Fig. 4.45

Taking moment of all the forces about A

$$R_B \times 8 + 1{,}600 \times 3 = 4{,}000 \times 5 + 2{,}000 \times 10$$

Solving $\quad R_B = 4{,}400$ kN

$\therefore \quad R_A = (\text{Total load}) - R_B = 7{,}200 - 4{,}000$

$$= 3{,}200 \text{ kN}$$

1. Shear Force

Shear force at C = –1,600 N

Shear force at $\text{A} = -1{,}600 + R_A = -1{,}600 + 3{,}200 = 1{,}600$ N

Shear force at $D = 1{,}600 - 4{,}000 = -2{,}400$ N

Shear force at $B = -2{,}400 + R_B = -2{,}400 + 4{,}200 = 2{,}000$ N

2. Bending Moment

Bending moment at C, and E are zero, i.e., $M_C = M_E = 0$

Bending moment at A, $M_A = -1{,}600 \times 3 = -4{,}800$ N·m

Bending moment at D, $M_D = -1{,}600 \times 8 = -3{,}200 \times 5 = 3{,}200$ N·m

Bending moment at B, $M_B = -2{,}000 \times 2 = -4{,}000$ N·m

3. Points of Contraflexure

O_1 lies between A and D. Let the point O_1 be at x m from A.

Bending moment at $O_1 = -1{,}600(3 + x) + R_A \times x$

$$= -1{,}600(3 + x) + 3{,}200x$$

At the point of contraflexure, the bending moment is zero

$$\therefore -1{,}600(3 + x) + 3{,}200x = 0$$

Solving $x = 3$ m from A.

O_2 lies between D and B. Let the point O_2 be at x m from B. Then

Bending moment at $O_2 = 2{,}000(x_1 + 2) - R_B \times x_1$

Again, letting this moment to zero $2{,}000(x_1 + 2) - 4{,}400x = 0$

Solving $x_1 = 1.66$ m from B.

Points of contraflexures are O_1 and O_2 which are located at 3 m from A and 1.66 m from B, respectively.

SALIENT POINTS

- A Beam is a horizontal structural member which is subjected to transverse loading.
- The load-bearing capacity of a beam varies with the type of support. Types of supports are
  - (i) simply supported or rolled support
  - (ii) hinged support or pinned support and
  - (iii) fixed support or rigid support or encastred support.
- Simply supported is the simplest support whose reaction is normal to the plane of the roller or the support.
- Pinned support can withstand force in any direction. The reaction is normal to the plane of the support and in addition it has a horizontal reaction along the axis of the beam.
- Fixed support is the one which prevents the beam not only from rotation but also linear movement or translation.
- Beams are classified according to their support conditions and not on any other parameter. Different types of beam are simply supported beam, cantilever beam, overhanging beam, dropped cantilever beam, continuous beam and fixed beam.

- Types of loading are: concentrated load or point load, uniformly distributed load, non-uniform load and couple or moment.
- Bending moment at a section is defined as the algebraic sum of the moments about the section of all the forces (including the reaction) acting on the beam either to the left or to the right of the section.
- Shear force at a section in a beam is the algebraic sum of all the forces including the reactions acting normal to the axis of the beam either to the left or to the right of the section.
- Bending moment is said to be positive (sagging) moment, at a section, when it is acting in an anticlockwise direction to the right and negative (hogging) moment when acting in the clockwise direction.
- Shear force having an upward direction to the right-hand side of a section or downwards to the left of the section will be taken as positive. Similarly, a negative shearing force will be the one that has a downward direction to the right of the section or upward direction to the left of the section.

QUESTIONS

1. A horizontal cantilever 6 m long beam carries a point load of 2 kN at the free end and an UDL of 1 kN/m over a length of 4 m, from the free end. Draw the shear force and bending moment diagrams.
2. A cantilever carries loads of 2 kN and 4 kN at 2 m and 6 m, respectively from the fixed end and an UDL of 10 kN/m over its entire length. Draw the shear force and bending moment diagrams.
3. A beam ABC is 12 m long is simply supported at B and C. AB = 3 m and BC = 9 m. It carries a point load of 25 kN at the free end A and UDL of 21 kN/m throughout the entire span. Draw shear force and bending moment diagrams marking all the salient values.
4. An overhanging beam ABC is simply supported at A and B over a span of 8 m and BC overhangs by 4 m. If the support span AB carries central concentrated load of 10 kN and overhanging span BC carries 2 kN/m completely. Draw SF and BM diagrams indicating the salient points.
5. Analyze the simply supported beam shown in Figure 4.46. Also, sketch the S.F. and B.M. diagrams.

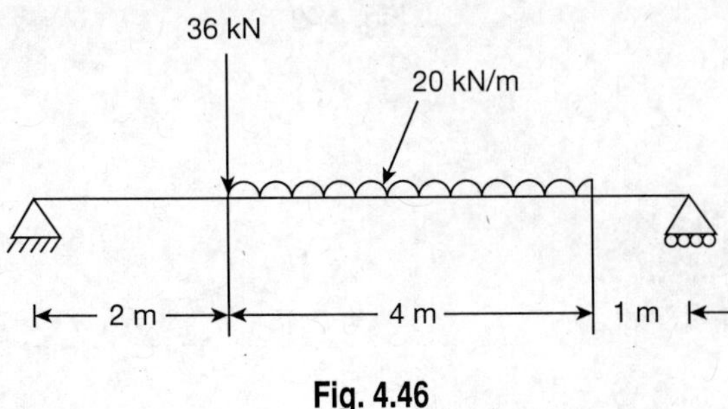

Fig. 4.46

6. Draw the shear force and bending moment diagrams for the loaded beam shown in Figure 4.47.

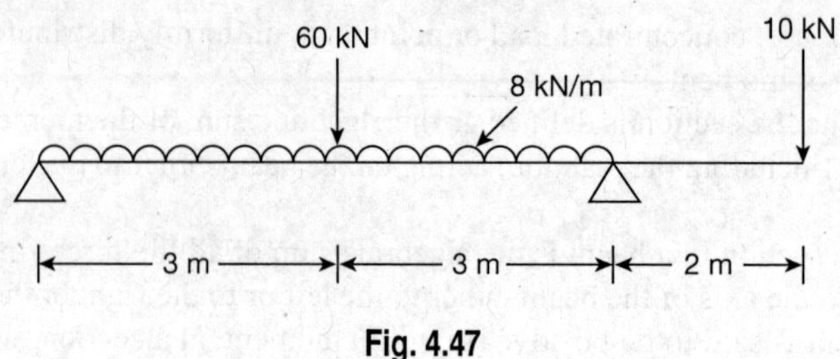

Fig. 4.47

7. Draw the shear force and bending moment diagrams for the beam shown in Figure 4.48.

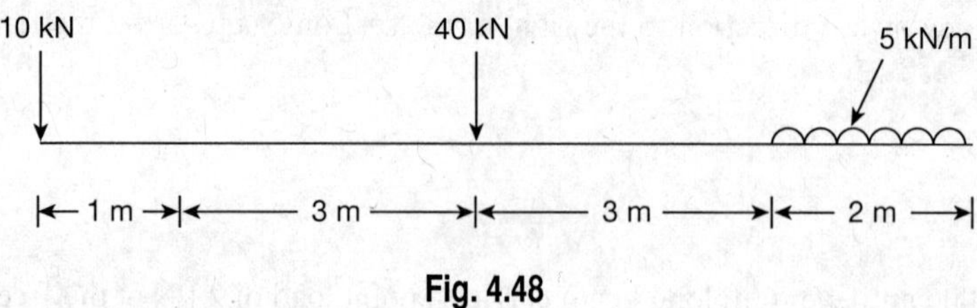

Fig. 4.48

8. Draw the shear force and bending moment diagrams for the triangularly loaded beam shown in Figure 4.49.

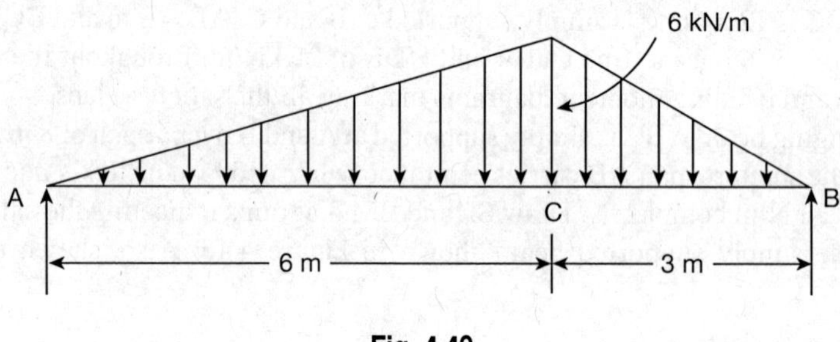

Fig. 4.49

5

Bending and Shear Stresses

LEARNING OBJECTIVES

5.1 THEORY OF SIMPLE BENDING

When a beam is subjected to a loading system or by a force couple acting on a plane passing through the axis, then the beam deforms. In simple terms, this axial deformation is called as *bending of a beam* (Figure 5.1).

When a beam is subjected to action of any load, then the fibres on one side of the beam are stretched and those on the other side are compressed. If the beam is simply supported, then the lower fibres will be under tension and the upper fibres are on compression. On the other hand, in a cantilever beam the upper fibres will be on tension and the lower fibres will be under compression.

When a loaded beam is unaccompanied by shear force and subjected only to a constant bending moment, then the beam is said to be in a *state of simple bending*.

When a beam is loaded longitudinally, stresses are induced in its cross section. For an efficient utility of any beam, it is necessary to establish a relationship between the radius of curvature to which the beam bends, the bending stresses, the bending moment and its cross-sectional dimension. The equation which connects these parameters is known as *bending equation* or the *flexural formula*.

Following are the assumptions made in the derivation of bending equation.

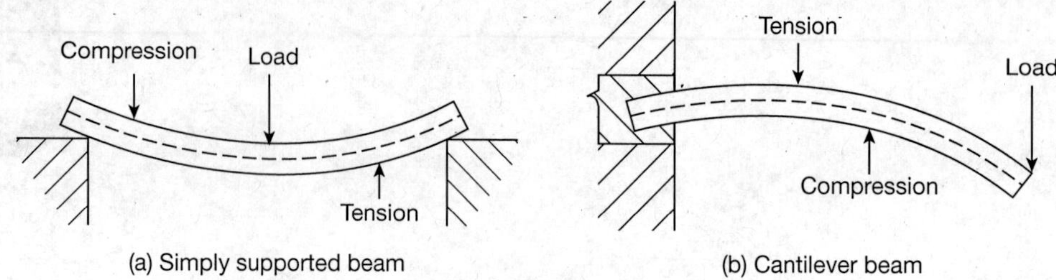

(a) Simply supported beam

(b) Cantilever beam

Fig. 5.1 Bending of beams.

(i) Transverse section which are plane before bending remain plane after bending. Any distortion due to shear is neglected.

(ii) Material of the beam is homogeneous and isotropic, i.e., it possesses the same elastic properties in all directions.

(iii) Stress induced is proportional to the strain and it no place the stress exceeds the elastic limit, i.e., obeys Hooke's law.

(iv) Each layer is free to contract or expand independently of the layer above or below it.

(v) The value of modulus of elasticity, E, is same for the fibre beam under compression or under tension.

(vi) The section of the beam is symmetrical about the plane of bending and the loads on the beam act in this plane.

In Figure 5.2, section AB and CD are separated by a distance dx. Due to the load W, section AB and CD have rotated relatively towards each other by an account $d\theta$. However, still they remain straight and undistorted as per the first assumption (Figure 5.2(a)).

Top fibre AC is shortened and the bottom fibre BD is lengthened. Somewhere in between these two fibres, one can locate a fibre GH whose length has not changed. Let a line $C'D'$ parallel to AB passing through H (Figure 5.2(b)) be drawn.

It could be observed that the fibre AC has shortened by a length CC' and is in compression and the fibre BD has been elongated by a length $D'D$ and is in tension.

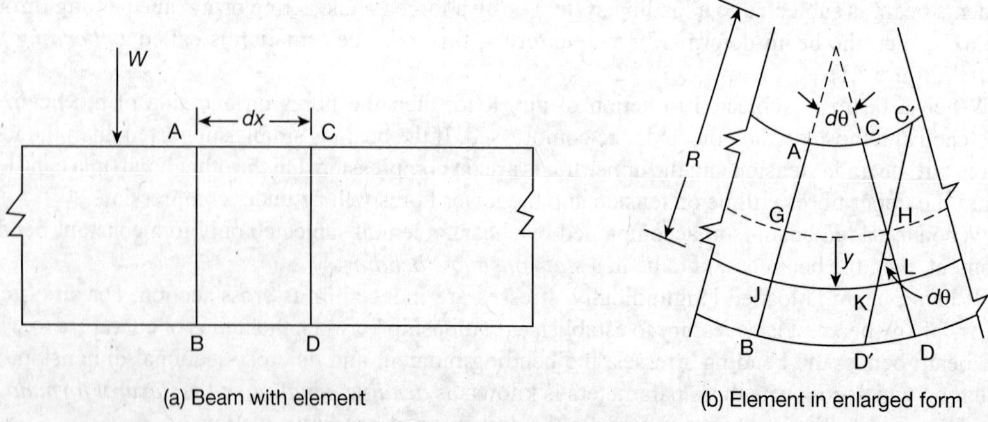

(a) Beam with element

(b) Element in enlarged form

Fig. 5.2 Section of a beam.

Since the fibre GH has not undergone any change in length, the plane containing the fibre GH is known as the *neutral plane* or the *neutral axis* (NA). The fibre in this plane is not subjected to any type of stress.

A fibre JK located at a distance y from the neutral axis is considered. The elongation of JK is KL.

Now, $\angle KHL = d\theta$

The elongated length, $KL = \delta = y d\theta$ (as $d\theta$ is very small)

Strain, $$e = \frac{\text{Change in length}}{\text{Original length}}$$

i.e., $$e = \frac{KL}{JK} = \frac{\delta}{L} = \frac{y d\theta}{GH}$$

where L is the original length.

When R is the radius of a curvature of the fibre JK, the curved length $GH = R d\theta$.

$\therefore$ $$e = \frac{y d\theta}{R d\theta}$$

i.e., $$e = \frac{y}{R}$$

As per the third assumption, the material obeys Hooke's law, then stress

$$p = Ee = \frac{Ey}{R}$$

$\therefore$ $$p = \frac{E}{R} y \qquad (5.1)$$

Since it is assumed that E is equal in compression and tension and the radius of curvature R of the neutral plane is independent of the location y of the fibre, Equation (5.1) indicates that the stress in any fibre varies directly with its location y from the neutral axis.

5.1.1 Neutral Axis

In the end view of the beam, consider an elemental arc (Figure 5.3) δa at a distance from the neutral axis. If p is the stress on the elemental area, then force on the elemental area is $\delta F = p_y \delta a$.

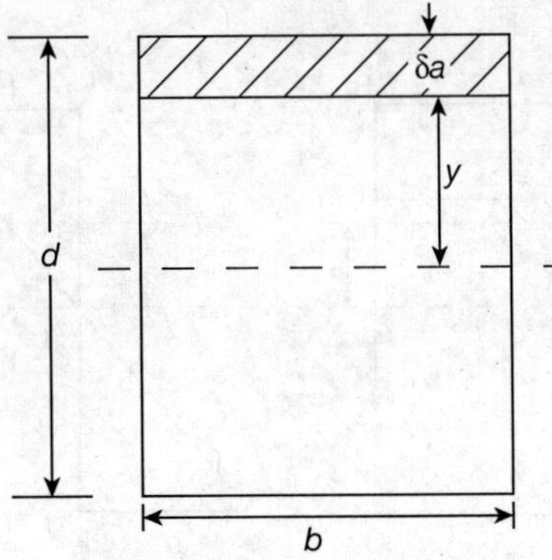

Fig. 5.3 Neutral axis.

Since the total force on the section is zero

$$F = \Sigma p_y \delta a = 0$$

But,
$$p_y = \frac{E}{R} y$$

$\therefore$
$$\Sigma p_y \delta a = \Sigma \frac{E}{R} y \cdot \delta a = \frac{E}{R} \Sigma y \cdot \delta a = 0$$

Since $\frac{E}{R}$ is a constant for a particular section, then $\frac{E}{R} \Sigma y \delta a = 0$. But $\Sigma y \delta a$ is the moment of the area about neutral axis and is equal to zero. Thus, neutral axis passes through the centroid of the section.

5.1.2 Relation Between Moment of Resistance and the Stress

Let us consider an elementary strip of the area a at a distance y from the neutral axis on which the intensity of stress is p (Figure 5.4).

Force on the elemental area = pa

Its moment about NA is,
$$pay = \frac{E}{R} a \cdot y^2$$

Total moment for the entire area of section $= \frac{E}{R} \Sigma a y^2$

$\Sigma a y^2$ or sum of all the elementary areas each multiplied by the square of its distance from NA (second moment of area) is termed as *moment of inertia* of the cross section about the axis and is denoted by I.

The algebraic sum of all the moments of elementary pulls and pushes about the neutral axis will be equal to the resultant couple or the *moment of resistance*, M.

For equilibrium condition, we have

$$M = \frac{E}{R} \Sigma a y^2 = \frac{E}{R} I \tag{5.2}$$

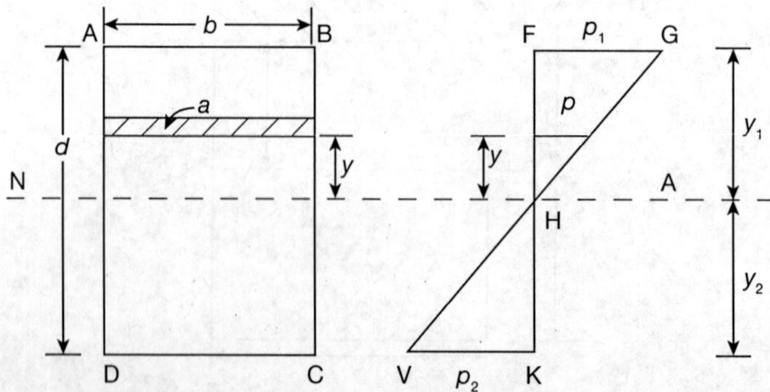

Fig. 5.4 Stress distribution in rectangular beam.

But
$$\frac{E}{R} = \frac{p}{y}$$

Then
$$M = \frac{p}{y} I \tag{5.3}$$

The moment of resistance produces a uniformly varying longitudinal stress, the intensity of which is p acting at a distance y from the neutral axis, i.e., $p = \frac{My}{I}$.

Let y_1, and y_2, be the distances from *NA* to the top and bottom edge of the cross section and p_1 and p_2 be the maximum stresses corresponding to y_1 and y_2, respectively then

$$M = \frac{p_1}{y_1} I = \frac{p_2}{y_2} I \tag{5.4}$$

Equation (5.4) can be used to find the moment of resistance to the boundary or the strength of the beam. Let R be the radius of curvature of the bent beam at the place under consideration and the curvature can hence be expressed as

$$\frac{1}{R} = \frac{p}{Ey} = \frac{M}{EI}$$

i.e.,
$$\frac{M}{I} = \frac{p}{y} = \frac{E}{R} \tag{5.5}$$

Equation (5.5) in known as *Bending equation.*
The stress p can also be represented by f. Then, the flexure equation will be

$$\frac{M}{I} = \frac{f}{y} = \frac{E}{R} \tag{5.6}$$

5.2 SECTION MODULUS

From Equation (5.6), let two terms be considered

$$\frac{M}{I} = \frac{f}{y}$$

It is known that the stress in a fibre is proportional to its distance from the neutral axis. If y_{max} is the distance of the extreme fibre from neutral axis then

$$M = f_{max} \times \frac{I}{y_{max}} = f_{max} Z \tag{5.7}$$

Here, Z represents *Section modulus* or *Modulus of section*, i.e., $Z = I / y_{max}$
Thus, moment of resistance of a section is

$$M = fz = pz$$

where f or p denotes the maximum stress, tensile or compressive in nature.

5.2.1 Section Modulus of a Rectangular Section

For symmetrical section, the neutral axis and centre of gravity of the cross section coincides.

Moment of inertia of a rectangular section of width b and depth d about an axis parallel to its breadth through its centre of gravity is $bd^3/12$.

The distance of the extreme fibre from the centre of gravity is $y_{max} = d/2$.

$$\therefore \quad Z = \frac{I}{y_{max}} = \frac{bd^3}{12} \div \frac{d}{2} = \frac{bd^2}{6}$$

i.e.,
$$Z = \frac{bd^2}{6} \tag{5.8}$$

Hence, the section modulus of a rectangular section is $bd^2/6$.

5.2.2 Section Modulus of a Circular Section

Moment of inertia of a circular section of diameter d about an axis through its centre (centre of gravity) is $\pi d^4/64$.

The distance of extreme fibre from centre of gravity, y_{max} is $\frac{d}{2}$.

Then
$$Z = \frac{I}{y_{max}} = \frac{\pi d^4}{64} \div \frac{d}{2} = \frac{\pi d^3}{32}$$

$$\therefore \quad Z = \frac{\pi d^3}{32} \tag{5.9}$$

Hence, section modulus of a circular section is $\frac{\pi d^3}{32}$.

5.2.3 Section Modulus for a Hollow Circular Section

Moment of inertia of a hollow circular section of external diameter, d_0, and internal diameter d_i, about an axis through its centre (centre of gravity) is $\frac{\pi}{4}\left(d_0^4 - d_i^4\right)$.

Distance of the extreme fibre from the centre of gravity is $\frac{d_0}{2}$.

$$\therefore \quad Z = \frac{I}{y_{max}} = \frac{\pi}{64}\left(d_0^4 - d_i^4\right) \div \frac{d_0}{2}$$

i.e.,
$$Z = \frac{\pi}{32}\left(\frac{d_0^4 - d_i^4}{d_0}\right) \tag{5.10}$$

Section modulus for a hollow circular section is $\frac{\pi}{32}\left(\frac{d_0^4 - d_i^4}{d_0}\right)$.

SOLVED PROBLEM 5.1

Calculate the maximum stress in a piece of rectangular steel strip 2.5 mm wide and 3 mm thick when it is bent round a drum of 3.0 m diameter. $E = 2\times10^5$ N/mm^2.

Solution:

Given data: Width = 2.5 mm, thickness = 3 mm and $E = 2\times10^5$ N/mm^2.

As the strip is bent round the drum, then the neutral axis is also bent. Let D be the diameter and t be the thickness.

Radius of the drum $= \dfrac{D+t}{2}$

$$= \frac{3,000+3}{2} = 1,501.5 \text{ mm}$$

Maximum distance of fibre, $y_{max} = \dfrac{t}{2} = \dfrac{3}{2} = 1.5$ mm

From the bending equation,

$$f_{max} = \frac{E}{R}\times y_{max}$$

$$\therefore \quad f_{max} = \frac{2\times10^5}{1,501.5}\times1.5 = 199.8 \text{ N/mm}^2$$

Maximum stress in the steel strip = 199.8 N/mm^2

SOLVED PROBLEM 5.2

A rectangular beam 250 mm deep is simply supported over a span of 4 m. If I is 8×10^6 mm^4 and bending stress should not exceed 120 MPa, then what will be the UDL the beam can carry?

Solution:

Given data: Depth, $d = 250$ mm, span = 4 m, $f_{max} = 120$ MPa and $I = 8\times10^6$ mm^4.
Let w be the uniformly distributed load the beam can carry.

$$y_{max} = \frac{d}{2} = \frac{250}{2} = 125 \text{ mm}$$

$$\text{Section modulus } Z = \frac{I}{y_{max}} = \frac{8\times10^6}{125} = 64\times10^3 \text{ mm}^3$$

$$\text{Moment of resistance } M = f_{max}\times Z$$

$$= 120\times64\times10^3 = 7,680\times10^3 \text{ N}\cdot\text{mm}$$

Bending moment of a simply supported beam subjected to UDL over the entire span is $wl^2/8$. Equating the bending moment to moment of resistance, we have

$$7,680\times10^3 = \frac{w\times4,000^2}{8}$$

Solving $w = 3.84$ N/mm

$= 3.84$ kN/m

i.e., the UDL the beam can carry is 3.84 kN/m.

SOLVED PROBLEM 5.3

A beam of depth equal to twice the width carries a load of 18 kN over the entire span of 350 cm. Determine the actual dimensions of the beam, if the maximum stress does not exceed 8 MPa.

Solution:

Given data: Load, $w = 18$ kN, span, $l = 350$ cm and $f_{max} = 8$ MPa.
Let b be the width of the beam. Then depth, $d = 2b$.

Section modulus of rectangular section $= \frac{bd^2}{6} = \frac{b(2b)^2}{6} = \frac{4b^3}{6} = \frac{2b^3}{3}$.

Maximum bending moment due to UDL on the entire span of simply supported beam is $\frac{wl^2}{8}$:

$$M_{max} = \frac{wl^2}{8} = \frac{Wl}{8} = \frac{(18\times10^3)\times3,500}{8}$$

$$= 7,875\times10^3 \text{ N}\cdot\text{mm}$$

$$\text{Maximum bending stress} = \frac{M_{max}}{Z}$$

$$8 = \frac{7,875\times10^3}{2b^3/3} = \frac{11.81\times10^3}{b^3}$$

Solving for b, $b = 113.86$ mm

Say, $b = 115$ mm

Then depth, $d = 2b = 230$ mm.

SOLVED PROBLEM 5.4

Calculate the cross-sectional dimensions of the strongest beam that can be cut out of a cylindrical log of wood of 100 cm diameter.

Solution:

Let b be the breadth and d be the depth of the beam.
The diagonal should be equal to the diameter of the log of wood

i.e., $$b^2 + d^2 = 100^2$$

$\therefore$ $$d^2 = 100^2 - b^2$$

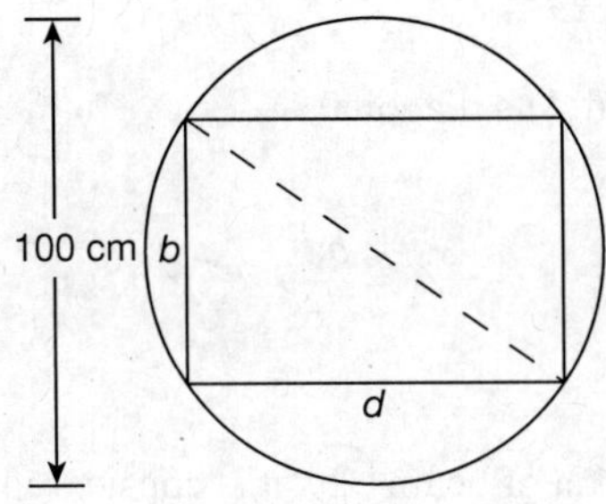

Fig. 5.5

The value of p depends on the material of the beam and it is constant for a particular material. Therefore, when Z is maximum, moment of resistance will be maximum.

Section modulus, $$Z = \frac{bd^2}{6} = \frac{1}{6}b(100^2 - b^2)$$

For Z to be maximum, $$\frac{dZ}{db} = 0$$

Differentiating, $$100^2 - 3b^2 = 0$$

$$\therefore \quad b = 57.7 \text{ cm}$$

Then, $$d = \sqrt{100^2 - 57.7^2} = 81.7 \text{ cm}$$

Dimension of the strongest section will be $577 \text{ mm} \times 817 \text{ mm}$.

SOLVED PROBLEM 5.5

Two beams of square cross section of same material are placed with two sides vertical and the other placed with the diagonal vertical. If the maximum bending stress is the same then compare the flexural strength.

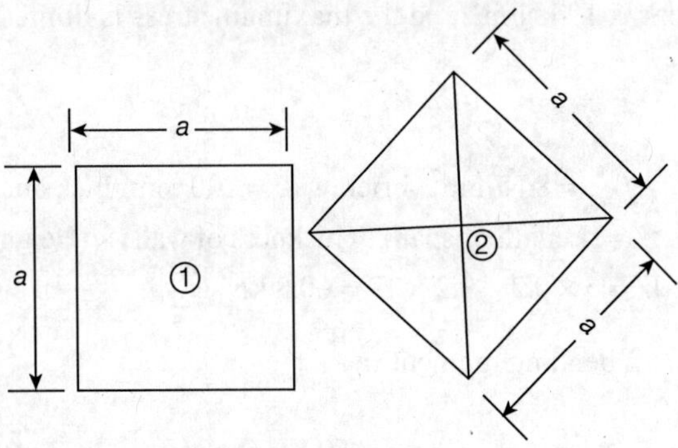

Fig. 5.6

Solution:

Let a be the side of the square and d its diagonal.

Then, $d = a\sqrt{2}$

Section modulus of beam (1) $\quad Z_1 = \dfrac{bd^2}{6} = \dfrac{a}{6} \times a^2$

$$= a^3/6$$

To the find the moment of inertia of beam (2), it is considered to be made up of two triangles attached back to back along their base.

Moment of inertia of triangle to its base $= \dfrac{bh^3}{12} = \dfrac{1}{12} d\left(\dfrac{d}{2}\right)^3 = \dfrac{d^4}{96}$

Moment of inertia of beam (2) $= \dfrac{2 \times d^4}{96} = \dfrac{d^4}{48}$

Section modulus of beam (2) $\quad Z_2 = \dfrac{I}{y_{max}} = \dfrac{d^4}{48} \times \dfrac{1}{d/2}$

i.e., $\quad Z_2 = \dfrac{(a\sqrt{2})^3}{24} = \dfrac{\sqrt{2}a^3}{12} = \dfrac{a^3}{6\sqrt{2}}$

Flexural ratio will be in the same ratio as of Z_1 and Z_2

Then, $\quad \dfrac{Z_1}{Z_2} = \dfrac{a^3}{6} \times \dfrac{6\sqrt{2}}{a^3} = \sqrt{2} = 1.414$

Hence, the first beam is 41.4% stronger than the second beam.

SOLVED PROBLEM 5.6

Find the dimensions of altimeter joist, span 5 m to carry a brick wall 200 mm thick and 3.2 m high. The weight of brick work is 19 kN/m^3 and the maximum stress is limited to 8 N/mm^2. The depth is to be twice the width.

Solution:

Given data: Span = 5 m, $f_{max} = 8$ N/mm^2, brick wall = 200 mm thick and 3.2 m high.

Total weight of wall, W = (Length of span) (Thickness of wall) × (Height of wall) (Density of brick)

$$W = 5 \times 0.2 \times 3.2 \times 19 = 60.8 \text{ kN}$$

$$\text{Maximum bending moment} = \frac{Wl}{8}$$

$$= \frac{60.8 \times 5}{8} = 38 \text{ kN} \cdot \text{m}$$

From bending equation

$$M = \frac{f}{y} \times I = \frac{f \times bd^3/12}{d/2} = \frac{8{,}000 \times bd^2}{6}$$

Equating moment of resistance to bending moment

$$\frac{8{,}000 \times bd^2}{6} = 38$$

$$\therefore \quad bd^2 = 4.75 \times 10^{-3}$$

But $d = 2b$

Then, $b \times (2b)^2 = 4.75 \times 10^{-3}$

Solving, $b = 192$ mm

$\therefore$ $d = 384$ mm

5.3 COMPOSITE BEAMS OR FLITCHED BEAMS

When a beam is made up of more than one material, then it is known as a composite beam. An RCC beam can be considered as a composite beam since it is made up of concrete and steel.

Similarly, wooden beam reinforced with steel strips or flitter form composite beams. Composite beams having flitched sections are known as flitched beams. Flitched section are stronger than pure wooden beams of same dimensions and are also economical. The reinforcing material should have a modulus of elasticity greater than that of the reinforced material. The steel plates are bolted or screwed to the timber beams. The connection is made perfectly such that there is no slipping between them. The composite beam behaves like a single monolithic beam. Different forms of flitched beams are shown in Figure 5.7.

5.3.1 Modular Ratio

It is assumed that the stress in wood and in steel at a distance y from the neutral axis be p_w and p_s, respectively. Further, there is no slip between the material but behave like a single monolithic beam. Then, the strain in wood and steel is the same. Then

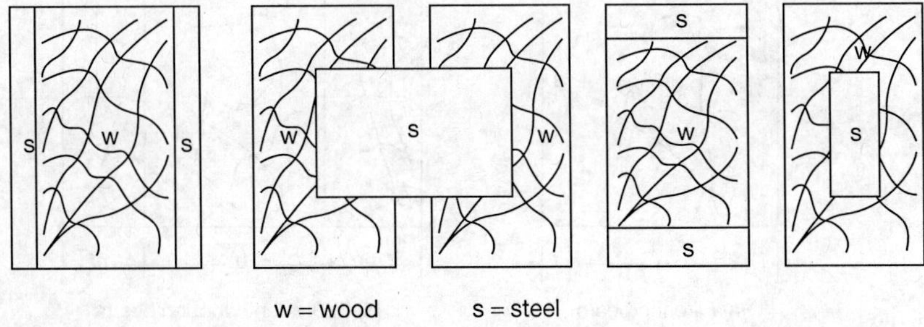

Fig. 5.7 Different forms of flitched beams.

Strain, $$e = \frac{p_w}{E_w} = \frac{p_s}{E_s}$$

where E_w and E_s are the modulus of elasticity of wood and steel, respectively.

$$\therefore \quad p_s = \frac{E_s}{E_w} \cdot p_w \tag{5.11}$$

$$p_s = mp_w \tag{5.12}$$

where $m = \frac{E_s}{E_w}$ which is known as the modular ratio.

5.3.2 Moment of resistance of the composite sections

1. Sections of Equal Depth

Let M_f be the moment of resistance of a flitched or composite section. Let M_w and M_s be the moments of resistance offered by wood and steel, respectively. Let b and d be the breadth and depth of the wooden section. Let t be the thickness and d be the depth of steel section (Figure 5.8(a)), then

$$M_f = M_w + M_s$$

If p_w and p_s be the extreme stresses in wood and steel and Z_w and Z_s be the modulus of sections of wood and steel, respectively then

$$M_f = p_w Z_w + p_s Z_s$$

$$= p_w \frac{bd^2}{6} + p_s \frac{2td^2}{6} \quad \text{(two steel plates of thickness)}$$

But $m = p_s/p_w$, i.e., $p_s = mp_w$

$$\therefore \quad M_f = p_w \frac{bd^2}{6} + mp_w \frac{2td^2}{6}$$

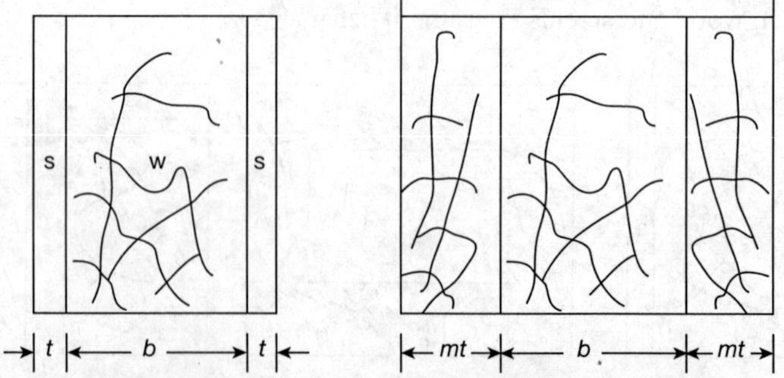

(a) Flitched beam (b) Equivalent wooden beam

Fig. 5.8 Composite beams of equal depth.

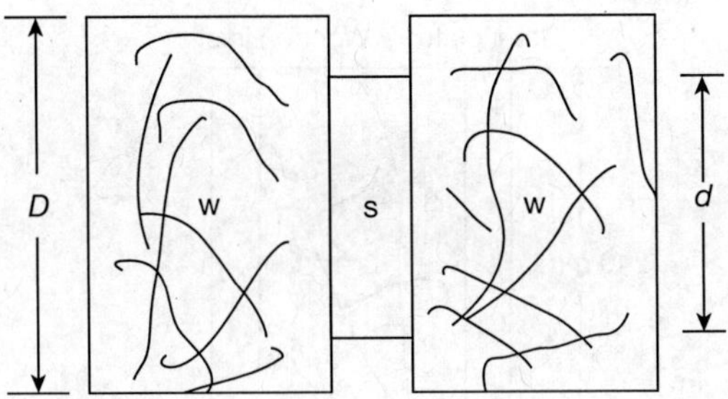

Fig. 5.9

i.e., $$M_f = \frac{p_w d^2}{6}(b + 2mt) \tag{5.13}$$

From the above relation, it is clearly seen that the moment of resistance of the flitched section M_f is nothing but the moment of resistance of a wooden section of width $(b + 2mt)$. This wooden section of width $(b + 2mt)$ and depth d is known as *equivalent wooden section* (Figure 5.8(b)).

2. Sections of Unequal Depths

For a flitched beam subjected to bending moment, the neutral axis and radius of curvature R is common for both wood and steel. Then, from bending equation

$$R = \frac{E_s I_s}{M_s} = \frac{E_w I_w}{M_w}$$

$$\therefore \quad \frac{M_s}{M_w} = \frac{E_s I_s}{E_w I_w} = \frac{E_s td^3}{E_w BD^3} \quad \left(\because I_s = \frac{td^3}{12} \text{ and } I_w = \frac{BD^3}{12}\right)$$

$$\frac{p_s td^2}{p_w BD^2} = \frac{E_s td^3}{E_w BD^3}$$

$$\frac{p_s}{p_w} = \frac{E_s}{E_w} \times \frac{d}{D}$$

$$\therefore \quad p_s = m\frac{d}{D}p_w \tag{5.14}$$

SOLVED PROBLEM 5.7

A flitched beam of a wooden joist is 100 mm wide and 240 mm deep and it is strengthened by steel plates 100 mm thick and 240 mm deep on either side of the joist. Determine the moment of resistance. The maximum stress in steel, p_s = 140 N/mm^2 and the maximum stress in wood, p_w = 7 N/mm^2.

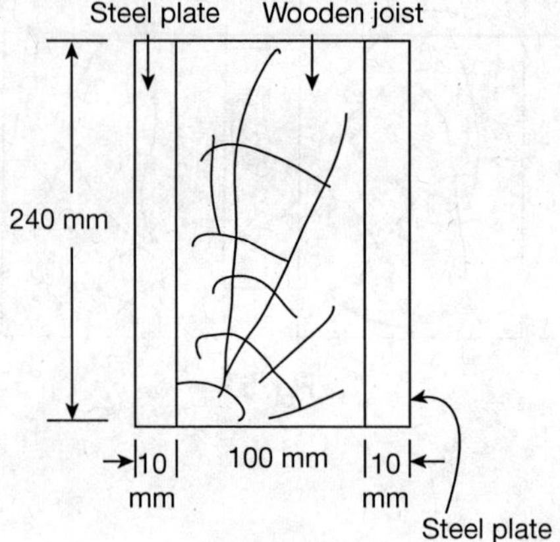

Fig. 5.10

Solution:

Given data: Wooden joist size = 100 mm × 240 mm, steel plates size e = 10 mm × 240 mm, p_s = 140 N/mm² and p_w = 7 N/mm².

Moment of resistance of wooden joist, $M_w = p_w Z_w$

$$= p_w \frac{bd^2}{6}$$

$$= 7 \times \frac{100 \times 240 \times 240}{6}$$

$$= 6{,}720 \times 10^3 \text{ N} \cdot \text{mm}$$

$$= 6{,}720 \text{ N}$$

Moment of resistance of steel plate, $M_s = p_s Z_s$

$$= p_s \frac{2td^2}{6}$$

$$= 140 \times \frac{2 \times 10 \times 240 \times 240}{6}$$

$$= 26{,}880 \times 10^3 \text{ N} \cdot \text{mm}$$

$$= 26{,}880 \text{ N} \cdot \text{m}$$

Moment of resistance of flitched beam $M_f = M_w + M_s = 6{,}720 + 2{,}880$

$$= 33{,}600 \text{ N} \cdot \text{m}$$

SOLVED PROBLEM 5.8

A flitched beam consists of two timber joists 100 mm wide and 240 mm deep with a steel plate 180 mm deep and 10 mm thick placed symmetrically between the timber joists and well clamped. Determine

(i) The maximum fibre stress in steel when the maximum fibre stress in wood is 80 kg/cm².
(ii) The combined moment of resistance if the modular ratio is 18.

Solution:
Given data: Joist size = 100 mm × 240 mm, steel plates: 180 mm × 10 mm, $f_{max} = 80$ kg/cm² and modular ratio = 18.

When the depth of steel plate and wooden joists are not equal, then maximum fibre stress in steel is given by the relation

$$p_s = mp_w \cdot \frac{d}{D}$$

where d is the depth of steel plate and D is the depth of timber joist

$$p_s = 18 \times 8 \times \frac{180}{240} = 108 \text{ N/mm}^2$$

Maximum resistance of timber joist

$$M_w = p_w \cdot Z_w$$

$$M_w = p_w \times 2 \times \frac{bd^2}{6}$$

$$= 8 \times 2 \times \frac{100 \times 240^2}{6}$$

$$= 15{,}360 \times 10^3 \text{ N} \cdot \text{mm}$$

$$= 15{,}360 \text{ N} \cdot \text{m}$$

Moment of resistance of steel plate

$$M_s = p_s \cdot Z_s$$

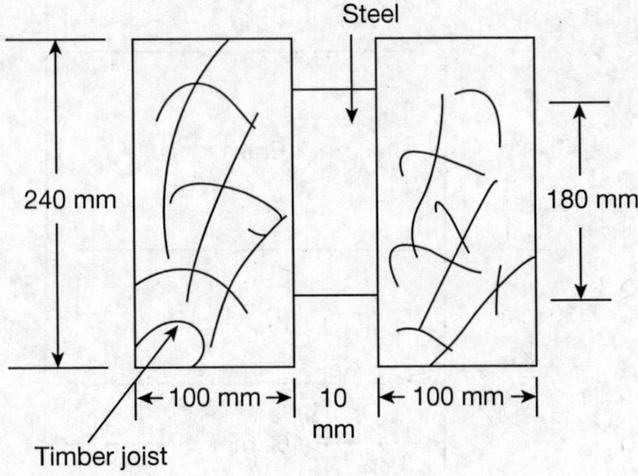

Fig. 5.11

$$= p_s \cdot \frac{bd^2}{6}$$

$$= 108 \times \frac{10 \times 180^2}{6}$$

$$= 5,832 \times 10^3 \text{ N} \cdot \text{mm}$$

$$= 5,832 \text{ N} \cdot \text{m}$$

Combined moment of resistance,

$$M_f = M_w + M_s$$

$$= 15,360 + 5,832$$

$$= 21,192 \text{ N} \cdot \text{m}$$

SOLVED PROBLEM 5.9

A steel plate of width 100 mm and depth 30 mm is placed symmetrically below a brass plate of width 100 mm and depth 40 mm to form a composite beam. Determine the moment of resistance when the plates are allowed to bend independently and when they are glued together to bend as a composite beam. $E_s/E_b = 2$. Allowable stress in steel and brass are 120 N/mm^2 and 60 N/mm^2, respectively.

Solution:

Given data: Steel plate: 100 mm × 30 mm, brass plate = 100 mm × 40 mm and $E_s/E_b = 2$.

$$I_s = \frac{100 \times 30^3}{12} = 22.5 \times 10^4 \text{ mm}^4$$

$$I_b = \frac{100 \times 40^3}{12} = 53.33 \times 10^4 \text{ mm}^4$$

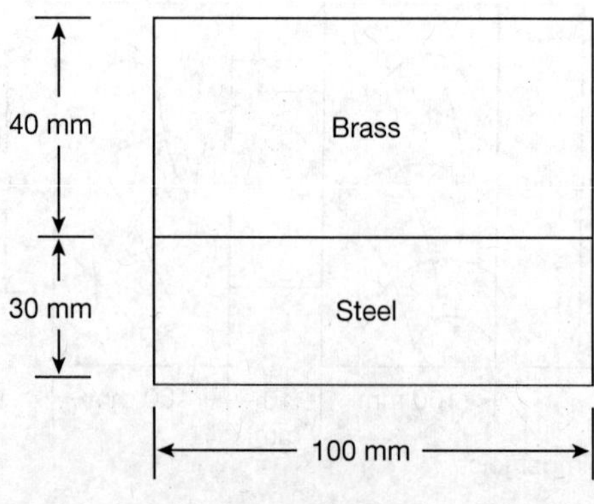

Fig. 5.12

When the materials act individually
Moment of resistance for steel section

$$M_s = p_s \frac{I_s}{y_s} = 120 \times \frac{22.5 \times 10^4}{15}$$
$$= 10 \times 10^4 \text{ N} \cdot \text{mm}$$

Moment of resistance of brass section

$$M_b = p_b \frac{I_b}{y_b} = 60 \times \frac{53.33 \times 10^4}{20}$$
$$= 180 \times 10^4 \text{ N} \cdot \text{mm}$$

Combined moment of resistance

$$M = M_s + M_b = (180 \times 10^4) + (160 \times 10^4)$$
$$= 340 \times 10^4 \text{ N} \cdot \text{mm}$$

When the materials act as a single beam

Using the relationship $\frac{E_s}{E_b} = 2$, the thickness of the steel section equivalent to 100 mm width of brass section is $\frac{100}{2} = 50$ mm (Figure 5.13).

Let $\bar{y}$ be the distance of NA from bottom of section.

$$\bar{y} = \frac{100 \times 30 \times 15 + \left[50 \times 40 \times \left(30 + \frac{40}{2} \right) \right]}{(100 \times 30) + (50 \times 40)}$$
$$\bar{y} = 29 \text{ mm}$$

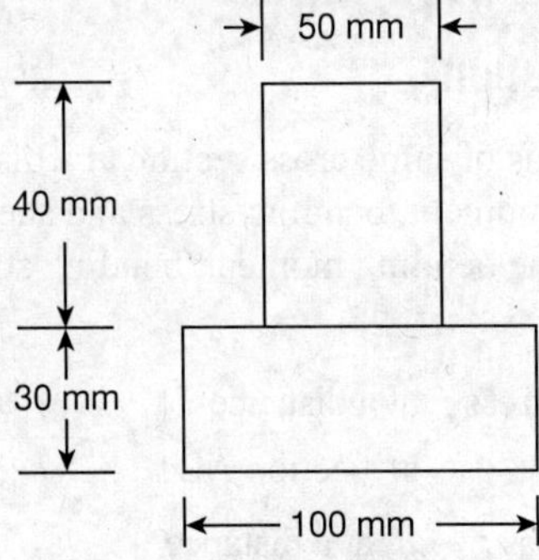

Fig. 5.13 Equivalent steel section.

Distance of extreme fibre from NA

$$y_{max} = 70 - 29 = 41 \text{ mm.}$$

Moment of inertia of equivalent section, I

$$= \frac{100 \times 30^3}{12} + [100 \times 30 \times (29-15)^2] + \frac{50 \times 40^3}{2} + [50 \times 40 \times (70-29-20)^2]$$

$$= (22.5 \times 10^4) + (58.8 \times 10^4) + (26.7 \times 10^4) + (88.2 \times 10^4)$$

$$= 196.2 \times 10^4 \text{ mm}^4$$

Moment of resistance of the section, $M_s = p_s \dfrac{I}{y_{max}}$

$$= 120 \times \frac{1{,}962 \times 10^4}{41} = 574 \times 10^4 \text{ N.mm.}$$

Moment of resistance when materials act individually $= 340 \times 10^4$ N·mm

Moment of resistance when materials act monolithically $= 574 \times 10^4$ N·mm

5.4 SHEAR STRESSES IN BEAMS

When a beam is subjected to bending moment, normal tensile and compressive stresses prevail along the cross section of the beam. In addition to the above stresses, transverse sections are subjected to vertical shearing stresses. Vertical shearing stresses are always accompanied by complementally shear stress of equal intensity in a perpendicular plane. Shear stress varies along longitudinal planes since the perpendicular section resists the shear forces.

Assumptions made for simple bending may also be adopted for their investigation. It is seen that transverse shear stress in beams generally increases towards the neutral axis from top to bottom of the section. As longitudinal stresses are almost negligible near the neutral axis, the width of the beam is generally kept small where the shear stress is maximum. When transverse stress increases to the ultimate value due to faulty design, then the beam may fail. It is said the failure has occurred due to *shear flow.*

5.4.1 Shear Stress Distribution

Let AB and CD be the two sections of same cross section at a distance of dx.

Let M, f and S be the bending moment, bending stress and shear force respectively at section AB.

Let $M + dM$, $f + df$, $S + dS$ be the bending moment, bending stress and shear force, respectively, at section CD (Figure 5.14).

$$\left.\begin{array}{r}\text{Bending stress } f \text{ at a distance } y\\ \text{at section AB}\end{array}\right\} f = \frac{My}{I}$$

$$\left.\begin{array}{r}\text{Bending stress } f + df \text{ at a distance}\\ y \text{ at section CD}\end{array}\right\} f + df = \frac{(M + dM)}{I} y$$

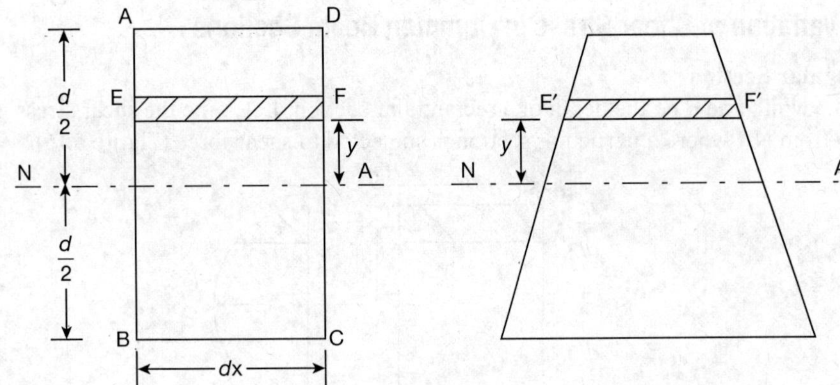

Fig. 5.14

Considering an elemental area dA, at this level the unbalanced force due to bending stress on the element will be

$$f = (f + df)\,dA - f dA$$

$$= \frac{M + dM}{I} dA - \frac{M.y}{I} \cdot dA$$

$$= \frac{dM}{I} \cdot y \cdot dA$$

Total unbalanced force above a distance y from neutral axis $= \int_y^{y_{\max}} \frac{dM}{I} \cdot y \cdot dA$

This will set up a horizontal shear stress τ at this level and if the width of the section is b at this level.

$$\tau\, b\, dx = \int_y^{y_{\max}} \frac{dM}{I} .y \cdot dA$$

Then,

$$\tau = \frac{dM}{dx} \frac{1}{Ib} \int_y^{y_{\max}} y \cdot dA$$

$$\tau = \frac{SA\bar{y}}{Ib}$$

i.e.,

$$\tau = \frac{SQ}{Ib}. \tag{5.15}$$

where S is the shear force at this section and Q is the moment of area $(A\,\bar{y})$ beyond the distance y from the neutral axis.

5.4.2 Variation of Shear Stress in Common Beam Sections

1. Rectangular Section

Let b be the width and d be the depth of a rectangular section. Let τ be the shear stress in a layer at distance y from NA where a particular section is subjected to shear force (Figure 5.15).

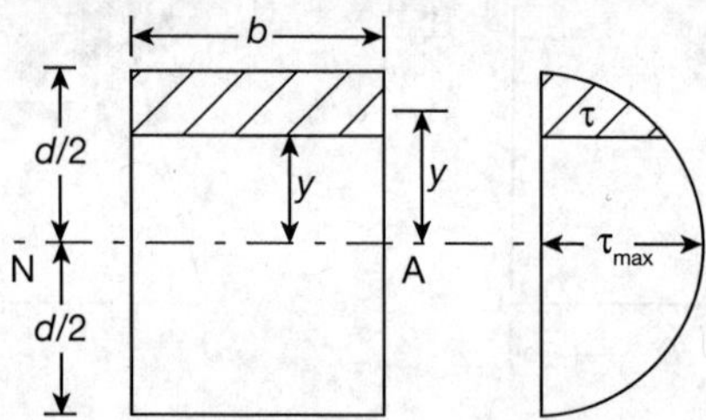

Fig. 5.15 Shear stress distribution in rectangular section.

Area of shaded portion

$$A = b \times \left(\frac{d}{2} - y\right)$$

Distance of centre of gravity of the shaded area from neutral axis.

$$\bar{y} = y + \frac{1}{2}\left(\frac{d}{2} - y\right) = y + \frac{d}{4} - \frac{y}{2}$$

i.e., $$\bar{y} = \frac{1}{2}\left(\frac{d}{2} + y\right)$$

$\therefore$ Shear stress, $$\tau = \frac{SQ}{Ib} = \frac{SA\bar{y}}{Ib}$$

$$= \frac{S \times b\left(\frac{d}{2} - y\right) \times \frac{1}{2}\left(\frac{d}{2} + y\right)}{\frac{bd^3}{12} b}$$

$$= \frac{S\left(\frac{bd^2}{8} - \frac{by^2}{2}\right)}{\frac{b^2 d^3}{12}}$$

$$\therefore \quad \tau = \frac{6S}{bd^3}\left(\frac{d^2}{4} - y^2\right) \tag{5.16}$$

Maximum value of τ is at the neutral axis, when $y = 0$.

or $$\tau_{max} = \frac{6S}{bd^3}\left(\frac{d^2}{4} - y^2\right)$$

i.e., $$\tau_{max} = \frac{3}{2}\frac{S}{bd}$$

But $\frac{S}{bd}$ is the mean shear stress, τ_{mean}

Hence, $$\tau_{max} = \frac{3}{2}\tau_{mean} \tag{5.17}$$

2. Solid Circular Section

A solid circular section of radius R and diameter D be considered. Let B be the length of the chord at a distance y from neutral axis (centre of circle).

Length of the chord, $$B = 2\sqrt{R^2 - y^2}$$

Elemental area of a strip of thickness dy is

$$dA = 2y\sqrt{R^2 - y^2} \cdot dy$$

Moment of dA about neutral axis is

$$y \cdot dA = 2y\sqrt{R^2 - y^2} \cdot dy$$

Moment of shaded area about neutral axis is

$$\int_y^R 2y\sqrt{R^2 - y^2} \cdot dy = \int_y^R y \cdot b \cdot dy$$

We know $$B = 2\sqrt{R^2 - y^2}$$

i.e., $$B^2 = 4\left(R^2 - y^2\right)$$

Differentiating on both sides

$$2B \cdot dB = -8 \cdot y \cdot dy$$

$$\therefore \quad y\,dy = -\frac{1}{4} B\,dB$$

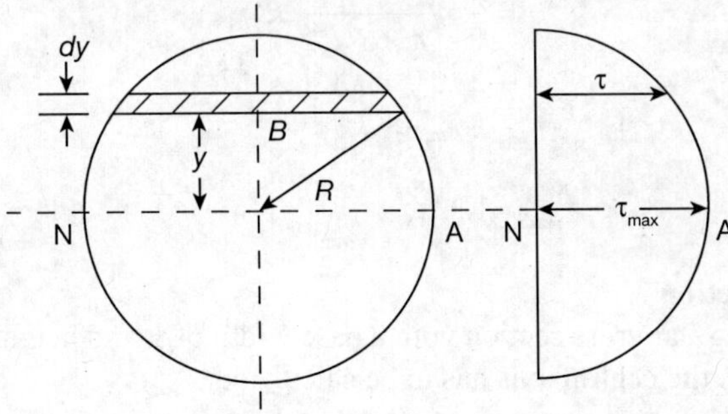

Fig. 5.16 Shear stress distribution in circular section.

When $y = 0$, then $B = D$
When $y = R$, then $B = 0$
Substituting this value in the shaded area equation and integrating

$$A\bar{y} = \int_y^R 2y\sqrt{R^2 - y^2} \cdot dy$$

$$= \int_y^R 2y\sqrt{R^2 - y^2} \times y \cdot dy$$

$$= \int_y^R B \times \left(-\frac{1}{4} B\, dB \right)$$

$$= \frac{1}{4}\int_0^B B^2\, dB = \frac{1}{4}\left[\frac{B^3}{3} \right]_0^B$$

$$= \frac{B^3}{12}$$

$$Q = A\bar{y} = \frac{B^3}{12} \times I = \frac{\pi d^4}{64}$$

$$\tau = \frac{SQ}{Ib} = \frac{SB^3}{12 \cdot I \cdot B} = \frac{SB^2}{12I}$$ (here the width b is replaced by B, the chord length).

$$\therefore \quad \tau = \frac{S}{12I}\left(2\sqrt{R^2 - y^2}\right)^2 = \frac{4S}{12I}\left(R^2 - y^2\right)$$

$$\therefore \quad \tau = \frac{S}{3I}\left(R^2 - y^2\right) \tag{5.18}$$

Since this is a second-degree equation in y the variation is parabolic.
When $y = R$, then $\tau = 0$ and
When $y = 0$ (i.e., at NA), τ will be maximum.

$$\tau_{max} = \frac{S}{3} \times \frac{64}{\pi d^4} \cdot R^2$$

Hence,

$$= \frac{S}{3} \times \frac{64}{\pi \cdot 16 \cdot R^4} \cdot R^2$$

$\therefore$

$$\tau_{max} = \frac{4}{3} \cdot \frac{S}{\pi R^2} = \frac{4}{3}\tau_{mean}$$

i.e.,

$$\therefore \tau_{max} = 1.33\tau_{av} \tag{5.19}$$

3. Triangular Cross Section

Consider a beam of triangular cross section with a base width of B and height H, the shear stress in a layer at a distance y from the central axis has to be calculated.

Height of shaded triangle, $h = \frac{2H}{3} - y$

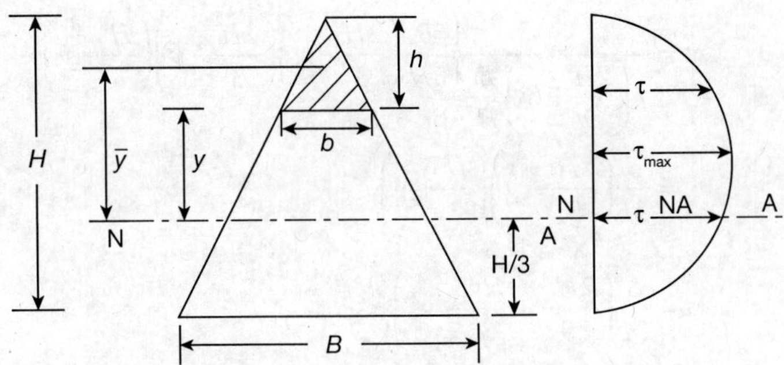

Fig. 5.17 Shear stress distribution in triangular section.

From similar triangles, $$\frac{h}{b} = \frac{H}{B}$$

i.e., $$b = \frac{h \cdot B}{H}$$

i.e., $$b = \left(\frac{2H}{3} - y\right)\frac{B}{H}$$

Distance of centre of gravity of the shaded triangle from NA $\bar{y} = y + \frac{h}{3}$

$$\bar{y} = y + \frac{1}{3}\left(\frac{2H}{3} - y\right)$$

$$\bar{y} = \frac{2}{3}\left(\frac{H}{3} + y\right)$$

∴ $$A\bar{y} = \left(\frac{1}{2}b \times h\right) \times \frac{2}{3}\left(\frac{H}{3} + y\right)$$

$$A\bar{y} = \frac{1}{2}\left(\frac{2H}{y} - y\right)\frac{B}{H}\left(\frac{2H}{y} - y\right) \times \frac{2}{3}\left(\frac{H}{3} + y\right)$$

i.e., $$= \frac{B}{3H}\left(\frac{2H}{3} - y\right)\left(\frac{2H}{3} - y\right)\left(\frac{H}{3} + y\right)$$

Shear stress, $$\tau = \frac{SQ}{Ib}$$

But $$Q = A\bar{y}; \quad I = \frac{BH^3}{16}$$

$$\therefore \qquad \tau = \frac{S}{I\left(\frac{2H}{3}-y\right)\frac{B}{H}}\left\{\frac{B}{3H}\left(\frac{2H}{3}-y\right)\left(\frac{2H}{3}-y\right)\left(\frac{H}{3}+y\right)\right\}$$

$$= \frac{S}{3I}\left(\frac{2H}{3}-y\right)\left(\frac{H}{3}+y\right)$$

i.e.,
$$\tau = \frac{S}{3I}\left(\frac{2H^2}{9}+\frac{Hy}{3}-y^2\right)$$

When $\frac{d\tau}{dy} = 0$, the shear stress is maximum.

Differentiating the above relation in τ and equating it to zero.

$$\frac{d\tau}{dy} = \frac{S}{3I}\left(\frac{H}{3}-y\right) = 0$$

As $\frac{S}{3I}$ cannot be equal to zero.

$$\frac{H}{3} - 2y = 0$$

i.e.,
$$y = \frac{H}{6}$$

At a distance $y = \frac{H}{6}$ from the neutral axis, the shear stress will be maximum.

$$\tau_{max} = \frac{S}{3I}\left\{\frac{2H^2}{9}+\frac{H}{3}\times\frac{H}{6}-\frac{H}{6}\times\frac{H}{6}\right\}$$

$$= \frac{S}{3I}\left(\frac{H^2}{4}\right)$$

$$\tau_{max} = \frac{SH^2}{12}\times\left(\frac{36}{BH^3}\right) = \frac{3S}{BH}$$

τ_{max} occurs at $\frac{H}{2}\left(=\frac{H}{3}+\frac{H}{6}\right)$ above the base of the triangle.

$$\tau_{av} = \frac{2S}{bh} \qquad \left(\because \quad \tau_{av} = \frac{\text{Shear stress}}{\text{Area of cross section}}\right)$$

$$\therefore \quad \tau_{max} = 1.5\tau_{av}$$

Shear stress at neutral axis, i.e., where $y = 0$

$$\tau_{av} = \frac{S}{I}\left(\frac{2H^2}{9}\right)$$

$$\tau_{av} = \frac{2SH^2}{3I} \times \frac{36}{DH^3} \tag{5.20}$$

$$\tau_{av} = \frac{8}{3BH}$$

4. I-Sections

An I-section of flange width B and overall depth D is considered, whose web width is b and depth d.

Shear stress distribution in flange.

Area of the shaded portion in the flange $A = B\left(\frac{D}{2} - y\right)$

Distance of the center of gravity of the shaded area from the neutral axis of the section

$$\bar{y} = \frac{1}{2}\left(\frac{D}{2} - y\right) + y$$

$$\bar{y} = \frac{1}{2}\left(\frac{D}{2} + y\right)$$

$$\therefore \qquad \tau = \frac{SQ}{Ib} = \frac{SA\bar{y}}{Ib}$$

$$= \frac{S}{IB} \cdot B\left(\frac{D}{2} - y\right) \times \frac{1}{2}\left(\frac{D}{2} + y\right)$$

$$= \frac{S}{2I}\left(\frac{D}{2} - y\right)\left(\frac{D}{2} + y\right)$$

$$\tau = \frac{S}{2I}\left(\frac{D^4}{4} - y^2\right)$$

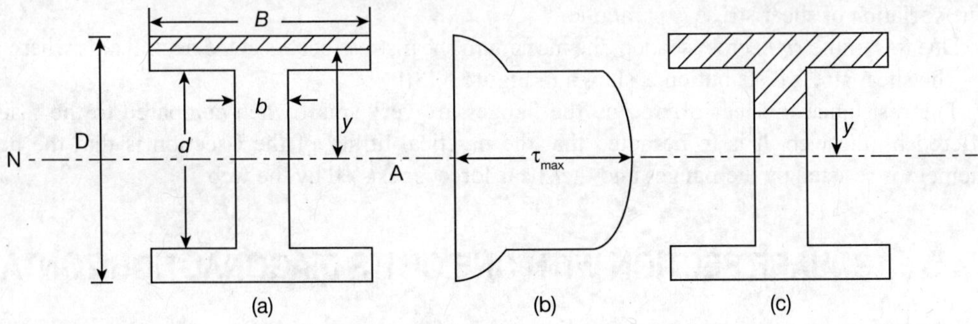

Fig. 5.18 Shear stress distribution in I-section.

This equation is a second-degree equation in y and hence the variation is parabolic.

At the junction of flange and web, $y = d/2$, then

$$\tau = \frac{S}{8I}\left(D^2 - d^2\right) \tag{5.21}$$

Shear stress distribution in web

At the junction of flange and web, the width of the I-section suddenly changes from B to b (Figure 5.18(c)).

Accordingly, the change in τ will be from $\frac{S}{8I}\left(D^2 - d^2\right)$ to $\frac{B}{b}\cdot\frac{S}{8I}\left(D^2 - d^2\right)$

Area of the shaded portion is the sum of the shaded area of the flange and the shaded area in web

$$A = B\left(\frac{D-d}{2}\right) + b\left(\frac{d}{2} - y\right)$$

Distance of the centre of gravity of the shaded area from neutral axis

$$\bar{y} = \frac{d}{2} + \frac{1}{2}\left(\frac{D}{2} - \frac{d}{2}\right) = \frac{1}{2}\left(\frac{D}{2} + \frac{d}{2}\right) \quad \text{(flange area)}$$

$$\bar{y} = y + \frac{1}{2}\left(\frac{d}{2} - y\right) = \frac{1}{2}\left(\frac{d}{2} + y\right) \quad \text{(web area)}$$

Moment of shaded area with reference to neutral axis is

$$A\bar{y} = B\left(\frac{D-d}{2}\right)\frac{1}{2}\left(\frac{D}{2} + \frac{d}{2}\right) + b\left(\frac{d}{2} - y\right)\frac{1}{2}\left(\frac{d}{2} + y\right)$$

$$= \frac{B}{8}\left(D^2 - d^2\right) + \frac{b}{2}\left(\frac{d^4}{4} - y^2\right)$$

Shear stress, $$\tau = \frac{SQ}{Ib} \times A\bar{y}$$

$$= \frac{S}{8Ib}\left[B\left(D^2 - d^2\right) + b\left(d^2 - 4y^2\right)\right] \tag{5.22}$$

The variation of shear stress is parabolic.

Due to symmetry in cross section, the maximum shear stress occurs at the neutral axis where $y = 0$. The shear stress distribution is shown in Figure 5.18(b).

The resistance to shear offered by the flanges are very small when compared to the resistance offered by the web. It is to be noted that the practical utility of the I-section is that the bending moment is resisted by the flanges and the shear force is resisted by the web.

5.5 SQUARE SECTION WITH ONE OF ITS DIAGONAL HORIZONTAL

A beam of square cross section with diagonal as D is considered. One of its diagonal is kept in horizontal position (Figure 5.19).

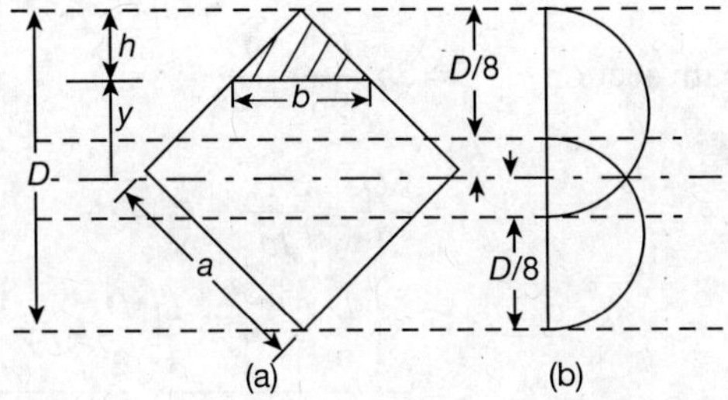

Fig. 5.19 Shear stress distribution in square section.

Let S be the shearing force to which the section is subjected to.

Height of the shaded triangle, $h = \frac{D}{2} - y$

Base of the shaded triangle, $b = 2\left(\frac{D}{2} - y\right)$

Area of the shaded triangle, $A = \frac{1}{2} bh$

i.e., $A = \frac{1}{2} \times 2\left(\frac{D}{2} - y\right) \times \left(\frac{D}{2} - y\right)$

i.e., $A = \left(\frac{D}{2} - y\right)^2$

Distance of the centre of gravity of the shaded triangle from neutral axis $\bar{y} = y + \frac{h}{3}$

i.e., $\bar{y} = y + \frac{1}{3}\left(\frac{D}{2} - y\right)$

$$= \left(\frac{D}{6} + \frac{2y}{3}\right)$$

$\therefore$ $\bar{y} = \frac{1}{3}\left(\frac{D}{2} + 2y\right)$

Here the moment of inertia of the cross-section is found out by considering them as two triangular sections attached at their base D.

Moment of criteria of a triangle with reference to the base = (Moment of inertia about NA) + (Area of triangle) $\times$ (Distance between NA and reference line)^2

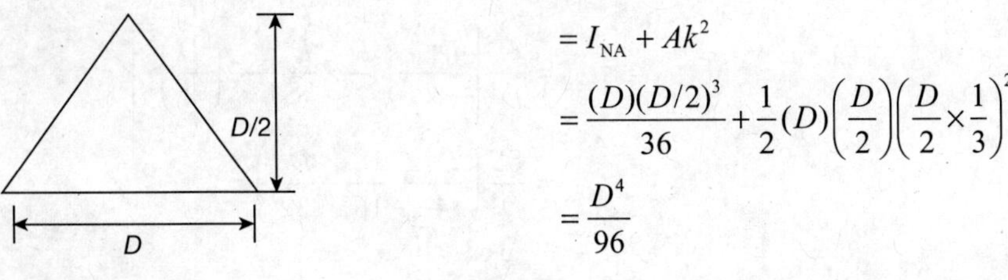

$$= I_{NA} + Ak^2$$

$$= \frac{(D)(D/2)^3}{36} + \frac{1}{2}(D)\left(\frac{D}{2}\right)\left(\frac{D}{2} \times \frac{1}{3}\right)^2$$

$$= \frac{D^4}{96}$$

Fig. 5.20

Moment of Inertia of the beam section $= 2 \times \dfrac{D^4}{96} = \dfrac{D^4}{48}$

Shear stress $\tau = \dfrac{SQ}{Ib} = \dfrac{SA\bar{y}}{Ib}$

$$= \frac{S\left(\frac{D}{2} - y\right)^2 \times \frac{1}{3}\left(\frac{D}{2} + 2y\right)}{\frac{D^4}{48} \times 2\left(\frac{D}{2} - y\right)}$$

$$= \frac{S\left(\frac{D}{2} - y\right)^2 \times \frac{1}{3}\left(\frac{D}{2} + 2y\right)}{48/2D^4}$$

$$= \frac{8S}{D^2}\left(\frac{D}{2} - y\right)\left(\frac{D}{2} + 2y\right)$$

$$\tau = \frac{8S}{D^4}\left(\frac{D^2}{4} + \frac{Dy}{2} - 2y^2\right) \quad (5.23)$$

The stress variation is parabolic.

To find the maximum shear, the above relation in τ is differentiated and equated to zero.

i.e., $\dfrac{d\tau}{dy} = 0$

$\therefore$ $\dfrac{8S}{D^4}\left(\dfrac{D}{2} - 4y\right) = 0$

As $\dfrac{8S}{D^4}$ cannot be zero,

$$\left(\frac{D}{2} - 4y\right) = 0$$

$\therefore$ $y = \dfrac{D}{8}$

Hence, τ_{max} will occur at a distance $\dfrac{D}{8}$ above and below neutral axis which is the horizontal diagonal.

$$\tau_{max} = \frac{8S}{D^4}\left[\frac{D^2}{4} + \left(\frac{D}{2} \times \frac{D}{8}\right) - 2\left(\frac{D}{8}\right)^2\right]$$

$$\tau_{max} = \frac{8S}{D^4}\left(\frac{D^2}{4} + \frac{D^2}{16} - \frac{D^2}{32}\right)$$

$$\tau_{max} = \frac{9S}{4D^2} \quad (5.24)$$

Area of section $= \frac{D^2}{2}$

$$\tau_{av} = \frac{S \times 2}{D^2}$$

$\therefore$ $$\frac{\tau_{max}}{\tau_{av}} = \frac{9S}{4D^2} \times \frac{D^2}{2S} = \frac{9}{8}$$

or $$\tau_{max} = 1.125\tau_{av}$$

i.e., τ_{max} is 12.5% more than τ_{av}

Shear stress at neutral axis, i.e., when $y = 0$

$$\tau_{NA} = \frac{8S}{D^4}\left(\frac{D^2}{4}\right) = \frac{2S}{D^2}$$

i.e., $$\tau_{NA} = \tau_{av}$$

The intensity of shear stress at neutral axis is equal to the average shear stress intensity.

5.6 BEAMS OF UNIFORM STRENGTH

In general, beams have uniform cross section throughout their length. When they are loaded, there is a variation in bending moment from section to section along the length. The stress in extreme outer fibre (top and bottom) also vary from section to section along their length. The extreme fibres can be loaded to the maximum capacity of permissible stress (say p_{max}), but they are loaded to less capacity. Hence, in beams of uniform cross section there is a considerable waste of materials.

When a beam is suitably designed such that the extreme fibres are loaded to the maximum permissble stress p_{max} by varying the cross section it will be known as a beam of uniform strength.

We know, $$p = \frac{My}{I} = \frac{M}{Z} \qquad \left(\because Z = \frac{I}{y}\right)$$

If at every point of the extreme fibre, stress is equal to p_{max} then the section modulus of beam at any section should be proportional to the bending moment at that section.

To obtain beams of uniform strength the sections of the beam may \be varied by

(i) keeping the width constant throughout and varying the depth,
(ii) keeping the depth constant throughout the length and varying the width,
(iii) by varying the width and depth in a suitable way and
(iv) circular beam of uniform strength can be made by varying diameter in such a way that $\frac{M}{Z}$ is a constant.

Carriage springs (leaf springs), tapered masts, electric pole of varying diameter, latticed girders, plate girders and box girders are examples of beam of uniform strength.

SOLVED PROBLEM 5.10

A beam is made up of a channel section 100 mm × 50 mm which has a uniform thickness of 15 mm. Draw the distribution of shear stress across a vertical section if the shearing force is 100 kN. Also find the ratio between the maximum shear stress and average shear stress.

Solution:

Given data: Channel section = 100 mm × 50 mm × 15 mm and shear force = 100 kN.

Moment of inertia of the channel section with reference to neutral axis,

$$I = \frac{50 \times 100^3}{12} - \frac{35 \times 70^3}{12}$$

$$I = 4.17 \times 10^6 - 1.00 \times 10^6$$

$$I = 3.17 \times 10^6 \text{ mm}^4$$

Shear stress at the flange section

Area of the flange = 50 × 15 = 750 mm².

Distance of centre of gravity of the flange section from *NA*.

$$\bar{y} = \frac{100}{2} - \frac{15}{2}$$

$$= 42.5 \text{ mm}$$

$$\text{Shear stress} \quad \tau = \frac{SQ}{Ib} = \frac{SA\bar{y}}{Ib}$$

$$= \frac{100 \times 10^3 \times 750 \times 42.5}{6.17 \times 10^6 \times 50}$$

$$= 10.33 \text{ N/mm}^2$$

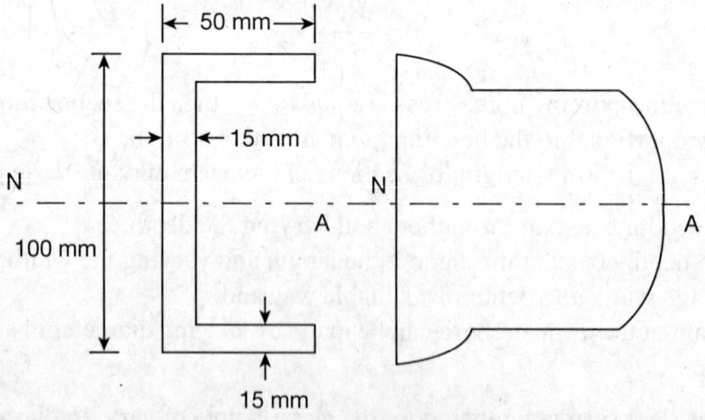

Fig. 5.21 Channel section.

Shear stress at the junction of flange and web

Shear stress at the junction of flange and web, $\tau = \dfrac{100 \times 10^3 \times 750 \times 42.5}{3.17 \times 10^6 \times 50}$

$= 20.11$

The values of 67.03 N/mm² can be cross checked by multiplying the shear stress at the angle with the width of the channel divided by the width of the web.

i.e., $\tau = 20.11 \times \dfrac{50}{15} = 67.03 \text{ N/mm}^2$

Shear stress at web

Area of flange $= 50 \times 15 = 750 \text{ mm}^2$

Area of web till neutral axis $= (50-15) \times 15$

$= 525 \text{ mm}^2$

Area of section till neutral axis $= 750 + 525$

$= 1{,}275 \text{ mm}^2$

Distance of centre of gravity from neutral axis, $\bar{y} = \left(50 - \dfrac{15}{2}\right) = 42.5$ mm (flange)

and $\bar{y} = \left(\dfrac{50-15}{2}\right) = 17.5$ mm (web)

Due to symmetry, the shear stress will be maximum at neutral axis.

$$\tau_{max} = \frac{QA\bar{y}}{Ib} = \frac{100 \times 10^3 \left[(750 \times 42.5) + (525 \times 17.5)\right]}{3.17 \times 10^6 \times 15}$$

Total area of the section $= (50 \times 100) - (35 \times 70)$

$= 2{,}550 \text{ mm}^2$

$$\tau_{max} = \frac{S}{\text{Area of section}} = \frac{100 \times 10^3}{2{,}550} = 39.22 \text{ N/mm}^2$$

Ratio between τ_{max} and τ_{av} is

$$\frac{\tau_{max}}{\tau_{av}} = \frac{86.36}{49.22} = 1.75$$

SOLVED PROBLEM 5.11

A simply supported beam of 6 m span is subjected to an uniformly distributed load of 15 kN/m over its entire length. The cross section of the beam is 20 cm wide and 30 cm deep. Sketch the variation of bending stress in the beam cross-section.

Solution:

Given data: Span = 6 m, UDL = 15 kN/m and beam cross-section = 20 cm × 30 cm

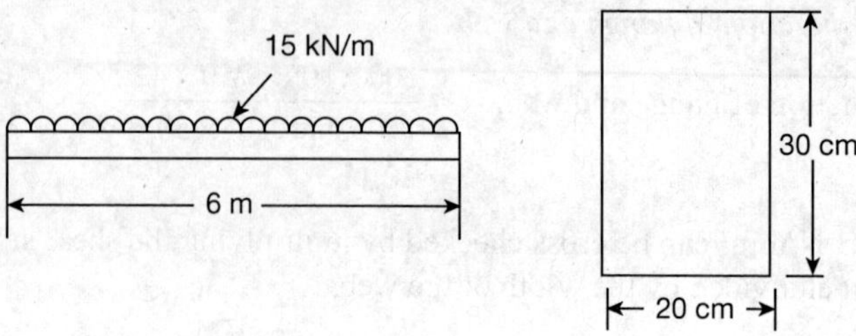

Fig. 5.22

Bending stress

Section modulus, $Z = \frac{bd^2}{6} = \frac{200 \times 300^2}{6} = 300 \times 10^4 \text{ mm}^3$

Maximum bending moment, $M = \frac{wl^2}{8} = \frac{15 \times 6^2}{8} = 67.5 \text{ kN} \cdot \text{m}$

Using bending equation

$$f = \frac{M}{Z}$$

$$= \frac{67.5 \times 10^6}{300 \times 10^4} = 22.5 \text{ N/mm}^2$$

Maximum compressive and tensile stresses occur at the outermost layer of the beam.

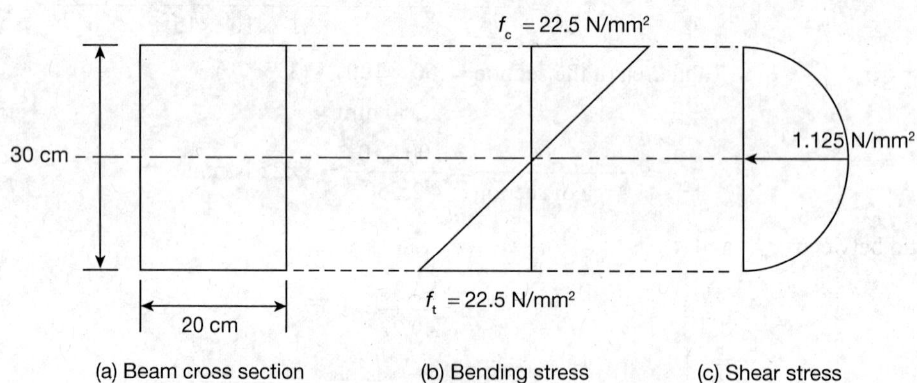

(a) Beam cross section (b) Bending stress (c) Shear stress

Fig. 5.23

Shear stress

Maximum shear force at the support $= \frac{Wl}{2}$

$$= 15 \times \frac{6}{2} = 45 \text{ kN.}$$

Maximum shear stress is at neutral axis $\tau_{max} = 1.5 \times \tau_{av}$

$$= 1.5 \times \frac{S}{b \times d}$$

$$\tau_{max} = 1.5 \times \frac{45}{200 \times 300} = 1.125 \text{ N/mm}^2$$

SOLVED PROBLEM 5.12

A T-section of simply supported beam has the following dimensions. Width of the flange = 100 mm, overall depth = 100 mm, and thickness of stem and flange = 20 mm.

(a) Determine the maximum stress in the beam when a bending moment at 1,200 N·m is acting on the section.

(b) Calculate the shear stress at the neutral axis and at the junction of web and the flange when a shear force of 50 kN acting on the beam.

Solution:

Given data: Width of the flange = 100 mm, overall depth = 100 mm and thickness = 20 mm.

Bending stress

Let $\bar{y}$ be the centre of gravity of the section from the bottom.

Then,
$$\bar{y} = \frac{A_1 y_1 + A_2 y_2}{A_1 + A_2}$$

$$\bar{y} = \frac{(100 \times 20) \times (80 + 20/2) + 80 \times 20 \times 80/2}{(100 \times 20) + (80 \times 20)}$$

$$= 67.78 \text{ mm}$$

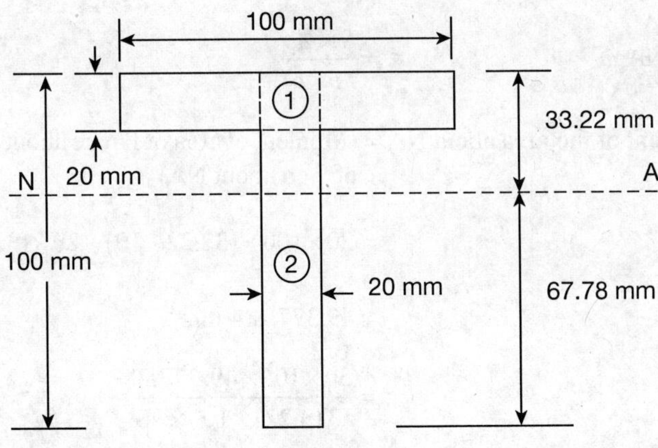

Fig. 5.24

∴ Neutral axis lies at a distance of 67.78 mm from the bottom face or 100 – 67.78 = 32.22 mm from the top face.

Now moment of inertia of the section about NA

$$I = I_1 + I_2$$

$$= \frac{b_1 d_1^3}{12} + a_1 h_1^2 + \frac{b_2 d_2^3}{12} + a_2 h_2^2$$

$$= \frac{100\times 20^3}{12} + 100\times 20\left(32.23 - \frac{20}{2}\right)^2 + \frac{20\times 80^3}{12} + (20\times 80)\left(67.77 - \frac{80}{2}\right)^2$$

$$= 3,142,222.4 \text{ mm}^4$$

Using the bending formula

$$f = \frac{M}{I}\times y$$

Maximum tensile and compressive stress for a simply supported beam occur at the extreme bottom and top fibre, respectively.

Maximum tensile stress, $f_t = \dfrac{1,200\times 10^3}{3,142,222.4}\times 67.78 = 25.885 \text{ N/mm}^2$

Maximum compressive stress, $f_c = \dfrac{1,200\times 10^3}{31,42,222.4}\times 32.22 = 12.30 \text{ N/mm}^2$

Shear stress

Shear force, $S = 50$ kN

Moment of inertia about NA, $I = 314.221\times 10^4 \text{ mm}^4$

The shear stress at the top edge of the flange and bottom of the web is zero.

Shear stress at NA, $\tau = \dfrac{S\times A\bar{y}}{I\times b}$

where $A\bar{y}$ = moment of the area about NA = (Moment of area of flange about NA) + (Moment of area of web about NA)

$$= 20\times 100\times (32.22 - 10) + 20\times (32.22 - 10)\times \frac{22.22}{2}$$

$$= 49,377.284 \text{ mm}^2$$

∴ $$\tau = \frac{50\times 10^3\times 49,377.28}{314.221\times 10^4\times 20}$$

$$= 39.285 \text{ N/mm}^2$$

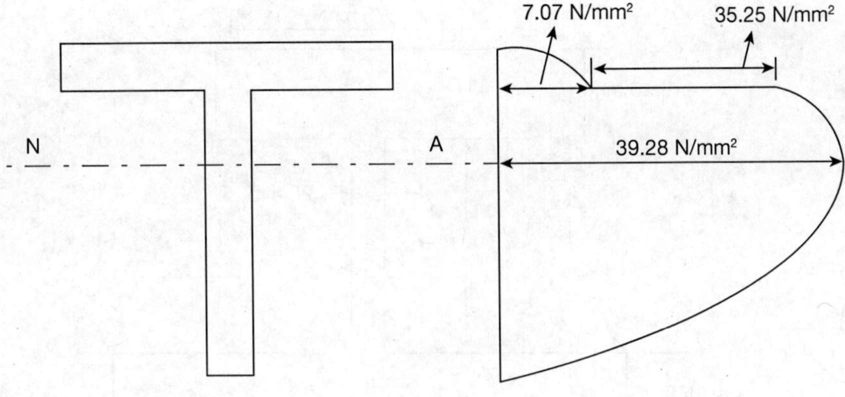

Fig. 5.25

Shear stress in the flange just at the junction of the flange and web

$$\tau = \frac{S \times A\bar{y}}{Ib}$$

$$= \frac{50 \times 10^3 \times (100 \times 200)\left(32.22 - \frac{20}{2}\right)}{314.221 \times 10^4 \times 100}$$

$$= 7.07 \text{ N/mm}^2$$

Shear stress at the web just at the junction of the web and flange.

$$= \frac{100}{20} \times 7.07 = 35.25 \text{ N/mm}^2.$$

SOLVED PROBLEM 5.13

A beam has an I-section with top flange 75 mm × 20 mm, web 100 mm × 20 mm and bottom flange 150 mm × 40 mm. If tensile stress is not to exceed 30 N/mm² and compressive stress 80 N/mm², what is the maximum uniformly distributed load the beam can carry over a simply supported span of 6 m?

Solution:

Given data: Top flange = 75 mm × 40 mm, web = 100 mm × 20 mm, bottom flange = 150 mm × 40 mm, $f_t = 30$ N/mm², $f = 80$ N/mm² and span = 6 m.

Let $\bar{y}$ be the distance of centroid (neutral axis) from bottom fibre.

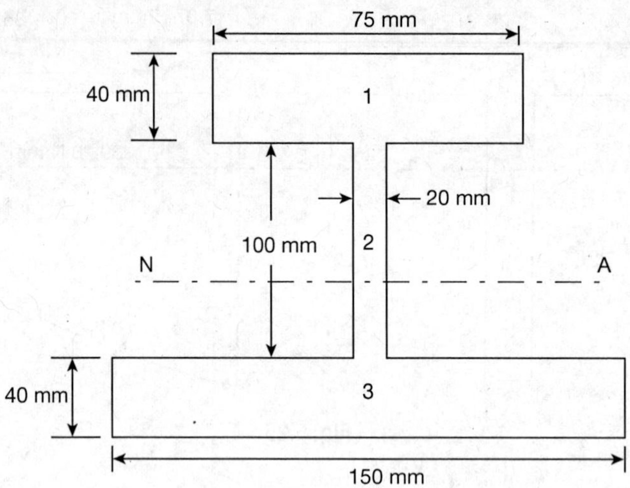

Fig. 5.26

$$\bar{y} = \frac{A_1 y_1 + A_2 y_2 + A_3 y_3}{A_1 + A_2 + A_3}$$

$$= \frac{75 \times 40 \times (100 + 40 + 40/2) + 20 \times 100 \times (40 + 100/2) + (40 \times 150 \times 10/2)}{75 \times 40 + 100 \times 20 + 150 \times 40}$$

$$= \frac{75 \times 40 \times (100 + 40 + 40/2) + 100 \times 20 \times (40 + 100/2) + (150 \times 40 \times 40/2)}{75 \times 40 + 100 \times 20 + 40 \times 150}$$

$= 62.73$ mm

$$I = \frac{1}{12} \times 70 \times 40^2 + 75 \times 40(160 - 70.91)^2 + \frac{1}{12} \times 20 \times 100^3 + 20 \times 100(90 - 70.91)^2$$

$$+ \frac{1}{12} \times 150 \times 40^3 + 150 \times 40 \times (20 - 70.91)^2$$

$= 58{,}134{,}670$ mm^4.

In the case of a simply supported beam, tension occurs at bottom and compression at top. Extreme distance of tension and compression fibres are

$$y_t = \bar{y} = 62.73 \text{ mm}$$

$$y_c = 180 - 62.73 = 117.27 \text{ mm}$$

Using the relation $$\frac{M}{I} = \frac{f}{y}$$

Moment carrying capacity considering tensile stress

$$= f_t \times \frac{I}{y_t}$$

$$= 30 \times \frac{I}{y_t}$$

$$= 30 \times \frac{58,134,670}{62.73} = 27,802,329.03 \text{ N} \cdot \text{mm}$$

$$= 27.8 \text{ kNm}$$

Moment carrying capacity considering compressive stress

$$= f_c \times \frac{I}{y_c}$$

$$= 80 \times \frac{58,134,670}{117.27}$$

$$= 39,658,681.67 \text{ Nmm}$$

$$= 39.65 \text{ kNm}$$

Actual moment carrying capacity of the section is smaller of the above two, i.e.,
Moment carrying capacity of the section = 25.9 kN m.
Maximum bending moment in a simply supported beam of span 6 m due to UDL of 10 kN/m

$$= w \times \frac{6^2}{8}$$

Equating to the moment carrying capacity

$$\frac{w \times 6^2}{8} = 24.59$$

$$\therefore\ w = 5.46 \text{ kN/m}$$

Maximum uniformly distributed load the beam can carry is 5.46 kN/m.

SOLVED PROBLEM 5.14

An I-Section with rectangular ends has the following dimensions: Flanges 100 mm × 10 mm and web 120 mm × 10 mm. If their section is subjected to a bending moment of 6 kN · m and a shearing force of 5 kN, find the maximum tensile and shear stress on it.

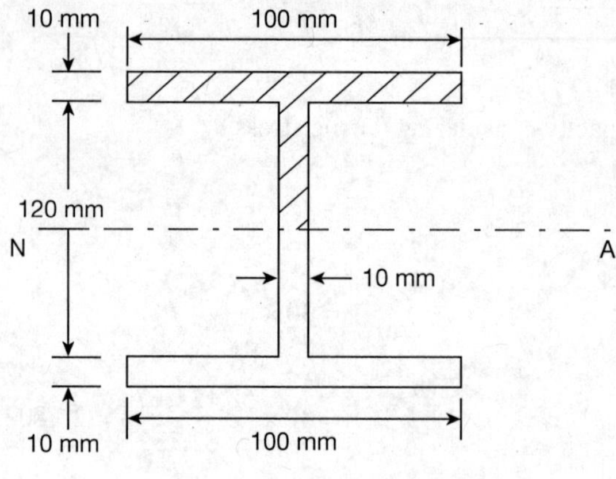

Fig. 5.27

Solution:

Given data: Flange = 100 mm × 10 mm, web = 120 mm × 10 mm, bending moment = 6 kN·m and shear force = 5 kN.

Moment of inertia of the section about NA,

$$I = \frac{100 \times 140^3}{12} - \frac{90 \times 120^3}{12}$$

$$= 9,906.666 \text{ mm}^4$$

$$I = 990.67 \times 10^{-8} \text{ m}^4$$

If f_t is the maximum tensile stress developed at outer face of the flange, then

$$f_t = \frac{M}{I} \times y_t$$

$$= \frac{6 \times 0.07}{990.67 \times 10^{-8}} \text{ kN/m}^2$$

$$= 42,395 \text{ kN/m}^2$$

Shear stress is obviously maximum in the web at NA we know that

$$\tau_{max} = \frac{SA\bar{y}}{Ib}$$

To find $A\bar{y}$, the shaded area above neutral axis is split into two rectangles 100 mm × 10 mm and 10 mm × 60 mm, then

$$A\bar{y} = (100 \times 10) \times (60 + 10/2) + (10 \times 60) \times \frac{60}{2}$$

$$= 83 \times 10^{-6} \text{ m}^3$$

$$\therefore \quad \tau_{max} = \frac{5 \times 83 \times 10^{-6}}{990.67 \times 10^{-8} \times 0.01}$$

$$= 4,189 \text{ kN/m}^2$$

SOLVED UNIVERSITY QUESTIONS

SOLVED PROBLEM 5.15

A steel plate of width 60 mm and thickness 10 mm is bent into a circular arc of radius 10 m. Determine the maximum stress induced and bending moment which will produce the maximum stress. Assume the modulus of elasticity of steel as 2×10^5 N/mm² (Anna Univ., June 2005, ME)

Solution:

Given data: b = 60 mm, thickness of plate, t = 10 mm, radius of curvature, R = 10 m and $E = 2 \times 10^5$ N/mm².

Let f_{max} be the maximum stress induced, and M be the bending moment.

Using bending equation,

$$f = \frac{E}{R} \times y$$

This equation gives the stress at a distance y from neutral axis. Stress will be maximum when y is maximum.

$$\therefore \quad y_{max} = \frac{t}{2} = \frac{10}{2} = 5 \text{ mm}$$

$$\therefore \quad f_{max} = \frac{E}{R} \times y_{max} = \frac{2 \times 10^5}{10 \times 10^3} \times 5 = 100 \text{ N/mm}^2$$

Using the relation,

$$M = \frac{E}{R} \times I$$

$$M = \frac{2 \times 10^5 \times 5{,}000}{10 \times 10^3}$$

$$M = 1 \times 10^5 \text{ N} \cdot \text{mm}$$

SOLVED PROBLEM 5.16

A uniform T-section beam is 100 mm wide and 150 mm deep with a flange thickness of 25 mm and a web thickness of 12 mm. If the limiting bending stress is 160 MPa in tension, find the maximum uniformly distributed load that the beam can carry over a simply supported span of 5 m. Also determine the corresponding maximum bending stress in compression (Anna Univ., Apr. 2008, ME).

Solution:

Given data: T-beam size: 100 mm × 150 mm, flange thickness = 25 mm, web thickness = 12 mm, f = 160 MPa and span = 5 m.

Centre of gravity of the section is found from

$$\bar{y} = \frac{A_1 y_1 + A_2 y_2}{A_1 + A_2}$$

$$= \frac{100 \times 25 \times \left(125 + \dfrac{25}{2}\right) + 125 \times 12 \times (125/2)}{100 \times 25 + 125 \times 12}$$

$$= 109.4 \text{ mm}$$

∴ Neutral axis at a distance of 109.44 mm from the bottom or 40.6 mm (150 – 109.4) from the top face.

Moment of inertia of the section about NA is

$$I = I_1 + I_2$$

$$= \frac{b_1 d_1^3}{12} + a_1 h_1^2 + \frac{b_2 d_2^3}{12} + a_2 h_1^2$$

$$I = \frac{100 \times 25^3}{12} + \left[(100 \times 25)\left(40.6 - \frac{25}{2}\right)^2\right] + \frac{12 \times 125^3}{12} + \left[(12 \times 125)\left(109.4 - \frac{125}{2}\right)^2\right]$$

$$I = 22.07 \times 10^6 \text{ mm}^4$$

For a simply supported beam, the maximum tensile stress occurs at the bottom and maximum compressive stress occurs at the top.

Let the uniformly distributed load be w N/m.

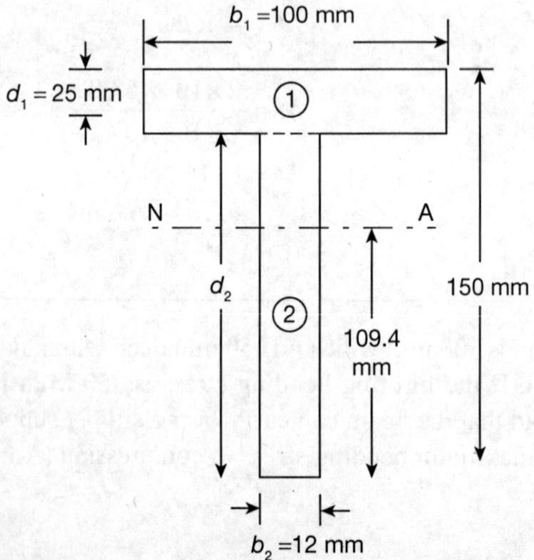

Fig. 5.28

Maximum bending moment, $M = \dfrac{wl^2}{8}$

$$= \frac{w \times 5^2}{8} = \frac{25}{8} w \text{ N}\cdot\text{m}$$

$$= 3.125 w \text{ N}\cdot\text{m}$$

But $M = \dfrac{f}{y} \times I$

$$= \frac{160}{109.4} \times 22.07 \times 10^6$$

$$= 32,277,879.3 \text{ Nm}$$

Equating both the moments, we have

$$3.125\, w = 32,277,879.3$$

$\therefore$ $w = 10,323.92$ N/m

Maximum compressive stress, f_c is

$$f_c = \frac{M}{I} \times y$$

$$= \frac{32,277,879.3}{22.07 \times 10^6} \times 40.6$$

$$= 59.40 \text{ N/mm}^2$$

SOLVED PROBLEM 5.17

A beam of rectangular cross section 50 mm wide and 150 mm deep is used as a cantilever 6 m long and subjected to a uniformly distributed load of 2 N/m over the centre length. Determine the bending stress at 50 mm from the top fibre, at the mid-span of the beam. Also, calculate the maximum bending stress (Anna Univ., Nov. 2008, ME).

Solution:

Given date: Rectangular cross section = 50 mm × 150 mm, span = 6 m and UDL = 2 kN/m.

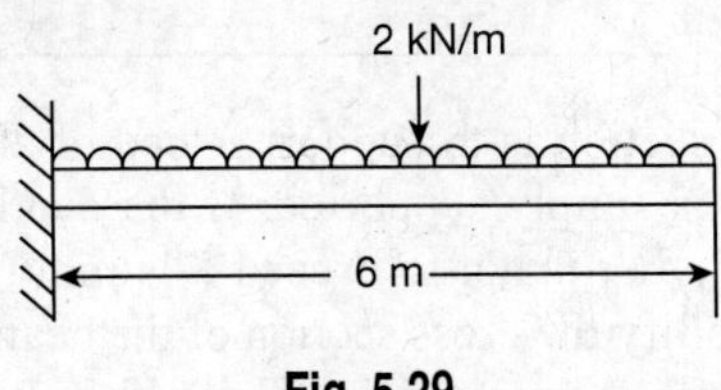

Fig. 5.29

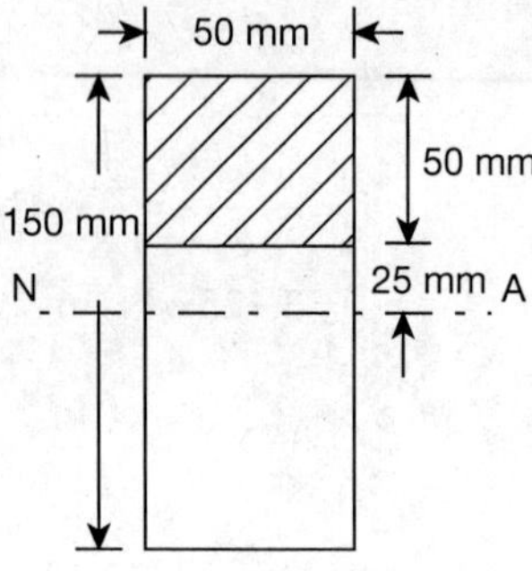

Fig. 5.30

Maximum bending moment $= w \times l \times \dfrac{l}{2}$

$$= 2 \times 6 \times \frac{6}{2} = 36 \text{ kN} \cdot \text{m}$$

Bending moment at mid-span $= 2 \times 3 \times \dfrac{3}{2} = 9 \text{ kN} \cdot \text{m}$

Moment of inertia, $I = \dfrac{bd^3}{12} = \dfrac{50 \times 150^3}{12}$

$$= 14.06 \times 10^6 \text{ mm}^4$$

$$f_{max} = \frac{M_{max}}{I} y_{max}$$

$$= \frac{36 \times 10^6}{14.06 \times 10^4} \times 75 = 192 \text{ N/mm}^2$$

Bending stress at 50 mm from the top fibre, $f_{25} = \dfrac{\text{Moment at mid-span}}{I} \times y$

$$= \frac{9 \times 10^6}{14.06 \times 10^6} \times 25$$

$$f_{25} = 16.00 \text{ N/mm}^2$$

SOLVED PROBLEM 5.18

A timber beam of rectangular section is to support a load of 20 kN uniformly distributed over a span of 3.6 m when the beam is simply supported. If the depth of the section is to be twice the breadth and the stress in the timber is not to exceed 7 N/mm², find the depth and breadth of the cross section. How will you modify the cross section of the beam, if it carries concentrated load of 30 kN placed at the mid-span with the same ratio of breadth to depth (Anna Univ., May 2006, ME).

Solution:

Given data: UDL = 20 kN/m, span = 3.6 m, depth = 2 × (twice the breadth), $f \ngtr 7$ N/mm² and concentrated load = 30 kN.

Dimension of beam with UDL

Maximum stress = 7 N/mm²

$$\text{Section modulus} = \frac{bd^2}{6} = \frac{2\times(2b)^2}{6} = \frac{2b^3}{6}\ \text{mm}^3$$

$$\left.\begin{array}{r}\text{Maximum bending moment of a simply}\\ \text{supported beam with UDL}\end{array}\right\} = \frac{wl^2}{8} = \frac{Wl}{8}$$

$$M_{\text{UDL}} = \frac{20\times1{,}000\times3.6}{8} = 9{,}000\ \text{N}\cdot\text{m}$$

$$M_{\text{UDL}} = 9{,}000\times10^3\ \text{N}\cdot\text{mm}$$

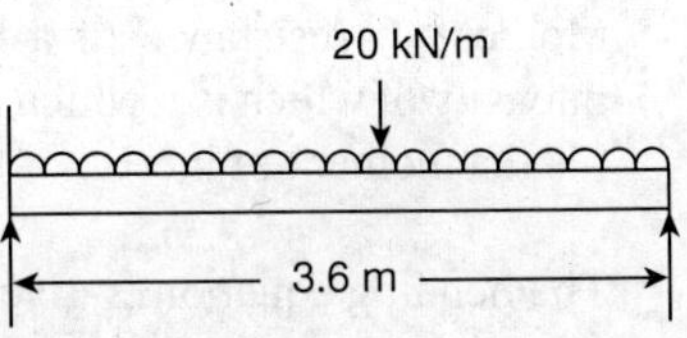

Fig. 5.31

$$M_{\text{UDL}} = f_{\max}\times Z$$

But,

$$9{,}000\times10^3 = 7\times\frac{2b^3}{6}$$

Solving,

$$b = 124.4\ \text{mm}$$

$$d = 2\times124.44\ \text{mm} = 248.8\ \text{mm}$$

Cross section of beam:

Width = 124.4 mm

Depth = 248.8 mm

Dimension of the beam with 30 kN central load

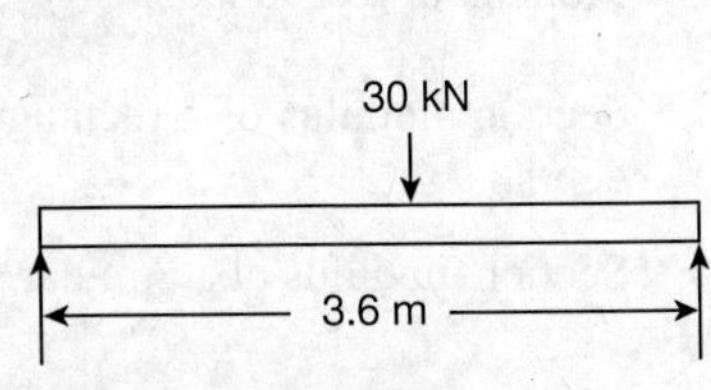

Fig. 5.32

$$\text{Maximum bending moment, } M_{\text{cL}} = \frac{Wl}{4}$$

$$= \frac{30\times10^3\times3.6}{4}$$

$$= 27{,}000\ \text{N}\cdot\text{m}$$

But

$$M_{\text{cL}} = f_{\max}\times Z$$

$$27{,}000 = 7\times\frac{2b^3}{6}$$

Solving,

$$b = 179.52\ \text{mm}$$

$$d = 2\times179.52\ \text{mm} = 359.04\ \text{mm}$$

Cross section of beam:

Width = 179.52 mm

Depth = 359.04 mm

SALIENT POINTS

- When a beam is subjected to a loading system or by a force couple acting on a plane passing through the axis, then the beam deforms. The deformation is called as bending of a beam.
- In a loaded cantilever beam, the upper fibres will be tensioned and lower fibres will be compressed.
- In a loaded simply supported beam, the lower fibres will be tensioned and the upper fibres are compressed.
- When a loaded beam is unaccompanied by shear force and subjected only to a constant bending moment, then the beam is said to be in a state of simple bending.
- When a beam is loaded longitudinally, stresses are induced in its cross section.
- The equation which connects radius of curvature of the beam, bending stresses, bending moment and its cross-sectional dimensions is known as bending equation.
- Neutral axis is an unstressed plane of a beam which will pass through the centroid of the section.
- Moment of resistance of a beam, M, produces a uniformly varying longitudinal stress the intensity of which is p which is acting at a distance y from the neutral axis, i.e., $p = My/I$ where I is the moment of inertia.
- The bending equation is given as $\frac{M}{I} = \frac{p}{y} = \frac{E}{R}$

  where E is the modulus of elasticity and R is the radius of curvature.
- Modulus of section or section modulus, $Z = \frac{I}{y_{max}}$
- Moment of resistance, $M = f \times Z$ where f is the stress.
- Section modulus of a rectangular section of a beam is $\frac{bd^2}{6}$ where b is the width and d is the depth.
- Section modulus of a circular section of a beam is $\pi d^3/32$ where d is the diameter of circle.
- Section modulus of a hollow circular section of a beam is $\frac{\pi}{32}\left(\frac{d_0^4 - d_i^4}{d_0}\right)$ where d_0 is the external diameter and d_i is the internal diameter.
- When a beam is made up of more than one material, then it is known as a composite beam or flitched beam.
- When a beam is subjected to bending moment, normal tensile and compressive stresses prevail along the cross section of the beam. In addition to the above stresses, transverse section are subjected to vertical shearing stresses. Vertical shear stresses are always accompanied by complementary shear stresses of equal intensity in a perpendicular plane.
- Shear stress varies along longitudinal planes since the perpendicular section resists the shear force.
- When transverse stress increases to the ultimate value due to faulty design, then the beam may fail. It is said that the failure has occurred due to shear flow.

QUESTIONS

1. What do you understand by moment of resistance of a beam subjected to a bending moment.?
2. Sketch the bending stress distribution in a simply supported beam and cantilever beam indicating the nature of stress.
3. What do you understand by modulus of a beam? Write down expression for section modulus of the following sections.
   (i) Rectangular, (ii) Circular section, and (iii) Hollow circular section.
4. What do you understand by shear stresses in beam? Show that for a rectangular section the maximum shear stress is 1.5 times the average stress.
5. A simply supported beam of span 10 m carries a central concentrated load of 20 kN. If the cross section of the beam is a rectangle 40 mm × 50 mm, what is the maximum shear stress set up in the beam?
6. A rectangular section 30 mm × 40 mm is subjected to a shear force of 12 kN. What is the maximum shear stress set up in a beam?
7. A timber beam is 125 mm wide and 150 mm deep. Two steel plates each 125 mm wide and 10 mm thick are provided. Case (i) at top and bottom and case (ii) at the sides. Determine the moment of resistance for each and compare them. $E_s = 2 \times 10^5$ N/mm^2, $E_w = 0.16 \times 10^5$ N/mm^2 and allowable stress in steel is 140 N/mm^2.
8. A beam is subjected to a bending stress of 5 N/mm^2 and the section modulus is 3,530 cm^3. What is the moment of resistance of the beam?
9. A brass strip of 75 mm wide and 30 mm thick is bent into an arc of radius 60 m. What is the maximum stress developed in the strip of $E_{brass} = 1 \times 10^5$ mN/m^2?
10. A rectangular timber beam of 3 m is simply supported at the ends and acted upon by a uniformly distributed load of 25 kN/m. If the depth of the beam is twice the width and the stress in timber not to exceed 7.5 N/mm^2 find the depth of the cross section. How will you modify the cross section if the beam is a cantilever beam?

6

Torsion

LEARNING OBJECTIVES

6.1 INTRODUCTION

In certain field applications, like in work-shops and factories, a turning force is applied to transmit energy by rotation. For example, this turning force is applied to the rim of a pulley or through a key to the shaft or at any required point from the axis of shaft. The product of this turning force and the corresponding distance is known as the torque. This distance is one between the point of application of the force and axis of the shaft. Torque is also referred to as turning moment or twisting moment. Thus, a member is considered to be under torsion if it transmits the moments of a couple in a direction normal to the plane of the couple.

Torsion is one of the four basic modes by which loads are transmitted from one portion of structural or machine member to another. Common simple examples of members subjected to torsion are: door knobs, screw drivers, electric motors, etc. Torsion induces shear stresses and strains in the material.

6.2 TORSION OF SOLID CIRCULAR SHAFT

6.2.1 Assumptions

In the development of a torsion formula for a circular shaft, the following assumptions are made:

(i) Material of the shaft is homogeneous throughout the length of the shaft.
(ii) Shaft is straight and of uniform circular cross section over its length.
(iii) Torsion is constant along the length of the shaft.
(iv) Cross section of the shaft which are plane before torsion remain plane after torsion.
(v) Radial lines remain radial during torsion.
(vi) Stresses induced during torsion are within the elastic limit.

The above assumptions are reasonably justified as long as the torque applied and the angle of twist are small.

The stresses induced at any point in the cross section of the shaft is one of pure shear.

6.2.2 Derivation

Let us consider a circular shaft be fixed at one end and a torque be applied at the other end along a plane perpendicular to the longitudinal axis of the beam (Figure 6.1).

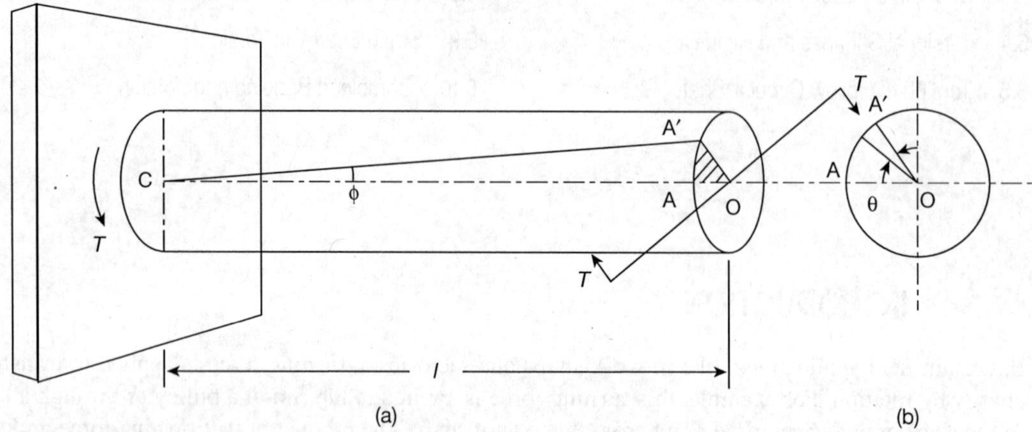

Fig. 6.1 Shaft under torque.

Let R be the radius of the circular shaft, l be the length of shaft and T be the torque applied.

Due to the application of the torque, every cross section of the shaft is subjected to shear stress. Thus, the line CA on the surface of the shaft had deformed to the position CA′ and the line OA to OA′ (Figure 6.1).

Let ∠ACA′ be ϕ and let ∠AOA′ be θ. It is known that shear strain is equal to deformation per unit length.

$$\therefore \quad \text{Shear strain at the outer surface} = \frac{\text{Distortion at the outer surface}}{\text{Length of the shaft}}$$

$$= \frac{AA'}{l} = \tan\phi \approx \phi \text{ (in radians) } (\because \phi \text{ is very small}),$$

But

$$AA' = OA \times \theta$$

i.e.,

$$AA' = R \times \theta \qquad \text{(where } \theta \text{ is in radians)}$$

$$\therefore \quad \phi = \frac{AA'}{l} = \frac{R\theta}{l} \tag{6.1}$$

If f_s is the intensity of shear stress on the outer surface and G is the modulus of rigidity of the shaft, then the shear strain

$$\phi = \frac{f_s}{G} \tag{6.2}$$

Comparing Equations (6.1) and (6.2) we have

$$\frac{f_s}{G} = \frac{R\theta}{l} \tag{6.3}$$

or

$$\frac{f_s}{R} = \frac{G\theta}{l} \tag{6.4}$$

For a given shaft subjected to torque T, the values of G, θ and l are constants. Thus, the shear induced is proportional to the radius R.

i.e.,

$$f_s \; \alpha \; R \tag{6.5}$$

or

$$\frac{f_s}{R} = \text{constant} \tag{6.6}$$

If q is the shear stress induced at a radius r from the centre of the shaft, then

$$\frac{f_s}{R} = \frac{q}{r} \tag{6.7}$$

Comparing Equations (6.4) and (6.7) we have

$$\frac{f_s}{R} = \frac{q}{r} = \frac{G\theta}{l} \tag{6.8}$$

It may be observed from Equation (6.5) that the shear stress has linear variation with radius at any cross section and the shear stress is maximum at the outer surface and zero at the axis of the shaft.

Thus, it may be stated that the intensity of shear stress induced at any point in the cross section of a circular shaft subjected to a couple is proportional to its distance from the centre.

6.3 STRENGTH OF A SOLID CIRCULAR SHAFT

Maximum torque or horse power that can be transmitted by a shaft is referred to as the strength of a shaft.

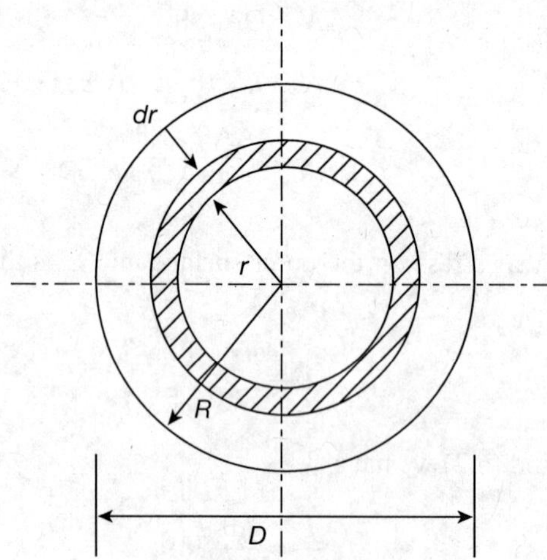

Fig. 6.2 Solid circular shaft.

Let a circular shaft subjected to a torque T be considered and let an elemental ring of area dA be taken (Figure 6.2). Then

$$dA = 2\pi r dr \tag{6.9}$$

From Equation (6.7), the shear stress at the radius r is given as

$$q = f_s \frac{r}{R} \tag{6.10}$$

Shear force on the elemental ring = (Shear stress at the radius r) × (Area of the elemental ring)

$$F = q\, dA$$

$$dF = \left(f_s \frac{r}{R} \right) \times (2\pi r dr)$$

$$dF = \left(\frac{f_s 2\pi r^2 dr}{R} \right)$$

Now, the torque (dT) due to this shear force on the elemental ring (dF)

= (Shear force on the elemental ring) (Distance of the ring from the centre)

$$= \left(\frac{f_s}{R} 2\pi r^2 dr \right) \times r$$

$$= \frac{f_s}{R} 2\pi r^3 dr \tag{6.11}$$

∴ Total torque, $$T = \int_0^R \frac{f_s}{R} 2\pi r^3 dr$$

i.e., $$= 2\pi \frac{f_s}{R}\left[\frac{r^4}{4}\right]_0^R$$

$$T = \frac{\pi}{16} f_s D^3 \tag{6.12}$$

6.4 TORSIONAL STIFFNESS AND RIGIDITY

In Equation (6.12) for torque, T, f_s may be substituted from Equation (6.4) and rewritten as

$$T = \frac{\pi}{16}\left(\frac{D}{2}\right)\frac{G\theta}{l} D^3$$

i.e., $$\frac{T}{\frac{\pi D^4}{32}} = \frac{G\theta}{l} \tag{6.13}$$

i.e., $$\frac{T}{I_p} = \frac{G\theta}{l} \tag{6.14}$$

where $I_p = \pi D^4/32$ is the polar moment of inertia of a circular cross section.
Thus, combining Equations (6.8) and (6.14) we have

$$\frac{T}{I_p} = \frac{q}{r} = \frac{G\theta}{l} \tag{6.15}$$

Thus, the torque required for unit twist, i.e., $T(\theta)$ is called the *torsional stiffness.*
From the above relation, $\theta = \frac{Tl}{GI_p}$

The quantity $\frac{GI_p}{l}$ is known as *torsional rigidity* and is represented by k.

From the above relation, $k = \frac{GI_p}{l} = \frac{T}{\theta}$

6.5 TORSION OF HOLLOW CIRCULAR SHAFT

Let a hollow circular shaft subjected to a torque T be considered (Figure 6.3). An elemental ring of area dA is considered. Then

$$dA = 2\pi r dr \tag{6.16}$$

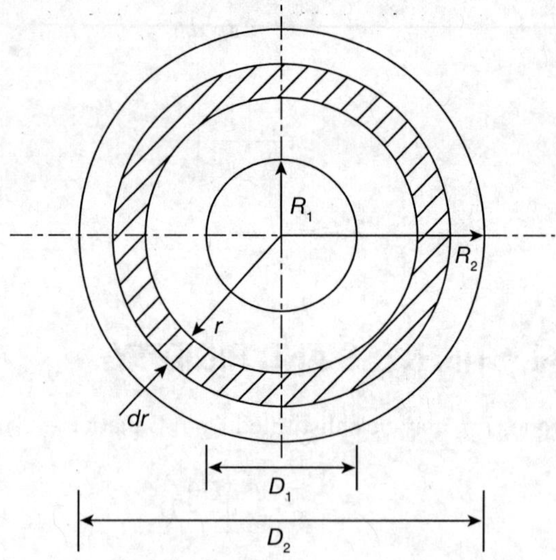

Fig. 6.3 Hollow circular shaft.

Let R_e and R_i be the external and internal radius and D_e and D_i be the external and internal diameter of the hollow shaft, respectively.

From Equation (6.7), the shear stress at the radius r is

$$q = \frac{f_s}{R_e} r \qquad (\because R = R_e)$$

Shear force dF as the elemental ring

$$= qdA$$

$$= \left(\frac{f_s}{R_e} r\right)(2\pi r dr)$$

$$= 2\pi \frac{f_s}{R_e} r^2 dr \tag{6.17}$$

Now the torque, dT, due to this shear force as the elemental ring

$$dT = \left(2\pi \frac{f_s}{R_e} r^2 dr\right) r$$

$$= 2\pi \frac{f_s}{R_e} r^3 dr$$

$\therefore$ Total torque,

$$T = \int_{R_i}^{R_e} 2\pi \frac{f_s}{R_e} r^3 dr$$

$$= 2\pi \frac{f_s}{R_e} \left[\frac{r^4}{4} \right]_{R_i}^{R_e}$$

i.e.,
$$T = 2\pi \frac{f_s}{R_e} \left[\frac{R_e^4 - R_i^4}{4} \right] \tag{6.18}$$

or
$$T = \frac{\pi}{16} f_s \left[\frac{D_e^4 - D_i^4}{D_e} \right] \tag{6.19}$$

6.6 POWER TRANSMITTED BY SHAFT

The main purpose of a shaft is to transmit its power to another member. Let a rotating shaft transmitting power from one of its ends to another be considered.

Let N be the revolutions per minute (rpm), T be the average torque N·m and ω be the angular speed of the shaft.

$$\text{Work done per minute} = (\text{Force}) \times (\text{Distance})$$
$$= (\text{Average torque}) \times (\text{Angular displacement})$$
$$= T \times 2\pi N/60$$

$\therefore$
$$\text{Power, } P = \frac{2\pi NT}{60} \text{ watts} \tag{6.20a}$$

also
$$P = T\omega \tag{6.20b}$$

where
$$\omega = \frac{2\pi N}{60}$$

6.7 POLAR MODULUS

Polar modulus is defined as the ratio of the polar moment of inertia to the radius of the shaft. It is also called as *torsional section modulus.* It is denoted by Z_p.

i.e.,
$$Z_p = \frac{I_p}{R} \tag{6.21}$$

1. For solid shaft

$$I_p = \frac{\pi}{32} D^4$$

$$Z_p = \frac{\frac{\pi}{32} D^4}{D/2} = \frac{\pi}{16} D^3 \tag{6.22}$$

2. For hollow shaft

$$Z_p = \frac{\frac{\pi}{32}\left(D_e^4 - D_i^4\right)}{R}$$

$$= \frac{\frac{\pi}{32}\left(D_e^4 - D_i^4\right)}{D_e/2}$$

$$= \frac{\pi}{16D_e}\left(D_e^4 - D_i^4\right) \quad (6.23)$$

SOLVED PROBLEM 6.1

Determine the diameter of a solid shaft which transmits 300 kW at 250 rpm. The maximum shear stress must not exceed 30 N/mm² and the twist should not be more than 1° in a shaft length of 2 m. $G = 1 \times 10^5$ N/mm².

Solution:

Given data: Power transmitted, P = 300 kW, speed, N = 250 rpm, f_s = 30 N/mm², l = 2 m, $G = 1 \times 10^5$ N/mm² and twist, $\theta = 1°$

Using,
$$P = \frac{2\pi NT}{60}$$

$$300 \times 1{,}000 = \frac{2 \times \pi \times 250 \times T}{60}$$

∴
$$T = \frac{300 \times 60 \times 1{,}000}{2 \times \pi \times 250} = 11{,}459 \text{ N}\cdot\text{m}$$

Shear stress consideration

Using,
$$T = \frac{\pi}{16} \times f_s \times D^3$$

$$11{,}459 = \frac{\pi}{16} \times 30 \times 10^6 \times D^3$$

Solving
$$D = 0.125 \text{ m}$$
$$= 125 \text{ mm}$$

Twist consideration

Using
$$\frac{T}{I_p} = \frac{G\theta}{l}$$

$$\frac{11{,}459}{\frac{\pi}{32} \times D^4} = 1 \times 10^{11} \times \frac{\pi}{180} \qquad \left(\because 1° = \frac{\pi}{180} \text{ radians}\right)$$

Solving $D = 0.1075$ m

i.e., $D = 107.5$ mm.

Suitable diameter of the shaft is the higher of the two values.

i.e., Diameter of the shaft = 125 mm.

SOLVED PROBLEM 6.2

Design a suitable diameter for a circular shaft required to transmit 75 kW at 180 rpm. The shear stress in the shaft is not to exceed 70 N/mm² and the maximum torque exceeds the mean by 40%.

Solution:

Given data: Power transmitted, $P = 75$ kW, speed, $N = 180$ rpm, $f_s = 70$ N/mm² and $T_{max} = 1.40\ T_{mean}$

Using
$$P = \frac{2\pi NT}{60}$$

$$75 \times 1{,}000 = \frac{2 \times \pi \times 180 \times T_{mean}}{60}$$

Solving
$$T_{mean} = 3{,}978\ \text{N}\cdot\text{m} = 3{,}978 \times 10^3\ \text{N}\cdot\text{mm}$$

$$T_{max} = 1.4 \times T_{mean} = 1.40 \times 3{,}978 \times 10^3\ \text{N}\cdot\text{mm}$$

$$T_{max} = 5{,}570 \times 10^3\ \text{N}\cdot\text{mm}$$

Using,
$$T = \frac{\pi}{16} \times f_s \times D^3$$

$$5{,}570 \times 10^3 = \frac{\pi}{16} \times 70 \times D^3$$

Solving,
$$D = 74.01\ \text{mm}$$

Diameter of circular shaft $= 74.01$ mm.

SOLVED PROBLEM 6.3

The working condition to be satisfied by a shaft transmitting power are (i) the shear stress must not exceed 45 N/mm² and (ii) that the shaft must not twist more than 1.5° on a length of 16 diameters. Compute the actual working stress and the diameter of the shaft to transmit 1 MW at 200 rpm. Take $G = 80{,}000$ N/mm².

Solution:

Given data: $f_s = 45$ N/mm², twist, $\theta = 1.5°$, power transmitted, $P = 1$ MW, speed, $N = 200$ rpm and $G = 80{,}000$ N/mm².

Shear stress condition

Using $$T = \frac{\pi}{16} f_s D^3$$

$$= \frac{\pi}{16} \times 45 \times D^3 = 8.83 D^3 \text{ N}\cdot\text{mm}$$

Twist condition

Using $$T = I_p \frac{G\theta}{l}$$

$$= \frac{\pi}{32} D^4 \times 80{,}000 \left(1.5 \times \frac{\pi}{180}\right) \frac{1}{16D}$$

$$= 12.85\, D^3 \text{ N}\cdot\text{mm}$$

Adopting smaller value of *T*, [i.e., condition (i)], it follows that the twist is less than the permitted by condition (ii)

Using, $$P = \frac{2\pi NT}{60}$$

$$1 \times 10^6 = \frac{2\pi \times 200 \times 8.83 D^3}{60 \times 1{,}000}$$

Solving $$D = 175.5 \text{ mm}$$

SOLVED PROBLEM 6.4

Determine the diameter of a solid shaft which will transmit 112 kW at 200 rpm. Also, determine the length of the shaft if the twist must not exceed 1.5° over the entire length. The maximum shear stress is limited to 50 N/mm². Take the modulus of rigidity = 8×10^4 N/mm².

Solution:

Given data: Power transmitted, $P = 112$ kW, speed, $N = 200$ rpm, twist, $\theta = 1.5°$, $f_s = 50$ N/mm² and $G = 8 \times 10^4$ N/mm².

Using $$P = \frac{2\pi NT}{60}$$

$$112 \times 1{,}000 = \frac{2\pi \times 200 \times T}{60}$$

Solving, $$T = 5{,}347.6 \text{ N}\cdot\text{m} = 5{,}347.6 \times 10^3 \text{ N}\cdot\text{mm}.$$

Using, $$T = \frac{\pi}{16} f_s D^3$$

$$5{,}347.6 \times 10^3 = \frac{\pi}{16} \times 50 \times D^3$$

Solving $$D = 81.56 \text{ mm}$$

Using $$\frac{f_s}{R} = \frac{G\theta}{l}$$

$$\frac{50}{\frac{81.56}{2}} = \frac{8\times10^4\times1.5\times\pi/180}{l}$$

Solving, length $$l = \frac{8\times10^4\times1.5\times\pi}{1.22\times180}$$

$$= 1,715.85 \text{ mm}$$

Length of shaft $$= 1,715.85 \text{ mm}$$

SOLVED PROBLEM 6.5

In a tensile test, a test piece of 25 mm diameter 200 mm gauge length is stretched by 0.0975 mm under a pull of 50 kN. In a torsion test, the same rod is twisted by 0.025 radian over a length of 200 mm when a torque of 400 kN·m was applied. Calculate Poisson's ratio and the three elastic modulus of the material.

Solution:

Given data: Diameter of bar = 25 mm, gauge length, l = 200 mm, δl = 0.0975 mm, tensile load, P = 50 kN, twist = 0.025 radians, rod length = 200 mm and T = 400 kN·m.

Tensile stress $$= \frac{\text{Load}}{\text{Area}} = \frac{50\times10^3}{\frac{\pi}{4}\times25^2} = 101.85 \text{ N/mm}^2$$

Tensile strain $$= \frac{0.0975}{200} = 4.875\times10^{-4}$$

∴ Young's modulus $$= \frac{101.85}{4.875\times10^{-4}} = 2.089\times10^5 \text{ N/mm}^2$$

Using $$\frac{T}{I_p} = \frac{G\theta}{l}$$

∴ $$G = \frac{Tl}{I_p} = \frac{(400\times10^3)\times200}{0.025\times\frac{\pi}{32}\times(25)^4} = 0.834\times10^5 \text{ N/mm}^2$$

Using $$E = 2G(1+1/m)$$

i.e., $$0.834\times10^5 = \frac{m\times2.089\times10^5}{2(m+1)}$$

$$1/m = 0.252$$

Using $$E = 3K(l - 2/m)$$

i.e., $$K = \frac{E}{3(1-2/m)} = \frac{2.089\times 10^5}{3(1-2\times 0.252)}$$

$$K = 1.404\times 10^5 \text{ N/mm}^2$$

SOLVED PROBLEM 6.6

A solid shaft of 20 cm diameter is used to transmit a required torque. Compute the maximum torque transmitted by the shaft if the maximum shear stress induced in the shaft is 50 N/mm². What would be the torque by the same shaft if a circular bore of 5 cm diameter is made centrally throughout the length of the shaft?

Solution:

Given data: Diameter of shaft = 20 cm, f_s = 50 N/mm² and diameter of bore = 5 cm.

Torque of solid circular shaft $T = \frac{\pi}{16} f_s D^3 = \frac{\pi}{16} \times \frac{50\times(20\times 10)^3}{1,000}$

$$= 78,539 \text{ N}\cdot\text{m}$$

Torque of hollow circular shaft $T = \frac{\pi}{16} f_s \left[\frac{D_e^4 - D_i^4}{D_e}\right]$

$$= \frac{\pi}{16}\times 50\times\left[\frac{(200)^4-(50)^4}{200}\right]\times\frac{1}{1,000}$$

$$= 78,233 \text{ N}\cdot\text{m}$$

SOLVED PROBLEM 6.7

A hollow shaft has an external diameter of 300 mm and a bore of 150 mm diameter. When transmitting power, it is found that the angle of twist is 0.5° in a length of 4 m. Find (i) the power transmitted if the speed is 300 rpm and (ii) the maximum shear stress in the shaft. $G = 0.8\times 10^5$ MPa.

Solution:

Given data: D_e = 300 mm, D_i = 150 mm, twist, θ = 0.5°, length, l = 4 m and speed, N = 300 rpm.

Using $$\frac{f_s}{R} = \frac{G\theta}{l}$$

i.e., $$f_s = \frac{RG\theta}{l} = \frac{300}{2}(0.8\times 10^5)\times\left(0.5\times\frac{\pi}{180}\right)\times\frac{1}{4\times 1,000}$$

$$= 26.17 \text{ N/mm}^2$$

Maximum shear stress = 26.17 N/mm²

Using
$$T = \frac{\pi}{16} \times f_s \left[\frac{D_e^4 - D_i^4}{D_e}\right]$$
$$= \frac{\pi}{16} \times 26.17 \left[\frac{300^4 - 150^4}{300}\right]$$
$$= 130{,}067.45\ \text{N}\cdot\text{m}$$

Power,
$$P = \frac{2\pi NT}{60}$$
$$= \frac{2 \times \pi \times 300 \times 130{,}067.45}{60 \times 1{,}000}$$
$$= 4{,}086.189\ \text{kW}$$

SOLVED PROBLEM 6.8

A shaft is required to transmit a power of 300 kW running at a speed of 120 rpm. If the shear strength of the shaft material is 70 N/mm², design a hollow shaft with inner diameter equal to 0.75 times the outer diameter.

Solution:

Given data: Power transmitted, $P = 300$ kW, speed $N = 120$ rpm, $f_s = 70$ N/mm² and $D_i = 0.75\ D_e$.

Using
$$P = \frac{2\pi NT}{60}$$
$$300 \times 10^3 = \frac{2 \times \pi \times 120 \times T}{60}$$

Solving
$$T = 23{,}873\ \text{N}\cdot\text{m}$$

i.e,
$$T = 23{,}873 \times 10^3\ \text{N}\cdot\text{mm}$$

For hollow shaft torque,
$$T = \frac{\pi}{16} \times f_s \left[\frac{D_e^4 - D_i^4}{D_e}\right]$$

i.e.,
$$23{,}873 \times 10^3 = \frac{\pi}{16} \times 70 \left[\frac{D_e^4 - (0.75 D_e)^4}{D_e}\right]$$
$$= \frac{\pi}{16} \times 70 \left[\frac{D_e^4 - 0.3164 D_e^4}{D_e}\right]$$
$$= \frac{\pi}{16} \times 70 \times 0.683 D_e^3$$

Solving
$$D_e = 136\ \text{mm}$$

Inner diameter,
$$D_i = 0.75 \times 136 = 102\ \text{mm}$$

SOLVED PROBLEM 6.9

The internal and external diameters of a hollow shaft is in the ratio of 2:3. The hollow shaft has to transmit a 400 kW power at 120 rpm. The maximum expected torque is 15% greater than the mean. The shear stress is not to exceed 50 N/mm^2 and the twist is limited to 2° over a length of 4 m. Estimate the design cross section of the shaft which would satisfy the shear stress and twist conditions. Take $G = 0.85 \times 10^5$ N/mm^2.

Solution:

Given data: $D_i : D_e :: 2:3$, power, $P = 400$ kW, speed, $N = 120$ rpm, $T_{max} = 1.15\ T_{mean}$, $f_s = 50$ N/mm^2, twist, $\theta = 2°$, length, $l = 4$ m and $G = 0.85 \times 10^5$ N/mm^2.

$$\frac{D_e}{D_i} = \frac{3}{2}$$

i.e.,

$$D_i = \frac{2}{3} D_e$$

Using,

$$P = \frac{2\pi NT}{60}$$

$$400 \times 1{,}000 = \frac{2\pi \times 120 \times T}{60}$$

Solving,

$$T = \frac{400 \times 1{,}000 \times 60}{2 \times \pi \times 120} = 31{,}830.98\ \text{N}\cdot\text{m}$$

$$\begin{aligned} T_{max} &= 1.15 T_{mean} \\ &= 1.15 \times 31{,}830.98 \\ &= 36{,}605.64\ \text{N}\cdot\text{m} \\ &= 36{,}605.64 \times 10^3\ \text{N}\cdot\text{mm} \end{aligned}$$

Shear stress condition

Using,

$$T = \frac{\pi}{16} f_s \left[\frac{D_e^4 - D_i^4}{D_1}\right]$$

$$36{,}605.64 \times 10^3 = \frac{\pi}{16} \times 50 \left[\frac{D_e^4 - (0.67 D_e)^4}{D_e}\right]$$

i.e.,

$$36{,}605.64 \times 10^3 = \frac{\pi}{16} \times 50 \times D_e^3 \times 0.798$$

Solving,

$$D_e = 167.2\ \text{mm} \approx 168.0\ \text{mm}$$

∴

$$D_i = \frac{2}{3} \times 168 = 112.0\ \text{mm}$$

Twist condition

Using,
$$\frac{T}{I_p} = \frac{G\theta}{l}$$

$\therefore$
$$I_p = \frac{Tl}{G\theta}$$

$$\frac{\pi}{32}\left[D_e^4 - D_i^4\right] = \frac{36,606.64\times10^3\times4\times1,000}{0.85\times10^5\times2\times\frac{\pi}{180}}$$

$\therefore$
$$\left[D_e^4 - (0.67D_e)^4\right] = \frac{36,606.64\times10^3\times4\times1,000\times32}{\pi\times0.85\times10^5\times2\times\frac{\pi}{180}}$$

Solving, $D_e = 158.5$ mm ≈ 159 mm

$\therefore$
$$D_i = \frac{2}{3}\times159 = 106 \text{ mm}$$

Design cross sections of hollow shaft is $D_e = 168$ mm and $D_i = 112$ mm which satisfy both the conditions.

SOLVED PROBLEM 6.10

A solid circular shaft 200 mm in diameter has the same cross-sectional area as a hollow circular shaft of the same material with inside diameter of 150 mm. For the same maximum shear stress, determine the ration of torque transmitted by the hollow shaft to that by the solid shaft. Also, compare the angle of twist in the above shaft for equal length and same maximum shear stress. From the above results, what will be your conclusion regarding strength and stiffness of two shafts?

Solution:

Given data: Solid shaft diameter = 200 mm, D_i of hollow shaft = 150 mm and $A_s = A_h$.
Let D_e be the outer diameter of the hollow shaft:

$$\text{As } A_s = A_h$$

$$\frac{\pi}{4}\times200^2 = \frac{\pi}{4}\left[D_e^2 - 150^2\right]$$

$\therefore$
$$D_e = 250 \text{ mm}$$

Torque for solid shaft,
$$T_s = \frac{\pi}{16}\times f_s\times D^3$$
$$= \frac{\pi}{16}\times f_s\times(200)^3$$

Torque for hollow shaft, $$T_h = \frac{\pi}{16} \times f_s \left[\frac{250^4 - 150^4}{250} \right]$$

$$\therefore \quad \frac{T_h}{T_s} = \frac{\frac{\pi}{16} \times f_s \left[\frac{250^4 - 150^4}{250} \right]}{\frac{\pi}{16} \times f_s \times 200^3}$$

i.e., $$\frac{T_g}{T_s} = 1{,}069 \approx 1.70$$

Using $$\frac{f_s}{R} = \frac{G\theta}{l}$$

i.e., $$\theta = \frac{f_s l}{RG}$$

As f_s, l and G are same for both the shafts

$$\frac{\theta_h}{\theta_s} = \left[1/(250/2)\right]/\left[1/(250/2)\right]$$

$$\frac{\theta_h}{\theta_s} = 1$$

Conclusion:

1. Hollow shaft is 1.7 times stronger than the solid shaft.
2. Hollow shaft has 1 times the angle of twist than that of solid shaft.
   Hence, hollow shaft is stronger and stiffer than solid shaft.

6.8 SHAFTS UNDER SPECIAL CONDITIONS

Under certain special field conditions, the shafts may have varying cross sections, may be made out of two or more materials and may be subjected to varying torque.

6.8.1 Shafts of Varying Sections

In the case of shafts of different lengths and of different diameters, the torque transmitted by individual sections are calculated first. The strength of the shaft is governed by the minimum value of these torques.

6.8.2 Shafts of Different Materials

Shafts made out of different materials are called as composite shafts which behave as a single shaft. In this type of shafts, one type of shaft is rigidly sleeved over another type of the shaft. The total torque transmitted by a composite shaft is the sum of the torque transmitted by each individual shaft. However, the angle of twist in each shaft is the same.

6.8.3 Shafts with Varying Torques

Sometimes, shafts are subjected to torques of different magnitudes at the ends and also at some locations in between the ends. Also, the power available on the shaft at one end is shared by the pulleys provided along the length of the shaft. In such cases, the combined twist is calculated due to different torques to which the shaft is subjected.

6.9 STRAIN ENERGY IN TORSION

Let a shaft of length l be under the action of torque T. Then, the work done in twisting shall be equal to the strain energy due to torsion (U). That is

$$U = \frac{1}{2}T\theta \qquad (6.24)$$

For a gradually applied torque (Figure 6.4), considering the maximum torque for circular shaft Equation (6.5) and representing θ in terms of maximum shear stress, G, radius and length of shaft Equation (6.8), then

$$U = \frac{1}{2}\left(\pi D^3 \frac{f_s}{16}\right) \times \left(2 f_s \frac{l}{GD}\right)$$

$$U = \left(\frac{f_s^2}{4G}\right) \times (\text{Volume}) \qquad (6.25)$$

Equation (6.25) gives the total strain energy over the whole shaft, for which the shear stress is varying from zero at the centre to f_s at the outside.

Thus, the maximum strain energy per unit volume for a solid shaft $= \dfrac{f_s^2}{2G}$ (6.26)

This maximum strain energy is called the *torsional resistance* of a solid shaft.

For a hollow circular shaft Equation (6.25) can be written as

$$U = \left(\frac{f_s^2}{4G}\right)\left(\frac{D_e^2 + D_i^2}{D_e^2}\right) \times (\text{Volume}) \qquad (6.27)$$

SOLVED PROBLEM 6.11

It is intended to replace a solid aluminium shaft to 80 cm long and 4 cm diameter by a tabular steel shaft of same length and outside diameter so that either shaft could carry the same magnitude of torque and have the same angle of twist over the entire length. Find the diameter of the bore of the tabular steel shaft. Modulus of rigidity of steel and aluminium are 8.0×10^4 N/mm² and 2.8×10^4 N/mm², respectively.

Solution:

Given data:Shaft diameter = 4 cm, length = 80 cm, $G_s = 8.0 \times 10^4$ N/mm² and $G_a = 2.8 \times 10^4$ N/mm².

Both the shafts have the same torque, angle of twist, length and external diameter.

Using $$\frac{T}{I_p} = \frac{G\theta}{l}$$

For aluminium solid shaft

$$\therefore \quad \theta_a = \frac{T_a l_a}{(I_p)_a G_a} = \frac{T_a l_a}{\frac{\pi}{32}(40)^4 \times 2.8 \times 10^4}$$

$\therefore$ For tubular steel shaft

$$\theta_s = \frac{T_s l_s}{(I_p)_s G_s} = \frac{T_s l_s}{\frac{\pi}{32}(40^4 - D_i^4) \times 8 \times 10^4}$$

Equating, $\theta_q = \theta_s$, we have

$$\frac{T_a l_a}{\frac{\pi}{32}(40)^4 \times 2.8 \times 10^4} = \frac{T_s l_s}{\frac{\pi}{32}(40^4 - D_i^4) \times 8 \times 10^4}$$

i.e., $$[(40)^4 - (D_i)^4] = \frac{40^4 \times 2.8}{8} \quad (\because T_a l_a = T_s l_s)$$

Solving $$D_i = 35.9 \text{ mm} \approx 36 \text{ mm}$$

Internal diameter of the tubular shaft = 36 mm

SOLVED PROBLEM 6.12

Two shafts A and B of the same material and of same length are subjected to same torque. Shaft A is of solid circular section, while shaft B is a hollow circular section whose internal diameter is 0.7 times the outside diameter. If the maximum shear stress in each shaft is to be the same, compare the weight of the two shafts.

Solution:

Given data: $T_s = T_h$, $D_i = 0.7D_e$ and $l_s = l_h$.

As torque is the same in both the shafts

$$\frac{\pi}{16} f_s D_s^3 = \frac{\pi}{16} f_s \left[\frac{D_e^4 - D_i^4}{D_e}\right]$$

i.e.,
$$D_s^3 = \left[\frac{D_e^4 - (0.7D_e)^4}{D_e}\right]$$

Solving,
$$D_e = 1.096\, D_s.$$

When the length and material are the same, weight will be proportional to sectional area:

$$\frac{W_{solid}}{W_{hollow}} = \frac{\frac{\pi D_e^2}{4}}{\frac{\pi}{4}(D_e^2 - D_i^2)} = \frac{D_s^2}{0.51D_e^2}$$

i.e.,
$$\frac{W_{solid}}{W_{hollow}} = \frac{D_s^2}{0.51(1.096D_s)^2} = 1.632$$

Hence, the weight of solid shaft is 1.632 times as that of hollow shaft.

SOLVED PROBLEM 6.13

A shaft of 400 mm length and 50 mm diameter is bored for a part of its length to a 20 mm diameter. Find the maximum power the shaft can transmit at a speed of 180 rpm with a limiting shear stress of 80 N/mm². Also, determine the length of the shaft that has to bear bore to 20 mm diameter, if the angle of twist is uniform in the entire length of the shaft.

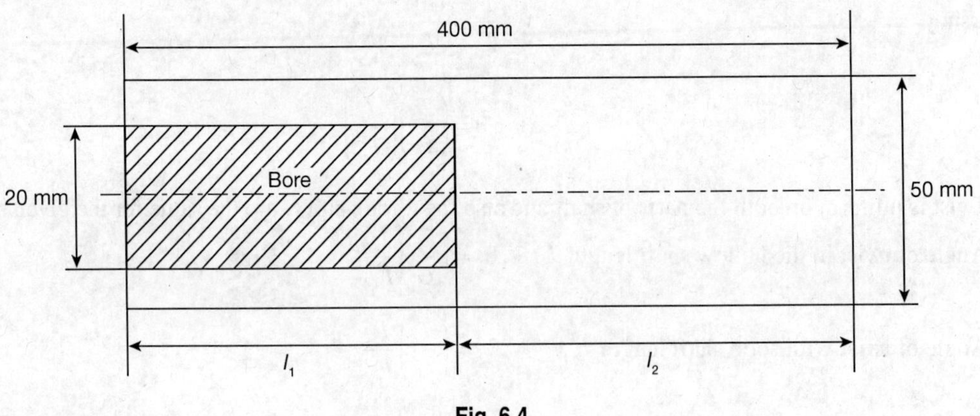

Fig. 6.4

Solution:

Given data: Length = 400 mm, diameter of shaft = 50 mm, diameter of bore = 20 mm, speed, N = 180 rpm and f_s = 80 N/mm².

Let l_1 be the length of the shaft that has been bored to 20 mm diameter and l_2 be the remaining solid length of shaft of 50 mm diameter.

Torque transmitted by hollow portion of shaft, $T_1 = \frac{\pi}{16} f_s \left[\frac{D_e^4 - D_i^4}{D_e} \right]$

$$T_1 = \frac{\pi}{16} \times 80 \left[\frac{50^4 - 20^4}{50} \right] \times \frac{1}{1{,}000}$$

i.e., $T_1 = 1{,}913.23$ N·m.

Torque transmitted by the solid portion of the shaft, $T_2 = \frac{\pi}{16} f_s D_e^3$

$$T_2 = \frac{\pi}{16} \times 80 \times 50^3$$

i.e., $T_2 = 1{,}963{,}495$ N·m.

Safe torque transmitted by the whole shaft is minimum of the torques T_1 and T_2.

∴ Safe torque, $T = 1{,}913.23$ N·m.

Using $P = \frac{2\pi NT}{60}$

$$= \frac{2 \times \pi \times 180 \times 1{,}913.23}{60 \times 1{,}000}$$

$$= 36.06 \text{ kW}$$

Using $\frac{T}{T_1} = \frac{G\theta}{l}$

i.e., $\theta = \left(\frac{T}{G}\right)\left(\frac{l}{I_p}\right)$

Twist is uniform on both the parts of shaft and also the torque and G are the same for the given shaft.

Angle of twist in the hollow shaft length l_1, $\theta_1 = \left(\frac{T}{G}\right)\frac{l_1}{I_{P_1}}$

Angle of twist with solid shaft length l_2, $\theta_2 = \left(\frac{T}{G}\right)\frac{l_2}{I_{P_2}}$

As $\theta_1 = \theta_2$

$$\frac{l_1}{I_{P_1}} = \frac{l_2}{I_{P_2}}$$

i.e., $$\frac{l_1}{\frac{\pi}{32}\left(50^4 - 20^4\right)} = \frac{l_2}{\frac{\pi}{32} 50^4}$$

i.e., $l_1 = 0.97 l_2$

But, $l_1 + l_2 = 500$ mm

Solving $l_1 = 246.19 \text{ mm} \approx 246 \text{ mm}$

i.e., Length of hollow portion $= 246$ mm

Length of solid portion $= 254$ mm

SOLVED PROBLEM 6.14

A stepped solid shaft of 2 m length consists of three lengths of diameter of 90 mm, 70 mm and 50 mm in sequence. If the angle of twist is the same for each section, compute the length of each section and the total twist. The maximum shear stress in the entire shaft is not to exceed 50 N/mm². Take $G = 8.0 \times 10^4$ N/mm².

Solution:

Given data: Total shaft length = 2 m, $D_1 = 90$ mm, $D_2 = 70$ mm, $D_3 = 50$ mm, $f_s = 50$ N/mm² and $G = 8.0 \times 10^4$ N/mm².

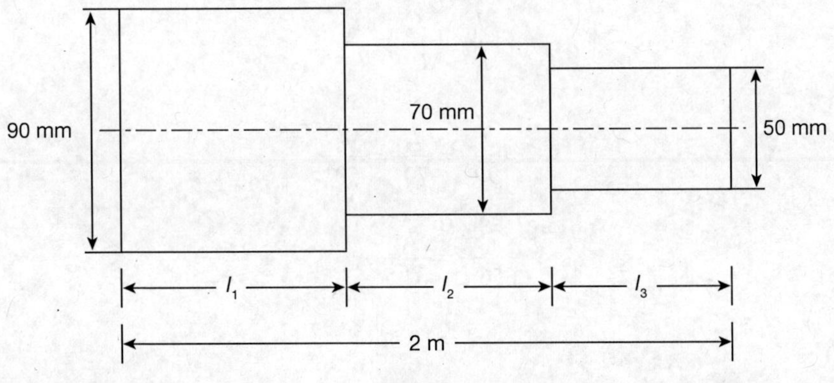

Fig. 6.5

Let l_1, l_2 and l_3 be the lengths corresponding to 90 mm, 70 mm and 50 mm diameters, respectively. Polar moment of inertia of each section is given as

$$I_{p_1} = \frac{\pi}{32} D_1^4 = \frac{\pi}{32} \times 90^4$$

$$I_{p_2} = \frac{\pi}{32} D_2^4 = \frac{\pi}{32} \times 70^4$$

$$I_{p_3} = \frac{\pi}{32} D_3^4 = \frac{\pi}{32} \times 50^4$$

Using $$\frac{T}{I_p} = \frac{G\theta}{l}$$

i.e., $$\theta = \left(\frac{T}{G}\right)\left(\frac{l}{I_p}\right)$$

$\therefore$ $$\theta_1 = \left(\frac{T}{G}\right)\left(\frac{l_1}{I_{p_1}}\right)$$

$$\theta_2 = \left(\frac{T}{G}\right)\left(\frac{l_2}{I_{p_2}}\right)$$

$$\theta_3 = \left(\frac{T}{G}\right)\left(\frac{l_3}{I_{p_3}}\right)$$

As the angle of twist is uniform over the entire length, $\theta_1 = \theta_2 = \theta_3$

i.e., $$\frac{l_1}{I_{p_1}} = \frac{l_2}{I_{p_2}} = \frac{l_3}{I_{p_3}}$$

i.e., $$\frac{l_1}{90^4} = \frac{l_2}{70^4} = \frac{l_3}{50^4}$$

i.e., $$\frac{l_1}{656.1} = \frac{l_3}{62.5}$$

i.e., $$l_1 = 10.5 l_3$$

Similarly, $$l_2 = \frac{240.1}{62.5} l_3 = 3.85 l_3$$

But, $$l_1 + l_2 + l_3 = 2{,}000$$

$$10.5 l_3 + 3.85 l_3 + l_3 = 2{,}000$$

$$\therefore l_3 = 130.3 \text{ mm}$$

$$l_2 = 501.7 \text{ mm}$$

$$l_1 = 1{,}368.2 \text{ mm}$$

SOLVED PROBLEM 6.15

A composite shaft consists of a 30 mm diameter steel shaft surrounded by a brass tube with an internal diameter of 30 mm. Both the shafts are fixed rigidly and share the applied torque equally. Find the thickness of the brass tube if $G_s = 8\times10^4$ N/mm^2 and that of brass, $G_b = 4\times10^4$ N/mm^2.

Solution:

Given data: Diameter steel shaft = 30 mm, internal diameter of brass tube = 30 mm, $G_s = 8\times10^4$ N/mm^2 and $G_b = 4\times10^4$ N/mm^2.

Let the outer diameter of the brass tube be D_e.

Then,
$$(I_p)_s = \frac{\pi\cdot30^4}{32}\ \text{mm}^4$$
$$(I_p)_b = \frac{\pi}{32}\left(D_b^4 - 30^4\right)\ \text{mm}^4$$

Using,
$$\frac{T}{I_P} = \frac{G\theta}{l}$$
$$T_s = (I_p)_s\frac{G_s\theta_s}{l_s}$$

and
$$T_b = (I_p)_b\frac{G_b\theta_b}{l_b}$$

As the shafts are rigidly fixed, the twist is equal and both the lengths are also equal, i.e., $\theta_s = \theta_b$, and $l_s = l_b$, Further, torques are equal, i.e., $T_s = T_b$. Then

$$(T_p)_s\frac{G_s\theta_s}{l_s} = (T_p)_b\frac{G_b\theta_b}{l_b}$$

i.e.,
$$\frac{\pi}{32}(30)^4\times8\times10^4 = \frac{\pi}{32}\left(D_e^4 - 30^4\right)4\times10^4$$

i.e.,
$$\left(D_e^4 - 30^4\right) = \frac{30^4\times8}{4}$$

Solution,
$$D_e = 39.48\ \text{mm}$$

∴ Thick of brass tube
$$= \frac{39.48-30}{2}\ \text{mm}$$
$$= 4.74\ \text{mm}$$

SOLVED PROBLEM 6.16

A composite shaft consists of a steel rod of 30 mm diameter enclosed in a copper tube of external diameter 50 mm and 10 mm thick. The shaft is required to transmit a torque of 1,000 N·m. Determine the shear stress developed in copper and steel, if both the shafts have equal length and are welded to a plate at each end, so that their twist are equal. Take $G_s = 2G_{cu}$.

Solution:

Given data: Steel rod diameter = 30 mm, copper tube, $D_e = 50$ mm, thickness = 10 mm, $T = 1{,}000$ N·m and $G_s = 2G_{cu}$.

Polar moment of inertia of steel, $(I_p)_s = \frac{\pi}{32} \times 30^4$

$= 79{,}521 \text{ mm}^4$

Polar moment of inertia of copper, $(I_p)_{cu} = \frac{\pi}{32}(50^4 - 30^4) \text{ mm}^4$

$= 53{,}400.75 \text{ mm}^4$

For a composite shaft the total torque $T = T_{cu} + T_s = 1{,}000 \times 10^3$ N·mm

Using, $\frac{T}{I_p} = \frac{G\theta}{l}$,

i.e., $\theta = \frac{T \times l}{GI_p}$

For steel rod, $\theta_s = \frac{T_s \times l_s}{G_s (I_p)_s}$

For copper tube, $\theta_{cu} = \frac{T_{cu} \times l_{cu}}{G_{cu} \times (I_p)_{cu}}$

But, $\theta_s = \theta_{cu}$

i.e., $\frac{T_s \times l_s}{G_s \times (I_p)_s} = \frac{T_{cu} \times l_{cu}}{G_{cu} \times (I_p)_{cu}}$

$\therefore$ $T_s = \frac{G_s}{G_{cu}} \times \frac{l_{cu}}{l_s} \times \frac{(I_p)_s}{(I_p)_{cu}} \times T_{cu}$

i.e., $= 2 \times 1 \times \frac{79{,}521}{534{,}070.71} \times T_{cu}$

i.e., $T_s = 0.298\, T_{cu}$

i.e., $T_{cu} + 0.298\, T_{cu} = 1{,}000 \times 10^3$

i.e., $T_{cu} = 770{,}416$ N·mm

$\therefore$ $T_s = 1{,}000 \times 10^3 - 770{,}416$

$T_s = 229{,}583$ N·mm

Stress developed in steel is got from $\dfrac{T_s}{I_p} = \dfrac{f_s}{R}$

i.e.,

$$f_s = \frac{T_s R}{I_p}$$

$$= \frac{229,583 \times 3,012}{\dfrac{\pi}{32} \times 30^4}$$

Solving,

$$f_s = 8,695.80 \text{ N/mm}^2$$

Stress developed in copper is

$$f_{cu} = \frac{T_{cu} \times \dfrac{D_e}{2}}{\left(I_p\right)_{cu}}$$

$$= \frac{770,416 \times 50/2}{534,070.75}$$

$$= 36.06 \text{ N/mm}^2$$

SOLVED PROBLEM 6.17

A circular shaft of 1,000 mm diameter and 2 m length is subjected to a twisting moment which creates a shear stress of 20 N/mm^2 at 30 mm from the axis of the shaft. Calculate the angle of twist and the strain energy stored in the shaft. Take $G = 8 \times 10^4$ N/mm^2.

Solution:

Given data: Shear stress at $r = 30$ m is 20 N/mm^2, diameter of shaft = 1,000 mm, length, $l = 2$ m, $f_s = 20$ N/mm^2 and $G = 8 \times 10^4$ N/mm^2.

$$f_s = \frac{q_s}{r} R = \frac{20}{30} \times \frac{100}{2} = 33.33 \text{ N/mm}^2$$

Using

$$T = \frac{\pi}{16} \times 33.33 \times 100^3 = 6.154 \times 10^6 \text{ N} \cdot \text{mm}$$

Angle of twist

Using

$$\frac{T}{I_p} = \frac{G\theta}{l}$$

i.e.,

$$\theta = \frac{Tl}{GI_p} = \frac{6.5 \times 10^6 \times 2 \times 1,000}{8 \times 10^4 \times \pi \times \dfrac{100^4}{32}}$$

$$= 0.016 \text{ radians}$$

Strain energy

Using,
$$U = \frac{f_s^2}{4G} \times V$$
$$= \frac{(33.33)^2}{4 \times 8 \times 10^4} \times \frac{\pi \times 100^2}{4} \times 2 \times 1,000$$
$$= 54,530 \text{ N} \cdot \text{mm}$$

6.10 COMBINED BENDING AND TORSION

A shaft transmitting torque or power is not only subjected to shear stresses but also to bending moment due to self-weight of the shaft, pulleys and couplings and due to the pulls exerted by belts and rope drives (Figure 6.5).

Thus, at any point in the shaft, the component of stresses are:

(i) the shear stress due to torsion,
(ii) the bending stresses (tensile or compressive) and
(iii) shear stress due to other forces causing bending.

Out of these three component of stresses, (i) and (ii) are important. Further discussion is made on these aspects.

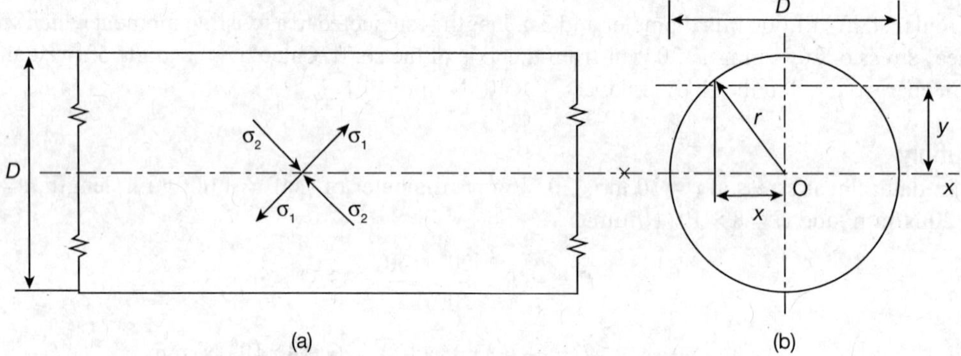

Fig. 6.6 Shaft under combined bending and torsion.

A shaft as shown in Figure 6.6 is considered. Let D be the diameter of the shaft.

Let M be the bending moment at the section of the shaft, q be the shear stress at a point produced by the torque, T and p be the bending stress at the same point produced by the bending moment M. Then, the shear stress as per Equation (6.15) is

$$q = \frac{T}{I_p} r$$

and the bending stress, $$p = \frac{M}{I} y \tag{6.28}$$

In shafts of short length, the bending stress and the shear stress are maximum at a point on the surface of the shaft, where $r = D/2$ and $y = D/2$.

The angle θ made by the plane of maximum shear with the normal cross section is given by

$$\tan 2\theta = \frac{2q}{p} \tag{6.29}$$

Let f_b be the maximum bending stress and f_s be the maximum shear stress

Then

$$f_b = \frac{M}{I} \times \frac{D}{2} = \frac{M}{\left(\dfrac{\pi D^4}{64}\right)} \times \frac{D}{2}$$

i.e.,
$$f_b = \frac{32}{\pi} \times \frac{M}{D^3} \tag{6.30}$$

and
$$f_s = \frac{I}{I_p} R = \frac{T}{\dfrac{\pi D^4}{32}} \times D/2$$

i.e.,
$$f_s = \frac{16}{\pi} \frac{T}{D^3} \tag{6.31}$$

Considering the maximum stress values, $q = f_s$ and $p = f_b$, Equation (6.28) reduces to

$$\tan 2\theta = \frac{2f_s}{f_b} = \frac{2 \times \left(\dfrac{16T}{\pi D^3}\right)}{\left(\dfrac{32}{\pi} \dfrac{M}{D^3}\right)}$$

i.e.,
$$\tan 2\theta = \frac{T}{M} \tag{6.32}$$

Major principal stress at any point,
$$\sigma_1 = \frac{p}{2} + \sqrt{\left(\frac{p}{2}\right)^2 + q^2} \tag{6.33a}$$

and minor principal stress at any point,
$$\sigma_2 = \frac{p}{2} - \sqrt{\left(\frac{p}{2}\right)^2 + q^2} \tag{6.34a}$$

Considering $q = f_s$ and $p = f_b$, then

$$\sigma_1 = \frac{f_b}{2} + \sqrt{\left(\frac{f_b}{2}\right)^2 + f_s^2} \tag{6.33b}$$

and $$\sigma_2 = \frac{f_b}{2} + \sqrt{\left(\frac{f_b}{2}\right)^2 + f_s^2} \quad (6.34b)$$

Substituting for f_b and f_s from Equations (6.30) and (6.31)

$$\sigma_1 = \frac{32M}{2\times\pi D^3} + \sqrt{\left(\frac{32M}{2\times\pi D^3}\right)^2 + \left(\frac{16T}{\pi D^3}\right)^2}$$

i.e., $$\sigma_1 = \frac{16}{\pi D^3} + \left(M + \sqrt{M^2 + T^2}\right) \quad (6.33c)$$

and $$\sigma_2 = \frac{32M}{2\pi D^3} - \sqrt{\left(\frac{32M}{2\times\pi D^3}\right)^2 + \left(\frac{16T}{\pi D^3}\right)^2}$$

i.e., $$\sigma_2 = \frac{16}{\pi D^3} - \left(M - \sqrt{M^2 + T^2}\right) \quad (6.34c)$$

Then

Maximum shear stress, $$T = \frac{\sigma_1 - \sigma_2}{2}$$

i.e., $$T = \frac{16}{\pi D^3}\sqrt{M^2 + T^2} \quad (6.35)$$

It is to be noted that one of the principal stresses will be of opposite sign.

In the case of very short shafts, the shear stresses will be dominating and the bending stresses will be comparatively very small. In such cases, the maximum principal stress may occur within the shaft.

In the case of hollow shafts, the major and minor principal stresses and the shear stresses may be written as

$$\sigma_1 = \frac{16D_e}{\pi\,[D_e^4 - D_i^4](M + \sqrt{M^2 + T^2})} \quad (6.36)$$

$$\sigma_2 = \frac{16D_e}{\pi\,[D_e^4 - D_i^4](M - \sqrt{M^2 + T^2})} \quad (6.37)$$

and $$T = \frac{16D_e}{\pi\,[D_e^4 - D_i^4]\sqrt{M^2 + T^2}} \quad (6.38)$$

For the design of solid or hollow shafts, the values of principal stresses, maximum shear stress and strain energy are needed.

SOLVED PROBLEM 6.18

The maximum allowable shear stress in a hollow shaft of external diameter equal to twice the internal diameter is 70 N/mm^2. Determine the diameter of the shaft if it is subjected to a torque of 4×10^6 N·mm and a bending moment of 3×10^6 N·mm

Solution:
Given data: $D_e = 2D_i, f_s = 70$ N/mm^2, $T = 4 \times 10^6$ N·mm and bending moment, $M = 3 \times 10^6$ N·mm.

Using
$$f_s = \frac{16D_e}{\pi(D_e^4 - D_i^4)}\sqrt{M^2 + T^2}$$

i.e.,
$$70 = \frac{16D_e}{\pi\left[D_e^4 - \left(\frac{D_e}{2}\right)^4\right]}\sqrt{(3\times10^6)^2 + (4\times10^6)^2}$$

Reducing,
$$D_e^3 = \frac{16\times10^6\times5\times16}{\pi\times15\times70}$$

i.e.,
$$D_e = 72.62 \text{ mm}$$

Hence,
$$D_i = 36.31 \text{ mm}$$

SOLVED UNIVERSITY QUESTIONS

SOLVED PROBLEM 6.19

Find the maximum shear stress induced in a solid circular shaft of diameter 200 mm when the shaft transmits 190 kW power at 200 rpm (Anna Univ., May 2005, ME).

Solution:
Given data: Power transmitted, $D = 190$ kW, speed, $M = 200$ rpm and diameter of shaft = 200 mm.

Power transmitted,
$$P = \frac{2\pi NT}{60}$$

i.e.,
$$190\times10^3 = \frac{2\pi\times200\times T}{60}$$

∴
$$T = 9{,}076.4 \text{ N}\cdot\text{m} = 9{,}076.4 \times 10^3 \text{ N}\cdot\text{mm}$$

Using the relation,
$$T = \frac{\pi}{16} f_s D^3$$

$$9{,}076.4\times10^3 = \frac{\pi}{16}\times f_s(200)^3$$

Solving $$f_s = 5.78\ \text{N/mm}^2$$

SOLVED PROBLEM 6.20

A solid shaft is to transmit 400 kW at 100 rpm. If the shear stress is not to exceed 80 N/mm², find the diameter of the shaft. If this shaft were to be replaced by a hollow shaft of the same material and length with an internal diameter of 0.6 times the external diameter, what percentage of saving is possible? (Anna Univ., Nov. 2005, ME).

Solution:

Given data: Power transmitted, $P = 400$ kW, $f_s = 80$ N/mm², D_i of shaft = 0.6 times D_e of shaft and speed, $N = 100$ rpm.

Solid shaft

$$\text{Power} = \frac{2\pi NT}{60}$$

$$400\times1{,}000 = \frac{2\times\pi\times100\times T}{60}$$

Solving $$T = 38{,}197.18\ \text{N}\cdot\text{m}$$

Using

$$T = \frac{\pi}{6} f_s\times D^3$$

$$D^3 = \frac{16\times38{,}197.18\times10^3}{\pi\times80}$$

Solving $$D = 134.5\ \text{mm}$$

Hollow shaft

$$T = \frac{\pi}{16} f_s\left[\frac{D_e^4 - D_i^4}{D_e}\right]$$

i.e.,

$$38{,}197.18\times10^3 = \frac{\pi}{16}\times80\times\left[\frac{D_e^4-(0.6D_e)^4}{D_e}\right]$$

Solving for D_e, $$D_e = 140.82\ \text{mm}$$

Let ρ be the density of the material and l be the length of the shaft.

Weight of solid shaft $$= \left(\frac{\pi D^2}{4}\right)l\times\rho = \frac{\pi D^2}{4}\rho\times l$$

Weight of hollow shaft $= \frac{\pi}{4}\left(D_e^2 - D_i^2\right)\rho \times l$

$$= \frac{\pi}{4} D_e^2 (1-0.6^2)^2 \rho \times l$$

$$= \frac{\pi}{4} D_e^2 \times 0.4 \rho \times l$$

% Saving in weight $= \frac{(\text{Wt. of solid shaft}) - (\text{Wt. of hollow shaft})}{(\text{Wt. of solid shaft})} \times 100$

$$= \frac{\frac{\pi \rho l}{4} D^2 - \frac{\pi}{4} \rho l (0.64 D_e^2)}{\frac{\pi}{4} \rho l D^2} \times 100$$

$$= \frac{(134.5)^2 - 0.64(140.82)^2}{134.5} \times 100$$

$$= 40.14\%$$

SOLVED PROBLEM 6.21

A hollow shaft is to transmit 300 kW at 80 rpm. If the shear stress is not to exceed 60 N/mm² and the internal diameter is 0.6 times the external diameter, find the diameter (Anna Univ., Dec. 2005, ME).

Solution:

Given data: Power transmitted, $P = 300$ kW, speed, $N = 80$ rpm, $f_s = 60$ N/mm² and $D_i = 0.6 D_e$.

Using the relation, $P = \frac{2\pi NT}{60}$

$$300 \times 1{,}000 = \frac{2 \times \pi \times 80 \times T}{60}$$

Solving, $T = 35{,}828$ N·m.

But, $T = \frac{\pi f_s}{16}\left[\frac{D_e^4 - D_i^4}{D_e}\right]$

i.e., $35{,}828 = \frac{\pi}{16} \times 60 \times \left[\frac{D_e^4 - (0.6 D_e)^4}{D_e}\right]$

Solving, $D_e = 14.24$ mm

Internal diameter = 8.56 mm

Internal diameter $= 0.6 \times D_e = 0.6 \times 15.17$

$$= 9.10 \text{ mm}$$

SOLVED PROBLEM 6.22

Calculate the power that can be transmitted at a speed of 300 rpm by a hollow steel shaft of 75 mm external diameter and 50 mm internal diameter, when permissible shear stress for the steel is 70 N/mm^2 and the maximum torque is 1.3 times the mean. Compare the strength of this hollow shaft with that of a solid shaft. Material, weight and length of the shafts are the same (Anna Univ., May 2006, ME).

Solution:

Given data: Speed, $N = 300$ rpm and $D_e = 75$ mm, $D_i = 50$ mm, $f_s = 70$ N/mm^2 and $T_{max} = 1.3T_{av}$

$$T_{max} = \frac{\pi}{16} \times f_s \times \left[\frac{D_e^4 - D_i^4}{D_e}\right]$$

$$= \frac{\pi}{16} \times 70 \times \left[\frac{75^4 - 50^4}{75}\right]$$

$$= 4.65 \times 10^6 \text{ N}\cdot\text{mm}$$

$$T_{max} = 1.3\ T_{mean}$$

∴

$$T_{mean} = \frac{T_{max}}{1.3} = \frac{4.65 \times 10^6}{1.3}$$

$$= 3.57 \text{ kN}\cdot\text{m}$$

Power,

$$P = \frac{2\pi N T_{mean}}{60}$$

$$= \frac{2 \times \pi \times 300 \times 3.57}{60}$$

∴ $$P = 112.15 \text{ kW}$$

As the material, weight and length are the same, then

Area of hollow shaft = Area of solid shaft

$$\frac{\pi}{4}\left[D_e^2 - D_i^2\right] = \frac{\pi}{4} \times D_s^2$$

where D_s is the diameter of solid shaft.

$$\frac{\pi}{4}\left[75^2 - 50^2\right] = \frac{\pi}{4} \times D_s^2$$

Solving $$D_s = 55.90 \text{ mm}$$

Torque transmitted by solid shaft, $$T_s = \frac{\pi}{16} \times f_s \times D_s^3$$

$$T_s = \frac{\pi}{16} \times 70 \times (50.90)^3$$

$$= 2.40 \times 10^6 \text{ N}\cdot\text{mm}$$

$$\therefore \quad \frac{\text{Torque of hollow shaft}}{\text{Torque of solid shaft}} = \frac{T_h}{T_s}$$

$$= \frac{4.65 \times 10^6}{2.40 \times 10^6}$$

i.e.,

$$T_h = 1.937 T_s$$

SOLVED PROBLEM 6.23

A hollow steel shaft of outside diameter 75 mm is transmitting a power of 300 kW at 200 rpm. Find the thickness of the shaft if the maximum shear stress is not to exceed 40 N/mm^2 (Anna Univ., Nov. 2006, ME).

Solution:

Given data: $D_e = 75$ mm, power transmitted, $P = 300$ kW, speed, $N = 200$ rpm and $f_s = 40$ N/mm^2.

Using

$$P = \frac{2\pi NT}{60}$$

$$300 = \frac{2 \times \pi \times 200 \times T}{60}$$

$$\therefore \quad T = 1.4324 \times 10^3 \text{ N}\cdot\text{m}$$

$$= 1.4324 \times 10^6 \text{ N}\cdot\text{mm}$$

$$T = \frac{\pi}{16} f_s \left[\frac{D_e^4 - D_i^4}{D_e} \right]$$

$$1.4324 \times 10^6 = \frac{\pi}{16} \times 40 \left[\frac{75^4 - D_i^4}{75} \right]$$

Solving

$$D_i = 65.10 \text{ mm}$$

$$\therefore \quad \text{Thickness of hollow shaft} = \frac{D_e - D_i}{2}$$

$$= \frac{75 - 65.10}{2}$$

$$\therefore \quad \text{Thickness} = 4.90 \text{ mm}$$

SOLVED PROBLEM 6.24

A steel shaft is required to transmit 75 kW power at 100 rpm and maximum twisting moment is 30% greater than the mean. Find the diameter of the shaft, if the maximum stress is 70 N/mm². Also, find the angle of twist in a 3 m length of the shaft. Assume the shear modulus as 90 kN/mm² (Anna Univ., May/June 2006, ME).

Solution:

Given data: Power transmitted, $P = 75$ kW, speed, $N = 100$ rpm, $T_{max} = 1.30T_{mean}$, $f_s = 70$ N/mm², $l = 3$ m and $G = 90$ kN/mm².

Using
$$P = \frac{2\pi N T_{mean}}{60}$$

$$T_{mean} = \frac{P \times 60}{2\pi N}$$

$$= \frac{75 \times 1{,}000 \times 60}{2 \times \pi \times 100}$$

$$= 7{,}161 \text{ N}\cdot\text{m}$$

$$T_{max} = 1.3T_{mean} = 1.3 \times 7{,}161 = 9{,}311 \text{ N}\cdot\text{m}$$

Now, using
$$T = \frac{\pi}{16} \times f_s \times 10^3$$

$$T_{max} = \frac{\pi}{16} \times 70 \times D^3 = 9{,}311 \times 1{,}000 \text{ N}\cdot\text{m}$$

Solving
$$D = 8.78 \text{ mm}$$

Using $\frac{f_s}{R} = \frac{G\theta}{L}$, we have

Angel of twist,
$$\theta = \frac{70 \times 3 \times 1{,}000}{\frac{8.78}{2} \times 90 \times 10^3}$$

$$= 0.53°$$

SOLVED PROBLEM 6.25

Find the diameter of a solid shaft to transmit 90 kW at 160 rpm, such that the shear stress is limited to 60 N/mm². The maximum torque is likely to exceed the mean torque by 20%. Also, find the permissible length of the shaft, if the twist is not to exceed 1 degree over the entire length. Take rigidity modulus as 0.8×10^5 N/mm² (Anna Univ., Nov. 2008, ME).

Solution:

Given data: Power transmitted, $P = 90$ kW, speed, $N = 160$ rpm, $f_s = 60$ N/mm², $T_{max} = 1.2\,T_{mean}$, twist, $\theta = 1°$ and $G = 0.8 \times 10^5$ N/mm².

Using
$$P = \frac{2\pi NT}{60}$$

$$90 \times 10^3 = \frac{2 \times \pi \times 160 \times T}{60}$$

i.e.,
$$T_{mean} = \frac{90 \times 10^3 \times 60}{2 \times \pi \times 160} = 5{,}371.47 \text{ N} \cdot \text{m}$$

i.e.,
$$T_{mean} = 5{,}371.47 \times 10^3 \text{ N} \cdot \text{mm}$$

$$T_{max} = 1.20 \times T_{mean}$$
$$= 1.20 \times 5{,}371.47 \times 10^3$$
$$= 6{,}445.775 \times 10^3 \text{ N} \cdot \text{mm}$$

Using
$$T_{max} = \frac{\pi}{16} \times f_s \times D^3$$

i.e.,
$$6{,}445.775 \times 10^3 = \frac{\pi}{16} \times 60 \times D^3$$

Solving,
$$D = 81.789 \text{ mm}$$

$$\frac{f_s}{R} = \frac{G\theta}{l}$$

Using
$$\frac{60}{\frac{81.782}{2}} = \frac{0.8 \times 10^5 \times \pi/180}{l}$$

Solving,
$$\text{length of shaft} = 951.66 \text{ mm}$$

SALIENT POINTS

- A member is considered to be under torsion if it transmits the moments of a couple in a direction normal to the plane of the couple.
- Torsion is one of the four basic modes by which loads are transmitted from one portion of a structural or machine member to another.
- The intensity of shear stress induced at any point in the cross section of a circular shaft subjected to a couple is proportional to its distance from the centre.
- The maximum torque or horse power that can be transmitted by a shaft is referred to as the strength of a shaft.
- Torque required for unit twist is called the torsional stiffness.
- The quantity $(GI_p)/l$ is known as the torsional rigidity and is represented by k, where G is the rigidity modulus, I_p is the polar moment of inertia and l is the length.

- Torque, T, transmitted by a solid circular shaft is given as

$$T = \frac{\pi}{16} f_s D^3$$

where f_s = intensity of shear stress
D = diameter of the shaft

- Torque, T, transmitted by a hollow circular shaft is given as

$$T = \frac{\pi}{16} f_s \left[\frac{D_e^4 - D_i^4}{D_e}\right]$$

where D_e = external diameter of the shaft
D_i = internal diameter of the shaft
f = intensity of shear stress.

- Power transmitted by a shaft, P is given as

$$P = \frac{2\pi NT}{60} \text{ watts}$$

where N = revolutions per minute
T = average or mean torque

also, $P = TW$

where $W = \dfrac{2\pi N}{60}$

- Polar modulus is defined as the ratio of the polar moment of inertia to the radius of the shaft and also known as torsional section modulus. It is denoted as Z_p

$$Z_p = I_p/R$$

where R is the radius of the shaft

- Strain energy U over a whole shaft for the condition that the shear stress is varying from zero at the centre to f_s at the outside is given as $U = \dfrac{(f_s^2)}{4G}(\text{Volume})$.
- Maximum strain energy per unit volume for a solid shaft, i.e., torsional resistance = $(f_s^2/2G)$.
- A shaft transmitting torque or power is not only subjected to shear stresses but also bending moment due to self-weight of the shaft and action of other accessories on the shaft.
- At any point as a shaft, the component of stresses are (i) shear stress due to torsion, (ii) bending stress (tensile or compressive) and (iii) shearing stresses due to forces causing bending.

QUESTIONS

1. What is stiffness of a shaft? Why a hollow shaft is preferred over a solid shaft?
2. Define torsional rigidity of a shaft. How a moment is applied to produce torque in a shaft?
3. Define polar moment of inertia and polar modulus. Give the polar modulus for solid circular shaft and a hollow shaft.
4. What is strain energy? How strain energy of a circular shaft is calculated?

5. How the power transmitted by a shaft is computed?
6. A solid shaft has to transmit 360 kW power at 120 rpm. The shaft must not be stressed beyond 70 MPa and must not twist more than 1° in a length of 4 m. Select the suitable diameter. $G = 80$ GPa.
7. A solid shaft of 2.5 m length of 4 cm diameter is rigidly fixed at the ends. At a distance of 1.5 m from one end, a twisting moment of 1.8 kN·m is applied. Compute (i) the fixing couples at the ends, (ii) the maximum shear stress and (iii) the angle of twist of section where the twisting moment was applied. Take $G = 8.2 \times 10^4$ N/mm^2.
8. A hollow shaft of diameter ratio 3/8 is required to transmit 600 kW at 120 rpm, the maximum torque being 20% greater than the mean. The shear stress is not to exceed 60 N/mm^2 and a twist in a length of 3 m is not to exceed 1.5 degrees. Calculate the maximum external diameter satisfying these conditions. Take modulus of rigidity = 0.84×10^5 N/mm^2.
9. A hollow shaft is 1 m long and has external diameter 50 mm. It has 20 mm internal diameter for a part of the length and 30 mm internal diameter for the rest of the length. If the maximum shear stress in it is not to exceed 8 N/mm^2, determine the maximum power transmitted by it at a speed of 300 rpm. If the twists produced in the two portions of the shafts are equal, find the length of two portions.
10. A hollow steel shaft 24 cm external and 16 cm internal diameter is to be replaced by a solid alloy shaft. If both the shafts should have the same polar modulus, find the diameter of the latter and the ratio of torsional rigidities. Take N for steel and $2N$ for alloy.
11. A hollow shaft of steel has an external diameter of 150 mm and 75 mm internal diameter. It transmits a power of 1,200 kW at 200 rpm. Find the shear stress developed at the inner and outer surfaces. Also, calculate the strain energy per metre length. Take $G = 84{,}000$ N/mm^2.
12. A tubular steel shaft of 10 cm external diameter and 5 cm internal diameter is to be replaced by a solid alloy shaft. Find the diameter of alloy shaft if the polar moment of inertia of both the shafts is equal. Also, determine the ratio of the torsional rigidity of the shaft. Take G for steel as 2 times as that of alloy.
13. A shaft (Figure 6.7) rotates at a speed of 250 rpm. Gear A is fed by a power of 50 hp and 30 hp and 20 hp are taken off by gear B and C, respectively. Calculate the maximum shear stress developed in the shaft and the angle of twist of the gear A relative to C. Take $G = 8.5 \times 10^4$ N/mm^2.

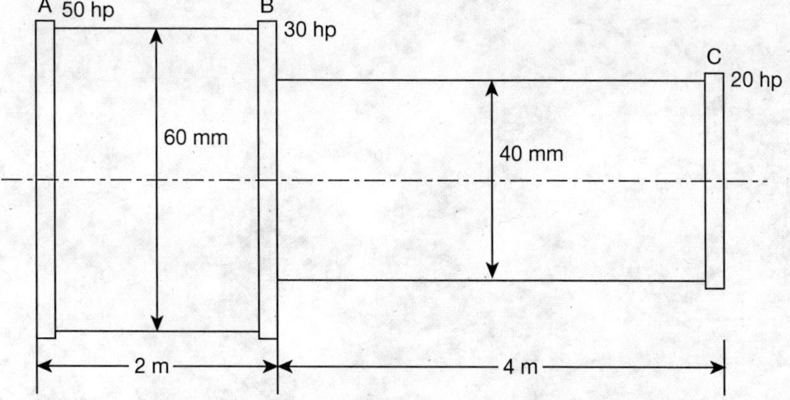

Fig. 6.7

14. A solid circular shaft has a diameter of d_1, at one end and d_2 at the other end. Derive an expression for the angle of twist over a fixed length l. Hence, compute the angle of twist if $d_1 = 8$ cm and $d_2 = 10$ cm and $l = 12$ cm.
15. A hollow shaft of 120 mm external diameter and 60 mm internal diameter transmits a power of 500 kW at 250 rpm. It is also subjected to a bending moment of 8 kN·m and an end thrust. Compute the end thrust, if the major principal stress is limited to 65 N/mm^2. Also, compute the minor principal stress.
16. A solid steel shaft of 40 mm diameter is fixed rigidly and concentrically in a bronze sleeve of 60 mm external diameter. Compute the angle of twist in a 1.2 m length of the composite shaft under the action of a torque of 600 N·m. Modulus of rigidity of steel is 8×10^4 N/mm^2 and that of the bronze is 4×10^4 N/mm^2.

7

Springs

LEARNING OBJECTIVES

7.1 INTRODUCTION

A spring is a device, made out of a particular material, which can undergo considerable angular and linear deformations without undergoing permanent distortion. Springs are used to absorb energy due to resilience which could be restored at the time of requirement. The quality of a spring is judged depending on its capacity in storing the greatest amount of energy for a given stress. *Stiffness* of a spring is defined as the load required to produce unit deflection.

7.2 TYPES OF SPRINGS

Springs may be broadly classified under two categories, viz., *bending springs* and *torsion springs.*

A spring which is subjected to only bending and the resilience is also due to the same is called the bending spring. The leaf, laminated or plate springs fall under this category. These springs are made out of plates and arranged in a particular pattern so as to attain the spring effect.

A spring which is subjected to a twisting moment or torsion and the resilience is only due to the same is called the *torsion spring.* Closely coiled helical springs fall under this group. These springs are made out of rod or wire and coiled in the form of a helix described on a right circular cylinder. The clock spring may be also grouped under this category.

An open-coiled helical spring may be considered under both the categories. Although there may be different forms of springs, the commonly used springs in engineering works are

(i) Leaf springs,
(ii) Helical springs and
(iii) Flat spiral springs.

7.3 LEAF SPRINGS

Leaf springs (also called as laminated springs) are commonly used in carriages such as cars, lorries and railway wagons. Shocks of vehicles give unpleasant feeling to the passengers and hence springs are used to absorb such shocks. The energy absorbed by such springs during a shock is immediately related without contributing for any useful work.

There are two main types of leaf springs, viz., the semi-elliptic leaf spring and the quarter-elliptic leaf spring.

7.3.1 Semi-elliptic Leaf Springs

This type of spring consists of a number of parallel metal strips having different lengths and of same width which are placed one over the other. The reason for such an arrangement is to provide uniform strength in the spring. These springs are fixed on the axis of a vehicle and their top plate is pinned at the ends to the chassis of the vehicle. Figure 7.1 shows the initial position of a spring before loading with a central deflection, δ.

The point load W acts at the centre of the lower most plate and is equally shared by the two ends of the top plate (Figure 7.1). When the load is fully applied all the plates become flat and the deflection δ disappears.

Let l be the span of the spring.

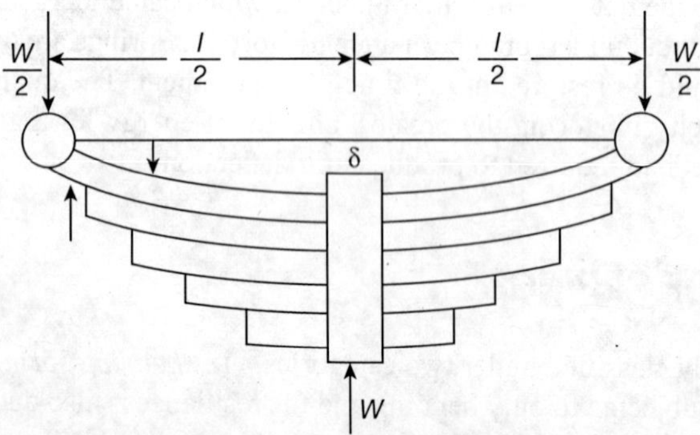

Fig. 7.1 Semi-elliptic leaf spring.

Then

$$\text{Bending moment at the centre} = (\text{Load at one end}) \times \frac{(\text{Span})}{2}$$

i.e.,

$$M = \frac{W}{2} \times \frac{l}{2} = \frac{Wl}{4}$$

As the plates have equal width b and thickness t, the moment of inertia of each plate

$$I = \frac{bt^3}{12}$$

It is known that,

$$\frac{M}{I} = \frac{f}{y}$$

i.e.,

$$M = \frac{f}{y} I \tag{7.1}$$

where f is the maximum bending stress in the plate
Here, $y = t/2$

$\therefore$

$$M = \frac{f}{y} I = \frac{f}{\frac{t}{2}} \times \frac{bt^3}{12} = \frac{fbt^2}{6} \tag{7.2a}$$

Total resisting moment for n number of plates $= n \times M$

$$= \frac{nfbt^2}{6} \tag{7.2b}$$

Equating max bending moment (Equation 7.1) to the total resisting moment (Equation 7.2)

$$\frac{Wl}{4} = \frac{nfbt^2}{6}$$

$\therefore$

$$f = \frac{6Wl}{4} \times \frac{1}{nbt^2}$$

i.e.,

$$f = \frac{3Wl}{2nbt^2} \tag{7.3}$$

Let us consider the central-line diagram of top plate (Figure 7.2).
Let R be the radius of the plate.
Then, from ΔACO,

$$AO^2 = AC^2 + CO^2$$

i.e.,

$$R^2 = \left(\frac{l}{2}\right)^2 + (R - \delta)^2$$

Rearranging

$$\delta = \frac{l^2}{8R} \tag{7.4}$$

Using the relation between bending stress, modulus of elasticity and radius of curvature R, is given by

$$\frac{f}{y} = \frac{E}{R} \tag{7.5}$$

$\therefore$

$$R = \frac{Ey}{f} = \frac{Et}{2f}$$

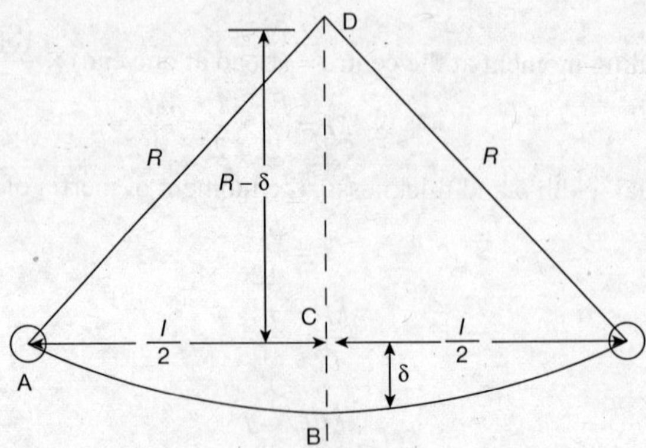

Fig. 7.2 Centre line diagram of top plate.

Substituting for R in Equation (7.4)

i.e.,

$$\delta = \frac{l^2}{8} \times \frac{2f}{Et}$$

$$\delta = \frac{fl^2}{4Et}$$

Substituting for f from Equation (7.4)

$$\delta = \frac{3Wl^3}{8Enbt^3}$$

Equations (7.6) and (7.7) give the central deflection of the spring. If δ_0 is the deflection corresponding to a load W_0 to make the spring flat then

$$\delta_0 = \frac{3W_0 l^3}{8Enbt^3}$$

$$W_0 = \frac{8Enbt^3\delta_0}{3l^3}$$

For a given material if f and E are known, then their equation may be used to fix proper relationship between the thickness and initial radius.

7.3.2 Quarter-Elliptic Leaf Springs

A quarter-elliptic leaf spring is half of a semi-elliptic spring (Figure 7.3). Let W be the load applied at the end of the spring to straighten the plate and δ be the corresponding deflection. Then

$$\frac{1}{R} = \frac{M}{EI}$$

where $M = Wl$, the bending moment at the fixed end and moment of inertia at the fixed end,

$$I = \frac{nbt^3}{12}.$$

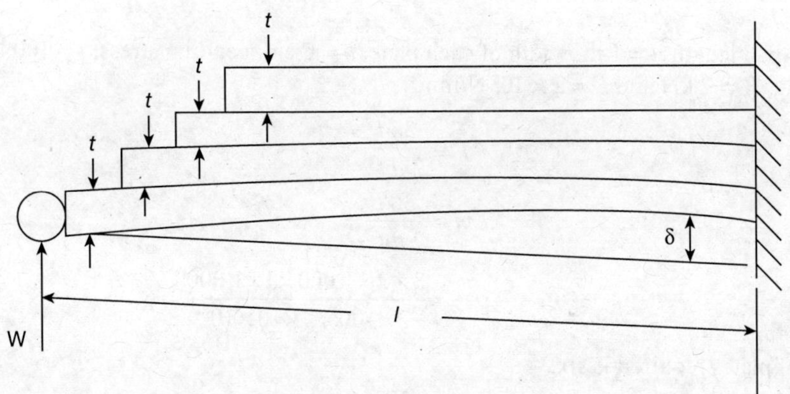

Fig. 7.3 Quarter-elliptic leaf spring.

$$\frac{1}{R} = \frac{12Wl}{nbt^3E}$$

For the properties of a circle $(2R - \delta)\,\delta = l \times l_0$

Neglecting the product of small quantities

$$2R\delta = l^2$$

i.e., $$\delta = \frac{l^2}{2R}$$

i.e., $$\delta = \frac{6Wl^2}{nbt^3E}$$

Maximum bending stress, $$f = \frac{M}{I}$$

$$= \frac{Wl}{\frac{1}{12} \times nbt^3} \times \frac{t}{2}$$

i.e., $$f = \frac{6Wl}{nbt^2}$$

It W_0 is the proof load and d_0 is the corresponding deflection, then $\delta_0 = \dfrac{6W_0l^2}{nbt^3E}$

or $$W_0 = \frac{nbt^3E}{6l^2}\delta_0$$

SOLVED PROBLEM 7.1

A laminated spring of 1 m long is made up of plates each 5 cm wide and 1 cm thick. If the bending stress in the plates is limited to 100 MPa, how many plates would be required to enable the spring to carry a central point load of 2 kN? What is the deflection under the load? $E = 2 \times 10^5$ N/mm^2.

Solution:

Given data: Span length, $l = 1$ m, width of each plate, $b = 5$ cm, bending stress, $f = 100$ MPa, central load on spring, $W = 2$ kN and $E = 2 \times 10^5$ N/mm².

Using
$$f = \frac{3Wl}{2nbt^2}$$
$$n = \frac{3Wl}{2fbt^2}$$
$$= \frac{3 \times 2 \times 1{,}000 \times 1 \times 1{,}000}{2 \times 100 \times 5 \times 10 \times 10^2} = 6$$

∴ Number of plates required is six.

Deflection,
$$\delta = \frac{3Wl^3}{8Enbt^3}$$
$$= \frac{3 \times 2 \times 1{,}000 \times 1{,}000^3}{8 \times 2 \times 10^5 \times 6 \times 50 \times 10^3}$$
$$= 12.5 \text{ mm}$$

Deflection under the load is 12.5 mm.

SOLVED PROBLEM 7.2

A laminated spring has 10 plates each of 40 mm width and 6 mm thick and the length of the longest plate is 500 mm. The bending stress is not to exceed 150 N/mm² and the central deflection is 8 mm. Compute the magnitude of the greatest central load that can be applied to the spring. $E = 2.2 \times 10^5$ N/mm².

Solution:

Given data: No. of plates = 10, width = 40 mm, thickness = 6 mm, longest plate length = 500 mm, pending stress = 150 N/mm², deflection = 8 mm and $E = 2.2 \times 10^5$ N/mm².

To satisfy the deflection conditions
$$W = \frac{8\delta Enbt^3}{3l^3}$$
$$W = \frac{8 \times 8 \times 2.2 \times 10^5 \times 10 \times 40 \times 6^3}{3 \times 500^3} = 3{,}244 \text{ N.}$$

To satisfy the bending stress condition.
$$W = \frac{2fnbt^2}{3l}$$
$$= \frac{2 \times 150 \times 10 \times 40 \times 6^2}{3 \times 500}$$

i.e.,
$$W = 2{,}880 \text{ N}$$

∴ The greatest load the spring can take is 2,880 N.

SOLVED PROBLEM 7.3

A laminated spring supported at the ends is to carry a central load of 7.5 kN. The horizontal distance between the supports is 900 mm. The central deflection is not to exceed 50 mm. The bending stress must not be greater than 140 N/mm^2. Plates are available in multiples of 8 mm for thickness and in multiples of 30 mm of width. Design the spring. $E = 210$ kN/mm^2.

Solution:

Given data: Load, W = 7.5 kN, E = 210 kN/mm^2, central deflection = 50 mm horizontal distance = 900 mm and bending stress f = 140 N/mm^2.

Using, $$\delta = \frac{3Wl^3}{8Enbt^3}$$

i.e., $$nbt^3 = \frac{3Wl^3}{8E\delta} \qquad \text{(i)}$$

Bending stress, $$f = \frac{3Wl}{2nbt^2}$$

i.e., $$nbt^2 = \frac{3Wl}{2f} \qquad \text{(ii)}$$

Dividing, Equation (ii) by Equation (i)

$$t = \frac{3Wl^3}{8E\delta} \times \frac{2f}{3Wl} = \frac{l^2 f}{4E\delta}$$

$$= \frac{900^2 \times 140}{4 \times 210 \times 1{,}000 \times 50} = 2.70 \text{ mm}$$

As the thickness of plate available is 8 mm, a thickness of 8 mm may be adopted. Now, for the condition of central deflection not to exceed 50 mm, from Equation (i)

$$nb = \frac{3Wl^3}{8E\delta t^3}$$

$$= \frac{3 \times 750 \times 900^3}{8 \times 210 \times 1{,}000 \times 50 \times 8^3} = 38.14 \text{ mm}$$

Now, for the condition of bending stress not to exceed 440 N/mm^2, from Equation (ii)

$$nb = \frac{3Wl}{2f \times t^2}$$

$$= \frac{3 \times 750 \times 900}{2 \times 140 \times 8^2} = 113.02 \text{ mm}$$

The width of plate available is 30 mm, hence 30 mm width may be adopted.

∴ Number of plates required to satisfy deflection condition

$$n = \frac{38.14}{b} = \frac{38.14}{30} = 1.27$$

∴ Number of plates required to satisfy deflection condition

$$n = \frac{113.02}{b} = \frac{113.02}{30} = 3.76$$

Hence, four plates of 30 mm width in provided, then both the deflection and bending stress conditions will be satisfied.

Hence,

$$\text{Thickness of plates} = 8 \text{ mm}$$
$$\text{Width of plates} = 30 \text{ mm}$$
$$\text{Number of plates} = 4.$$

SOLVED PROBLEM 7.4

A laminated quarter-elliptic steel spring has a span of 0.70 m and is required to carry a load of 10 kN. The static deflection is not to exceed 75 mm and the bending stress not to exceed 300 N (mm)^2. Plates are available in multiples of 1 mm and 4 mm for width. Estimate suitable values for width, thickness and number of plates and also calculate the radius to which the plates should be beat. Assume suitable width–thickness ratio. Take $E = 208{,}000$ N/mm^2.

Solution:

Given data: Span = 0.70 m, load = 10 kN, deflection = 75 mm, bending stress, $f = 300$ N/mm^2, plate sizes = 1 mm, $E = 208{,}000$ N/mm^2 and width of plate = 4 mm.

Using
$$f = \frac{6Wl}{nbt^2} \quad \text{(i)}$$

and
$$\delta = \frac{6Wl^3}{nbt^3 E} \quad \text{(ii)}$$

Dividing Equation (i) by Equation (ii), we have

∴
$$\frac{f}{\delta} = \frac{Et}{l^2}$$

$$t = \frac{fl^2}{\delta E} = \frac{300 \times (700)^2}{75 \times 208{,}000} = 9.42 \text{ mm}$$

As only 1 mm thick plates are available, 10 plates may be provided, i.e., $t = 10$ mm.

Assume a width to thickness ratio of 12:1.

∴ Width of plate = $12t = 12 \times 10 = 120$ mm

As only 4 mm plates are available, 30 plates may be provided.

Using
$$n = \frac{6Wl}{fbt^2} = \frac{6 \times 10 \times 1{,}000 \times 700}{300 \times 120 \times 10^2} = 11.66$$

and using
$$n = \frac{\delta Wl^3}{bt^3 E\delta} = \frac{6 \times 10 \times 1{,}000 \times 700^3}{120 \times 10^3 \times 208{,}000 \times 75} = 10.99$$

Hence, 12 plates of 120 mm width with 10 mm thickness are provided:

$$f = \frac{6Wl}{nbt^2}$$

$$= \frac{6 \times 10 \times 1{,}000 \times 700}{12 \times 120 \times 10^2} = 291.67 \text{ N/mm}^2$$

Using
$$\frac{f}{y} = \frac{E}{R} = \frac{M}{I}$$

$$R = \frac{E}{f} \times y = \frac{208{,}000}{291.67} \times \frac{10}{2}$$

$$R = 3{,}565.7 \text{ mm} = 3{,}566 \text{ mm}.$$

Check for deflection

$$\delta = \frac{l^2}{2R} = \frac{700^2}{2 \times 3{,}566} = 68.70 \text{ mm}$$

which is less than 75 mm and hence safe.

SOLVED PROBLEM 7.5

A leaf spring of semi-elliptic type has 11 plates each of 9 cm wide and 1.5 cm thick. The length of the spring is 1.5 m. The plates are made of steel having a proof stress (bending) of 650 MN/m². To what radius should the plates be bent initially? From what height can a load of 600 N fall on to the centre of the spring if maximum stress is to be one half of the proof stress? $E = 2.0 \times 10^6$ N/mm².

Solution:

Given data: No. of plates = 11, width = 9.0 cm, thickness = 1.5 cm, length of spring, $l = 1.5$ m, bending stress, $f = 650$ MN/m² and $E = 2.0 \times 10^6$ N/mm².

Using
$$\frac{M}{I} = \frac{f}{y} = \frac{E}{R_c}$$

$$R_c = \frac{E \times y}{f} = \frac{E \times \frac{t}{2}}{f} = \frac{2 \times 10^5 \times 1.5 \times 10/2}{650}$$

$$= 2{,}307 \text{ mm} = 2.31 \text{ m}$$

where R_C is the radius of curvature.

The stress in the second case is half the proof stress, i.e., 650/2 = 325 N/mm².

i.e, Let W be the equivalent static load which will produce the above stress.

i.e.,
$$f = \frac{3Wl}{2nbt^2}$$

$\therefore$
$$W = \frac{2nbt^2 f}{3l} = \frac{2 \times 11 \times 90 \times (15)^2 \times 325}{3 \times 1{,}500}$$

$$= 32{,}175 \text{ N}.$$

Deflection under this load

$$\delta = \frac{3Wl^3}{8Enbt^3} = \frac{3\times 32{,}175\times(1{,}500)^3}{8\times 2\times 10^5\times 11\times 90\times 15^3}$$
$$= 61 \text{ mm.}$$

Let P be the following weight

$\therefore$ $$P(h+\delta) = \text{Energy stored in the spring due to static load}$$

i.e., $$P(h+\delta) = W\times\delta$$

$$600(h+\delta) = \frac{1}{2}\times 32{,}175\times 61$$

$\therefore$ $$h = 1{,}574 \text{ mm.}$$
$$\text{Height of fall} = 1{,}574 \text{ mm.}$$

7.4 HELICAL SPRINGS

Helical springs are made out of thick wires coiled into a helix. They are of two types, viz., (i) close-coiled helical springs and (ii) open-coiled helical springs.

In close-coiled helical spring, the wire is coiled very closely and so it is assumed that each turn of the coil is practically a plane perpendicular to the axis of the helix. The close-coiled spring is a torsion spring and the bending stress is negligible.

In open-coiled helical spring, the wire is not coiled closely but with large gap between the two consecutive turns as a result of which there is bending stress also. Thus, open-coiled helical springs belong to the categories of torsion and bending springs.

7.4.1 Close-coiled Helical Springs Subjected to Axial Force

Let us consider a spring with radius of the coil as R, mean diameter of the coil as D and the diameter of the wire as d (Figure 7.4(a)).

The spring is equivalent to a shaft of diameter d and length l (Figure 7.4(b)).
where l is the total length of the wire.

$$l = 2\pi Rn$$

where n is the number of turns or coils.

Using the relationship $$\frac{T}{I_p} = \frac{G\theta}{l}$$

$$\theta = \frac{Tl}{GI_p} = \frac{WRl}{G\dfrac{\pi d^4}{32}}$$

i.e., $$\theta = \frac{32WRl}{\pi d^4 G} \tag{7.9}$$

Substituting for $l = 2\pi Rn$

$$\theta = \frac{64WR^2n}{Gd^4} \text{ radians} \tag{7.10}$$

Thus, the free end has twisted through an angle θ, given by Equation (7.10) and consequently the free end undergoes an axial movement or deflection, $R\theta$.

Let δ be the axial movement or deflection,

then
$$\delta = R\theta = R \times \frac{64WR^2n}{Gd^4} \tag{7.11}$$

i.e.,
$$\delta = \frac{64WR^3n}{Gd^4} \tag{7.12}$$

Stiffness s of the spring is given as

i.e.,
$$s = \frac{W}{\delta} \tag{7.13}$$

$$s = \frac{Gd^4}{64nR^3}$$

Shear stress is induced at any section, distant R from the axis, both due to torsion ($T = WR$) and the shear force caused due to axial load (W).

Let f_{st} be the shear stress due to torque and f_{ss} be the shear stress due to shear force.

$$f_{st} = \frac{T}{I_p} R$$

i.e.,
$$= \frac{T}{\frac{\pi d^4}{32}} \times \frac{d}{2} = \frac{(WR)}{\frac{\pi d^3}{16}} \tag{7.14}$$

$$f_{st} = \frac{16WR}{\pi d^3}$$

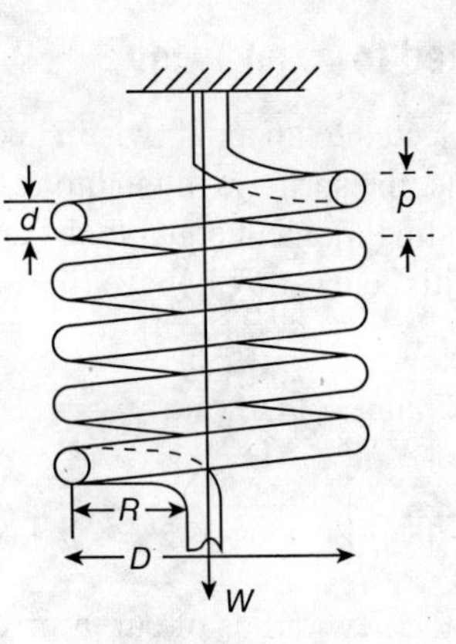

(a) Spring with load

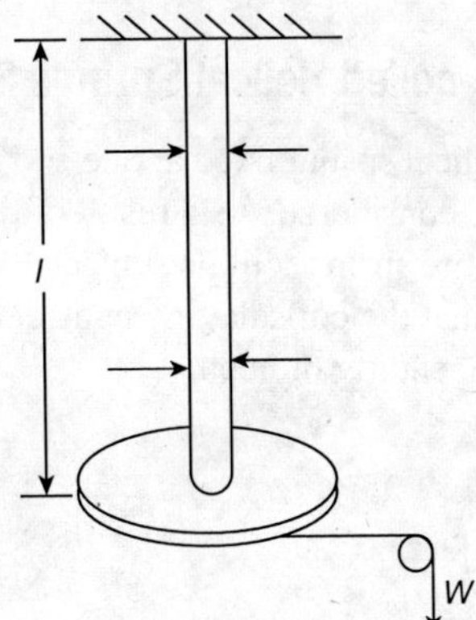

(b) Equivalent shaft

Fig. 7.4 Close-coiled helical spring under axial force.

and $$f_{ss} = \frac{\text{Load}}{\text{Area of cross section}} = \frac{W}{\pi\frac{d^2}{4}} \tag{7.15}$$

i.e., $$f_{ss} = \frac{4W}{\pi d^2}$$

∴ Total maximum shear stress

$$(f_s)_{max} = \frac{16WR}{\pi d^3} + \frac{4W}{\pi d^2}$$

$$(f_s)_{max} = \frac{16WR}{\pi d^3}\left(1 + \frac{d}{4R}\right) \tag{7.16}$$

∴ $$(f_s)_{max} = \frac{16WR}{\pi d^3} \tag{7.17}$$

It is to be noted that each section of the coil is under torsion but there are small quantities of bending and shear stresses which being small are usually neglected.

Strain energy stored by the spring,

$$U = \frac{f_s^2}{4G} \times (\text{Volume})$$

i.e., $$U = \frac{(16WR)^2}{\pi d^3} \times \frac{1}{4G}\left(\frac{\pi d^2}{4} \times 2\pi Rn\right) \tag{7.18}$$

$$= \frac{32W^2R^2}{Gd^4} \times R.n$$

i.e., $$U = \frac{32W^2R^2}{Gd^4} \times n$$

7.4.2 Close-coiled Helical Springs Subjected to Axial Twist

A close-coiled helical spring fixed at one end and subjected to an axial twisting couple M (Figure 7.5) at the other end is considered. As a result of the couple, the spring will tend to unwind or wind up the spring and the entire spring will be subjected to a bending moment equal to the applied torque M.

It is assumed that the coils having considerable initial curvature behave like a beam of zero initial curvature. Then, bending moment

$$M = EI \times (\text{Change of curvature})$$

i.e., $$M = EI\left(\frac{1}{R'} - \frac{1}{R}\right) \tag{7.19}$$

where R is the original radius of curvature and R' is the new radius of curvature.

The above equation refers to a condition that the moment causes an increase in curvature (i.e., $R' < R$).

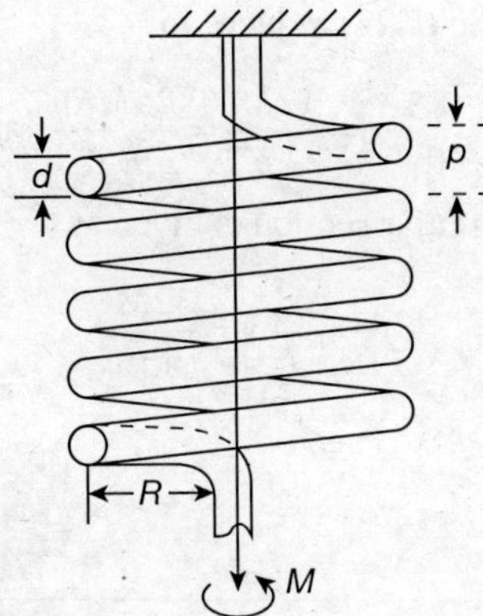

Fig. 7.5 Close coiled helical spring under axial twist.

and $$\frac{1}{R} = \frac{2\pi n}{l}$$

and $$\frac{1}{R'} = \frac{2\pi n'}{l}$$

where n' is the changed number of coils or turns.

Substituting for R and R' in Equation (7.19)

$$M = EI\left\{\frac{2\pi n'}{l} - \frac{2\pi n}{l}\right\} \tag{7.20}$$

i.e., $$M = \frac{2\pi EI}{l}(n' - n)$$

Total twist ϕ of the free end (in radians) is given as

$$\phi = 2\pi(n' - n)\ \frac{Ml}{EI} \tag{7.21}$$

The above result may also be obtained from the strain energy consideration. Thus

$$\frac{1}{2}M\phi = \frac{1}{2EI}\int_0^l M^2\,dx = \frac{1}{2}\frac{M^2 l}{EI}$$

i.e., $$\phi = \frac{Ml}{EI}$$

The change of curvature or angle of bend per unit length is

$$\frac{d\phi}{dl} = \frac{1}{R'} = \frac{M}{EI}$$

For a solid circular section, equation is reduced to

$$\phi = \frac{64Ml}{\pi Ed^4} = \frac{128MRn}{Ed^4} \text{ radians} \tag{7.22}$$

For a solid square section, Equation (7.22) reduces to

$$\phi = \frac{12Ml}{Eb^4} = \frac{24MRn}{Eb^4} \text{ radians} \tag{7.23}$$

Maximum bending stress f_b for circular section is given as

$$f_b = \frac{M}{I} y = \frac{32M}{\pi d^3} \tag{7.24}$$

Stiffness for torsion

$$= \frac{M}{\phi} \tag{7.25}$$

SOLVED PROBLEM 7.6

A close-coiled helical spring of 8 mm diameter wire with 12 coils of mean diameter 100 mm carries an axial load of 400 N. Find the shear stress induced and the deflection caused. What is the strain energy stored? Take $G = 8 \times 10^4$ N/mm^2.

Solution:

Given data: Load on spring $W = 400$ N, diameter of wire, $d = 8$ mm, No. of coils, $n = 12$ and mean diameter of coil, $D = 100$ mm.

Using

$$(f_s)_{max} = \frac{16WR}{\pi d^3}$$

$$= \frac{16 \times 400 \times 100}{\pi \times (8)^3 \times 2} = 199 \text{ N/mm}^2$$

and

$$\delta = \frac{64WR^3 n}{Gd^4} = \frac{64 \times 400 \times \left(\frac{100}{2}\right)^3 12}{8 \times 10^4 \times 8^4}$$

$$= 117.2 \text{ mm}$$

$$\text{Energy stored} = \frac{1}{2} W\delta$$

$$= \frac{1}{2} \times 400 \times 117.2 \text{ mm}$$

$$= 23{,}440 \text{ N} \cdot \text{mm.}$$

SOLVED PROBLEM 7.7

A close-coiled helical spring made out of 8 mm diameter wire has 18 coils. Each coil is of 80 mm mean diameter. If the maximum allowable stress in the spring is 140 MPa, determine the maximum allowable load on the spring, elongation of the spring and stiffness of the spring, $G = 8.2 \times 10^4$ N/mm^2.

Solution:

Given data: Diameter of wire, $d = 8$ mm, No. of coils, $n = 18$, mean diameter of coil, $D = 80$ mm and allowable stress = 140 MPa.

Using
$$(f_s)_{max} = \frac{16WR}{\pi d^3}$$

$$\therefore \quad W = \frac{f_s \times \pi \times d^3}{16R} = \frac{140 \times \pi \times 8^3}{16 \times 40} = 351 \text{ N.}$$

$\therefore$ Maximum allowable load = 351 N.

Elongation of the spring,
$$\delta = \frac{64WR^3n}{Gd^4}$$

$$= \frac{64 \times 335 \times (80/2)^3 \times 18}{8.2 \times 10^4 \times (8)^4}$$

$$= 73.5 \text{ mm}$$

Stiffness of the spring,
$$s = \frac{Gd^4}{64nR^3}$$

$$= \frac{8.2 \times 10^4 \times 8^4}{64 \times 18\left(\frac{80}{2}\right)^3}$$

$$= 4.56 \text{ N/mm}$$

SOLVED PROBLEM 7.8

A close-coiled helical spring made of circular wire is required to absorb 1,000 N·mm of energy for a deflection of 10 mm and the stress not exceeding 40 N/mm². Determine a suitable diameter and length of wire, given the mean coil diameter as 80 mm. Take N (or G) = 0.85×10^5 N/mm².

Solution:

Given data: Energy stored = 1,000 N·mm, deflection = 10 mm, allowable stress = 40 N/mm², mean coil diameter, $D = 80$ mm and N or $G = 0.85 \times 10^5$ N/mm².

Energy stored
$$= \frac{1}{2}W\delta$$

$$1{,}000 = \frac{1}{2} \times W \times 10$$

$$\therefore \quad W = \frac{1{,}000 \times 2}{10} = 200 \text{ N}$$

Using $$(f_s)_{max} = \frac{16WR}{\pi d^3}$$

$$40 = \frac{16\times 200\times\left(\frac{80}{2}\right)}{\pi d^3}$$

Simplifying $$d^3 = 1{,}019.1 \text{ mm}^3$$

Solving, $$d = 10.27 \text{ mm.}$$

Using $$\delta = \frac{64WR^3}{Gd^4}n$$

$\therefore$ $$n = \frac{\delta G d^4}{64WR^3}$$

$$= \frac{10\times 0.85\times 10^5\times(10.27)^4}{64\times 200\times\left(\frac{80}{2}\right)^3}$$

$$= 11.54 \approx 12$$

Length of wire, $$l = 2\pi nR$$

$$= 2\times\pi\times 12\times\frac{80}{2}$$

$$= 3{,}014.4 \text{ mm}$$

SOLVED PROBLEM 7.9

A close-coiled helical spring has a stiffness of 5 N/mm. Its length when fully comprised with adjacent coils touching each other is 40 cm. The modulus of rigidity of the material of the spring = 0.8×10^5 N/mm^2. Determine the wire diameter and mean coil diameter if their ratio is 1/10. What is the corresponding maximum shear stress in the spring?

Solution:

Given data: Stiffness = 5 N/mm, solid length = 40 cm, $G = 0.8\times 10^5$ N/mm^2 and diameter ratio = 110.

Solid length $$= n\times d$$

$$400 = n\times d$$

$\therefore$ $$n = \frac{400}{d}$$

Using $$s = \frac{Gd^4}{64nR^3}$$

$$s = \frac{0.8\times 10^5\times d^4}{64\times R^3\times n}$$

$$\therefore \quad d^4 = \frac{5\times 64\times R^3 \times n}{0.8\times 10^5} \quad \text{(i)}$$

$$d^4 = 0.004R^3 \times n$$

Substituting $n = \frac{400}{d}$ in Equation (i)

$$d^4 = 0.004R^3 \times \frac{400}{d}$$

$$d^5 = 1.6R^3$$

But, mean coil radius, $R = D/2$

$$\therefore \quad d^5 = 1.6\left(\frac{D}{2}\right)^3 = 0.2D^3$$

i.e.,

$$\frac{d^5}{D^3} = 0.2$$

or

$$\frac{d^3}{D^3}\times d^2 = 0.2$$

Substituting

$$\frac{d}{D} = \frac{1}{10}$$

Solving,

$$(1/10)^3.d^2 = 0.2$$

$$d = 10 \text{ mm}$$

$$\therefore \quad D = 10 \times d = 10 \times 10 = 100 \text{ mm}$$

$$n = \frac{400}{d} = \frac{400}{10} = 40$$

$$\therefore \quad \delta = 2 \times n = 2 \times 40 = 80 \text{ mm}$$

Using,

$$s = \frac{W}{\delta}$$

$$\therefore \quad W = s \times \delta = 5 \times 80 = 400 \text{ N}$$

Using,

$$f_s = \frac{16WR}{\pi d^3}$$

$$= \frac{16\times 400\times 100/2}{\pi\times 10^3}$$

$$= 101 \text{ N/mm}^2$$

SOLVED PROBLEM 7.10

A close-coiled helical spring of circular cross section has a mean coil diameter of 80 mm. The coil has developed an angular rotation of 90° when subjected to a torque of 7.0 N·m and when an axial load of 250 N applied the spring showed an elongation of 120 mm. Find the Poisson's ratio.

Solution:

Given data: Mean coil diameter = 80 mm, angular rotation = 90°, torque = 7.0 N·m, axial load = 250 N and $\delta = 120$ mm.

Using,
$$\phi = \frac{64Ml}{\pi E d^4}$$

i.e.,
$$d^4 = \frac{64Ml}{\pi E \phi} \quad \text{(i)}$$

Using,
$$\delta = \frac{64WR^3n}{4d^4}$$

i.e.,
$$d^4 = \frac{64WR^3n}{\delta G} \quad \text{(ii)}$$

Equating (i) and (ii)

$$\frac{64M \times 2\pi Rn}{\pi E\phi} = \frac{64WR^3n}{\delta G} \quad (\because l = 2\pi RM)$$

i.e.,
$$\frac{M}{W} = \frac{R^2E\phi}{2\pi\delta G} \quad \text{(iii)}$$

we know,
$$E = 2G(1+\mu)$$

i.e.,
$$(1+\mu) = \frac{E}{2G}$$

Equation (iii) can be written as

$$\frac{M}{W} = R^2\frac{\phi}{\delta}(1+\mu)$$

$$(1+\mu) = \frac{M}{W}\frac{\delta}{R^2\phi}$$

$$(1+\mu) = \frac{7 \times 1{,}000 \times 120}{250 \times \left(\frac{80}{2}\right)^2 \times 90 \times \frac{\pi}{180}}$$

$$= 1.3375$$

∴ Poisson's ratio, $\mu = 1.3375 - 1.0 = 0.3375$

SOLVED PROBLEM 7.11

A closely-coiled helical spring made of wire of 6 mm in diameter and having an inside diameter of 50 mm joins two shafts. The effective number of coils between the shafts is 16 and 0.750 kW is transmitted through the spring at 1,000 rpm. Calculate the relative axial twist in degrees between the ends of spring and also the intensity of bearing stress in the material. $E = 2 \times 10^5$ N/mm^2.

Solution:

Given data: Wire diameter = 6 mm, $n = 16$, power transmitted = 0.750 kW, speed, $N = 1{,}000$ rpm and $E = 2 \times 10^5$ N/mm^2.

Mean diameter of coil, $d = 0.05 + 0.06 = 0.056$ m

$$R = \frac{0.056}{2} = 0.028 \text{ m}$$

Using
$$P = \frac{2\pi NT}{60}$$

$$0.75 \times 1{,}000 = \frac{2\pi \times M \times N}{60} \qquad (\because M = T)$$

or
$$M = \frac{0.750 \times 1{,}000 \times 60}{2\pi \times 1{,}000} = 7.16 \text{ N}\cdot\text{m}$$

Using
$$\phi = \frac{128MRn}{Ed^4}$$

$$= \frac{128 \times 7.16 \times 0.028 \times 16 \times 1{,}000 \times 1{,}000}{2 \times 10^5 \times (6)^4}$$

$$= 1.58 \text{ radians}$$

$$= 90.5°$$

∴
$$f = \frac{32M}{\pi d^3} = \frac{32 \times 7 \times 10^3}{\pi \times (6)^3}$$

$$= 330 \text{ N/mm}^2$$

SOLVED PROBLEM 7.12

In a compound helical spring, the inner spring is arranged within and concentric with the outer one but is 9 mm shorter. The outer spring has 10 coils of mean diameter 24 mm and the wire diameter is 3 mm. Find the stiffness of the inner spring if an axial load of 150 N, causes the outer one to compress 18 mm. If the radial clearance between the springs is 1.5 mm, find the wire diameter of the inner spring when it has eight coils. $N = 77{,}000$ N/mm^2.

Solution:

Given data: Outer spring-coils = 10, outer spring-mean diameter = 24 mm, outer spring-wire diameter = 3 mm, axial load = 150 N and $G = 77{,}000$ N/mm^2.

Since the inner spring is shorter than the outer spring by 9 mm, the outer spring compresses first, before both the springs start taking the load.

Load taken by the outer spring to compress 18 mm } $W_{\text{outer}} = \dfrac{Gd^4\delta}{64R^3n}$

i.e.,
$$W_{\text{outer}} = \frac{77{,}000 \times 3^4 \times 18}{64 \times 12^3 \times 10} = 101.5 \text{ N.}$$

Load taken by the inner spring $W_{inner} = 150 - 101.5 = 48.5$ N

Compression, $\delta = 18 - 9 = 9$ mm.

∴ Stiffness of inner spring, $s_{inner} = \frac{W_{inner}}{\delta} = \frac{48.5}{9} = 5.39$ N/mm

Let d be the wire diameter of the inner spring. Now, mean coil radius of outer spring = 24/2 = 12 mm

Diameter of outer spring wire = 3 mm

Inner coil radius of outer spring $= 12 - \frac{3}{2} = 10.5$ mm

Radial clearance between the two springs = 1.5 mm

∴ Outer radius of inner spring = 10.5 – 1.5 = 9 mm

Mean coil radius, $R_{inner} = \left(9 - \frac{d}{2}\right)$ mm

Now

$$W_{inner} = \frac{77{,}000 \times d^4 \times 9}{64 \times \left(9 - \frac{d}{2}\right)^3 \times 10}$$

$$48.5 = \frac{1{,}564 d^4}{\left(9 - \frac{d}{2}\right)^3}$$

i.e.,

$$\left(9 - \frac{d}{2}\right)^3 = \frac{1{,}564}{48.5} = 32.24 d^4$$

$$(18 - d)^3 = 257.94 d^4$$

Solving $d = 2.06$ mm

Wire diameter of inner spring = 2.06 mm.

SOLVED PROBLEM 7.13

A closely-coiled helical spring of steel 8 mm in diameter having 12 complete turns with a mean diameter of 100 mm is subjected to an axial load of 25 N. Determine (i) the deflection of the spring, (ii) maximum shear stress in the wire and (iii) stiffness of the spring. Take $G = 80$ kN/mm^2.

Solution:

Given data: Diameter of wire, $d = 8$ mm, $n = 12$, mean diameter of coil, $D = 100$ mm, axial load = 25 N and $G = 80$ kN/mm^2.

Shear stress, $f_s = \frac{16WR}{\pi d^3}$

$$= \frac{16 \times 25 \times 100/2}{\pi \times (80)^3} = 12.43 \text{ N/mm}^2$$

Deflection, $\delta = \dfrac{64WR^3n}{Gd^4}$

$$= \frac{64 \times 25 \times (100/2)^3 \times 12}{8 \times 10^3 \times 8^3}$$

Stiffness, $\delta = 58.59$ mm

$$s = \frac{W}{\delta} = \frac{2s}{58.59} = 0.4267 \text{ N/mm}$$

SOLVED PROBLEM 7.14

A close-coiled helical spring is to have a stiffness of 1 kN/m of compression under a maximum load of 45 kN and maximum shear stress of 126 MPa. The solid length of the spring is to be 45 mm. Find the diameter of the wire and mean diameter of the coil required. $G = 42 \times 10^3$ N/mm^2.

Solution:

Given data: stiffness, $s = 1$ kN/m, load, $W = 45$ N, maximum shear stress, $f_s = 126 \times 10^6$ N/mm^2 and $G = 42 \times 10^3$ N/mm^2.

Stiffness, $s = \dfrac{W}{\delta}$ and $\delta = \dfrac{64WR^3n}{Gd^4}$

Substituting for δ and multiply by $\dfrac{d}{d}$

$$s = W \times \frac{Gd^4}{64WR^3n} \times \frac{d}{d}$$

$$1 = \frac{42 \times 10^3 \times d^5}{64 \times R^3 \times nd} = \frac{42 \times 10^3 \times d^5}{64 \times R^3 \times 45}$$

i.e., $\dfrac{R^3}{d^5} = 14.585$

Maximum shear stress, $f_s = \dfrac{16WR}{\pi d^3}$

$$126 = \frac{16 \times 45 \times R}{\pi d^3}$$

i.e., $R = 0.5497d^3$

$$\frac{(0.5497d^3)^3}{d^5} = 14.585$$

i.e., $d^4 = 87.79$

$$\therefore \quad d = 3.06 \text{ mm}$$

$$R = 0.5497 \times 3.06^3$$

$$R = 15.766 \text{ mm}$$

7.4.3 Open-coiled Helical Spring Subjected to Axial Force

An open-coiled helical spring, as shown in Figure 7.6(a), subjected to axial force is considered. In Figure 7.6(b) one limb of the coil at any point is shown with an angle of inclination of α to the horizontal. Then, the total length

$$l = 2\pi Rn \sec \alpha$$

where R is the mean radius of each coil and n the total number of turns.

The moment about the axis $u - u$ is WR. This moment is resolved into two components.

(i) A moment T in the plane $x - x$, causing a twisting action and

(ii) A moment M normal to the plane $x - x$ causing a bending action. Thus

$$T = WR \cos \alpha \tag{7.26}$$

and

$$M = WR \sin \alpha \tag{7.27}$$

The angle of torsion θ in the plane $x - x$ is given as

$$\theta = \frac{Tl}{GI_p}$$

Work done against torsional stress

$$= \frac{1}{2} T\theta = \frac{T^2 l}{2GI_p}$$

i.e.,

$$T\theta = \frac{W^2 R^2 \cos^2 \alpha}{GI_p} \tag{7.28}$$

Work done against bending stress

$$= \frac{M^2 l}{2EI}$$

$$= \frac{(W^2 R^2 \sin^2 \alpha) l}{2EI} \tag{7.29}$$

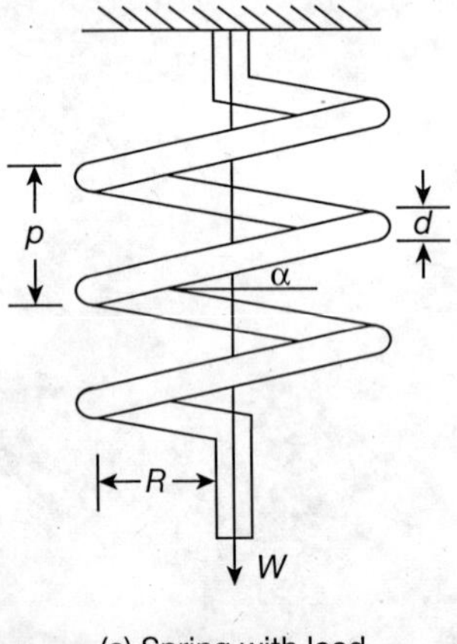

(a) Spring with load

(b) One LIMB of spring

Fig. 7.6 Open-coiled helical spring under axial force.

$\therefore$ Total work done
$$= \frac{\left(W^2R^2\cos^2\alpha\right)l}{2GI_p} + \frac{\left(W^2R^2\sin^2\alpha\right)l}{2EI} \tag{7.30}$$

Work done by the force
$$W = \frac{1}{2}W\delta \tag{7.31}$$

where δ is the axial extension.

Equating Equations (7.30) and (7.31)

$$\frac{1}{2}W\delta = \frac{\left(W^2R^2\cos^2\alpha\right)l}{2GI_p} + \frac{\left(W^2R^2\sin^2\alpha\right)l}{2EI}$$

$$= \frac{W^2R^2l}{2}\left(\frac{\cos^2\alpha}{GI_p} + \frac{\sin^2\alpha}{EI}\right)$$

i.e.,
$$\delta = W^2R^2l\left(\frac{\cos^2\alpha}{GI_p} + \frac{\sin^2\alpha}{EI}\right)$$

Substituting for $l = 2\pi Rn\sec\alpha$

and
$$J = 2I = \frac{\pi}{32}d^4 \quad \text{(for circular section)}$$

$$\delta = \frac{64WR^3n\sec\alpha}{d^4}\left(\frac{\cos^2\alpha}{G} + \frac{2\sin^2\alpha}{E}\right) \tag{7.33}$$

When $\alpha = 0$, Equation (7.32) reduces to that of the close-coiled helical spring subjected to axial force.

Now,

Bending stress,
$$f_b = \frac{M}{Z} = \frac{32WR\sin\alpha}{\pi d^3} \tag{7.34}$$

and shear stress,
$$f_s = \frac{16I}{\pi d^3} + \frac{W\sin\alpha}{\frac{\pi}{4}d^4}$$

i.e.,
$$f_s = \frac{16WR\cos\alpha}{\pi d^3}\left(l + \frac{d\tan\alpha}{4R}\right) \tag{7.35}$$

Neglecting the term $\frac{d\tan\alpha}{4R}$ which is usually of the range 0.01 to 0.02, then

$$f_s = \frac{16WR\cos\alpha}{\pi d^3} \tag{7.36}$$

7.4.4 Open-coiled Helical Springs Subjected to Axial Twist

Let the spring be subjected to an axial torque M. The torque M is considered positive if there is an increase in curvature or an increase in φ.

Resolving the moment along $x - x$ and $y - y$ axis,

We have

$M_{xx} = M \cos \alpha$, causing bending

$M_{yy} = M \sin \alpha$, causing torsion

Considering the strain energy equation

$$\frac{1}{2} M\phi = \frac{1}{2}\frac{(M\cos^2\alpha)l}{EI} + \frac{1}{2}\frac{(M\sin^2\alpha)l}{GI_p}$$

i.e.,

$$\phi = Ml\left(\frac{\cos^2\alpha}{EI} + \frac{\sin^2\alpha}{GI_p}\right)$$

Substituting for l, I and I_p for a circular section

$$\phi = \frac{64 MRn\sec\alpha}{d^4}\left(\frac{2\cos^2\alpha}{E} + \frac{\sin^2\alpha}{G}\right) \tag{7.38}$$

When $\alpha = 0$, Equations (6.38) reduces to that of close-coiled spring subjected to axial twist.

The axial extension caused by torque M is obtained by resolving the rotations. Thus, for circular section.

$$\delta = MRI\sin\alpha\cos\alpha\left(\frac{I}{GI_p} - \frac{l}{EI}\right)$$

i.e.,

$$= \frac{32MRl}{\pi d^4}\sin\alpha\cos\alpha\left(\frac{1}{G} - \frac{2}{E}\right) \tag{7.40}$$

$$\delta = \frac{64MR^2}{d^4} n\sin\alpha\left(\frac{1}{G} - \frac{2}{E}\right)$$

SOLVED PROBLEM 7.15

An open-coiled helical mild-steel spring has a wire diameter of 12 mm and 12 coils with a mean diameter of 90 mm. The angle of helix is 20°. Compute the deflection under an axial load of 250 N and the intensities of stresses developed in the wire. Take $G = 0.85 \times 10^5$ N/mm² and $E = 2.2 \times 10^5$ N/mm².

Solution:

Given data: Wire diameter, $d = 12$ mm, $n = 12$, mean diameter $D = 90$ mm, angle of helix, $\alpha = 20°$, axial load, $W = 250$ N, $G = 0.85 \times 10^5$ N/mm² and $E = 2.2 \times 10^5$ N/mm².

Deflection, $$\delta = \frac{64WR^3 n\sec\alpha}{d^4}\left(\frac{\cos^2\alpha}{G} + \frac{2\sin^2\alpha}{E}\right)$$

$$\delta = \frac{64\times 250\times\left(\frac{50}{2}\right)^3 \times 12\times\sec 20°}{(12)^4}\times\left(\frac{\cos^2 20°}{0.85\times 10^5} + \frac{2\sin^2 20°}{2.2\times 10^5}\right)$$

$$\delta = 8.98(1.039 + 0.106) = 10.28 \text{ mm}$$

Bending stress, $$f_b = \frac{32WR\sin\alpha}{\pi d^3}$$

$$f_b = \frac{32\times 250\times\left(\frac{90}{2}\right)\sin 20°}{\pi\times 12^3} = 22.96 \text{ N/mm}^2$$

Shear stress, $$f_s = \frac{16WR\cos\alpha}{\pi d^3}$$

$$= \frac{16\times 250\left(\frac{90}{2}\right)\cos 20°}{\pi\times 12^3}$$

$$f_s = 31.31 \text{ N/mm}^2$$

SOLVED PROBLEM 7.16

An open-coiled spring has 12 coils, when subjected to an axial load, the stresses due to bending and twisting inducted in the wire are 160 N/mm^2 and 180 N/mm^2, respectively. Calculate the maximum permissible axial load and the wire diameter for a maximum extension of 20 mm. The mean diameter of the coil is approximately equal to ten times that of the diameter of the wire. Take $E = 2.05 \times 10^5$ N/mm^2 and $G = 8.1 \times 10^4$ N/mm^2.

Solution:

Given data: $n = 12$, bending stress = 160 N/mm^2, twisting = 180 N/mm^2, maximum extension = 20 mm, $D = 10d$, $E = 2.05 \times 10^5$ N/mm^2 and $G = 8.1 \times 10^4$ N/mm^2.

Bending stress, $$f_b = \frac{32WR\sin\alpha}{\pi d^3} \quad \text{(i)}$$

Shear stress, $$f_s = \frac{16WR\cos\alpha}{\pi d^3} \quad \text{(ii)}$$

Dividing Equation (i) by Equation (ii)

$$\frac{f_b}{f_s} = \frac{32WR\sin\alpha \times \pi d^3}{\pi d^3 \times 16WR\cos\alpha}$$

i.e., $$\tan\alpha = \frac{f_b}{2f_s} = \frac{160}{2\times 180} = 0.44$$

∴ $$\alpha = 23.96°$$

Deflection, $$\delta = \frac{64WR^3 \sec\alpha}{d^4}\left(\frac{\cos^2\alpha}{G} + \frac{2\sin^2\alpha}{E}\right)$$

$$20 = \frac{64W\left(10\times\frac{d}{2}\right)\sec 23.96}{d^4}\left(\frac{\cos^2 23.96}{G} + \frac{2\sin^2 23.96}{E}\right)$$

$$20 = \frac{W}{d}\times 105{,}052.46\left(\frac{0.835}{0.81\times 10^5} + \frac{0.33}{2.05\times 10^5}\right)$$

$$20 = \frac{W}{d}\times 1.27$$

From Equation (i)

$$160 = \frac{32W\left(\frac{Wd}{2}\right)\sin 23.96°}{\pi d^2}$$

$$160 = 20.69\frac{Wd}{d^3} = 20.69\times\left(\frac{W}{d}\right)\frac{1}{d}$$

∴ Solving, $$d = 2.04 \text{ mm}$$

$$180 = \frac{16W\left(\frac{Wd}{2}\right)\cos 23.96°}{\pi d^3}$$

$$180 = 23.28\frac{Wd}{d^3}$$

$$= 23.28\left(\frac{W}{d}\right)\frac{1}{d}$$

$$d = \frac{23.28}{180}\times\frac{20}{1.27}$$

$$d = 2.04 \text{ mm}$$

SOLVED PROBLEM 7.17

An open-coiled helical spring made of steel rod 12 mm diameter has 10 coils of mean diameter 76 mm. The angle of helix is 15° . The spring is subjected to an axial torque of 10 N·m. Calculate the angle of rotation of the axis of the coil and the axial deflection. Take $E = 2.1 \times 10^2$ N/mm² and G (or N) = 0.84×10^5 N/mm².

Solution:

Given data: Diameter of rod, $d = 12$ mm, $n = 10$, mean diameter $D = 76$ mm, angle of helix = 15°, axial torque = 10 N·m, $E = 2.1 \times 10^5$ N/mm² and $G = 0.84 \times 10^5$ N/mm².

Angle of rotation,
$$\phi = \frac{64 MRn \sec\alpha}{d^4}\left(\frac{2\cos^2\alpha}{E} + \frac{\sin^2\alpha}{G}\right)$$

$$\phi = \frac{64 \times 10 \times 1{,}000 \times \left(\frac{76}{2}\right) \times 10 \times \sec 15°}{12^4}\left(\frac{2\cos^2 15°}{2.1 \times 10^5} + \frac{\sin^2 15°}{0.84 \times 10^5}\right)$$

$$\phi = \frac{12{,}142 \times 0.9688}{10^5} = 0.1175 \text{ rad.} = 6.44°$$

Axial deflection,
$$\delta = \frac{64\, MR^2}{d^4} n \sin\alpha\left(\frac{1}{G} - \frac{2}{E}\right)$$

$$\delta = \frac{64 \times 10 \times 1{,}000\left(\frac{76}{2}\right)^2 10 \sin 15°}{12^4} \times \left(\frac{1}{0.84 \times 10^5} - \frac{2}{2.1 \times 10^5}\right)$$

$$\delta = 0.275 \text{ mm}$$

7.5 FLAT SPIRAL SPRINGS

This type of spring consists of a uniform thin strip wound into a spiral in one plane, and pinned at its outer end. Winding of the spring is done by applying a torque to a spindle attached to the centre of the spiral.

Let T be the torque tending to wind up the spring. Let T_x and T_y be the components of reaction at the outer end of the spring O.

Taking moments about the spindle axis,

$$T = T_{xy} R \tag{7.41}$$

where R is the maximum radius of the spiral.

$$U = \int \frac{\left(T_y x - T_x y\right)^2 ds}{2EI}$$

Substituting for T_y from Equation (7.39)

$$U = \int \frac{\left(\left(\frac{T}{R}\right)x - T_x y\right) ds}{2EI}$$

As O is the fixed point $\dfrac{\partial U}{\partial x} = 0$, then

$$T_x = \left(\frac{T}{R}\right)\frac{\int xyds}{\int y^2 ds} = 0 \qquad \text{(by symmetry)}$$

Then,
$$\theta = \frac{\partial U}{\partial T}$$

i.e.,
$$\theta = \left(\frac{2T}{R^2}\right)\int \frac{x^2 ds}{2EI}$$

Treating the spiral as a uniform disc $\int x^2 ds$ may be approximately taken as $\left(\dfrac{R^2}{4} + R^2\right)1$, then,

$$\theta = 1.25\frac{Tl}{EI}$$

$$\text{Strain energy} = \frac{1}{2}T\theta$$

$$= \frac{1.25T^2 l}{2EI}$$

Maximum bending moment at the left-hand edge $= y \times 2R = 2T$

$$\text{Maximum stress } = \frac{2T}{Z}$$

$$= \frac{12T}{bt^2} \qquad (7.41)$$

where b is the width and t the thickness of the spiral material.

SOLVED PROBLEM 7.18

A flat spring is 2.0 m long, 5 mm wide and 0.3 mm thick. The maximum stress included at a point of greatest bending moment is 780 N/mm^2. Find the torque, the angle of rotation and the work stored, in the spring. Take $E = 2.05 \times 10^5$ N/mm^2.

Solution:

Given data: Length, $l = 2.0$ m, width = 5 mm, thickness = 0.3 mm, maximum stress = 780 N/mm^2 and $E = 2.05 \times 10^5$ N/mm^2.

$$\text{Maximum stress, } f = 12T$$

$$780 = \frac{12 \times T}{5 \times (0.3)^2}$$

Solving,
$$T = \frac{780 \times 5 \times (0.3)^2}{12} = 29.25 \text{ N} \cdot \text{mm}$$

Angle of rotation,
$$\theta = \frac{1.25Tl}{EI}$$

$$= \frac{1.25 \times 29.25 \times 2 \times 1,000}{2.05 \times 10^5 \left(\frac{1}{2} \times 5 \times 0.3^3 \right)}$$

$$= 5.285 \text{ radians}$$

Work stored in the spring $= \frac{1}{2} T\theta$

$$= \frac{1}{2} \times 29.25 \times 5.285$$

$$= 77.286 \text{ Nmm.}$$

7.6 BUFFER SPRINGS

Buffer spring is mostly used in the railway wagons. The shock between two colliding bodies may be softened or cushioned by means of buffers. The purpose of the buffers is to increase the duration of impact, by allowing considerable local deformation of the colliding bodies and thus to reduce the magnitude of the force which acts between the bodies during impact. Energy is absorbed by the buffer spring during the interval of time required for the speeds of the colliding bodies to the equalized, and is returned, either wholly or impact, during the remainder of the period of impact.

7.6.1 Design of Buffer Spring

Let d be the diameter of the spring wire.
Kinetic energy absorbed by the buffer spring $= \frac{1}{2} mv^2$

$$= \frac{W}{g} v^2 \quad (7.42)$$

where m = mass of wagon
W_w = weight of wagon
Let w be the equivalent load which when applied gradually on each spring causes a deflection.
∴ Energy stored in the springs.

$$= \frac{1}{2} \times W \times \delta \quad (7.43)$$

Equating Equations (7.42) and (7.43) we get the value of W.

Torque $T = W \times \frac{D}{2}$

We also know that the torque transmitted by the spring, $T = \frac{\pi}{16} f_s d^3$

Let n be the number of active turns of the spring coil

Then
$$\delta = \frac{64WR^3 n}{Gd^4}$$

i.e.,
$$\delta = \frac{8WD^3 n}{Gd^4} \tag{7.44}$$

7.6.2 Solid Length of Buffer Spring

When the compression spring is compressed until the coils come in contact with each other, then the spring is said to be solid. The solid length of a spring is the product of the total number of coils and the diameter of wire.

$$L_s = n \cdot d. \tag{7.45}$$

7.6.3 Free Length of Buffer Spring

Free length of spring = (Solid length) + (Maximum compression) + (Clearance between adjacent coils)

$$= nd + \delta_{max} + 0.15\delta_{max} \tag{7.46}$$

Pitch of coil
$$p = \frac{\text{Free length}}{n-1} \tag{7.47}$$

SOLVED PROBLEM 7.19

A rail wagon of mass 2 tonnes is moving with a velocity of 2 m/s. It is brought to rest by two buffers with springs of 300 mm diameter. The maximum deflection of spring is 250 mm. The allowable shear stress in the spring material is 600 MPa. Design the spring for buffers.

Solution:

Given data: Wagon weight $W = 2$ tonnes, velocity = 2 m/s, diameter of spring = 300 mm, deflection, $\delta = 250$ mm and allowable shear stress = 600 MPa.

Let d be the diameter of the spring wire.

$$\text{Kinetic energy of the wagon} = \frac{1}{2}mv^2$$
$$= \frac{1}{2} \times 2{,}000(2)^2 = 40{,}000 \text{ kg} \cdot \text{m}$$
$$= 40 \times 10^6 \text{ N} \cdot \text{mm} \tag{i}$$

Let W be the equivalent load which when applied gradually on each spring causes a deflection of 250 mm twice there are two springs.

Energy stored in the springs
$$= \frac{1}{2} \times W \times \delta \times 2$$
$$= W \times \delta = W \times 250 = 250\, W \text{ N} \cdot \text{mm} \tag{ii}$$

Equating Equation (i) and Equation (ii)

$$W = 40 \times \frac{10^6}{250} = 160 \times 10^3 \text{ N.}$$

Torque transmitted by the spring,

$$T = W \times R = 160 \times 10^3 \times \frac{300}{2} = 24 \times 10^6 \text{ N} \cdot \text{mm}$$

Torque transmitted by the spring is also

i.e.,
$$T = \frac{\pi}{16} f_s d^3$$

i.e.,
$$24 \times 10^6 = \frac{\pi}{16} \times 600 \times d^3$$

$$d = 58.84 \text{ mm}$$

Let n be the number of turns of the spring

$$\delta = \frac{64WR^3 n}{Gd^4}$$

$$= \frac{64 \times W \left(\frac{D}{2}\right)^3 n}{Gd^4}$$

$$= \frac{8 \times WD^3 n}{Gd^4}$$

$$250 = \frac{8 \times 160 \times 10^3 \times (300)^3 n}{84 \times 10^3 \times (58.84)^4}$$

Solving $n = 7.28$ say 8.

$$\text{Free length of the spring} = nd + \delta + 0.15\delta$$

$$= 8 \times 60 + 250 + 0.15 \times 250$$

$$= 767.5 \text{ mm}$$

$$\text{Pitch of coil} = \frac{\text{Free length}}{(n-1)}$$

$$= \frac{767.5}{8-1}$$

$$= 109.6 \text{ mm}$$

It may be noted that for square and ground ends the total number of turns is increased by 2, i.e., $n' = n + 2$.

SOLVED PROBLEM 7.20

A wagon weighing 2,000 kg and moving at 0.69 m/s has to be brought to rest by a buffer. Compute the number of springs that would be required in the buffer stop to absorb the energy of motion during a compression of 15 cm. Each spring has 15 coils, made of 2 cm wire, the mean diameter of the coils being 20 cm and $G = 0.8 \times 10^5$ N/mm². Also, determine the stiffness of spring.

Solution:

Given data: Mean diameter of coil = 20 cm, wagon weight, W = 2,000 kg, velocity, v = 0.69 m/s, deflection, δ = 15 cm, number of coils, n = 15, $G = 0.8 \times 10^5$ N/mm² and wire diameter = 2 cm.

$$\text{Mean radius of coil,} \quad R = \frac{20}{2} = 10 \text{ cm} = 100 \text{ mm}$$

Deflection, $\delta = \dfrac{64WR^3n}{Gd^4}$ where W = axial load

$$15 = \frac{64 \times W \times 10^3 \times 15}{0.8 \times 10^5 \times 2^4}$$

Solving, $W = 200$ kg.

$$\text{Kinetic energy absorbed by the buffer springs} = \frac{1}{2} \times W \times v^2$$

$$= \frac{1}{2} \times \frac{W}{g} \times v^2$$

$$= \frac{1}{2} \times \frac{2{,}000}{9.81} \times (0.69)^2$$

$$= 48.52 \text{ kg·m}$$

i.e., $= 4{,}852$ kg·cm (i)

$$\text{Energy stored in } n \text{ springs} = n \times \frac{1}{2} \times W \times \delta$$

Equating Equations (i) and (ii)

$$4{,}852 = n \times \frac{1}{2} \times W \times \delta$$

Solving $n = 3.23$

Number of springs required = 4.

$$\text{Stiffness of spring} = \frac{W}{\delta}$$

$$= \frac{200}{15}$$

$$= 13.33 \text{ kg/cm}$$

7.7 SPRINGS IN SERIES

Figure 7.7 shows two springs of stiffness s_1 and s_2 is arranged in series. Each spring will be subjected to load W and the total extension produced will be the sum of the extensions of two springs.

Total extension

$$\delta = \delta_1 + \delta_2$$

i.e.,

$$\frac{W}{s} = \frac{W}{s_1} + \frac{W}{s_2} \tag{7.48}$$

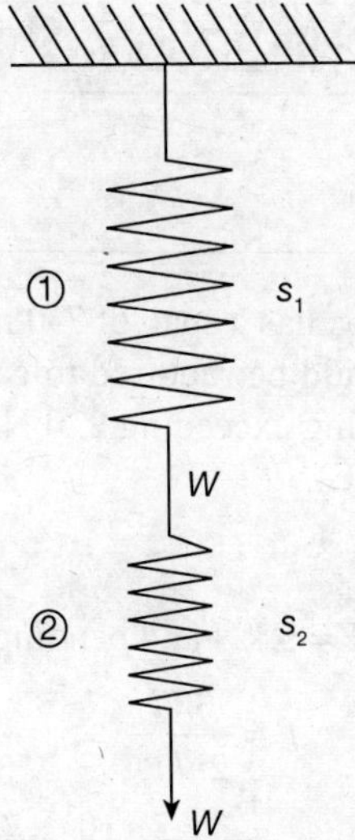

Fig. 7.7 Springs in series.

7.8 SPRINGS IN PARALLEL

When the springs are connected in parallel (Figure 7.8), the total load W is divided between the two springs, while the extension of each spring is the same.

$$W = W_1 + W_2$$

$$\delta.s = \delta s_1 + \delta s_2 \tag{7.49}$$

i.e.,

$$s = s_1 + s_2$$

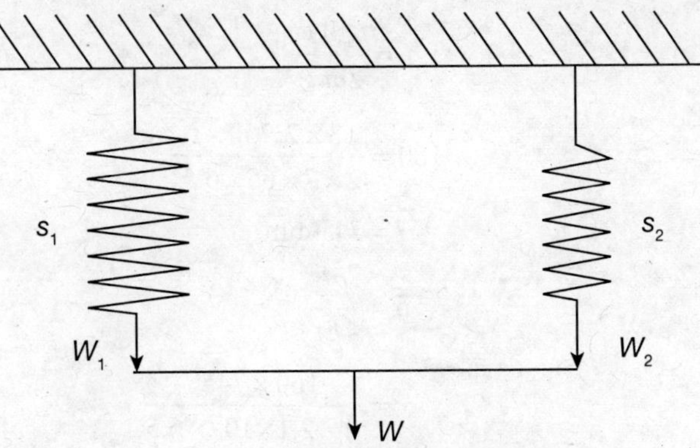

Fig. 7.8 Springs in parallel.

SOLVED UNIVERSITY QUESTIONS

SOLVED PROBLEM 7.21

A laminated spring of semi-elliptical type has a span of 750 mm and is built up of leaves 8 mm thick and 35 mm wide. How many leaves would be required to carry a central load of 5 kN without the bending stress in the material of the spring exceeding 220 N/mm². Assume modulus of elasticity of 200 N/mm² (Anna Univ., May 2006, EEE).

Solution:
Given data: Length, $l = 750$ mm, load, $W = 5$ kN and bending stress, $f = 220$ N/mm².

Bending stress,

$$f = \frac{3Wl}{2nbt^2}$$

$$220 = \frac{3 \times 5 \times 10^3 \times 750}{2 \times n \times 35 \times 8^2}$$

Solving,

$$n = 11$$

Number of leaves required = 11.

SOLVED PROBLEM 7.22

A leaf spring is to be made of seven steel plates 6.5 cm wide and 6.5 mm thick. Calculate the length of the spring, so that it may carry a central load of 2,750 N, the stress being limited to 16,000 N/cm². Also, calculate the deflection at the centre of the spring. Take $E = 2.1 \times 10^7$ N/cm² (Anna Univ., Dec. 2006, E&E).

Solution:
Given data: Central load, $W = 2{,}750$ N, no. of plates, $n = 7$, width of plate = 6.5 cm = 65 mm, thickness, $t = 6.5$ mm, bending stress, $f = 1{,}600$ kN/m² and $E = 2.1 \times 10^7$ N/m².

Bending stress,

$$f = \frac{3Wl}{2nbt^2}$$

i.e.,

$$160 = \frac{3 \times 2{,}750 \times l}{2 \times 7 \times 65 \times 6.5^2}$$

Solving,

$$l = 745 \text{ mm}$$

Deflection,

$$\delta = \frac{fl^2}{4Et}$$

$$= \frac{160 \times 745^2}{4 \times 2.1 \times 10^5 \times 6.5}$$

$$\delta = 16.26 \text{ mm}$$

SOLVED PROBLEM 7.23

A close-coiled spring is to carry an axial load of 1 kN. Its mean coil diameter is to be 10 times that of wire diameter. Calculate the diameter if the maximum shear stress in the material of the spring is 90 N/mm² (Anna Univ., May 2005, EEE).

Solution:

Given data: Axial load, $\frac{W}{2} = 1\text{ kN}$, $\frac{d}{D} = \frac{1}{10}$ and maximum shear stress, $f_s = 90$ N/mm².

Shear stress,
$$f_s = \frac{16WR}{\pi d^3}$$

$$90 = \frac{16 \times 1 \times 1{,}000 \times \left(\frac{D}{2}\right)}{\pi d^3}$$

$$90 = \frac{16 \times 1{,}000 \times \left(\frac{10d}{2}\right)}{\pi d^3}$$

$$90 \times \pi d^3 = 16{,}000 \times 5d$$

Solving,
$$d = 16.88 \text{ mm}$$

$$\text{Mean coil diameter} = 10 \times d = 10 \times 16.88$$
$$= 168 \text{ mm}$$

SOLVED PROBLEDM 7.24

A close-coiled helical spring is to have a stiffness of 90 N/mm, compression with a maximum load of 45 N and maximum shear stress of 12 N/mm². The solid length of the spring (i.e., coils are touching) is 45 mm. Find the diameter and its number of coils. $G = 40$ kN/mm² (Anna Univ., Nov. 2005, EEI).

Solution:

Given data: Stiffness, $s = 900$ N/m, $G = 40$ kN/mm², load, $W = 45$ N and shear stress, $f_s = 120$ N/mm².

$$\text{Deflection, } \delta = \frac{64WR^3 n}{Gd^4}$$

and
$$s = \frac{W}{\delta}$$

$$0.9 = \frac{40 \times 10^3 \times d^4}{64 \times R^3 n}$$

i.e.,
$$d^4 = \left[\frac{0.9 \times 64}{40 \times 10^3}\right] R^3 n$$

and
$$\text{shear stress, } f_s = \frac{16WR}{\pi d^3}$$

$$120 = \frac{16 \times 45 \times R}{\pi d^3}$$

i.e.,
$$R = \frac{120 \times \pi \times d^3}{16 \times 45 \times R}$$

i.e.,
$$R = 0.52d^3$$

Solid length (when the coils are touching)

$$= nd = 45$$

$$n = \frac{45}{d}$$

Substituting the values of R and n in Equation (i)

$$d^4 = \left[\frac{0.9 \times 64}{40 \times 10^3}\right] \times \left(0.52d^3\right)^3 \times \frac{45}{d}$$

$\therefore$
$$d = 3.24 \text{ mm}$$

$$\text{Mean coil radius, } R = 0.52d^3$$

$$= 0.52 \times (3.24)^3$$

$$= 17.68 \text{ mm}$$

$$\text{Number of coils, } n = \frac{45}{d}$$

$$= \frac{45}{3.24} = 13.88$$

$$= 14$$

SOLVED PROBLEM 7.25

A helical spring in which the mean diameter of the coil is eight times the wire diameter is to be designed to absorb 0.2 kN·m of energy with an extension of 100 mm. The maximum shear stress is not to exceed 125 N/mm^2. Determine the mean diameter of the spring, diameter of wire and the number of turns. Also find the load with which an extension of 40 mm could be produced in the spring. Assume $G = 84$ kN/mm^2 (Anna Univ., June 2007, EEE).

Solution:

Given data: $D = 8d$, energy = 0.20 kN·m, $\delta = 100$ mm, shear stress = 125 N/mm^2 and $G = 84$ kN/mm^2.

Strain energy,
$$U = \frac{f_s^2}{4G} \times (\text{Volume of the spring})$$

$$U = \frac{f_s^2}{4G} \times \frac{\pi d^2}{4} \times 2\pi R n$$

$$0.2 \times 10^6 = \frac{125^2}{4 \times 84 \times 10^3} \times \frac{\pi}{4} \times d^2 \times 2\pi \times (4d) \times n$$

$$d^3 n = 217,881$$

$$n = \frac{217,881}{d^3}$$

Shear stress, $f_s = \dfrac{16WR}{\pi d^3}$

$$125 = \frac{16W \times 4d}{\pi d^3}$$

i.e.,
$$d^2 = \frac{64W}{\pi \times 125} = 0.163W$$

$$\delta = \frac{64WR^3 n}{Gd^4}$$

Deflection,
$$100 = \frac{64W(4d)^3 \dfrac{217,881}{d^3}}{84 \times 10^3 \times (0.163W)^2}$$

Solving, $W = 3,997.65$ N.

$d^2 = 0.163 \times W = 0.163 \times 3,997.65 = 651.5$

$\therefore$ $d = 25.53$ mm

$\therefore$ $R = 4d = 4 \times 25.53 = 102.12$ mm

$D = 102.12 \times 2 = 204.24$ mm

$$n = \frac{217,881}{d^3} = \frac{217,881}{(25.53)^3} = 13.09$$

$$\delta = \frac{64WR^3 n}{Gd^4}$$

Deflection,
$$40 = \frac{64 \times W \times 102.12^3 \times 13.09}{80 \times 10^3 \times (25.53)^4}$$

Solving Load, $W = 1,523.7$ N.

SOLVED PROBLEM 7.26

A closely-coiled helical spring of round steel wire 10 mm in diameter having 10 complete turns with a mean diameter of 12 cm is subjected to an axial load of 250 N. Determine (i) deflection of the spring, (ii) maximum shear stress in the wire and (iii) stiffness of the spring and take $G = 0.8 \times 10^5$ N/mm^2 (Anna Univ., May 2007, ME).

Solution:

Given data: Diameter of wire, d = 10 mm, axial load = 250 N, mean diameter, D = 12 cm, $G = 0.8 \times 10^5$ N/mm² and mean radius, $R = \frac{12}{2} = 6$ cm = 60 mm

Deflection,
$$\delta = \frac{64WR^3}{Gd^4}n$$

$$\delta = \frac{64 \times 250 \times 60^3 \times 10}{0.8 \times 10^5 \times 10^4}$$

$$\delta = 43.2 \text{ mm}$$

Shear stress,
$$f_s = \frac{16WR}{\pi d^3}$$

$$= \frac{16 \times 250 \times 60}{\pi \times 10^3} = 76.43 \text{ N/mm}^2$$

Stiffness of spring,
$$s = \frac{W}{\delta} = \frac{250}{43.2} = 5.78 \text{ N/mm}$$

SOLVED PROBLEM 7.27

A close-coiled helical spring is required to absorb 2,250 joules of energy. Determine the diameter of the wire, the mean coil diameter of the spring and the number of coils necessary if (i) the maximum stress is not to exceed 400 MPa, (ii) the maximum compression of the spring is limited to 250 mm and (iii) the mean diameter of the spring is eight times the wire diameter. For the spring material, rigidity modulus is 70 GPa (Anna Univ., May 2008, ME).

Solution:

Given data: Energy = 2,250 joules, G = 70 GPa, maximum stress = 400 MPa, δ = 250 mm and $D = 8d$.

$$\text{Energy} = \frac{1}{2} \times W \times \delta$$

$$2{,}250 \times 10^3 = \frac{1}{2} \times W \times 250$$

∴
$$W = 18{,}000 \text{ N}$$

Maximum shear stress,
$$f_s = \frac{16WR}{\pi d^3}$$

i.e.,
$$400 = \frac{16 \times W \times 4d}{\pi d^3}$$

$$400 = \frac{16 \times 18{,}000 \times 4d}{\pi d^3}$$

$$d^2 = \frac{16 \times 18{,}000 \times 4}{400 \times \pi}$$

Solving, $d = 30.2$ mm

Mean diameter of spring $= 8d$

$= 8 \times 30.2 = 241.6$ mm

Deflection, $$\delta = \frac{64WR^3}{Gd^4} n = \frac{64 \times 18,000 \times 120.83^3 \times n}{70 \times 10^3 \times 30.24}$$

Equating δ to 250 and solving for n, $n = 7.2$.

SALIENT POINTS

- A spring is a device, made out of particular material, which can undergo considerable angular and linear deformation without undergoing permanent distortion.
- Stiffness of a spring is a measure of its capacity and is defined as the load required to produce unit deflection.
- A spring which is subjected to only bending and the resilience is also due to the same is called the bending spring.
- A spring which is subjected to a twisting moment or torsion and the resilence is only due to the same is called the torsion spring.
- Leaf springs (also called as laminated springs) are commonly used in carriages such as cars, lorries and railway wagons.
- Semi-elliptic leaf spring consists of a number of parallel metal strips having different lengths and same width which are placed are over the other.
- Quarter-elliptic leaf spring is half of a semi-elliptic spring.
- Helical springs are made up of thick wires called into a helix. They are of two types, viz., (i) close-coiled and (ii) open-coiled.
- In close-coiled helical spring, the wire is coiled very closely and so it is assumed that each turn is practically a plane perpendicular to the axis of the helix. An example of close-coiled springs is the torsion spring.
- In open-coiled helical spring, the wire is not coiled closely but with large gap between the two consecutive turns as a result of which there is bending stress also. Examples are torsion and bending springs.
- Flat spiral spring consists of a uniform thin strip wound into a spiral in one plane, and pinned at its outer end. Winding of the spring is done by applying a torque to a spindle attached to the centre of the spiral.
- Buffer springs are mostly used in railway wagons. The shock between two colliding bodies may be softened or cushioned by means of buffers.

QUESTIONS

1. What are leaf springs? Desive an expression for maximum bending stress for quarter-elliptic leaf springs.
2. Distinguish between close-coiled and open-coiled helical springs. Compare their deformation when both are subjected to an axial load.

3. What are buffer springs? Give their field applications.
4. Derive an expression for strain energy stored by a close-coiled helical spring subjected to axial thrust.
5. A leaf spring 750 mm long is required to carry a central load of 8 kN. If the central deflection is not to exceed 20 mm and the bending stress is not to be greater than 200 N/mm^2, determine the thickness, width and number of plates. Assume the width of the plates is 12 times, their thickness and modulus of elasticity of the springs material as 200 kN/mm^2 (Anna Univ. Question).
6. A quarter-elliptic laminated spring is 0.8 m long and is to carry a load 12.8 kN. The maximum deflection and the bending stress are restricted to 65 mm and 325 N/mm^2, respectively. Estimate suitable values of the number and size of plates required. Assume a suitable width to thickness ratio. Plates are available in multiples of 5 mm in width and 1.5 mm in thickness. $E = 2.2 \times 10^5$ N/mm^2.
7. A close-coiled helical spring made out of a 10 mm diameter steel bar has 12 complete coils, each of mean diameter of 100 mm. Calculate the stress induced in the section of rod, the deflection under the pull and the amount of energy stored in the spring during the extension. It is subjected to an axial pull of 200 N. Modulus of rigidity is 0.84×10^5 N/mm^2 (Anna Univ. Question).
8. A close-coiled helical spring of 100 mm mean diameter is made up of 6 mm diameter steel rod. The load is 250 N. Determine the following:
   (i) shear stress developed in the spring,
   (ii) stiffness of the spring,
   (iii) Deflection of the spring under the load and
   (iv) If another similar spring is attached to the above spring in series, determine the effective stiffness and deflection under the 250 N at the end of the second spring (Anna. Univ. Question).
9. A close-coiled helical spring is required to carry a maximum load of 800 N and to have a stiffness of 25 N/mm. The mean diameter is 60 mm. The allowable shear stress is 100 N/mm^2. Find the suitable diameter of the wire to make the spring and the approximate number of turns required (Anna Univ. Question).
10. A close-coiled helical spring, made of 6 mm diameter steel wire has 20 coils each of 100 mm mean diameter. The spring is subjected to an axial load of 70 N. Calculate (i) maximum shear stress induced (ii) the deflection and (iii) stiffness. Assume the modulus of rigidity of spring material as 84 kN/mm^2 (Anna Univ. Question).
11. An open-coiled helical spring has an helix angle of 28° with a wire diameter of 6 mm. Find the mean radius of the spring to give a vertical displacement of 20 mm and an angular rotation of the load end of 0.020 radius under an axial load of 35 N.
12. An open-coiled helical spring has coils having 16 turns with 10 mm diameter and a mean radius of 20 mm and of an helix angle of 25°. Compute the deflection and maximum stresses when subjected to an axial load of 280 N. If the axial load is replaced by an axial torque of 10 N·m, calculate the angle of rotation about the axis of the coil and the axial deflection. Take $E = 200{,}000$ N/mm^2 and $G = 80{,}000$ N/mm^2.
13. A flat steel spring has the following dimensions: width = 10 cm, thickness = 0.5 mm and length = 3.5 m. If the stress in the spring is not to exceed 500 N/mm^2, determine (i) the maximum

turning moment which could be applied, (ii) number of turns and (iii) energy stored. Take $E = 200{,}000$ N/mm^2.

14. Two springs with stiffness $f = 4.65$ N/mm and 5.56 N/mm are connected in series. Calculate the deflection if the load applied is 300 N. If the springs are connected in parallel what is the change in deformation?

8

Deflection of Beams

LEARNING OBJECTIVES

- 8.1 Introduction
- 8.2 Slope Deflection and Radius of Curvature
- 8.3 Methods of Determination of Slope and Deflection
- 8.4 Double Integration Method
- 8.5 Mecaulay's Method
- 8.6 Moment Area Method
- 8.7 Conjugate Beam Method

8.1 INTRODUCTION

When a beam is subjected to some type of loading, the beam deflects from its original position. The amount by which a beam deflects depends upon its support conditions, cross section of the beam and the bending moment.

Strength and stiffness are the two main design criteria for a beam. As per the strength criterion the beam should be strong enough to resist the bending stresses and shear stresses. According to stiffness criterion, the beam should be stiff enough to resist the deflection beyond a permissible limit.

When a load is placed on a beam, the beam deflects or sags as shown in Figure 8.1.

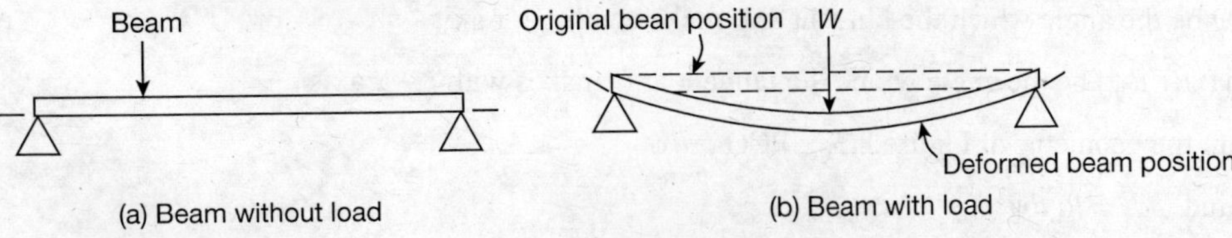

Fig. 8.1 Beam under load.

Under the load, the neutral axis becomes a curved line and is called *elastic curve* (Figure 8.2). The deflection 'y' is the vertical difference between a point on elastic curve and the unloaded neutral axis.

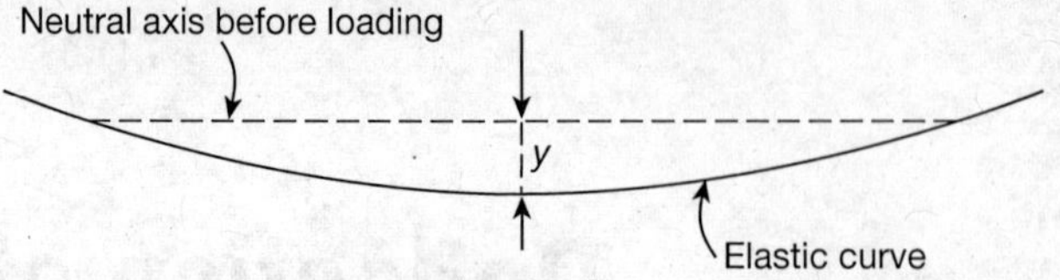

Fig. 8.2 Elastic curve.

The study of deflection is not only important as a design criterion but also as only means of determining bending moment (BM) and shear force (SF). in certain cases of beams, e.g., continuous beams and fixed beams, etc.

8.2 SLOPE DEFLECTION AND RADIUS OF CURVATURE

A small portion PQ of the beam, bent into an arc is considered (Figure 8.3).

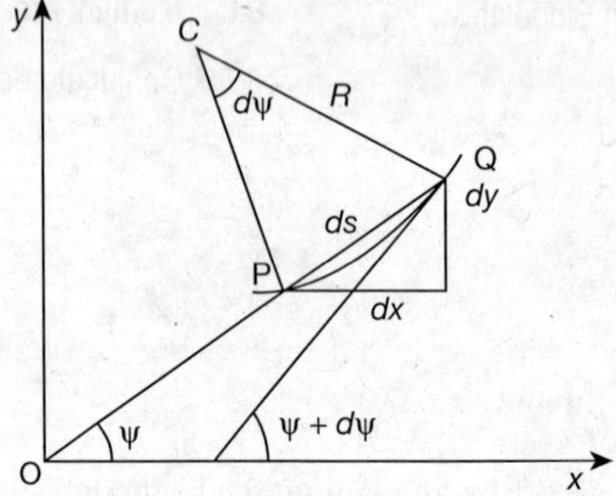

Fig. 8.3 Deflection of element.

Let ds be the elemental length PQ of the beam

R be the radius of the arc

C be the centre of the arc (into which the beam has been bent)

ψ be the angle which the tangent at P makes with $x - x$ axis

and $(\psi + d\psi)$ be the angle which the tangent at Q makes with $x - x$ axis

From the geometry of Figure 8.3, $\angle PCQ = d\psi$

and $ds = R.\ d\psi$

$$\therefore \qquad R = \frac{ds}{d\psi} = \frac{dx}{d\psi} \text{ (assuming } ds = dx\text{)}$$

or
$$\frac{1}{R} = \frac{d\psi}{dx} \tag{8.1}$$

If x and y are the coordinates of point P, then

$$\tan\psi = \frac{dy}{dx}$$

Since ψ is a very small angle, hence considering $\tan\psi = \psi$

$$\psi = \frac{dy}{dx}$$

Differentiating Equation (8.1) with respect to x, we get

$$\frac{d\psi}{dx} = \frac{d^2y}{dx^2}$$

i.e.,
$$\frac{1}{R} = \frac{d^2y}{dx^2} \tag{8.2}$$

From the bending equation,

$$\frac{M}{I} = \frac{E}{R}$$

i.e.,
$$\frac{M}{I} = E \times \frac{1}{R}$$

Substituting Equation (8.2) in the bending equation

$$M = EI\frac{d^2y}{dx^2} \tag{8.3}$$

Equation (8.3) is based only on BM. The effect of shear force being very small as compared to the BM it is neglected.

Differentiating Equation (8.3) with respect to x, we get

$$\frac{dM}{dx} = EI\frac{d^3y}{dx^3} \tag{8.4a}$$

But shear force,
$$S = \frac{dM}{dx}$$

i.e.,
$$S = EI\frac{d^3y}{dx^3} \tag{8.4b}$$

Differentiating Equation (8.4b) again with respect to x, we get

$$\frac{ds}{dx} = EI\frac{d^4y}{dx^4} \tag{8.5a}$$

But $\dfrac{ds}{dx} = w$ (the rate of loading)

i.e.,

$$w = EI\frac{d^4y}{dx^4} \tag{8.5b}$$

Hence, the relationships between curvature, slope, deflection, etc. at a section is given by

$$\text{Deflection} = y$$

$$\text{Slope} = \frac{dy}{dx}$$

$$\text{Bending moment} = EI\frac{d^2y}{dx^2}$$

$$\text{Shear force} = EI\frac{d^3y}{dx^3}$$

$$\text{Rate of loading} = EI\frac{d^4y}{dx^4}$$

8.3 METHODS OF DETERMINATION OF SLOPE AND DEFLECTION

Following are the important methods which are used for finding out the slope and deflection at a section in a loaded beam:

(i) Double integration method
(ii) Moment–area method
(iii) Mecaulay's method
(iv) Conjugate beam method.

The first two methods are suitable for single load, whereas the third one is suitable for several loads. The last method is a modified form of moment–area method and is conveniently used for finding out the slope and deflection of cantilever and simply supported beams with varying flexural rigidities.

8.4 DOUBLE INTEGRATION METHOD

In this method, the first moment, M at any distance x from one of the supports, is written with the sagging moment as positive.

Then

$$EI\frac{d^2y}{dx^2} = M$$

Integrating,

$$EI\frac{dy}{dx} = \int_0^x Mdx + C_1$$

and $$EIy = \int_0^x \int_0^x M dx + C_1 x + C_2$$

where C_1 and C_2 are constants.

Constants C_1 and C_2 are found by making use of boundary conditions.

(a) At simply supported/roller ends, deflection $y = 0$

(b) At fixed ends, deflection $y = 0$ and slope $\frac{dy}{dx} = 0$

(c) At point of symmetry $\frac{dy}{dx} = 0$

8.4.1 Simply Supported Beam Carrying a Point Load at the Centre

A simply supported beam AB of length l and carrying a point load W at the centre of beam C is considered (Figure 8.4). As the load is symmetrically applied, the reactions R_A and R_B will be equal.

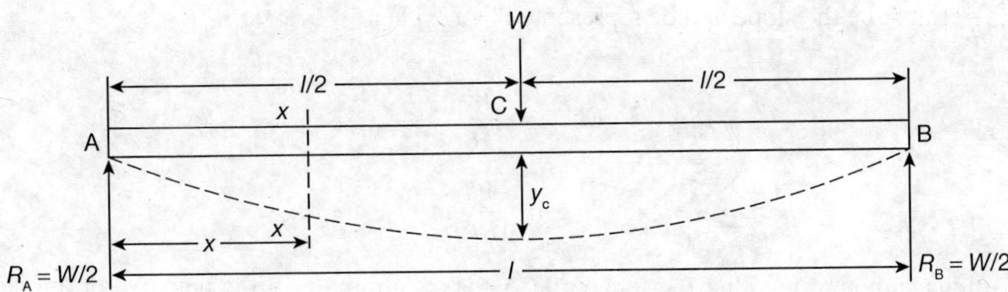

Fig. 8.4 Simply supported beam with central point load.

Now, $$R_A = R_B = \frac{W}{2}$$

A section $x - x$ at a distance x from A is considered.

Further, BM at this section is given as $M = R_A \times x$

i.e., $$M = \frac{W}{2} x$$

Then $$EI \frac{d^2 y}{dx^2} = \frac{Wx}{2}$$

Integrating the above equation $$EI \frac{dy}{dx} = \frac{W}{2} \times \frac{x^2}{2} + C_1$$

where C_1 is a constant of integration. It is known that at $x = \frac{l}{2}$, the slope $\frac{dy}{dx} = 0$

As the maximum BM is at the centre, hence slope at the centre will be zero. Substituting these values in the above equation

$$0 = \frac{W}{4} \times \left(\frac{l}{2}\right)^2 + C_1$$

i.e.,
$$C_1 = \frac{-Wl^2}{16}$$

Substituting the value of C_1 in the above slope equation

$$EI\frac{dy}{dx} = \frac{Wx^2}{4} - \frac{Wl^2}{16} \tag{8.6}$$

Equation (8.6) is known as *slope equation.* The slope at any point on the beam by substituting the value of x can be found. Slope is maximum at A and B.

1. Slope at A

Substituting $x = 0$ in Equation (8.6)

$$EI\left(\frac{dy}{dx}\right)_{\text{at A}} = \frac{W}{4}\times 0 - \frac{Wl^2}{16}$$

Let $\left(\frac{dy}{dx}\right)$ at A be the slope and be represented by i_A. Then

$$EI\ i_A = \frac{-Wl^2}{16}$$

i.e.,
$$i_A = \frac{-Wl^2}{16EI} \tag{8.7a}$$

The minus sign signifies that the tangent at A makes an angle in the negative or anticlockwise direction.

$$i_A = \frac{Wl^2}{16EI} \tag{8.7b}$$

The slope at B will be similar to slope at A since the load is symmetrically applied on the beam.

i.e.,
$$i_B = i_A = -\frac{Wl^2}{16EI}$$

2. Deflection

As deflection at any point is obtained by integrating the slope equation, we get

$$EIy = \frac{W}{4}\cdot\frac{x^3}{3} - \frac{Wl^2x}{16} + C_2 \tag{8.8a}$$

where C_2 is the second constant of integration.

At $x = 0$, $y = 0$. Substituting these values in Equation (8.8a), we get $C_2 = 0$. Then

$$EIy = \frac{Wx^3}{12} - \frac{Wl^2}{16}x \tag{8.8b}$$

Equation (8.8b) is known as the *deflection equation.*

Deflection at any point on the beam can be obtained by substituting the value of x. The deflection is maximum at centre, y_C, point C, where $x = l/2$.

$$EI \cdot y_c = \frac{W}{12}\left(\frac{l}{2}\right)^3 - \frac{Wl^2}{16}\left(\frac{l}{2}\right)$$

$$= \frac{Wl^3}{96} - \frac{Wl^3}{32}$$

$$= -\frac{Wl^3}{48}$$

$$y_c = -\frac{Wl^3}{48EI} \quad (8.9)$$

Minus sign signifies that the deflection is downwards.

8.4.2 Simply Supported Beam with a Uniformly Distributed Load

A simply supported beam AB of length l and carrying an uniformly distributed load of w per unit length over the entire length is considered (Figure 8.5).

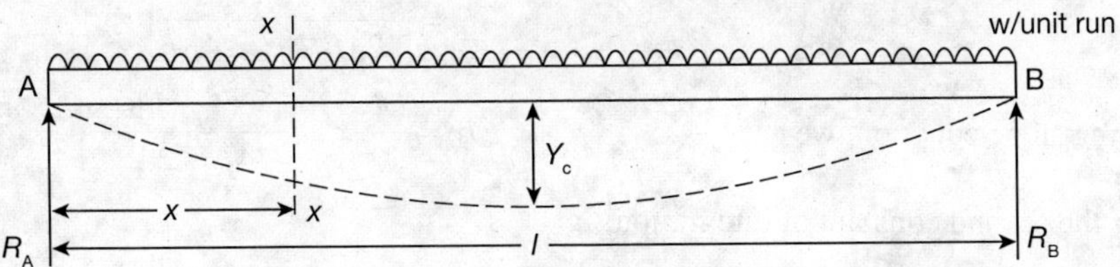

Fig. 8.5 Simply supported beam with UDL.

Reactions at A and B will be equal and given as $R_A = \frac{wl}{2}$ and $R_B = \frac{wl}{2}$

A section $x-x$ at a distance x from A is considered. Further BM at this section is given as

$$M_x = R_A x - w \times x \times \frac{x}{2}$$

i.e.,

$$EI\frac{d^2y}{dx^2} = \frac{wl}{2}x - \frac{wx^2}{2}$$

Integrating the above equation

$$EI\frac{dy}{dx} = \frac{wl}{2} \times \frac{x^2}{2} - \frac{wx^3}{2 \times 3} + C_1$$

i.e.,

$$EI\frac{dy}{dx} = wl\frac{x^2}{4} - \frac{wx^3}{6} + C_1 \quad (8.10a)$$

where C_1 is a constant of integration.

At $x = \frac{l}{2}$, $\quad \frac{dy}{dx} = 0$

$$\therefore \qquad 0 = \frac{wl}{4}\left(\frac{l}{2}\right)^2 - \frac{w}{6}\left(\frac{l}{2}\right)^3 + C_1$$

i.e.,
$$0 = \frac{wl^3}{16} - \frac{wl^3}{48} + C_1$$

i.e.,
$$C_1 = \frac{wl^3}{24}$$

Hence,
$$EI\frac{dy}{dx} = \frac{wlx^2}{4} - \frac{wx^3}{6} - \frac{wl^3}{24} \tag{8.10b}$$

Equation (8.10b) is the slope equation.

Slope at A is obtained by substituting $x = 0$ as $EI\ i_A = -\dfrac{wl^3}{24}$

i.e.,
$$i_A = -\frac{wl^3}{24EI} = -\frac{wl^2}{24EI}$$

Similarly
$$i_B = -\frac{wl^3}{24EI} = -\frac{wl^2}{24EI} \tag{8.10c}$$

Integrating slope equation, we get
$$EIy = \frac{wl}{12}x^3 - \frac{wx^4}{24} - \frac{wl^3x}{24} + C_2$$

where C_2 is the second constant of integration.

At $x = 0$, $y = 0$, then $C_2 = 0$

i.e.,
$$EIy = \frac{wlx^3}{12} - \frac{wx^4}{24} - \frac{wl^3}{24}x \tag{8.11}$$

Equation (8.11) is the deflection equation. Deflection at mid-span (y_{max}) is got by substituting $x = y_2$ in the above equation:

$$EIy_{max} = \frac{wl}{12}\left(\frac{l}{2}\right)^3 - \frac{w}{24}\left(\frac{l}{2}\right)^4 - \frac{w}{24}l^3\left(\frac{l}{2}\right)$$

i.e.,
$$EIy_{max} = \frac{wl^4}{96} - \frac{wl^4}{384} - \frac{wl^4}{48}$$

i.e.,
$$y_{max} = -\frac{5wl^4}{384EI} \tag{8.12}$$

8.4.3 Simply Supported Beam with a Gradually Varying Load

A simply supported beam AB of length l and carrying a gradually varying load from zero at B to w per unit run at A is considered (Figure 8.6).

From the geometry of the Figure 8.6, the reactions at A and B are given as

$$R_A = \frac{wl}{3} \text{ and } R_B = \frac{wl}{6}.$$

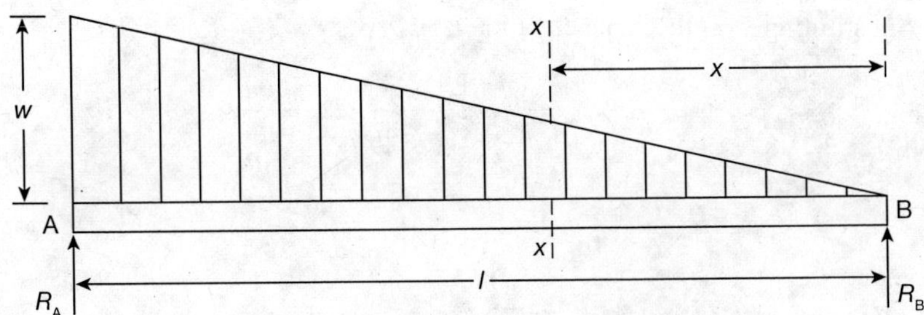

Fig. 8.6 Simply supported beam with varying (triangular) load.

A section $x - x$ at a section x from B is considered.

Now, BM at this section is given by $M_x = R_B x - \dfrac{wx}{l} \times \dfrac{x}{2} \times \dfrac{x}{3}$

$$M_x = \frac{wlx}{6} - \frac{wx^3}{6l}$$

$$\therefore \qquad EI\frac{d^2y}{dx^2} = \frac{wlx}{6} - \frac{wx^3}{6l} \qquad (8.13)$$

Integrating the above equation

$$EI\frac{dy}{dx} = \frac{wlx^2}{6\times 2} - \frac{wx^4}{6\times 4l} + C_1 = \frac{wlx^2}{12} - \frac{wx^4}{24l} + C_1 \qquad (8.14a)$$

where C_1 is a constant of integration.

Integrating the slope equation once again.

$$EIy = \frac{wlx^3}{36} - \frac{wx^5}{120l} + C_1x + C_2 \qquad (8.14b)$$

where C_2 is the second constant of integration. This is the deflection equation.

At $x = 0$, $y = 0$

$$\therefore \qquad C_2 = 0$$

Also, at $x = l$, $y = 0$

$$\therefore \qquad 0 = \frac{wl}{36} \times l^3 - \frac{w \times l^5}{120l} + C_1 l$$

i.e.,

$$C_1 = \frac{7wl^3}{360}$$

Now, substituting the value of C_1 in Equation (8.14a)

$$EI\frac{dy}{dx} = \frac{wlx^2}{12} - \frac{wx^4}{24l} - \frac{7wl^3}{360} \qquad (8.15a)$$

Equation (8.15a) is the slope equation for the given load condition.

Slope at A: Substituting $x = l$ the slope equation is

$$EI \times i_A = \frac{wl}{12} \times l^2 - \frac{w}{24l} \times l^4 - \frac{7wl^3}{360}$$

$$= \frac{wl^3}{45}$$

$$\therefore \quad i_A = \frac{wl^3}{45EI}$$

Slope at B: Substituting $x = 0$ the slope equation is

$$EI \times i_B = -\frac{7wl^3}{360}$$

$$i_B = -\frac{7wl^3}{360EI}$$

Now, substituting the value of C_1 in Equation (8.14b), we get

$$EI \times y = \frac{wlx^3}{36} - \frac{wx^5}{120l} - \frac{7wl^3 x}{360}$$

i.e.,

$$y = \frac{1}{EI}\left(\frac{wlx^3}{36} - \frac{wx^5}{120l} - \frac{7wl^3 x}{360}\right) \quad (8.15b)$$

Deflection at the centre of the span is obtained by substituting $x = l/2$:

$$y_c = \frac{1}{EI}\left[\frac{wl}{36}\left(\frac{l}{2}\right)^3 - \frac{w}{120l}\left(\frac{l}{2}\right)^5 - \frac{7wl^3}{360}\left(\frac{l}{2}\right)\right] \quad (8.15c)$$

i.e.,

$$y_c = -0.0065\frac{wl^4}{EI}$$

Maximum deflection will occur where slope of the beam is zero.

$\therefore$ Equating the slope equation to zero

$$\frac{wlx^2}{12} - \frac{wx^4}{24l} - \frac{7wl^3}{360} = 0$$

Solving $x = 0.519l$.

Now, substituting the value of x in the deflection equation

$$y_{max} = \frac{1}{EI}\left[\frac{wl}{36}(0.519l)^3 - \frac{w}{120l}(0.519l) - \frac{7wl}{360}(0.519l)\right]$$

i.e.,

$$y_{max} = 0.00652\frac{wl^4}{EI} \quad (8.16)$$

SOLVED PROBLEM 8.1

A wooden beam 4 m long, simply supported at its ends is carrying a point load of 7.25 kN at its centre. The cross section of the beam is 140 mm wide and 240 mm deep. If E for the beam is 6×10^3 N/mm^2, find the (i) deflection at the centre and (ii) slopes at the supports.

Solution:

Given data: Length, $l = 4$ m, point load, $W = 7.25$ kN, beam cross section = 140 mm × 240 mm and $E = 6 \times 10^3$ N/mm^2.

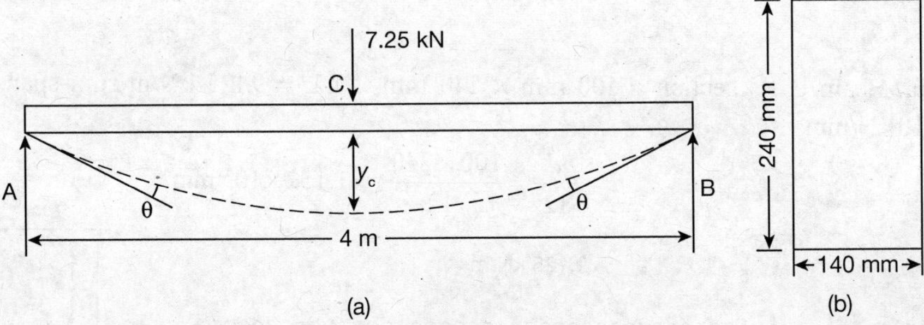

Fig. 8.7

Let y_c be the deflection at the centre and i_A and i_B be the slopes at the support.

Moment of inertia,

$$I = \frac{bd^3}{12}$$

$$= \frac{140 \times 240^3}{12}$$

i.e.,

$$I = 1.612 \times 10^8 \text{ mm}^4$$

(i) Deflection at the centre

Central deflection,

$$y_c = \frac{wl^3}{48EI}$$

$$= \frac{7.25 \times 1{,}000 \times (4 \times 1{,}000)^3}{48 \times 6 \times 10^3 \times 1.612 \times 10^8}$$

$$= 9.989 \text{ mm}$$

$$= 10 \text{ mm.}$$

(ii) Slope at the support

Slope at A, i_A = slope at B, i_B as the load is only a central load.

∴

$$i_A = i_B = \frac{Wl^2}{16EI}$$

$$= \frac{7.25 \times 1{,}000 \times (4 \times 1{,}000)^2}{16 \times 6 \times 10^3 \times 1.612 \times 10^8} \text{ radians}$$

$$= 7.4958 \times 10^4 \times \frac{180}{\pi} \text{ degrees}$$

$$= 0.429°$$

SOLVED PROBLEM 8.2

A beam of uniform rectangular section 100 mm wide and 240 mm deep is simply supported at its ends. It carries a uniformly distributed load of 9.125 kN/m run over the entire span of 4 m. Find (i) deflection of the centre, if $E = 1.1 \times 10^4$ N/mm^2 and (ii) slope at the supports.

Solution:

Given data: Beam cross section = 100 mm × 240 mm, UDL = 9.125 kN/m run, span, l = 4 m, $E = 1.1 \times 10^4$ N/mm^2

$$I = \frac{bd^3}{12} = \frac{100 \times 240^3}{12} = 1.152 \times 10^8 \text{ mm}^4$$

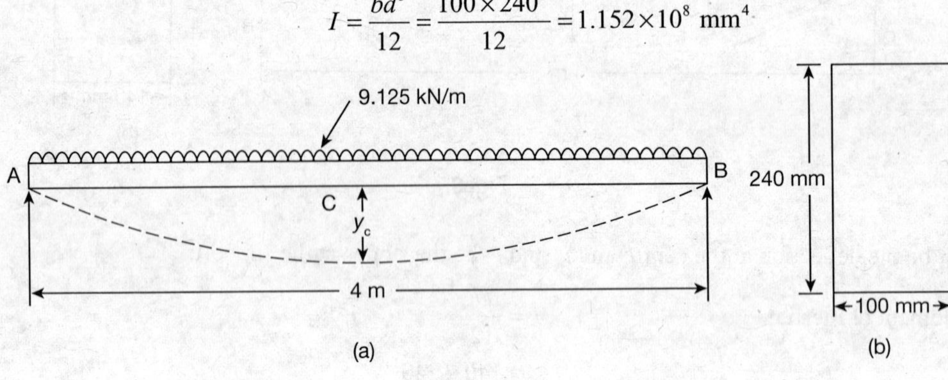

Fig. 8.8

Let i_A be the slope at the support and y_c be the maximum deflection.

(i) Slopes at the supports of the beam

$$i_A = i_B = -\frac{wl^3}{24EI}$$

$$= -\frac{9.125 \times 1{,}000 \times 4{,}000^3}{24 \times 1.1 \times 10^4 \times 1.152 \times 10^6}$$

$$= -0.0192 \text{ radians.}$$

(ii) Maximum deflection

Maximum deflection is at the centre

$$\therefore \quad y_c = \frac{5}{384}\frac{wl^4}{EI}$$

$$= \frac{5}{384} \times \frac{9{,}125 \times 4{,}000^4}{1.1 \times 10^4 \times 1.152 \times 10^8}$$

$$= 24.003 \text{ mm.}$$

SOLVED PROBLEM 8.3

A beam of length 4.8 m and of uniform rectangular section is simply supported at its ends. It carries a uniformly distributed load of 9.375 kN/m run over the entire length. Calculate the width and depth of the beam if permissible bending stress is = 7 N/mm² and maximum deflection is not to exceed 0.95 cm. Take $E = 1.05\times10^4$ N/mm².

Solution:

Given data: Length, $l = 4.8$ m, UDL = 9.375 kN, bending stress, $f = 7$ N/mm², $E = 1.05\times10^4$ N/mm² and central deflection, $y_c = 0.95$ cm.

Let b be the width of the beam in mm and d be the depth of the beam in mm.

Moment of inertia, $$I = \frac{bd^3}{12}$$

Central deflection, $$y_c = \frac{5}{384}\times\frac{wl^4}{EI}$$

i.e., $$9.5 = \frac{5}{384}\times\frac{9.375\times4{,}800^4}{1.05\times10^4\times\dfrac{bd^3}{12}}$$

Rearranging, $$bd^3 = \frac{5}{384}\times\frac{9.375\times4{,}800^4\times12}{1.05\times10^4\times9.5}$$

$$bd^3 = 818.52\times10^7 \text{ mm}^4 \qquad \text{(i)}$$

For a simply supported beam with UDL, the maximum bending moment,

$$M = \frac{wl^2}{8} = \frac{9.375\times4{,}800^2}{8} = 2{,}700{,}000 \text{ N}\cdot\text{mm}$$

Using bending equation, $$M = \frac{fI}{y}$$

i.e., $$2{,}700{,}000 = \frac{7\times bd^3/12}{(d/2)}$$

Rearranging, $$bd^2 = \frac{2{,}700{,}000\times12}{14}$$

$$bd^2 = 231{,}42{,}857.14 \text{ mm}^3 \qquad \text{(ii)}$$

Dividing Equation (i) by Equation (ii), we get

$$d = \frac{818.52\times10^7}{23{,}142{,}857.14} = 353 \text{ mm}$$

Substituting this value of d in Equation (ii), we get

$$b\times(353)^2 = 23{,}142{,}857.14$$

i.e.,
$$b = \frac{23,142,857.14}{(353)^2}$$
$$b = 185.72 \text{ mm}$$

Cross section of the beam is 185.72 mm × 353 mm

SOLVED PROBLEM 8.4

A steel girder of 6 m length acting as a beam carries a uniformly distributed load of 3.5 kN/m run throughout its length. If $I = 30 \times 10^{-6}$ m^4, find the slope and deflection (under this load) in the beam at a distance of 1.8 m from one end. Take $G = 200$ GN/m^2.

Solution:

Given data: Length, $l = 6$ m, UDL = 3.5 N/m run, $I = 30 \times 10^{-6}$ m^4, $G = 200$ GN/m^2 and $x = 1.8$ m.

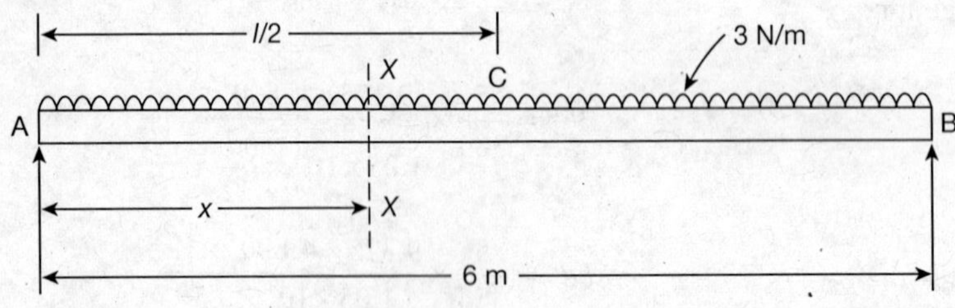

Fig. 8.9

A section $x - x$ at a distance x from A is considered.

Now, BM at this section,
$$M_x = \frac{wl}{2}x - \frac{wx^2}{2}$$

i.e.,
$$EI\frac{d^2y}{dx^2} = \frac{wl}{2}x - \frac{wx^2}{2} \qquad \text{(i)}$$

Integrating Equation (i), we have
$$EI\frac{dy}{dx} = \frac{wl}{4}x^2 - \frac{wx^3}{6} + C_1 \qquad \text{(iia)}$$

where C_1 is a constant of integration.

When $x = \frac{l}{2}, \frac{dy}{dx} = 0$, then
$$0 = \frac{wl}{4}\left(\frac{l}{2}\right)^2 - \frac{w}{6}\left(\frac{l}{2}\right)^3 + C_1$$
$$= \frac{wl^3}{16} - \frac{wl^3}{48} + C_1$$

i.e., $$C_1 = -\frac{wl^3}{24}$$

$\therefore$ $$EI\frac{dy}{dx} = \frac{wl}{4}x^2 - \frac{wx^3}{6} - \frac{wl^3}{24} \qquad \text{(iib)}$$

Integrating the slope equation [Equation (iib)] we have

$$EIy = \frac{wlx^3}{12} - \frac{wx^4}{24} - \frac{wl^3}{24}x + C_2 \qquad \text{(iiia)}$$

where C_2 is a constant of integration

When $x = 0,\ y = 0$ $\quad\therefore\ C_2 = 0,$

Then

$$EIy = \frac{wlx^3}{12} - \frac{wx^4}{24} - \frac{wl^3}{24}x \qquad \text{(iiib)}$$

Slope at $x = 1.8$ m: substituting $x = 1.8$ m in Equation (iib)

$$EI\frac{dy}{dx} = \frac{3,500 \times 6 \times 1.8^2}{4} - \frac{3,500 \times 1.8^3}{6} - \frac{3,500 \times 6^3}{24}$$
$$= -18,173.2$$

$\therefore$ $$\frac{dy}{dx} = i_{1.8} = -\frac{18,173.2}{200 \times 10^9 \times 30 \times 10^{-6}}$$

i.e., $$i_{1.8} = -0.00302 \text{ radians}$$

Deflection at $x = 1.8$ m: substituting $x = 1.8$ m in Equation (iiib)

$$EIy_{1.8} = \frac{3,500 \times 6 \times 1.8^3}{12} - \frac{3,500 \times 1.8^4}{24} - \frac{3,500 \times 6^3 \times 1.8}{24}$$
$$= -48,779.6$$

$$y_{1.8} = -\frac{48,779.6}{200 \times 10^9 \times 30 \times 10^{-6}}$$
$$y_{1.8} = -8.13 \text{ mm}$$

Slope at $x = 1.8$ m is -0.0303 radian.

Deflection at $x = 1.8$ m is -8.13 mm.

8.4.4 Cantilever Beam with a Point Load at the Free End

A cantilever beam AB of length l and carrying a point load W at the free end is considered (Figure 8.10). A section $x-x$ at a distance x from the fixed and A is considered.

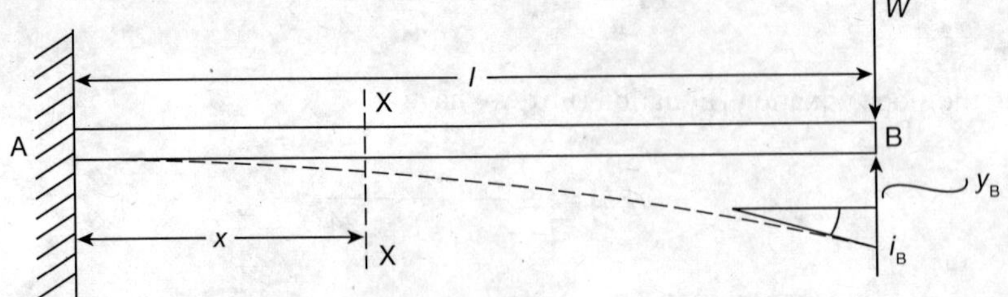

Fig. 8.10 Cantilever beam with point load at free end.

Now, BM at the section $x-x$ is given by

$$M_x = -W(l-x)$$

Minus sign is assigned as it is a hogging moment

$$\therefore \quad EI\frac{d^2y}{dx^2} = -W(l-x) \tag{8.17}$$

Integrating the above equation, we get

$$EI\frac{dy}{dx} = -W\left(lx - \frac{x^2}{2}\right) + C_1$$

$$EI\frac{dy}{dx} = -Wlx + \frac{Wx^2}{2} + C_1 \tag{8.18a}$$

where C_1 is a constant of integration.

At the fixed end, A: $x = 0$ and $\frac{dy}{dx} = 0$

$$\therefore \quad C_1 = 0$$

$$\therefore \quad EI\frac{dy}{dx} = -Wlx + \frac{Wx^2}{2} \tag{8.18b}$$

Equation (8.18c) is the slope equation

$$\frac{dy}{dx} = -\frac{1}{EI}W\left[lx - \frac{x^2}{2}\right] \tag{8.18c}$$

Slope at B: making $x = l$, we have

$$i_B = \frac{dy}{dx} = -\frac{1}{EI}W\left[l \times l - \frac{l^2}{2}\right]$$

i.e., $$i_B = -\frac{Wl^2}{2EI} \tag{8.19}$$

Integrating the slope equation we get the deflection equation:

i.e., $$EIy = -W\left[l\frac{x^2}{2} - \frac{x^3}{6}\right] + C_2 \tag{8.20a}$$

where C_2 is a constant of integration.

At the fixed end, A: $x = 0$ and $y = 0$ $C_2 = 0$

∴ $$EIy = -W\left[\frac{lx^2}{2} - \frac{x^3}{6}\right] \tag{8.20b}$$

∴ $$y = -\frac{W}{EI}\left[\frac{lx^2}{2} - \frac{x^3}{6}\right] \tag{8.20c}$$

Deflection at B is obtained by substituting $x = l$,

$$y_B = -\frac{W}{EI}\left[l \times \frac{l^2}{2} - \frac{l^3}{6}\right]$$

$$= -\frac{Wl^3}{3EI}$$

Negative sign signifies that the deflection is downwards.

8.4.5 Cantilever Beam with a Uniformly Distributed Load

A cantilever beam AB of length l, carrying a uniformly distributed load of w per unit run is considered (Figure 8.11). A section $x - x$ at a distance x from the fixed end A is considered.

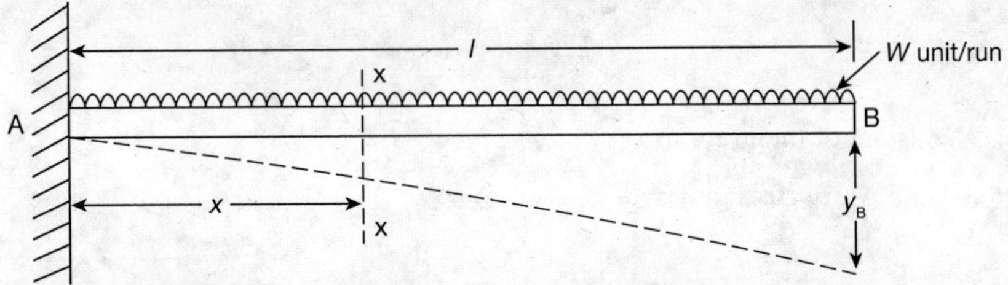

Fig. 8.11 Cantilever beam with uniformly distributed load.

Now, BM at the section is given by

$$M_x = -w(l - x) \times \frac{(l - x)}{2}$$

Minus sign is due to hogging BM

$$EI\frac{d^2y}{dx^2} = -\frac{w}{2}(l-x)^2 \tag{8.21}$$

Integrating the above equation, we get

$$EI\frac{dy}{dx} = -\frac{w}{2}\left(\frac{l-x}{3}\right)^3 \times (-1) + C_1$$

$$EI\frac{dy}{dx} = \frac{w}{6}(l-x)^3 + C_1 \tag{8.22a}$$

where C_1 is a constant of integration.

At the fixed end A, $x = 0$ and $\frac{dy}{dx} = 0$

$$\therefore \quad C_1 = -\frac{wl^3}{6}$$

$$\therefore \quad EI\frac{dy}{dx} = \frac{1}{6}(l-x)^3 - \frac{wl^3}{6} \tag{8.22b}$$

Slope at B: substituting $x = l$, we have

$$EIi_B = -\frac{wl^3}{6}$$

i.e.,

$$i_B = -\frac{wl^3}{6EI} \tag{8.23}$$

By integrating Equation (8.22b) we get

$$EIy = -\frac{w}{24}(l-x)^4 - \frac{wl^3}{6}x + C_2 \tag{8.24a}$$

where C_2 is a constant of integration.

At the fixed end A: $x = 0$ and $y = 0$.

$$\therefore \quad C_2 = \frac{wl^4}{24}$$

$$\therefore \quad EIy = -\frac{w}{24}(l-x)^4 - \frac{wl^3x}{6} + \frac{wl^4}{24} \tag{8.24b}$$

i.e.,

$$y = -\frac{1}{EI}\left[\frac{w}{24}(l-x)^4 + \frac{wl^3x}{6} - \frac{wl^4}{24}\right] \tag{8.24c}$$

Deflect at B: substituting $x = l$, we obtain from Equation (8.24c)

$$y_B = -\frac{wl^4}{8EI} \tag{8.25}$$

Negative sign signifies a downward deflection.

8.4.6 Cantilever Beam with a Moment Applied at the Free End

A cantilever beam AB of length l with a moment M applied at the free end is considered. A section $x-x$ at a distance x from the fixed end is considered (Figure 8.12).

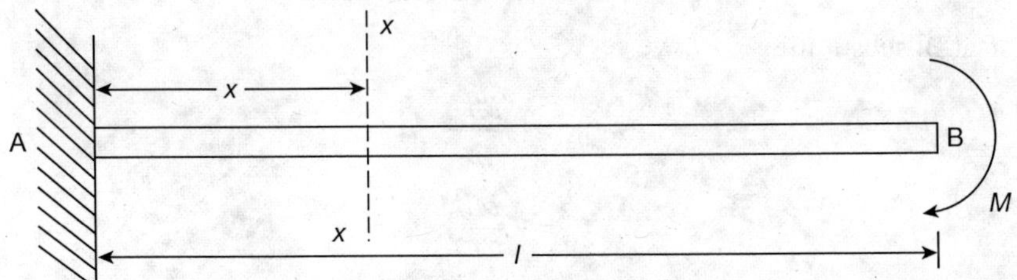

Fig. 8.12 Cantilever beam with a moment at free end.

Now, BM at this section is given by

$$M_x = -M$$

$$EI\frac{d^2y}{dx^2} = -M \tag{8.26}$$

Integrating Equation (8.26), we obtain

$$EI\frac{dy}{dx} = -M_x + C_1 \tag{8.27a}$$

where C_1 is a constant of integration.

At A, $x = 0$ and $\frac{dy}{dx} = 0$

$\therefore$ $$C_1 = 0$$

i.e., $$EI\frac{dy}{dx} = -Mx \tag{8.27b}$$

Slope at B: substituting $x = l$, we get

$$EI\frac{dy}{dx} = -Ml \tag{8.27c}$$

i.e., $$EI\, i_B = -Ml \tag{8.27d}$$

i.e., $$i_B = -\frac{Ml}{EI} \tag{8.27e}$$

Deflection equation is obtained by integrating Equation (8.27b).

i.e.,
$$EI \cdot y = -M\frac{x^2}{2} + C_2 \tag{8.28a}$$

where C_2 is a constant of integration

When $x = 0, y = 0$

$$\therefore \quad C_2 = 0$$

Hence
$$EI \cdot y = -M\frac{x^2}{2} \tag{8.28b}$$

Deflection at B: substituting $x = l$, we get

$$EI \cdot y_B = -\frac{Ml^2}{2}$$

i.e.,
$$y_B = -\frac{Ml^2}{2EI} \tag{8.28c}$$

8.4.7 Cantilever Beam with a Gradually Varying Load

A cantilever beam AB of length l and carrying a gradually varying load from zero at B to w per unit run at A is considered (Figure 8.13).

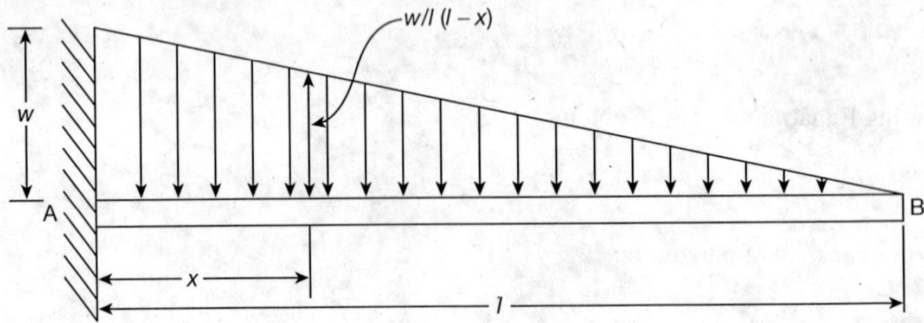

Fig. 8.13 Cantilever beam with varying (triangular) load.

A section $x - x$ at a distance x from the fixed end is considered.

Now, BM at this section is given by

$$M_x = -\frac{1}{2}(l-x)\times\frac{w}{l}(l-x)\times l - \frac{x}{3}$$

$$M_x = -\frac{w}{6l}(l-x)^3$$

$$\therefore \quad EI\frac{d^2y}{dx^2} = -\frac{w(l-x)^3}{6l} \tag{8.29}$$

Integrating Equation (8.29), the slope equation is obtained:

i.e,
$$EI\frac{dy}{dx} = -\frac{w(l-x)^4}{6l \times 4} + C_1 \quad (8.$$

where C_1 is a constant of integration.

When $x = 0$, $\frac{dy}{dx} = 0$

i.e.,
$$0 = -\frac{wl^3}{24} + C_1$$

∴
$$C_1 = \frac{wl^3}{24}$$

∴
$$EI\frac{dy}{dx} = -\frac{w(l-x)^4}{24l} + \frac{wl^3}{24} \quad (8.30b)$$

Equation (8.30b) is the slope equation for the given loading condition.
Slope B: substituting $x = l$, we get

$$EI\frac{dy}{dx} = \frac{wl^3}{24} \quad (8.30c)$$

i.e.,
$$i_B = \frac{wl^3}{24} \quad (8.30d)$$

By integrating the slope equation, deflection equation is obtained.

i.e.,
$$EI \cdot y = -\frac{w(l-x)^5}{120l} - \frac{wl^3}{24}x + C_2 \quad (8.31a)$$

where C_2 is the constant of integration
When $x = 0$, $y = 0$

i.e.,
$$0 = \frac{wl^4}{120}$$

Then,
$$EIy = \frac{w(l-x)^5}{120} - \frac{wl^3}{24}x + \frac{wl^4}{120} \quad (8.31b)$$

Deflection at B: substituting $x = l$, we get

$$EIy_B = -\frac{wl^4}{24} + \frac{wl^4}{120}$$

i.e.,
$$y_B = \frac{wl^4}{30EI} \quad (8.31c)$$

Cantilever Beam with a Point Load at a Distance from Fixed End

ıtilever beam AB of length l and carrying a point load w at C at a distance l_1 from the fixed end ‿nsidered (Figure 8.11).

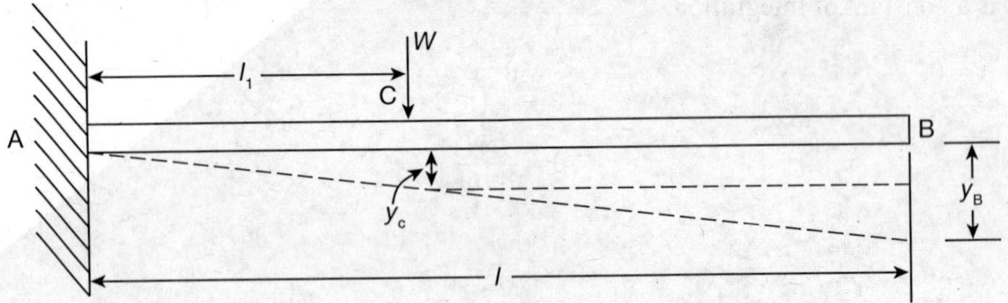

Fig. 8.14 Cantilever beam with a point load at a distance from fixed end.

Let i_c be the slope at point C.

y_c and y_b be the deflections at C and B, respectively.

The portion AC of the cantilever may be considered similar to a cantilever beam, with load at the free end:

$$\therefore \quad i_c = \frac{Wl_1^2}{2EI} \tag{8.32}$$

and

$$y_c = \frac{Wl_1^3}{3EI} \tag{8.33}$$

The beam will bend only between A and C, but from C to B it will remain straight since BM between C and B is zero.

Since the portion CB of the cantilever is straight.

$$i_B = i_C$$

or

$$i_C = i_B = \frac{Wl^2}{2EI}$$

From the geometry of Figure 8.11, we have

$$y_B = y_c + i_c(l - l_1)$$

i.e.,

$$y_B = \frac{Wl_1^3}{3EI} + \frac{Wl_1^3}{2EI}(l - l_1) \tag{8.34}$$

For a particular case of $l_1 = l/2$

$$y_B = \frac{W}{3EI}\left(\frac{l}{2}\right)^3 + \frac{W}{2EI}\left(\frac{l}{2}\right)^2 \times \frac{l}{2}$$

i.e.,
$$y_B = \frac{5Wl^3}{48EI} \tag{8.35}$$

SOLVED PROBLEM 8.5

A cantilever beam of length 3 m carries a point load of 60 kN at a distance of 2 m from the fixed end. If $E = 2 \times 10^5$ N/mm² and $I = 10^8$ mm⁴, find (i) slope at the free end and (ii) deflection at the free end.

Solution:

Given data: Length, $l = 3$ m, point load, $w = 60$ kN, $l_1 = 2$ m, $E = 2 \times 10^5$ N/mm², and $I = 10^8$ mm⁴.

(i) Slope at the free end

$$i_B = \frac{Wl^2}{2EI}$$
$$= \frac{60 \times 1,000 \times (2 \times 1,000)^2}{2 \times 2 \times 10^5 \times 10^8}$$

Hence, slope at free end $i_B = 0.006$ radians.

(ii) Deflection at the free end

$$y_B = y_c + i_c(l - l_1)$$
$$= \frac{Wl_1^3}{3EI} + \frac{Wl_1^2}{2EI}(l - l_1)$$
$$= \frac{50,000 \times 2,000^3}{3 \times 2 \times 10^5 \times 10^8} + \frac{50,000 \times 2,000^2}{2 \times 2 \times 10^5 \times 10^8} \times (3,000 - 2,000)$$
$$= 11.6 \text{ mm.}$$

Hence deflection at the free end = 11.6 mm.

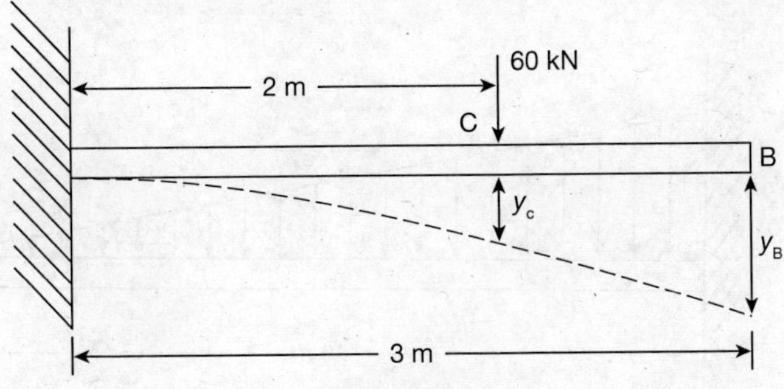

Fig. 8.15

SOLVED PROBLEM 8.6

A cantilever beam of length 3 m carries a uniformly distributed load over the entire length. If the slope at the free end is 0.01777 radians, find the deflection at the free end.

Solution:

Given data: Length, $l = 3$ m and slope at free end = 0.01777 radians.

Let y_B be the deflection at the free end.

$$y_B = -\frac{wl^3}{8EI} \quad \text{(i)}$$

and

$$i_B = -\frac{wl^2}{6EI} \quad \text{(ii)}$$

$$0.01777 = -\frac{wl^2}{6EI}$$

i.e.,

$$0.10662 = -\frac{wl^2}{EI} \quad \text{(iii)}$$

Substituting Equation (iii) in Equation (i)

$$y_B = -\frac{wl^3}{8EI} = -\frac{wl^2}{8EI} \times \frac{l}{8}$$

$$= 0.10662 \times \frac{3 \times 1,000}{8}$$

$$= 39.88 \text{ mm}$$

∴ Deflection at the free end = 40 mm.

SOLVED PROBLEM 8.7

A cantilever beam of length 4 m carries a uniformly distributed load of zero intensity at the free end and 50 kN/m at the fixed end. If $E = 2.1 \times 10^5$ N/mm^2 and $I = 10^8$ mm^4, find the slope and deflection at the free end.

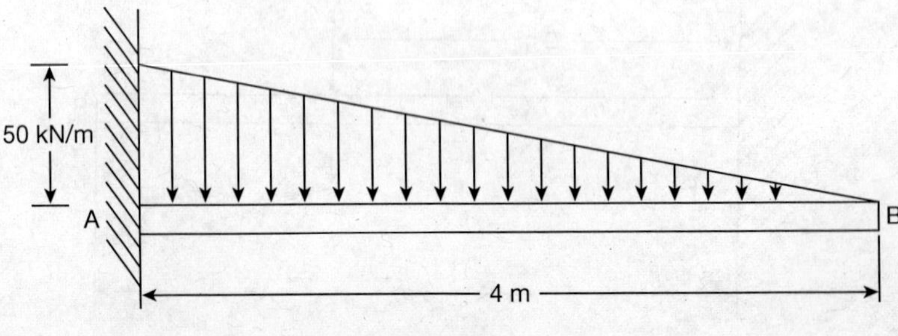

Fig. 8.16

Solution:

Given data: Length, $l = 4$ m, $w = 50$ kN/m, $E = 2.1\times10^5$ N/mm^2 and $I = 10^8$ mm^4.

Let i_B be the slope at the free end and y_B be the deflection at the free end.

(i) Slope at the free end

$$i_B = \frac{wl^3}{24EI}$$

$$= \frac{50\times(4,000)^3}{24\times2.1\times10^5\times10^8}$$

$$= 0.0067 \text{ radians}$$

(ii) Deflection at the free end

$$y_B = \frac{wl^4}{30EI}$$

$$= \frac{50\times(4,000)^4}{30\times2.1\times10^5\times10^8}$$

$$= 20.31 \text{ mm.}$$

SOLVED PROBLEM 8.8

A cantilever of 3 m length and of uniform rectangular cross section 150 mm wide and 300 mm deep is loaded with 30 kN load at its free end. In addition to this, it carries a UDL of 20 kN per meter run over its length. Calculate the maximum deflection and maximum slope of the beam.

Solution:

Given data: Length, l = 3 m, cross section = 150 mm × 300 mm, point load = 30 kN, and UDL = 20 kN/m run.

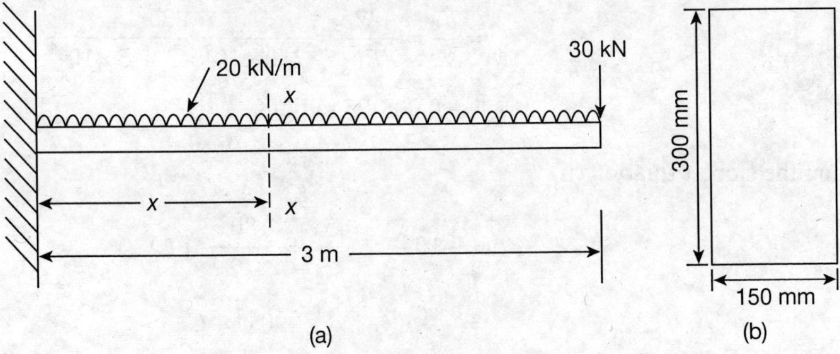

Fig. 8.17

Moment of inertia, $I = \dfrac{bd^3}{12} = \dfrac{150 \times 300^3}{12}$

$= 337.5 \times 10^{-6}\ \text{m}^4$

A section $x - x$ at a distance x from the fixed end is considered.

Bending moment at this section

$$M_x = -30(3-x) - \frac{20(3-x)^2}{2}$$

$$= -90 + 30x - 90 - 10x^2 + 60x$$

$$= -10x^2 + 90x - 180$$

$$\therefore \quad EI\frac{d^2y}{dx^2} = M_x = -10x^2 + 90x + 180 \qquad \text{(i)}$$

Integrating Equation (i), we get

$$EI\frac{dy}{dx} = \frac{10x^3}{3} + 45x^2 - 180x + C_1$$

where C_1 is the constant of integration.

When $x = 0$, $y = 0$

$$\therefore \quad C_1 = 0$$

$$\therefore \quad EI\frac{dy}{dx} = -\frac{10x^3}{3} + 45x^2 - 180x \qquad \text{(ii)}$$

Maximum slope is at the free end B

Substituting $x = 3$, we get

$$EI\frac{dy}{dx} = \frac{10 \times 3^3}{3} + 45 \times 3^2 - 180 \times 3$$

$$= -225$$

$$\therefore \quad i_A = \frac{dy}{dx} = \frac{-225}{EI} = \frac{-225}{210 \times 10^9 \times 337.5 \times 10^{-6}}$$

i.e., $i_A = -0.003175$ radians

Integrating the slope Equation (ii)

$$EIy = -10\frac{x^4}{12} + \frac{45x^3}{3} - \frac{180x^2}{2} + C_2$$

$$= -\frac{5}{6}x^4 + 15x^3 - 90x^2 + C_2$$

where C_2 is the constant of integration

When $x = 0, y = 0$

$$\therefore \quad C_2 = 0$$

$$\therefore \quad EIy = -\frac{5}{6}x^4 + 15x^3 - 90x^2 \qquad \text{(iii)}$$

Maximum deflection at the free end B

Substituting $x = 3$, we get

$$EIy_{max} = -\frac{5\times 3^4}{6} + 15\times 3^3 - 90\times 3^2$$

$$= -472.5$$

$$\therefore \quad y_{max} = -\frac{472.5}{I} = \frac{-472.5\times 10^3}{(210\times 10^9 \times 337.5\times 10^{-6})}$$

i.e.,

$$y_{max} = 6.67 \text{ mm.}$$

8.5 MECAULAY'S METHOD

For discontinuous uniformly distributed load and point loads at various points in a beam, a single equation of M_x will not be sufficient to determine the slope and deflection everywhere in the beam.

In such situation's Mecaulay's method may be used. In Mecaulay's method, a single equation of moment is formed for all loads on the beam in such a way that the constants of integration apply equally to all positions of the beam.

For example, let us consider a simply supported beam of span l which carries point loads W_1 and W_2 at a' and b' from the supports as shown in Figure 8.18.

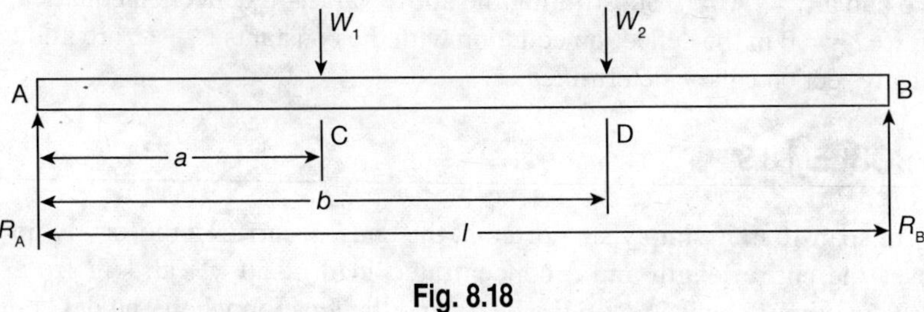

Fig. 8.18

Let R_A and R_B be the reactions at A and B, respectively.

Then, BM is given as

$$M_x = R_A x - \text{ [for portion of AC]}$$

$$M_x = R_A x - W_1(x-a) - \text{[for portion of CD]}$$

$$M_x = R_A x - W_1(x-a) - W_2(x-b) - \text{ [for portion of DB]}$$

In general, at any section the BM is given by

$$M_x = EI\frac{d^2y}{dx^2} = R_A x[-W_1(x-a)] - W_2(x-b) \tag{8.36}$$

Integrating Equation (8.36), we obtain the slope equation

$$EI\frac{dy}{dx} = R_A\frac{x^2}{2} + C_1\frac{W_1(x-a)^2}{2} - \frac{W_2(x-b)^2}{2} \tag{8.37}$$

Following points are to be noted in this regard:

(i) The constant of integration has to be written after the first term of the above equation.

(ii) The quantity $(x-a)$ should be integrated as $\dfrac{(x-a)^2}{2}$ and not as $\dfrac{x^2}{2} - ax$

(iii) The constant C_1 is valid for all values of x.

(iv) Integrating the slope equation, we obtain the deflection equation

$$ELy = R_A\frac{x^3}{6} + C_1 + C_2\left[-\frac{W_1(x-a)^3}{6}\right] - \frac{W_2(x-b)^3}{6}.$$

(v) $(x-a)^2$ has been integrated to $\dfrac{(x-a)^2}{6}$ and $(x-b)^2$ has been integrated to $\dfrac{(x-b)^3}{6}$.

(vi) Constant C_2 is written after C_1x. The constant C_1 is valid for all values of x.

(vii) Constants C_1 and C_2 are found from the boundary conditions. In this case, the boundary conditions are:

at $x = 0, y = 0$ and at $x = l, y = 0$. Substituting the above values in deflection equation, we get $C_2 = 0$. Substituting $x = l, y = 0$ in the deflection equation with the constants $C_1 + C_2$ are known, the slope and deflection at any section can be determined.

SOLVED PROBLEM 8.9

A beam AB of span 10 m is simply supported at the ends. It carries a uniformly distributed load of 25 kN/m over the entire length and a concentrated load of 50 kN at 4 m from the support A. Determine the maximum deflection in the beam and the location where the deflection occurs. Take $E = 200\times10^6$ kN/m^2, $I = 75\times10^{-6}$ m^4 using Mecaulay's method.

Solution:

Given data: Span = 10 m, UDL = 25 kN/m, point load = 50 kN, $E = 200\times10^6$ kN/m^2 and $I = 75\times10^{-6}$ m^4.

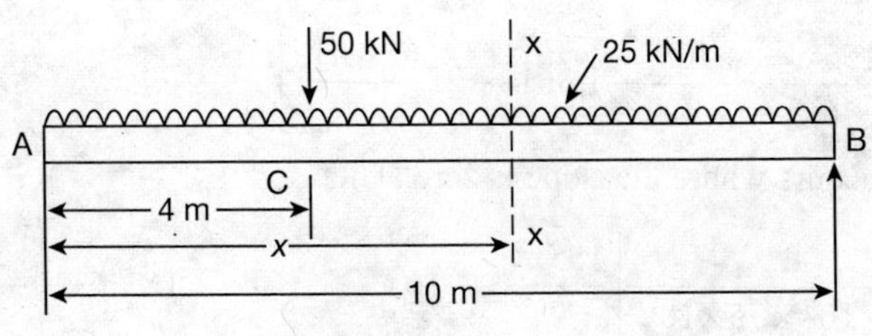

Fig. 8.19

While using Macaulay's method, the section $x-x$ is to be taken in the last portion of the beam.

Taking moments about A and equating clockwise and anticlockwise moments

$$R_B \times 10 = 50\times 4 + 25\times 10\times \frac{10}{2}$$

$\therefore$ $$R_B = 145 \text{ kN}$$

$$R_A + R_B = 50 + 25\times 10 = 300 \text{ kN}$$

$\therefore$ $$R_A = 300 - 145 = 155 \text{ kN}$$

A section $x-x$ at a distance x from B is considered.

Now, BM at the section $x-x$ is given as

$$M_x = EI\cdot\frac{d^2y}{dx^2} = 155x - 25\times x\times\frac{x}{2} - 50(x-4) \quad \text{(i)}$$

Integrating Equation (i), we have

$$EI\frac{dy}{dx} = \frac{155x^2}{2} - 25\frac{x^3}{6} + C_1 - \frac{50(x-4)^2}{2}$$

i.e., $$EI\frac{dy}{dx} = \frac{155x^2}{2} - 4.16x^3 + C_1 - 50\frac{(x-4)^2}{2} \quad \text{(iia)}$$

Integrating Equation (ii), we have

$$EIy = 155\frac{x^3}{6} - \frac{4.16}{4}x^4 + C_1x + C_2 - 25\frac{(x-4)^3}{3} \quad \text{(iiia)}$$

When $x = 0, y = 0$

$\therefore$ $$C_2 = 0$$

When $x = 10, y = 0;$ $$0 = \frac{155\times 10^3}{6} - \frac{4.16}{4}10^4 + 10C_1 - \frac{25(10-4)^3}{3}$$

i.e., $$C_1 = 1{,}361$$

Hence, the slope and deflection equations are

$$\frac{dy}{dx} = \frac{1}{EI}\left[\frac{155x^2}{2} - 4.16x^3 - 1{,}361x - 25(x-4)^2\right] \quad \text{(iib)}$$

and $$y = \frac{1}{EI}\left[\frac{155x^3}{6} - \frac{4.16x^4}{4} - 1{,}361x - 25\frac{(x-4)^3}{3}\right] \qquad \text{(iiib)}$$

Maximum deflection occurs where the slope is zero. Then

$$\frac{dy}{dx} = \frac{1}{200\times10^6\times75\times10^{-6}}\left[\frac{155x^2}{2} - 4.16x^3 - 1{,}361 - 25(x-4)^2\right] = 0$$

By trial and error, x is found as 4.72 m.

That is the maximum deflection occurs at a distance of 4.72 m from A. Substituting $x = 4.72$ m the deflection equation,

$$y_{\max} = \frac{-1}{200\times10^6\times75\times10^{-6}}\left[\frac{155(4.72)^3}{6} - \frac{4.16(4.72)^4}{4} - 1{,}361\times4.72 - 25\frac{(4.72-4)^3}{3}\right]$$

i.e., $y_{\max} = 2.66$ mm downwards.

SOLVED PROBLEM 8.10

A horizontal beam AB is freely supported at A and B, 8 m apart, and carries a uniformly distributed load of 10 kN/m run. A clockwise moment of 160 kN · m is applied to the beam at a point C, 3 m from the left-hand support A. Calculate the slope of the beam at C, if $EI = 10$ MN · m^2 by using Mecauley', method.

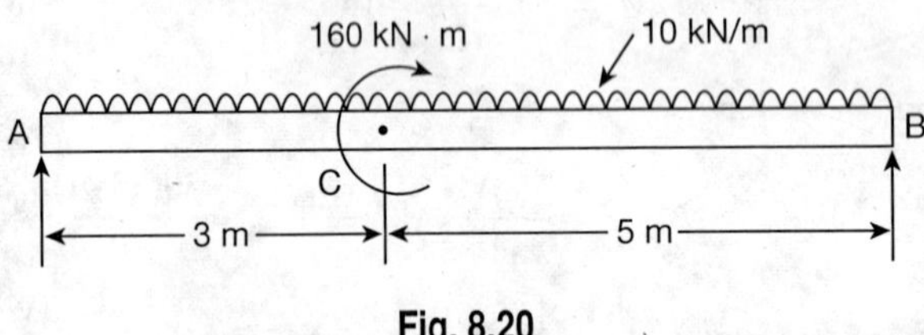

Fig. 8.20

Solution:

Given data: Span = 8 m, UDL = 10 kN/m, moment = 150 kN · m and $EI = 10$ MN · m^2.

Taking moment about A,

$$R_B \times 8 = w\times8\times\frac{8}{2} + 160$$

i.e., $$R_B = \frac{480}{8} = 60 \text{ kN}$$

$$R_A = (10\times8) - 60 = 20 \text{ kN}.$$

Taking A as origin, and using MeCaulay's method the BM at any section $x - x$, at a distance x from A,

$$EI\frac{d^2y}{dx^2} = -20x \;\vdots\; +10x\times\frac{x}{2} \;\vdots\; -160(x-3)$$

$$EI\frac{d^2y}{dx^2} = -20x \,\vdots + \frac{10x^2}{2} \,\vdots -160(x-3) \qquad \text{(i)}$$

Integrating Equation (i) we have

$$EI\frac{dy}{dx} = -20\frac{x^2}{2} + C_1 \,\vdots + \frac{10x^3}{6} \,\vdots -160(x-3)$$

i.e,

$$EI\frac{dy}{dx} = -10x^2 + C_1 \,\vdots + \frac{5x^2}{3} \,\vdots -160(x-3) \qquad \text{(ii)}$$

Integrating Equation (ii) we get

$$EIy = \frac{-10x^3}{3} + C_1x + C_2 \,\vdots + \frac{5x^4}{12} \,\vdots -160\frac{(x-3)^2}{2} \qquad \text{(iii)}$$

When $x = 0$, $y = 0$ and so $C_2 = 0$

$\therefore$

$$0 = \frac{10\times 80^3}{3} + C_1\times 8 + \frac{5\times 8^4}{12} - \frac{160\times 5^2}{2}$$

Solving $\qquad C_1 = 250$

Substituting the above values in Equation (ii), we have

$$EI\frac{dy}{dx} = -10x^2 + 250 + \frac{5x^3}{3} - 160(x-3)$$

For slope at C substituting $x = 3$ in the above equation up to C is

$$EI\frac{dy}{dx} = -10\times 3^2 + 250 + \frac{5\times 3^3}{3}$$

i.e.,

$$i_c = \frac{205}{4\times 10^3} = 0.00512 \text{ radians}$$

SOLVED PROBLEM 8.11

A 9 m long beam simply supported over a span at 8 m with a overhang of 1 m on left-hand side carries load, as shown in Figure 8.21. Determine the magnitude of the slope at A and D and deflection under the load of 40 kN in the beam $E = 200$ GPa and $I = 3{,}600$ cm^4.

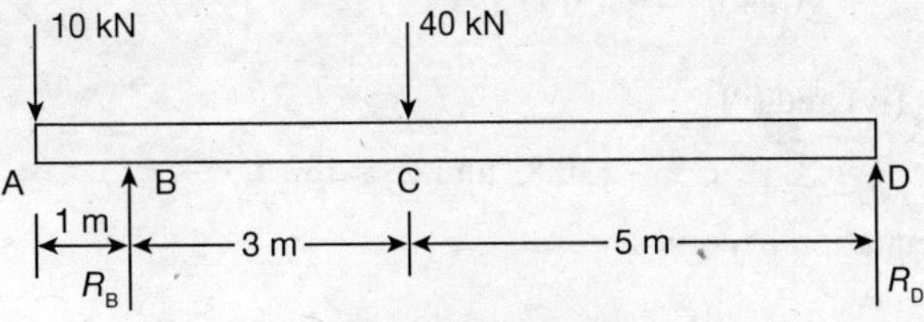

Fig. 8.21

Solution:

Given data: Length, $l = 9$ m, span $= 8$ m, overhang $= 1$ m, point loads $= 10$ kN and 40 kN, $E = 200$ GPa and $I = 3{,}600$ cm^4.

Taking moment about B,

$$R_D \times 8 + 10 \times 1 = 40 \times 3$$

$$R_D = 13.75 \text{ kN}$$

$$R_B + R_D = 10 + 40 = 50 \text{ kN}$$

$$\therefore \quad R_B = 50 - 13.75 = 36.25 \text{ kN.}$$

A section $x - x$ at a distance x from the end A in the last portion of the beam, i.e., CD is considered.

Now, BM at this section

$$M_x = -10x + \{R_B(x-1)\} - 40(x-4)$$

i.e.,
$$EI\frac{d^2y}{dx^2} = -10x + \vdots\, 36.25(x-4)\,\vdots - 40(x-4) \qquad \text{(i)}$$

Integrating Equation (i), slope equation is obtained as

$$EI\frac{dy}{dx} = -10\frac{x^2}{2} + C_1 + \vdots\, \frac{36.25}{2}(x-1)^2\,\vdots - \frac{40}{2}(x-4)^3 \qquad \text{(ii)}$$

Integrating Equation (ii), deflection equation is obtained

$$EIy = -\frac{5}{3}x^3 + C_1x + C_2 + \vdots\, \frac{36.25}{2}(x-1)^3\,\vdots - \frac{20}{3}(x-4)^3$$

When $x = 1$ m, $y = 0$ and $x = 9$ m, $y = 0$.

Substituting these values

$$0 = -\frac{5}{3}x^3 + C_1 + C_2$$

and
$$0 = -\frac{5}{3} \times 9^3 + \frac{36.25}{6}(9-1)^3 - \frac{20}{3}(9-4)^3 + 9C_1 + C_2$$

Then,
$$C_1 + C_2 = \frac{5}{3} \qquad \text{(iv)}$$

and
$$9C_1 + C_2 = -1{,}045 \qquad \text{(v)}$$

Solving Equations (iv) and (v)

$$C_1 = -130.83 \text{ and } C_2 = 132.5$$

Hence, the slope equation becomes

$$EI\frac{dy}{dx} = -5x^2 - 130.83\,\vdots + 18.125(x-1)\,\vdots - 20(x-4)^2$$

At $x = 0$, i.e., at end A

$$EI\frac{dy}{dx} = -130.833$$

So slope, $$i_A = -\frac{130.833}{200\times10^6\times3,600\times10^{-8}} = -0.0187 \text{ radians}$$

At $x = 9$ m, at end D

$$EI\frac{dy}{dx} = -5\times9^2 - 130.833 + 19.125(9-1)^2 - 20(90-4)^2$$

$$= 124.167$$

$$\therefore \quad i_D = \frac{124.167}{EI} = \frac{124.167}{200\times10^6\times3,600\times10^{-8}} = 0.0172 \text{ radians}$$

Equation for deflection

$$EIy = -\frac{5x^3}{3} - 130.833x + 132.5 \;\vdots + \frac{36.25}{3}(x-1)^3 \;\vdots - \frac{20}{3}(x-4)^3$$

Under the load of 40 kN, $x = 4$.
Substituting this value in the above equation

$$EIy_c = -\frac{5}{3}\times4^3 + \frac{36.25}{3}(3)^2 - 130.833\times4 + 132.5$$

$$= -388.75$$

$$\therefore \quad y_c = -\frac{170.919}{EI} = -\frac{170.919}{200\times10^6\times3,600\times10^{-8}} = -0.05399 \text{ mm}$$

SOLVED PROBLEM 8.12

A beam AB of span 8 m is simply supported at the ends A and B which is loaded as shown in Figure 8.22. If $E = 200$ kN/mm^2 and $I = 43\times10^8$ mm^4, determine (i) maximum deflection and (ii) slope at the end A.

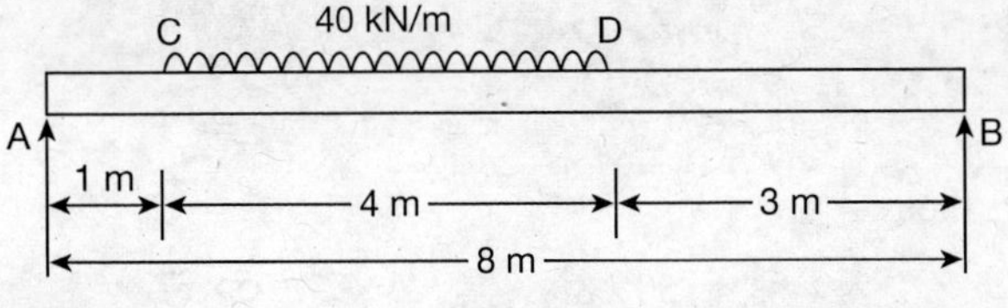

Fig. 8.22

Taking moments about the end A

$$R_B\times8 = 40\times4\times\left(1+\frac{4}{2}\right)$$

$$R_B = 60 \text{ kN}$$

$$R_B \times 8 = 40 \times 4 \times \left(1 + \frac{4}{2}\right)$$

$$R_B = 60 \text{ kN}$$

In order that the general expression for the BM at any section may be expressed in the form suitable for application at MeCeulay's method, let us assume a uniformly distributed load of 40 kN/m to be extended up to B. This assumed load be compensated by applying an equal and upward load of 40 kN/m over the span DB as shown in Figure 8.23.

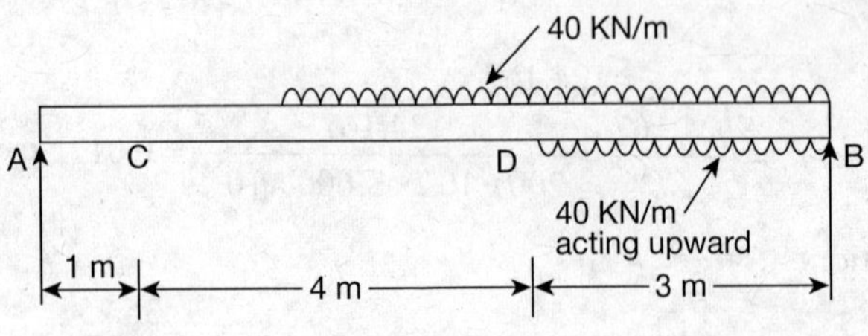

Fig. 8.23

Taking A as the origin, the BM at any section $x - x$ at a distance x from A is given as

$$EI = \frac{d^2y}{dx^2} = 100x \,\vdots\, \frac{40(x-1)^2}{2} \,\vdots\, + \frac{40(x-5)^2}{2}$$

$$EI = \frac{d^2y}{dx^2} = 100x + \,\vdots\, 20(x-1)^2 \,\vdots\, + 20(x-5)^2 \qquad \text{(i)}$$

Integrating Equation (i), we have

$$EI = \frac{dy}{dx} = 50x^2 + C_1 \,\vdots\, \frac{-20(x-1)^3}{3} \,\vdots\, + \frac{20}{3}(x-5)^3 \qquad \text{(ii)}$$

Integrating Equation (ii) we get

$$Ely = \frac{50x^2}{3} + C_1x + C_2 \,\vdots\, \frac{-5}{3}(x-1)^4 \,\vdots\, + \frac{5}{3}(x-5)^4$$

When $x = 0, y = 0$

$$\therefore \quad C_2 = 0$$

When $x = 8, y = 0,$ then

$$0 = \frac{50 \times 8^3}{3} + 8C_1 - \frac{5 \times 7^4}{3} + \frac{5 \times 3^4}{3}$$

$$C_1 = -583.33$$

Hence, slope and deflection equations are rewritten as

$$EI\frac{dy}{dx} = 50x^2 - 583.33 \,\vdots\, \frac{-20(x-1)^3}{3} \,\vdots\, + \frac{20}{3}(x-5)^3 \qquad \text{(iib)}$$

and $$EIy = \frac{50x^x}{3} - 583.33x \Big| \frac{-5}{3}(x-1)^4 \Big| \frac{5}{3}(x-5)^4 \qquad \text{(iiib)}$$

Maximum deflection, y_{max}

Let us assume that the deflection will be maximum between C and D. Equating the slope to zero.

$$EI\frac{dy}{dx} = 50x^2 - 583.33 - \frac{20}{3}(x-1)^3 = 0$$

Solving the above equation by trial and error, we get $x = 3.83$ m

Substituting $x = 3.83$ m in the deflection equation we have

$$EI\, y_{max} = \frac{50x^2}{3} - 583.33\times - \frac{5}{3}(x-1)^4$$

$$= \frac{50}{3}\times 3.83^2 - 583.33\times 3.83 - \frac{5}{3}(2.83)^4$$

$$= -1,403.78$$

$$\therefore \quad y_{max} = \frac{-1,403.78}{200\times 10^6 \times 43\times 10^{-4}} \times 1,000 \text{ mm}$$

$$= -16.32 \text{ mm}$$

Maximum deflection is downwards with 16.32 mm.

Slope at the end $A(i_A)$

For slope at A, substituting $x = 0$ in the slope equation

$$EI\frac{dy}{dx} = -583.33$$

$$\therefore \quad i_A = -\frac{583.33}{EI}$$

$$= \frac{-583.33}{200\times 10^6 \times 4.3\times 10^{-4}}$$

$$= -0.00678 \text{ radian}$$

i.e., $$i_A = -0.006 + 8\times\frac{180}{\pi}$$

$$= -0.388°$$

SOLVED PROBLEM 8.13

A beam of length 6 m is simply supported at its ends and carries two point loads of 48 kN and 40 kN at a distance 1 m and 3 m, respectively, from the left support. Find (i) deflection under each load, (ii) maximum deflection and (iii) the point at which maximum deflection occurs. Given $E = 2\times 10^5$ N/mm^2 and $I = 85\times 10^6$ mm^4. Use MeCaulay's method.

Solution:

Given data: Length of beam, $l = 69$ m, two point loads = 48 kN. 40 kN. $E = 2\times10^5$ N/mm^2 and $I = 85\times10^6$ mm^4.

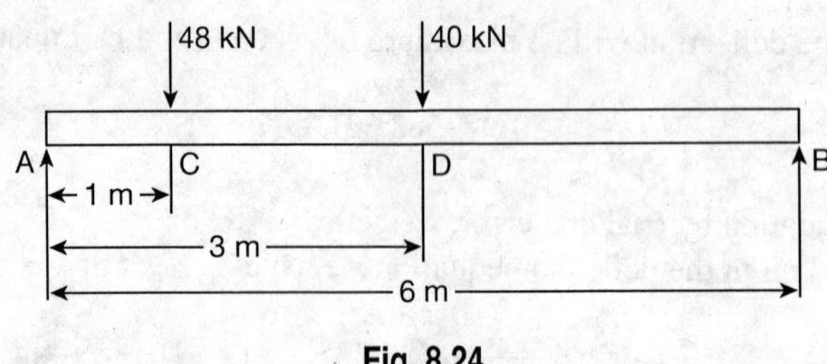

Fig. 8.24

Taking moment about B, we get

$$R_A \times 6 = 48\times5 + 40\times3$$

$$R_A = 60 \text{ kN}$$

$\therefore$

$$R_B = (48+40) - 60 = 28 \text{ kN}$$

A section $x-x$ is the last part of the beam at a distance x from the left support is considered. Now BM at this section is given by

$$EI\frac{d^2y}{dx^2} = R_A x - 48 \vdots -48(x-1) \vdots -40(x-3)$$

$$EI\frac{d^2y}{dx^2} = 60x - 48(x-1) - 40(x-3) \quad \text{(i)}$$

Integrating Equation (i) we get

$$EI\frac{dy}{dx} = \frac{60x^2}{2} + C_1 \vdots \frac{-48(x-1)^2}{2} - 20(x-3)^2 \vdots$$

i.e.,

$$EI\frac{dy}{dx} = 30x^2 + C_1 \vdots -24(x-1)^2 \vdots -20(x-3)^2 \quad \text{(iia)}$$

Integrating Equation (iia), we get

$$EIy = \frac{30x^3}{3} + C_1x + C_2 \vdots -\frac{24(x-1)^3}{3} \vdots -\frac{20(x-3)^3}{3}$$

$$EIy = 10x^3 + C_1x + C_2 \vdots -8(x-1)^3 \vdots -\left[\frac{20(x-3)^3}{2}\right] \quad \text{(iiia)}$$

To find the value of constants C_1 and C_2, we use two boundary conditions, i.e.,

At $x = 0$ y_0 and

At $x = 6$ m $y = 0$

Substituting the boundary conditions,

i.e., $x = 0$ in Equation (iiia) and considering the equation up to first dotted line we get

$$0 = 0 + 0 + C_2$$

$$\therefore \quad C_2 = 0$$

Substituting the second boundary condition

i.e., $x = 6$ m $y = 0$ in Equation (iiia) and considering the complete equation, we get

$$0 = 10 \times 6^3 + C_1 \times 6 + 0 - 8(6-1)^3 - \frac{20}{3}(6-3)^3$$

$$0 = 980 + 6C_1$$

$$\therefore \quad C_1 = \frac{980}{6} = -166.33$$

Substituting the values of C_1 and C_2 in Equation (iiia), we get

$$EIy = 10x^3 - 166.33 - 8\ (x-1)^2 - \frac{20}{3}(x-3)^3 \qquad \text{(iiib)}$$

Deflection under first load (Point C)

Deflection under first load point *C* is obtained by substituting $x = 1$ in Equation (iiib) up to the first dotted line *C* as the point *C* lies in the first part of the beam, we get,

$$EIy_c = 10 \times 1^3 - 163.33 \times 1 = -153.33 \text{ kN/m}^3$$

i.e.,

$$y_c = -\frac{153.33 \times 10^{12}}{2 \times 10^5 \times 85 \times 10^6}$$

$$= -9.013 \text{ mm}$$

Negative sign shows that the deflection is downwards.

Deflection at second load (Point D)

This is obtained by substituting $x = 3$ m in Equation (iiib) up to the second dotted line (as the point D lies in the second part of the beam) we get,

$$EIy_D = 10 \times 3^3 - 163.3 \times 3 - 8\ (3-1)^3 = -283.99 \times 10^{12}$$

$$y_D = \frac{-283.99 \times 10^{12}}{2 \times 10^5 \times 85 \times 10^6}$$

i.e.,

$$y_D = -16.77 \text{ mm}$$

Maximum Deflection

The deflection is likely to be maximum between C and D. For maximum deflection $\frac{dy}{dx}$ should be zero, i.e., slope is zero.

Hence, Equation (iia) is equated to zero up to the second dotted line.

i.e.,
$$30x^2 + C_1 - 24(x-1)^2 = 0$$

$$30x^2 - 163.33 - 24(x^2 + 1 - 2x) = 0$$

i.e.,
$$6x^2 + 48x - 187.33 = 0$$

Solving, $x = 2.87$ m (neglecting negative root).

Now substituting, $x = 2.87$ m in Equation (iiib) up to the second dotted line we get the maximum deflection as

$$EIy_{max} = 10 \times 2.87^3 - 163.33 \times 2.87 - 8(2.87-1)^3$$

$$= -284.67 \text{ kN m}^3 = -284.67 \times 10^{12} \text{ N/mm}^3$$

$$y_{max} = \frac{-284.67 \times 10^{12}}{2 \times 10^5 \times 85 \times 10^6}$$

i.e.,
$$y_{max} = -16.745 \text{ mm}$$

8.6 MOMENT AREA METHOD

This is a semigraphical method of dealing with the problem of deflection of beams subjected to bending. Figure 8.25 shows a beam AB carrying some type of loading and the corresponding BM.

Let the beam bend to a shape of A, P, Q, B.

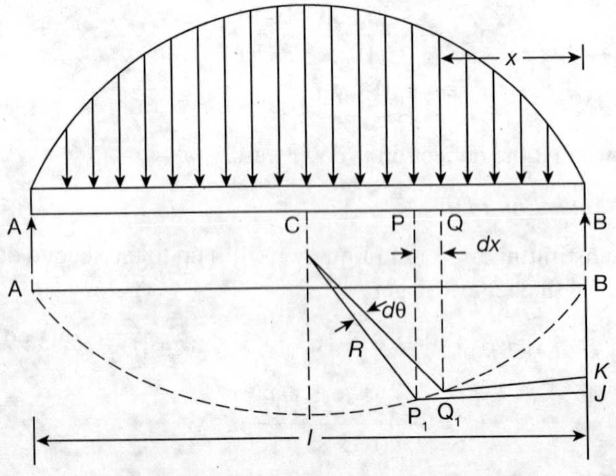

Fig. 8.25 Moment area method.

An element PQ of length dx at a distance x from B is considered.

Let R be the radius of curvature of deflected part of the beam.

Let $d\theta$ be the angle subtended by the arc $P_1 Q_1$, at the centre in radians and M be the BM between P and Q.

From the geometry of the bent up beam, we have $P_1Q_1 = Rd\theta$.

Substituting $P_1Q_1 = dx$, we have

$$dx = Rd\theta$$

i.e.,

$$d\theta = \frac{dx}{R} \tag{8.38}$$

But, for a loaded beam, we know

$$\frac{M}{I} = \frac{E}{R}$$

or

$$R = \frac{EI}{M}$$

Substituting this value of R in Equation (8.38), we have

$$d\theta = dx \cdot \frac{M}{EI} = \frac{Mdx}{EI} \tag{8.39}$$

Total change of slope from A to B may be found by integrating Equation (8.39) between the limits 0 to l.

$$\theta = \int_0^l \frac{Mdx}{EI} = \frac{1}{EI}\int_0^l Mdx \tag{8.40}$$

But Mdx represents the area of BMD of length dx.

Hence, $\int_0^l Mdx$ represents the area of BMD between A and B.

$$\theta = \frac{l}{EI} \text{ [Area of BMD over the entire span]}$$

$$\theta = \frac{A}{EI}$$

Tangents are drawn at P_1 and Q_1. Let the tangents meet at J and K on a vertical line drawn through B. The tangents at P_1 and Q_1 also meet at an angle $d\theta$ and

$$JK = x \cdot d\theta = x \cdot \frac{Mdx}{EI} = \frac{Mxdx}{EI}$$

The total intercept may be found out by integrating the above equation between the limits 0 and l.

$$y = \int_0^l \frac{M \cdot dx}{EI} = \frac{1}{EI}\int_0^l M \cdot x \cdot dx \tag{8.41}$$

But $x.M.dx$ represents the moment of the area of the BMD of length dx about point B.

Hence, $\int_0^l M.x.dx$ represents the moment of the area of BMD between B and A.

$$y = \frac{1}{EI} \times \bar{x} \times A = \frac{A\bar{x}}{EI} \tag{8.42}$$

where $\bar{x}$ = distance of centre of gravity of the area from A to B.

The results given by equations for slope and for deflection are known as Mohr's theorems I and II.

1. Mohr's Theorem I

Theorem I states that the change of slope between any two points is equal to the net area of the BMD between these points divided by *EI*.

2. Mohr's Theorem II

Theorem II states that the deflection between any two points is equal to the moment of the area of the BMD between the two points about the reference line divided by *EI*.

8.6.1 Simply Supported Beam Carrying a Point Load at the Centre

A simply supported beam AB of length l, carrying a point load w at the centre of the beam is considered (Figure 8.26).

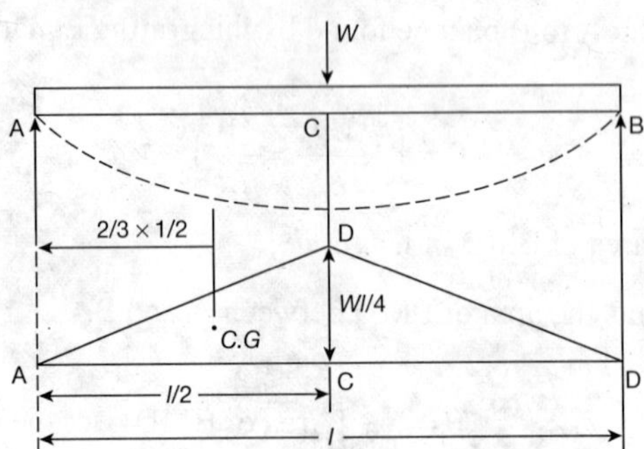

Fig. 8.26 Simply supported beam carrying a central point load.

The BMD is shown in Figure 8.26.

Using Mohr's theorem for slope, we get slope at A $(i_A) = \dfrac{\text{Area of BMD between A and C}}{EI}$

$$= \frac{\text{Area of } \Delta\text{ACD}}{EI}$$

i.e.,

$$i_A = \frac{\frac{1}{2} \times \frac{l}{2} \times \frac{Wl}{4}}{EI}$$

$$i_A = \frac{Wl^2}{16EI} \tag{8.43}$$

Using Mohr's theorem for deflection, we get

$$y = \frac{A\bar{x}}{EI}$$

where A = Area of BMD between A and C.

$$= \frac{Wl^2}{16}$$

and x = distance of $C.G.$ of area $\quad A = \frac{2}{3} \times \frac{l}{2} = \frac{l}{3}$

$$\therefore \quad y_c = \frac{\frac{Wl}{16} \times \frac{l}{3}}{EI}$$

i.e.,

$$y_c = \frac{Wl^2}{48EI} \tag{8.44}$$

8.6.2 Simply Supported Beam Carrying an Uniformly Distributed Load

A simply supported beam AB of length l carrying an uniformly distributed load of w per unit length over the entire span is considered (Figure 8.27).

Using Mohr's theorem for slope, we get

$$\text{Slope at A } (i_A) = \frac{\text{Area of BMD between A and C}}{EI}$$

$$= \frac{\text{Area of parabola ACD}}{EI}$$

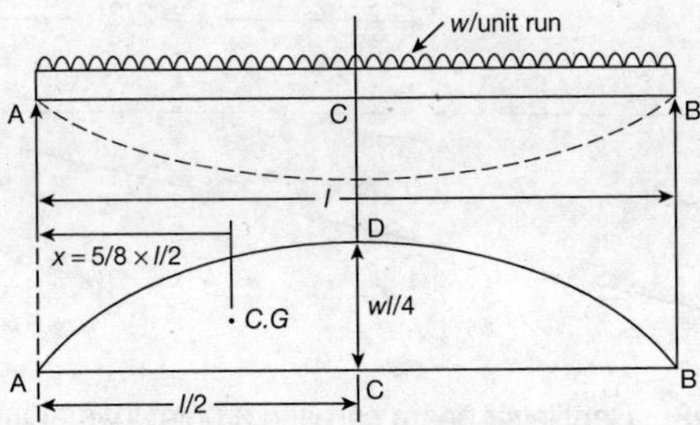

Fig. 8.27 Simply supported beam carrying an uniformly distributed load.

$$i_A = \frac{2/3 \times AC \times AD}{EI}$$

$$= 2/3 \times \frac{l}{2} \times \frac{wl^2}{8}$$

i.e., $$i_A = \frac{wl^3}{24EI} \quad (8.45)$$

Using Mohr's theorem for deflection, we get $$y_c = \frac{A\bar{x}}{EI}$$

where A = area of BMD between A and C. $$= \frac{wl^3}{24}$$

and $\bar{x}$ = distance of centre of gravity of area A from Point A.

$$= \frac{5l}{16}$$

i.e., $$y_c = \frac{\frac{wl^3}{24} \times \frac{5l}{16}}{EI}$$

i.e., $$y_c = \frac{5wl^4}{384EI} \quad (8.46)$$

8.6.3 Cantilever Beam Carrying a Point Load at the Free End

A cantilever beam AB carrying a point load at the free end is considered (Figure 8.28).

The BM will be zero at B and Wl at A. The variation of BM between A and B is linear (Figure 8.28).

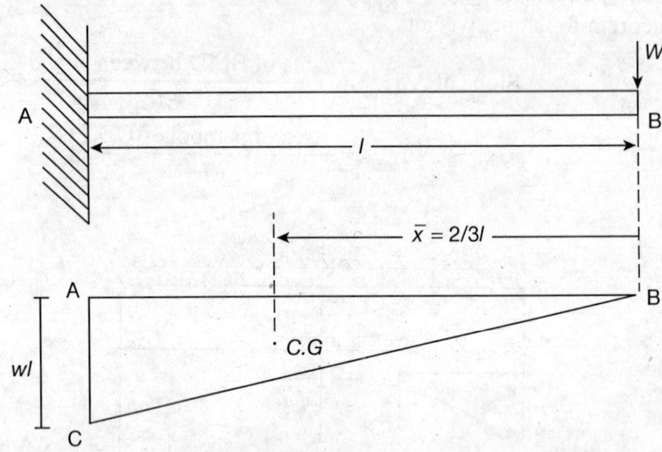

Fig. 8.28 Cantilever beam carrying a point load at the free end.

At the fixed end A, the slope and deflection are zero.

Using Mohr's theorem for slope, we get slope at B

$$i_B = \frac{A}{EI}$$

where A = Area of BMD between A and B

$$= \frac{1}{2} \times AB \times AC$$

$$= \frac{Wl^2}{2EI}.$$

$$\therefore \qquad i_B = \frac{Wl^2}{2EI} \qquad (8.47)$$

Using Mohr's theorem for deflection, we get $\quad y_B = \frac{A\bar{x}}{EI}$

where $\bar{x}$ = distance of centre of gravity of area of BMD for $B = \frac{2}{3}l$.

$$\therefore \qquad y_B = \frac{\frac{Wl^2}{2} \times \frac{2l}{3}}{EI} = \frac{Wl^3}{3EI} \qquad (8.48)$$

8.6.4 Cantilever Beam Carrying an Uniformly Distributed Load

A cantilever beam AB of length l carrying a uniformly distributed load w/unit run over the entire span is considered (Figure 8.29).

The BM is zero at B and $\frac{wl^2}{2}$ at A. The variation of BM between A and B is parabolic as shown in Figure 8.29.

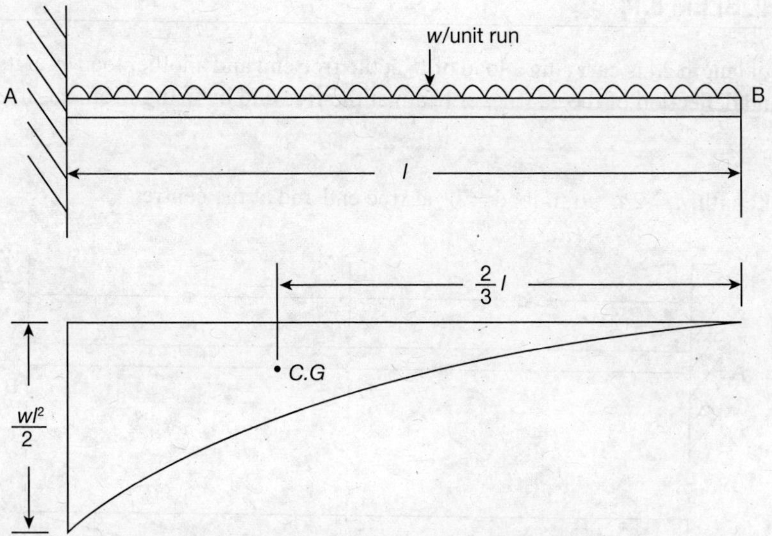

Fig. 8.29 Cantilever beam carrying a UDL.

Using Mohr's theorem for slope, we get,

$$\text{Slope at B} \quad i_B = \frac{A}{EI}$$

where A = Area of BMD

$$= \frac{1}{3} \times AB \times AC$$

$$= \frac{1}{3} \times l \times \frac{wl^2}{2} = \frac{wl^3}{6}$$

$$\therefore \quad i_B = \frac{wl^3}{6EI} \tag{8.49}$$

Using Mohr's theorem for deflection, we get $y_B = \frac{A\bar{x}}{EI}$

where $\bar{x}$ = distance of centre of gravity from B $= \frac{3}{4} wl$

$$\therefore \quad y_B = \frac{wl^3}{6} \times \frac{3}{4} \frac{l}{EI}$$

i.e.,

$$y_B = \frac{wl^4}{8EI} \tag{8.50}$$

SOLVED PROBLEM 8.14

A cantilever of length $2a$ is carrying a load of W at the free end and another load W at its centre. Determine slope and deflection of the cantilever beam at the free end by using Moment-area method.

Solution:

Given data: Length, $l = 2a$, point load = W at free end and at the centre.

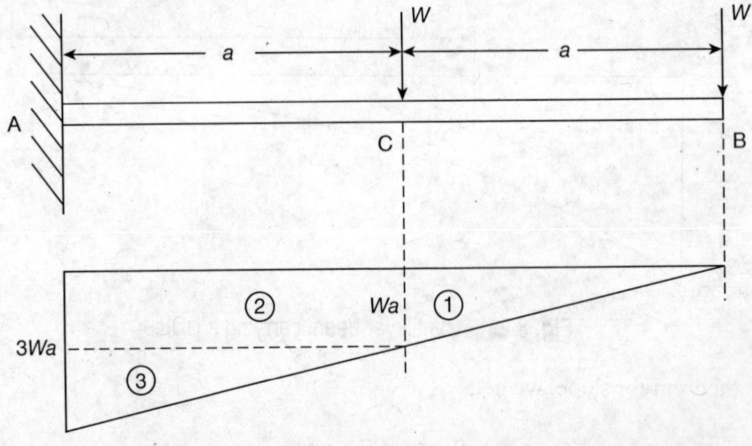

Fig. 8.30

Now, BM at B, $M_B = 0$

Further, BM at C, $M_C = -Wa$

Area of BMD of 1, $A_1 = \frac{1}{2} Wa \times a = \frac{Wa^2}{2}$

Area of BMD of 2, $A_2 = Wa \times a = Wa^2$

Area of BMD of 3, $A_3 = \frac{1}{2} \times 2Wa \times a = Wa^2$

Total area of BMD,

$$A = A_1 + A_2 + A_3$$

$$= \frac{Wa^2}{2} + Wa^2 + Wa^2$$

$$= \frac{5}{2} Wa^2$$

Using Mohr's theorem for slope we get slope at B, $i_B = \frac{A}{EI}$

$$i_B = \frac{5Wa^2}{EI}$$

Using Mohr's theorem for deflection, we get $y_B = \frac{A\bar{x}}{EI}$

Total moment of BMD about B,

$$A\bar{x} = A_1\bar{x}_1 + \bar{A}_2\bar{x}_2 + A_3\bar{x}_3$$

$$= \frac{Wa^2}{2} \times \frac{29}{3} + Wa^2 \times \frac{3a}{2} \times Wa^2 \times \frac{5a}{3}$$

$$= \frac{7Wa^3}{2}$$

$\therefore$

$$y_B = \frac{7Wa^3}{2EI}$$

8.7 CONJUGATE BEAM METHOD

Conjugate beam method is the modified moment–area method. This method is based on the construction of a conjugate beam, defined as an imaginary beam of length equal to that of the original beam and loaded with an elastic weight M/EI, where M is the BM of the actual beam. This method is specially useful for simply supported and cantilever beams with varying flexural rigidities.

This method is based on the following two theorems:

1. Conjugate beam Theorem I

Shear force at any section of the conjugate beam is equal to the slope of the elastic curve at the corresponding section of the actual beam.

2. Conjugate beam Theorem II

Bending moment at any section of the conjugate beam is equal to the deflection of the elastic curve at the corresponding section of actual beam.

8.7.1 Relation Between Real Beam and Conjugate Beam

The relation between given beam (actual beam) and the corresponding conjugate beam for different end conditions are given in Table 8.1.

Table 8.1 Companion of real and conjugate beams

Sl No.	Real (Actual) Beam	Conjugate Beam
1.	(a) Slope at any section (b) Deflection at any section (c) Loading as per the given system	(a) Shear force at the corresponding section (b) Bending moment at the corresponding section (c) $\frac{M}{EI}$ diagram is the loading system.
2.	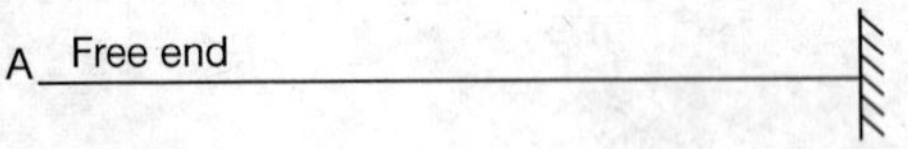 A Free end Slope and deflection exist	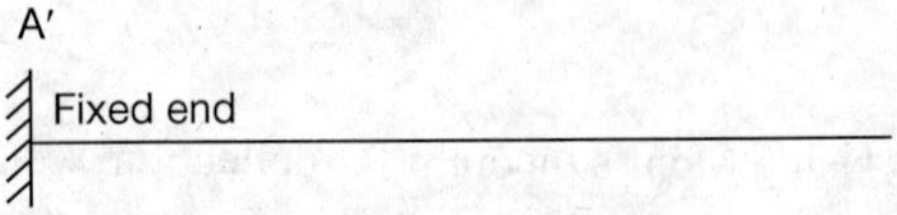 A′ Fixed end Shear force and bending moment exist.
3.	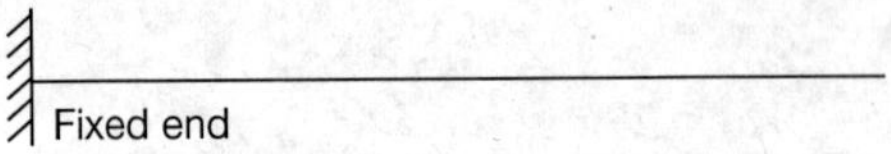 Fixed end Slope = 0 and deflection = 0	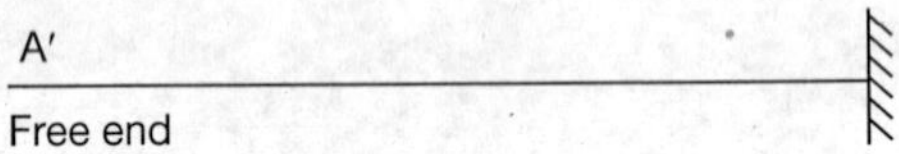 A′ Free end Shear force = 0; *BM* = 0
4.	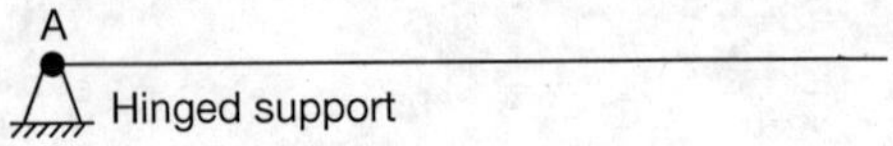 A Hinged support Slope exists but deflection = 0	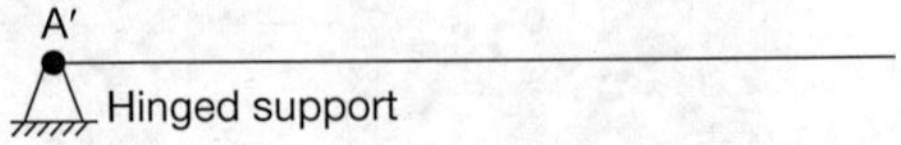 A′ Hinged support Shear force exists but *BM* = 0
5.	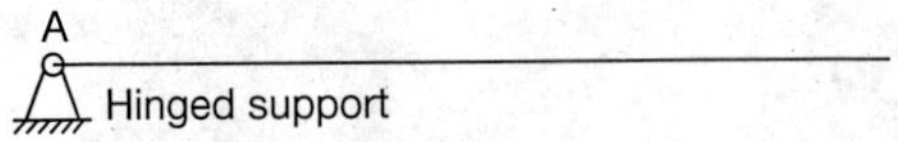 A Hinged support Slope exists but deflection = 0	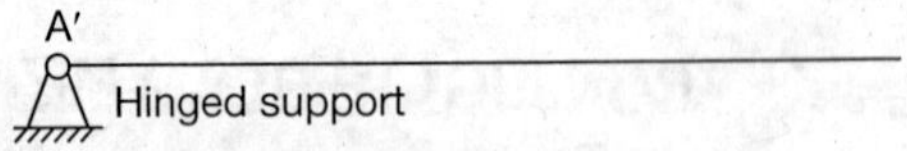 A′ Hinged support Shear force exists but deflection = 0

8.7.2 Simply Supported Beam Carrying a Point Load at Mid-Span

A simply supported beam AB of length l carrying a point load W at the mid-span is considered (Figure 8.31(a)).

The BMD is shown in Figure 8.31(b). Now, the conjugate beam A'B'C' is drawn with central ordinates as $\frac{Wl}{4EI}$ (Figure 8.31(c)).

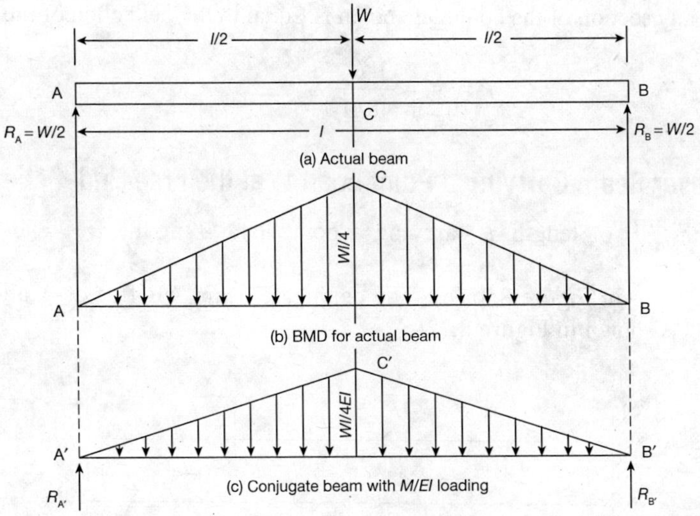

Fig. 8.31 Simply supported beam with point load at mid-span.

Total load on the conjugate beam = Area of the load diagram $= \frac{1}{2} \times l \times \frac{Wl}{4EI} = \frac{Wl^2}{8EI}$

$\therefore$ Reaction at each support for the conjugate beam

$$R'_A = R'_B = \frac{1}{2} \times \frac{Wl^2}{8EI} = \frac{Wl^2}{16EI}$$

But, shear force at any section of the conjugate beam is equal to the slope of the real beam.

Hence, i_A = Slope at the end A of the real beam

$$i_A = -\frac{Wl^2}{16EI} \tag{8.51}$$

Similarly,

$$i_B = -\frac{Wl^2}{16EI} \tag{8.52}$$

Deflection at C for the given beam

= BM at C for the conjugate beam

$$= R'_A \times \frac{l}{2} - \left(\frac{1}{2} \times \frac{l}{2} \times \frac{Wl}{4EI}\right)\left(\frac{1}{3} \times \frac{l}{2}\right)$$

$$= \frac{Wl^3}{48EI}$$

But the BM at any section of the conjugate beam is Equal to the deflection of the real beam

Hence, $$y_c = \frac{Wl^3}{48EI} \tag{8.53}$$

8.7.3 Cantilever Beam Carrying a Point Load *W* at the Free End

A cantilever beam AB of length l carrying a point load W at the free end B is considered (Figure 8.32).

The BMD is shown in Figure 8.32(b). The conjugate beam AB free at A and fixed at B and loaded with BMD is shown in Figure 8.32(c).

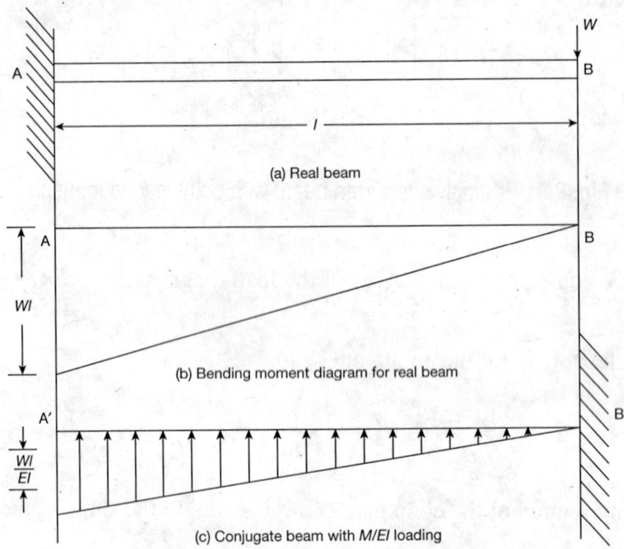

Fig. 8.32 Cantilever beam with point load at free end.

Slope at B for real beam = BM at B′ for the conjugate beam

$$\therefore \quad i_B = \frac{1}{2} l \frac{Wl}{EI} = \frac{Wl^2}{2EI} \tag{8.54}$$

Deflection at B for the given beam = BM at B′ for the conjugate beam.

$$= \frac{1}{2} l \frac{Wl}{EI} \times \frac{2}{3} l$$

$$y_B = \frac{Wl^3}{3EI} \tag{8.55}$$

SOLVED PROBLEM 8.15

A simply supported beam AB of span 10 m carries a point load of 15 kN, 8 m from end A. The value of I of the left half of the beam is $4I$ and that of right half is I. Find (i) slope at end A, (ii) deflection at the mid-span and (iii) maximum deflection. $I = 8 \times 10^5$ mm^4 and E = 200 × 10^6 kN/m^2. Use conjugate beam method.

Solution:

Given data: Span, l = 10 m, point load, W = 15 kN, $(I)_{\text{left}} = 4I$, $(I)_{\text{right}} = I$, $I = 8 \times 10^5$ mm^4 and $E = 200 \times 10^6$ kN/m^2.

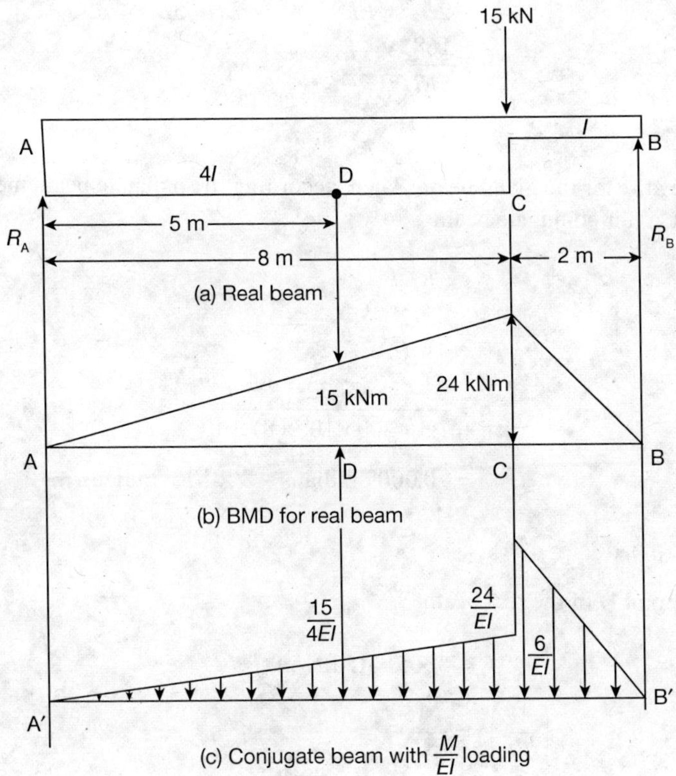

Fig. 8.33

Taking moments of forces about A

$$R_B \times 10 = 15 \times 8$$

$$\therefore \quad R_B = 12 \text{ kN}$$

$$R_A = 15 - 12 = 3 \text{ kN}.$$

Further, BM at C,

$$M_C = 12 \times 2 = 24 \text{ kN} \cdot \text{m}.$$

Conjugate beam can be constructed by dividing BM at any section by the product of E and I.

Let R'_A and R'_B be the reactions at the ends of the conjugate beam.
Taking moments of forces about the point B′, we get

$$R'_A \times 10 = \left[\left(\frac{1}{2}\times 8\times\frac{6}{EI}\right)\times\left(\frac{8}{3}\times 2\right)+\left(\frac{1}{2}\times 2\times\frac{24}{EI}\right)\times\frac{2}{3}\times 2\right]$$

i.e.,

$$R'_A = \frac{72}{5EI}$$

$$R_B{}' = \frac{1}{2}\times 8\times\frac{6}{EI}+\frac{1}{2}\times 2\times\frac{24}{EI}-\frac{72}{5EI}$$

$$R_B{}' = \frac{168}{5EI}$$

Slope at the end A

Let i_A be the slope at A for the given beam. Then, according to conjugate beam method,
i_A = shear force at A′ for conjugate beam

$$= -R'_A$$

$$= -\frac{72}{8EI}$$

$$= -\frac{72}{5\times 200\times 10^6\times 8\times 10^5}$$

$$= -0.0009 \text{ radians} = 9\times 10^{-14} \text{ radians}$$

Deflection at mid-span

The BMD of the point D of the real beam,

$$M_D = 3\times 5 = 15 \text{ kN/m}$$

$$\therefore \quad \left(\frac{M}{EI}\right)_D = \frac{15}{(4I)E} = \frac{15}{4EI}$$

$$\therefore \quad y_D = M'_D \text{ for the conjugate beam}$$

$$= R'_A\times 5-\frac{1}{2}\times 5\times\frac{15}{4EI}\times\frac{5}{3}$$

$$= \frac{56.37}{EI}$$

$$= \frac{56.37}{200\times 10^6\times 8\times 10^5}\times 10^3 \text{ mm}$$

$$= 3.52\times 10^{-7} \text{ mm}$$

SOLVED PROBLEM 8.16

Using conjugate beam method for the beam shown in Figure 8.34, find the slopes and deflections at A, B, C and D. Given $E = 2 \times 10^2$ kN/mm² and $I = 200 \times 10^{10}$ mm⁴. Neglect the weight of the beam.

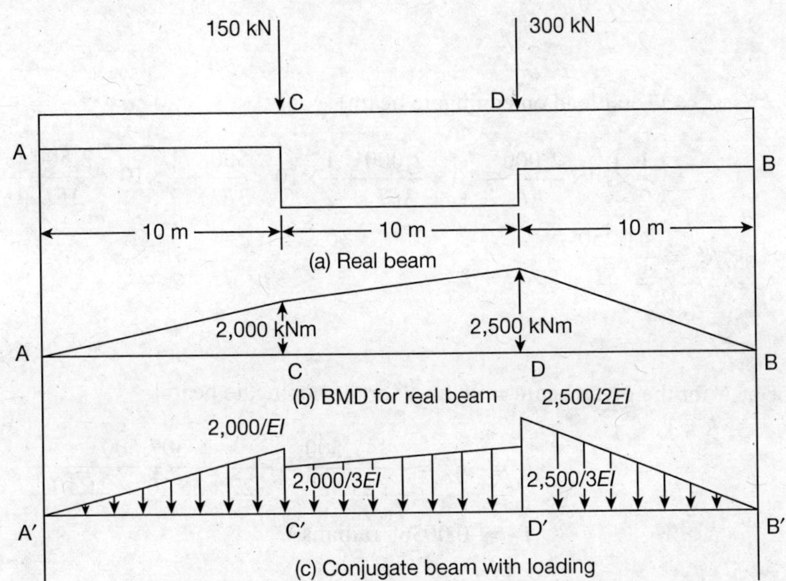

Fig. 8.34

Taking moments about A

$$R_B \times 30 = 150 \times 10 + 300 \times 20 = 7{,}500$$

i.e.,

$$R_B = \frac{7{,}500}{30} = 250 \text{ kN}$$

$$R_A = 150 + 300 - 250 = 200 \text{ kN}$$

BM at C,

$$M_C = 200 \times 10 = 2{,}000 \text{ kN} \cdot \text{m}.$$

BM at D,

$$M_D = 250 \times 10 = 2{,}500 \text{ kN} \cdot \text{m}.$$

Let R'_A and R'_B be the reactions for conjugate beam.

Taking moments of all the forces acting on conjugate beam about A'

$$R'_B \times 30 = \left(\frac{1}{2}\times 10\times \frac{2,000}{EI}\right)\times\left(\frac{2}{3}\times 10\right)+\left(10\times\frac{2,000}{EI}\right)\times 15$$

$$+\left(\frac{1}{2}\times 10\times\frac{500}{3EI}\right)\times\left(\frac{20}{3}+10\right)+\left(\frac{1}{2}\times\frac{2,500}{2EI}\times 10\right)\times\left(20+\frac{10}{3}\right)$$

$$=\frac{293,750}{27EI}$$

$$R'_A = \text{(Total load on conjugate beam)} - R'_B$$

$$=\left(\frac{1}{2}\times 10\times\frac{2,000}{EI}+10\times\frac{2,000}{3EI}+\frac{1}{2}\times 10\times\frac{500}{3EI}+\frac{1}{2}\times 10\times\frac{2,500}{2EI}\right)-\frac{23,750}{EI}$$

$$R'_A = \frac{641,250}{27EI}$$

Slopes

Slope at A for the given beam = SF at A for the conjugate beam.

$$i_A = -R'_A = -\frac{347,500}{27EI} = -\frac{397,500}{27\times 2\times 10^8\times 2\times 10^{-2}}$$

$$i_A = -0.00368 \text{ radians.}$$

Slope at C for the given beam = SF at C for the conjugate beam.

$$=\frac{347,500}{27EI}+\frac{1}{2}\times 10\times\frac{2,000}{EI}$$

$$=\frac{77,500}{27EI}=\frac{77,500}{27\times 2\times 10^8\times 2\times 10^{-2}}$$

$$= -0.0007176 \text{ radians}$$

Slope at D for the given beam = SF at D' for the conjugate beam

$$= R_B{}' - \frac{1}{2}\times 10\times\frac{2,500}{27EI}$$

$$=\frac{293,750}{27EI}-\frac{12,500}{27EI}=\frac{281,250}{27EI}$$

$$=\frac{281,250}{27\times 2\times 10^8\times 2\times 10^{-2}}$$

$$= -0.0026042 \text{ radians}$$

Slope at B for the given beam = SF at B' for the conjugate beam.

$$= +R_B{}' = \frac{293,750}{27EI}$$

$$= \frac{293,750}{27\times 2\times 10^8 \times 2\times 10^{-2}}$$

$$= -0.00272 \text{ radians}$$

Deflections

Deflection at A of the given beam = BM at A' for conjugate beam

i.e., $y_A = 0$

Deflection at B for the given beam = BM at B' for conjugate beam $y_B = 0$

Deflection at C for the given beam = BM at C' for the conjugate beam.

$$y_C = R_A' \times 10 - \frac{1}{2}\times 10 \times \frac{2,000}{EI}\times\frac{10}{3}$$

$$= \frac{3,475,000}{27EI} - \frac{100,000}{3EI} = \frac{3,475,000}{27EI} - \frac{900,000}{27EI}$$

$$= \frac{2,575,000}{27EI}$$

$$= \frac{2,575,000}{27\times 2\times 10^8 \times 2\times 10^{-2}}$$

$$= 0.02384 \text{ m} = 23.84 \text{ mm}$$

Deflection at D for the given beam = BM at D' for the conjugate beam

$$y_D = R_B{}' \times 10 - \frac{1}{2}\times 10 \times \frac{2,500}{2EI}\times\frac{10}{3}$$

$$= \frac{2,937,500}{27EI} - \frac{62,500}{3EI} = \frac{2,375,000}{27EI}$$

$$= \frac{2,375,000}{27\times 2\times 10^5 \times 2\times 10^{-2}}$$

$$= 0.02199 \text{ m} = 21.99 \text{ mm}$$

SOLVED PROBLEM 8.17

A cantilever 2 m long carries a load of 20 kN at a distance of 1 m from the fixed end and a load of 15 kN at the free end. Determine the deflection at the free end. Use conjugate beam method. Take $E = 200\times 10^4 \text{ kN/m}^2$, $I = 15\times 10^6 \text{ m}^4$.

Solution:

Given data: Span, $l = 2$ m, point loads = 20 kN and 15 kN, $E = 200\times 10^4 \text{ kN/m}^2$, and $I = 15\times 10^6 \text{ m}^4$.

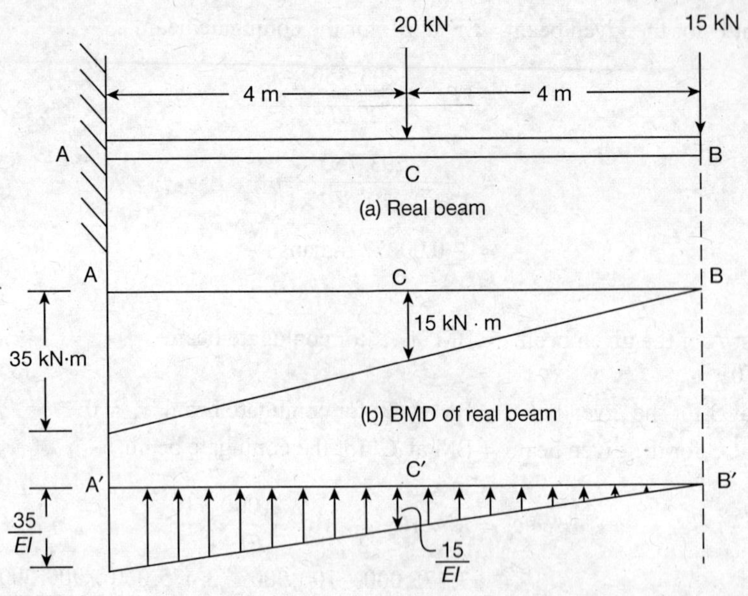

Fig. 8.35

Deflection at free end of given beam

$$y_B = \text{BM at } B' \text{ of conjugate beam}$$

$$y_B = \frac{15}{EI}\times 1\times\left(\frac{1}{2}\times 1\right)+\frac{1}{2}\times\frac{20}{EI}\left(\frac{2}{3}\times 1+1\right)+\left(\frac{1}{2}\times 1\times\frac{15}{EI}\times\frac{2}{3}\right)$$

$$= -\frac{1}{EI}\left[22.5+\frac{100}{6}+\frac{15}{3}\right]$$

$$= -\frac{1}{200\times 10^4\times 15\times 10^6}\times\frac{265}{6} = 00$$

SOLVED UNIVERSITY QUESTIONS

SOLVED PROBLEM 8.18

A beam is simply supported at its ends over a span of 10 m and carries two concentrated loads of 100 kN and 60 kN at distances of 2 m and 5 m, respectively from the left support. Calculate (i) slope at the left support and (ii) slope and deflection under the 100 kN load. Assume $EI = 36\times 10^4$ kNm^2 (Anna Univ., June 2006, ME).

Solution:

Given data: Span, $l = 10$, point load = 100 kN and 60 kN, $EI = 36\times10^4$ kNm^2.

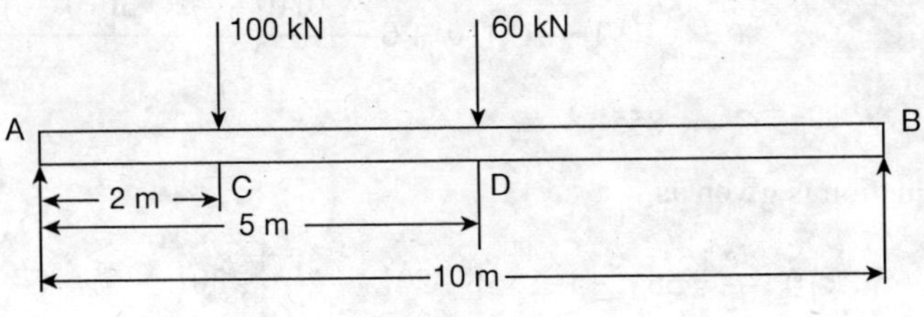

Fig. 8.36

Taking moments about A,

$$R_B \times 10 = 60\times5+100\times2$$

$$\therefore \quad R_B = 50 \text{ kN}$$

$$R_A = 160-5 = 110 \text{ kN}$$

A section $x-x$ in the last part of the beam (DB) at a distance x from A is considered. Now, BM of this section is

$$EI\frac{d^2y}{dx^2} = R_A \times x \,\vdots\, -100(x-2) \,\vdots\, -60(x-5)$$

$$EI\frac{d^2y}{dx^2} = 110\times x \,\vdots\, -100(x-2) \,\vdots\, -60(x-5)$$

Integrating to get the slope equation

$$EI\frac{dy}{dx} = 100\times\frac{x^2}{2}+C_1 \,\vdots\, \frac{-100(x-2)^2}{2} \,\vdots\, -60\frac{(x-5)^2}{2}$$

i.e., $$EI\frac{dy}{dx} = 50x^2+C_1 \,\vdots\, -50(x-2)^2 \,\vdots\, -30(x-5)^2 \quad \text{(i)}$$

Integrating to get the deflection equation

$$Ely = 50\frac{x^3}{3}+C_1x+C_2 \,\vdots\, \frac{-50(x-2)^3}{3} \,\vdots\, -30\frac{(x-5)^3}{3} \quad \text{(ii)}$$

Applying the boundary condition:

(i) At $x = 0$, $y = 0$ and

(ii) at $x = 10$, $y = 0$.

Substituting $x = 0$ and $y = 0$ in Equation (ii)

$$0 = C_2 = 0$$

$$\therefore \quad C_2 = 0$$

Substituting $x = 10$ and $y = 0$ in Equation (ii)

$$0 = \frac{55}{3}(10)^3 + C_1(10) + 0 - \frac{50(10-2)^2}{3} - \frac{30(0-5)^3}{3}$$

$$\therefore \qquad C_1 = -855$$

Now the slope equation is given as

$$36 \times 10^4 \frac{dy}{dx} = 55x^2 + (-855) \,\vdots\, -50(x-2)^2 \,\vdots\, -30(x-5)^2 \qquad \text{(iii)}$$

(i) Slope at the left support

Substituting $x = 0$ in Equation (ii) up to first dotted line

i.e., $$36 \times 10^4 \frac{dy}{dx} = 55(0)^2 - 855$$

i.e., $$\frac{dy}{dx} = -2.375 \times 10^{-3} \text{ radians}$$

i.e., $$i_A = -2.375 \times 10^{-3} \text{ radians}$$

(ii) Slope at 100 kN load

Substituting $x = 2$ m in Equation (iii) up to second dotted line

$$36 \times 10^4 \frac{dy}{dx} = 55(2)^2 - 855 - 50(1-2)^2$$

i.e., $$36 \times 10^4 \frac{dy}{dx} = -635$$

i.e., $$\frac{dy}{dx} = -1.76 \times 10^{-3} \text{ radians}$$

i.e., $$i_C = -1.76 \times 10^{-3} \text{ radians}$$

(iii) Deflection at 100 kN load

Substituting $x = 2$ m in Equation (ii) up to second dotted line.

$$36 \times 10^4 y = \frac{55x^3}{3} + (-855)x - \frac{50(x-2)^3}{3}$$

$$36 \times 10^4 y = \frac{55 \times 2^3}{3} + (-855) \times 2 - \frac{50(2-2)^3}{3}$$

i.e., $$y = -4.34 \text{ mm}$$

Negative sign shows the download deflection.

SOLVED PROBLEM 8.19

Find the maximum deflection of the beam shown in Figure 8.37. $EI = 1 \times 10^{11}$ kN/mm^2. Use MeCaulay's method (Anna Univ., Nov. 2006, ME).

Solution:

Given data: Moment = 300 kNm and $EI = 1 \times 10$ kN $\cdot$ mm^2.

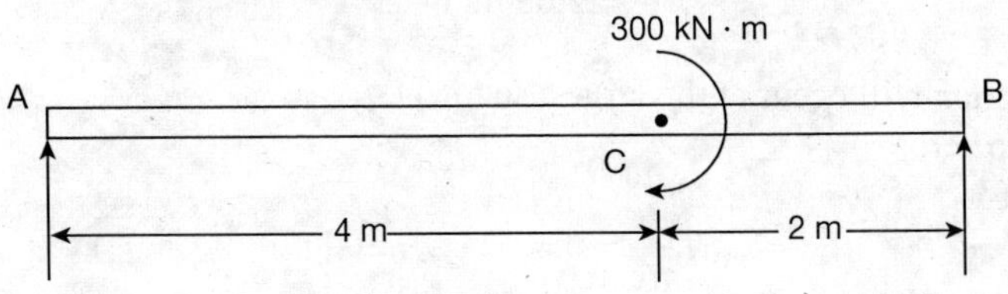

Fig. 8.37

Taking moment A about A

$$R_B \times 6 = 300$$

$\therefore$
$$R_B = 50 \text{ kN}(\uparrow)$$

$\therefore$
$$R_A = \text{Total load } - R_B.$$
$$= 0 - 50 = -50 \text{ kN}(\downarrow)$$

Further, BM at any section x from A, $EI\dfrac{d^2y}{dx^2} = -50x + 300(x-4)^\circ$

Integrating the above equation,

$$EI\frac{dy}{dx} = -25x^2 + C_1 \,\vdots\, 300(x-4)^1 \,\vdots$$

Integrating again

$$EI_y = \frac{-25x^3}{3} + C_1x + C_2 \,\vdots\, 300\frac{(x-4)^2}{2} \,\vdots$$

At $x = 0$, $y = 0$,

$\therefore$
$$C_2 = 0$$

At $x = 6$ m, $y = 0$

$$0 = \frac{-25}{3} \times 6^3 + 6C_1 + 150 \times 2^2$$

i.e.,
$$C_1 = 200$$

Deflection at C

i.e., at $x = 4$ m

$$EIy_C = \frac{-25}{3} \times 4^3 + 200 \times 4$$

i.e.,
$$EIy_c = 266.67$$

$$y_c = \frac{266.67}{1 \times 10^4}$$
$$y_c = 2.66 \text{ mm}$$

Maximum deflection will occur at the largest segment,
Equating slope to zero

$$EI\frac{dy}{dx} = 0 = -25x^2 + 200$$

i.e.,
$$x = 2\sqrt{2} \text{ m}$$

$\therefore$
$$EIy_{max} = -\frac{25}{3}(2\sqrt{2})^3 + 200 \times 2\sqrt{2}$$
$$= \frac{800\sqrt{2}}{3}$$

i.e.,
$$y_{max} = \frac{800\sqrt{2}}{3 \times 10^4} = 3.77 \text{ mm}$$

SALIENT POINTS

- When a beam is subjected to same type of loadings the beam deflects from its original position.
- Strength and stiffness are the two main design criteria for a beam.
- As per the strength criterion of the beam design, it should be strong enough to resist the bending stresses and shear stresses. As per the stiffness criterion, the beam should be stiff enough not to deflect more than the permissible limit.
- Elastic curve is the deformed position of the neutral axis due to loading.
- At a section of a loaded beam the following properties may be needed.
  Deflection = y

  Slope $= dy/dx$

  Bending moment $= EI\frac{d^2y}{dx^2}$

  Shear force $= EI\frac{d^3y}{dx^3}$

  Rate of loading $= EI\frac{d^4y}{dx^4}$

- In MeCaulay's method, a single equation of moment is formed for all loads on the beam in such a way that the constants of integration apply equally to all positions of the beam.
- Mohr's theorem I states that the change of slope between any two points is equal to the net area of the BMD between these points divided by the product of modulus of elasticity and moment of inertia.
- Mohr's theorem II states that the total deflection between any two points is equal to the moment of the area of BMD between the two points about the reference line divided by the product of modulus of elasticity and moment of inertia.
- Conjugate beam Theorem I states that the shear force at any section of the conjugate beam is equal to the slope of the elastic curve at the corresponding section of the actual beam.
- Conjugate beam Theorem II states that the BM at any section of the conjugate beam is equal to the deflection of the elastic curve at the corresponding section of the actual beam.

QUESTIONS

1. Explain MeCaulay's method for finding out the slopes and deflection of a beam.
2. What is the central deflection of a simply supported beam of span l when two equal moments of M_0 applied at ends?
3. Derive a relationship between deflection, BM and flexural rigidity for a beam.
4. Explain conjugate beam method. Draw the conjugate beam for a cantilever beam fixed at right end and loaded with central point load.
5. State moment–area theorems. How they are different from conjugate beam theorems?
6. Discuss double integration method of finding deflection of beams.
7. Using double integration method, find the deflection under the loads of the beam shown in Figure 8.38.

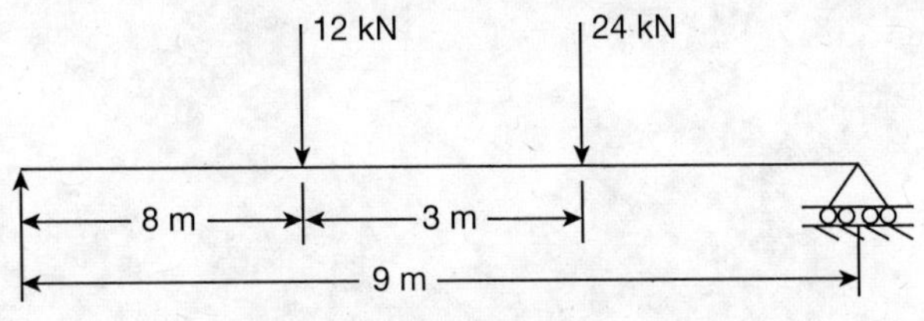

Fig. 8.38

8. A beam of span 8 m and flexural rigidity as $8\times10^4\ \text{kN}\cdot\text{m}^2$ is subjected to a clockwise couple of $600\ \text{kN}\cdot\text{m}$ at a distance of 4 m from the left end. Find (i) the deflection at the point of application of the couple and (ii) the maximum deflection using MeCaulay's method.
9. An overhanging beam ABC of 6 m long is supported at A and B such that AB = 5 m. It is loaded with point load of 7.5 kN at the end. If $E = 200\times10^6\ \text{kN/m}^2$ and $I = 10\times10^{-6}\ \text{m}^4$, determine (i) deflection at the point C (ii) maximum deflection between A and B using MaCaulay's method.
10. A steel cantilever of 2.5 m long is loaded with UDL of 10 kN per metre over a length of 1.5 m from the fixed end. Determine the slope and deflection at the free end. $I = 9{,}500\ \text{cm}^4$ and $E = 210\ \text{GN/m}^2$. Use moment–area method.

11. A beam 4 m long, simply supported at its ends, is carrying a point load (w) at its centre. If the slope at the ends of the beam is not to exceed 1°, find the deflection at the centre of the beam using moment–area method.
12. A cantilever of length 3 m is carrying a point load of 40 kN at a distance of 2 m from the fixed end. If $I = 10^8$ mm^4 and $E = 2\times10^5$ N/mm^2, find (i) slope at the free end and (ii) deflection at the free end using conjugate beam method.
13. A simply supported beam of 5 m span carries a UDL of 15 kN/m over the entire span and in addition carries a point load of 40 kN at the centre of the span. Calculate the slope at the ends and maximum deflection of the beam. Take $E = 200$ GN/m^2 and $I = 5{,}000$ cm^4. Use conjugate beam method.
14. A cantilever of 2.5 m effective length carries a load of 25 kN at its free end. If the deflection at the free end is not to exceed 40 mm, what must be the I value of the section of the cantilever. Take $E = 210$ GN/m^2. Adopt any suitable method justify your choice.

9

Thin Cylinders and Shells

LEARNING OBJECTIVES

9.1 INTRODUCTION

Fluids, gases or dry materials with or without pressures are stored in specially made vessels. The walls of these vessels are in the form of shells with certain thickness but adequate to sustain the load and pressure. The shapes of such vessels are in the form of cylinders and spheres. Steam boilers, reservoirs, pressure tanks, storing fluids (gases or liquids), working chamber of engines, compressed air receivers, etc. are the common examples of shells.

Generally, the walls of such vessels are thin compared to their diameters. If the ratio of the thickness of the wall of the shell to internal diameter is less than 1/20, it is known as a *thin shell*. In such cases, pressure is uniformly distributed through the thickness of wall and the stress is computed easily. However, if the wall of shell is thick, the pressures are no longer uniformly distributed and the problem becomes complex.

9.2 THIN CYLINDRICAL SHELLS UNDER INTERNAL PRESSURE

Shells may have different shapes such as cylindrical, cylindrical with conical bottom, hemispherical, spherical, etc. In this section, thin cylindrical shells under internal pressure is only dealt.

In a cylindrical shell due to the internal pressure *p*, stresses are developed on the walls of the cylinder on the cross section along the axis, on the cross section perpendicular to the axis and in radial direction (Figure 9.1)

Stress acting along the circumference of the cylinder is called *circumferential stress* or *hoop stress.* Stress acting along the longitudinal direction (i.e., along the length of the cylinder) is called *longitudinal stress* and stress acting along the radial direction is called *radial stress.* All these stresses are tensile stresses and are principal stresses.

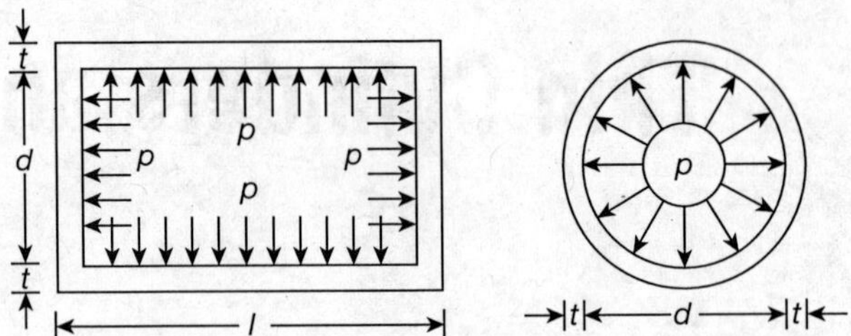

Fig. 9.1 Cylinder under internal pressure.

If the ratio of thickness to internal diameter is less than 1/20, it is reasonably assumed that the hoop and longitudinal stresses are constant over the thickness and the radial stress is small and is neglected.

The pressure of the fluid acting vertically upwards and downwards (due to setting up of hoop stress) on the thin cylinder may cause a burst leading to a failure of the type shown in Figure 9.2(a). The pressure of the fluid acting on the ends (due to setting up of longitudinal stress) of the thin cylinder may cause a burst leading to a failure of the type shown in Figure 9.2(b).

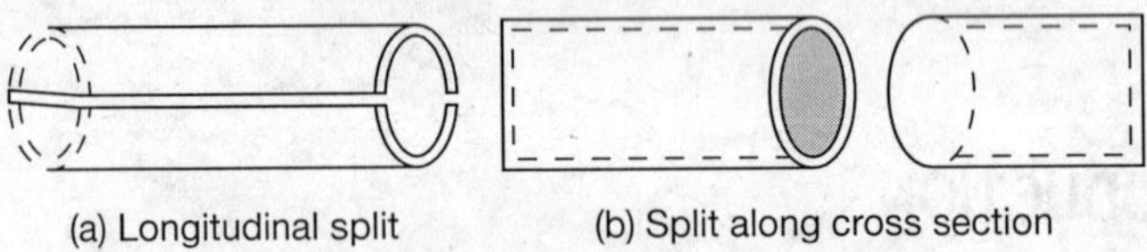

Fig. 9.2 Types of cylinder failure.

9.2.1 Circumferential or Hoop Stress

A thin cylindrical shell of length *l*, with internal diameter *d*, and shell thickness *t*, subjected to an internal pressure *p* is considered in Figure 9.3(a).

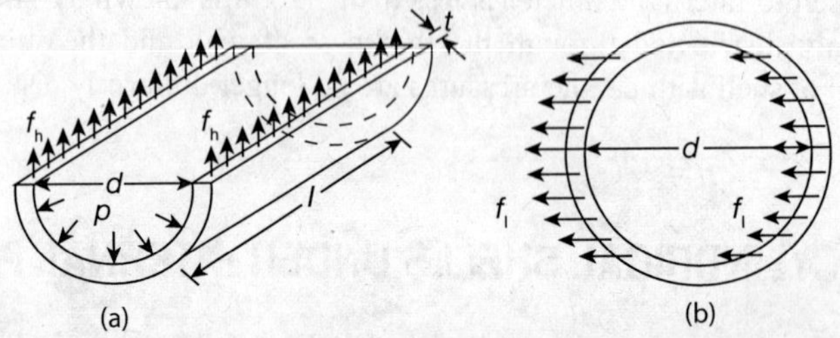

Fig. 9.3 Development of stresses in cylinder.

Splitting or bursting of the cylinder, as shown in Figure 9.2(a), may occur when the force due to fluid pressure is more than the resisting force developed due to hoop stress, f_h.

$$\text{Force due to fluid pressure} = p \times (\text{projected horizontal area on which } p \text{ acts})$$
$$= p \times d \times l$$
$$= pdl \tag{9.1}$$
$$\text{Resisting stress due to hoop stress} = f_h \times (\text{area over which } f_h \text{ is acting})$$
$$= f_h(l \times t + l \times t)$$
$$= 2f_h lt \tag{9.2}$$

For equilibrium Equation (9.1) should be equal to Equation (9.2)

i.e.,
$$pdl = 2f_h lt$$

i.e.,
$$f_h = \frac{pd}{2t} \tag{9.3}$$

The stress f_h is tensile.

9.2.2 Longitudinal Stress

Splitting or bursting of the cylinder, as shown in the Figure 9.2(b), may occur when the force due to fluid pressure acting on the ends of the cylinder is more than the resisting force developed due to longitudinal stress, f_l.

$$\text{Force due to fluid pressure} = p \times (\text{projected vertical area on which } p \text{ acts})$$
$$= p\frac{\pi}{4}d^2 \tag{9.4}$$
$$\text{Resisting force due to longitudinal stress} = f_l \times (\text{area on which } f_l \text{ is acting})$$
$$= f_l \pi dt \tag{9.5}$$

For equilibrium Equation (9.4) should be equal to Equation (9.5)

i.e.,
$$p\frac{\pi}{4}d^2 = f_l \pi dt$$

i.e.,
$$f_l = \frac{p\pi d^2}{4\pi dt}$$

∴
$$f_l = \frac{pd}{4t} \tag{9.6a}$$

The stress f_l is also tensile.

Comparing Equation (9.6a) with Equation (9.3), we have

$$f_l = \frac{1}{2}f_h \tag{9.6b}$$

i.e., Longitudinal stress is half of hoop stress.

If an axial pull P is applied the internal pressure will decrease, then the longitudinal stress is given as,

$$f_l = \frac{P - \dfrac{\pi d^2}{4}p}{\pi d} \tag{9.6c}$$

9.2.3 Maximum Shear Stress

As stated earlier the hoop stress f_h, acting circumferentially is a principal stress and the longitudinal stress acting parallel to the axis (i.e., perpendicular to plane on which f_h is acting) is also a principal stress.

∴ Maximum shear stress,

$$q_{max} = \frac{f_h - f_l}{2}$$

i.e.,

$$= \frac{\left(\frac{pd}{2t} - \frac{pd}{4t}\right)}{2} \tag{9.7}$$

$$q_{max} = \frac{pd}{8t}$$

9.2.4 Change in Dimensions of a Thin Cylinder

Let δd, δl and δV be the change in diameter, change in length and change in volume caused in the cylinder due to internal pressure.

Let e_h and e_l be the hoop strain and longitudinal strain, respectively. As f_h and f_l are the principal stresses, the *hoop strain* is given as

$$e_h = \frac{f_h}{E} - \frac{f_l}{mE}$$

Substituting for f_h and f_l from Equations (9.3) and (9.6), respectively

$$e_h = \frac{pd}{2tE} - \frac{pd}{4tmE}$$

i.e.,

$$e_h = \frac{pd}{2tE}\left[1 - \frac{1}{2m}\right] \tag{9.8}$$

We know,

$$e_h = \frac{\text{Change in circumference due to pressure}}{\text{Original circumference}}$$

$$= \frac{(\text{Final circumference}) - (\text{Original circumference})}{\text{Original circumference}}$$

i.e.,

$$e_h = \frac{\pi(d + \delta d) - \pi d}{\pi d}$$

i.e.,

$$e_h = \frac{\delta d}{d} \tag{9.9}$$

Equating the values of e_h from Equations (9.8) and (9.9), we have

$$\frac{\delta d}{d} = \frac{pd}{2tE}\left[1 - \frac{1}{2m}\right] \tag{9.10}$$

∴ Change in diameter, $$\delta d = \frac{pd^2}{2tE}\left[1 - \frac{1}{2m}\right]$$

Longitudinal strain is given as

$$e_l = \frac{f_l}{E} - \frac{f_h}{mE}$$

Substituting again for f_h and f_l we have

$$e_l = \frac{pd}{4tE} - \frac{pd}{2tmE}$$

i.e.,
$$e_l = \frac{pd}{2tE}\left[\frac{1}{2} - \frac{1}{m}\right] \quad (9.11)$$

Longitudinal strain is also given as

i.e.,
$$e_l = \frac{\text{Change in length}}{\text{Original length}} \quad (9.12)$$

$$e_l = \frac{\delta l}{l}$$

Equating values of e_l from Equations (9.11) and (9.12)

$$\frac{\delta l}{l} = \frac{pd}{2tE}\left[\frac{1}{2} - \frac{1}{m}\right]$$

i.e., Change in length, $$\delta l = \frac{pdl}{2Et}\left[\frac{1}{2} - \frac{1}{m}\right] \quad (9.13)$$

Thus the volumetric strain

$$e_v = \frac{\delta V}{V} = \frac{\text{Change in volume}}{\text{Original volume}} \quad (9.14)$$

Original volume, $$V = \frac{\pi d^2}{4} l$$

$$\text{Final volume} = (\text{Final cross section}) \times (\text{Final length})$$

$$= \frac{\pi}{4}(d + \delta d)^2 \times (l + \delta l) \quad (9.15)$$

$$= \frac{\pi}{4}(d^2 l + 2dl\delta d + \delta l d^2)$$

The small quantities such as $(\delta d)^2 l$, $\delta l(\delta d)^2$ and $2d\delta d\delta l$ are neglected
Change in volume,

$$\delta V = \frac{\pi}{4}(d^2 l + 2dl\delta d + \delta l d^2) - \frac{\pi d^2}{4} l \quad (9.16)$$

$$\delta V = \frac{\pi}{4}(2dl\delta d + \delta l d^2).$$

Considering Equations (9.14) and (9.16)
Volumetric strain,

$$e_V = \frac{\delta V}{V} = \frac{\frac{\pi}{4}(2dl\delta d + \delta l d^2)}{\frac{\pi d^2}{4} l} \quad (9.17)$$

i.e.,

$$e_V = \frac{2\delta d}{d} + \frac{\delta l}{l} \quad (9.18)$$

Substituting for $\frac{\delta d}{d}$ and $\frac{\delta l}{l}$ from Equation (9.7) from Equations (9.10) and (9.13), we have

$$e_V = 2 \times \frac{pd}{2tE}\left(1 - \frac{1}{2m}\right) + \frac{pd}{4tE}\left(\frac{1}{2} - \frac{1}{m}\right)$$

Rearranging,

$$e_V = \frac{\delta V}{V} = \frac{pd}{2Et}\left(\frac{5}{2} - \frac{2}{m}\right) \quad (9.19)$$

∴ Change in volume [from Equation (9.19)]

$$\delta V = V\left[\frac{pd}{4tE}\left(\frac{5}{2} - \frac{2}{m}\right)\right]$$

also from Equation (9.18)

$$\delta V = V(2e_h + e_l)$$

9.2.5 Thin Cylindrical Shells with Hemispherical Ends

Let t_1 and t_2 be the thickness of the shell of the cylinder and that of the hemispherical ends, respectively, with the same diameter, d (Figure 9.4).

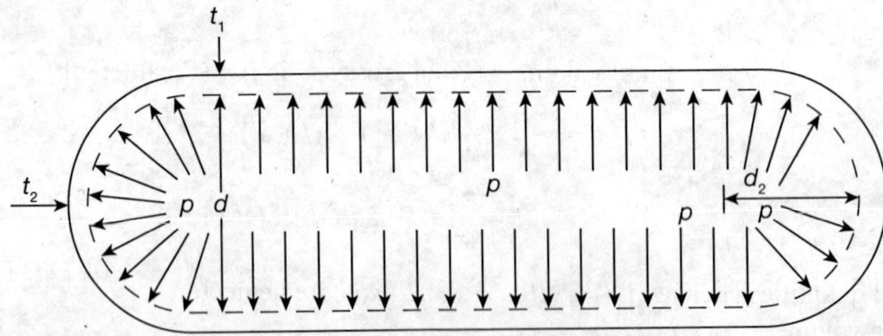

Fig. 9.4 Cylinder with hemispherical ends.

If the internal pressure is p, then Hoop stress, $f_h = \frac{pd}{2t_1}$ (9.20) and longitudinal stress, $$f_l = \frac{pd}{4t_1} \tag{9.21}$$

Then, hoop strain

$$e_h = \left(\frac{1}{E}\right)\left(f_h - \frac{f_l}{m}\right)$$

Substituting for f_h and f_l from Equations (9.20) and (9.21).

$$e_h = \left(\frac{1}{E}\right)\left(\frac{pd}{2t_1} - \frac{pd}{4mt_1}\right)$$

i.e., $$e_h = \frac{pd}{2Et_1}\left(1 - \frac{1}{2m}\right) \tag{9.22}$$

For the hemispherical ends (discussed in Section 9.3)

Hoop stress, $$(f_h)_{\text{ends}} = \frac{pd}{4t_2}$$

and

Hoop strain, $$(e_h)_{\text{ends}} = \frac{1}{E}\left[(f_h)_{\text{ends}} - \frac{(f_h)_{\text{ends}}}{m}\right]$$

i.e., $$(e_h)_{\text{ends}} = \frac{pd}{4Et_2\left(1 - \frac{1}{m}\right)} \tag{9.23}$$

For no distortion of the junction under pressure.

$$(e_h)_{\text{ends}} = e_h$$

i.e., $$\frac{pd}{2Et_1}\left(1 - \frac{1}{2m}\right) = \frac{pd}{4Et_2}\left(1 - \frac{1}{m}\right)$$

i.e., $$\frac{2m-1}{4mt_1} = \frac{m-1}{4mt_2}$$

i.e., $$\frac{t_2}{t_1} = \frac{(m-1)}{(2m-1)} \tag{9.24}$$

This shows that $\frac{t_2}{t_1}$ ratio is less than one. Hence, the maximum shear stress is to occur at the ends.

For equal stress in the cylindrical portion and at the ends $\frac{t_2}{t_1}$ should be equal to 0.5.

9.2.6 Design Aspects of Thin Cylindrical Shells

If f_a is the allowable tensile stress, then the major principal stress acting at the thin cylindrical shell should be less than f_a. Hence, the major principal stress for flat ends is f_h.

$$\therefore \qquad f_h \le f_a \tag{9.25}$$

$$\text{i.e.,} \qquad \frac{pd}{2t} \le f_a \tag{9.26}$$

Thus, for a cylindrical shell of given diameter d and flat ends, under an internal pressure of p, the thickness of the shell.

$$t_1 \ge \frac{pd}{2f_a} \tag{9.27}$$

In the case of hemispherical ends the maximum shear stress occurs at the ends and hence, t_2 is found first from

$$t_2 \ge \frac{pd}{2f_a} \tag{9.28}$$

Then, from a suitable ratio $\frac{t_2}{t_1}$, the thickness t_1 is found. For an ideal condition of equal stress throughout the body of the vessel the ratio $\frac{t_2}{t_1} = 0.50$ may be adopted.

9.2.7 Wire Winding of Thin Cylinders

It is shown that the hoop stress is twice the longitudinal stress when a thin cylinder is under fluid pressure. Hence, the failure of the thin cylinder may be anticipated to occur due to hoop stress. The maximum fluid pressure is restricted to a particular value for a given cylinder such that the hoop stress does not exceed the allowable pressure.

In the case of cylinders with high internal fluid pressure, such as that in a gun tube, the increase in thickness of the tube will be uneconomical and some method of reducing the hoop stresses has to be made. This is achieved by reinforcing the cylinder by either providing a wire winding or providing a compound tube (discussed elsewhere).

Steel wire or rectangular strip in wound under tension which puts the shell of the cylinder under an initial compressive stress. The bursting force due to internal pressure is possibly resisted by the shell and wire section. That is, the net effect of the initial compressive stress on the shell and those due to internal fluid pressure is to make the resultant stress less. Thus, the resultant stress in the shell will be hoop (tensile) stress due to the fluid pressure minus the initial compressive stress.

9.2.8 Rotational Stresses in Thin Cylindrical Shells

Let a cylinder of mean radius r and thickness t rotating at an angular velocity w about its axis be considered (Figure 9.5).

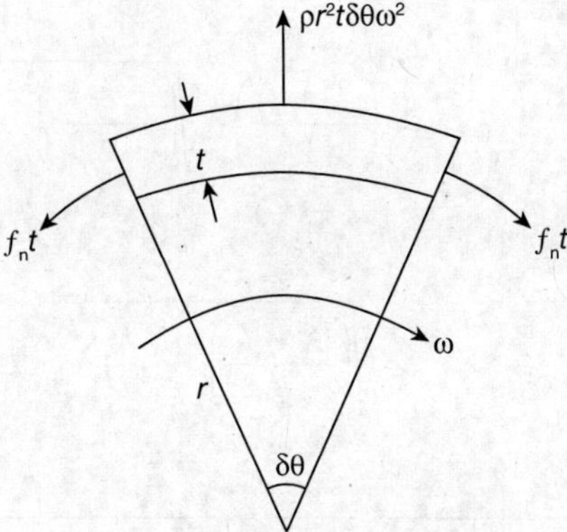

Fig. 9.5 Rotational forces.

The centrifugal force on the walls will produce a hoop stress f_h which may be assumed to be constant.

The centrifugal force on an element of unit length axially (Figure 9.5)

$$= (\rho r\delta\theta t)r\omega^2$$

where δ is the density of the material of the shell.

Resolving radially we have,

$$3f_h \sin\frac{1}{2}\delta\theta = \rho r^2\omega^2 t\delta\theta$$

or

$$f_h = \rho r^2\omega^2 \qquad \left(\because \sin\frac{1}{2}\delta\theta\frac{1}{2}\delta\theta\right)$$

where ρ is in kg/m³

∴ r is in m

and ω is in radians/sec.

9.2.9 Thin Cylindrical Shells Under Internal Pressure and Torque

It has been shown that in a cylindrical shell subjected to internal fluid pressure p, the stress developed are hoop stress, f_h and the longitudinal stress, f_l. If in addition to internal fluid pressure, a torque also acts in the cylindrical shell, then shear stresses will also be set up in the shell material.

Thus, at any point in the material there will be two tensile stresses acting along two perpendicular planes accompanied by a shear stress, f_s.

Here, the hoop stress is a tensile stress and longitudinal stress is also a tensile stress but the torque produces shear stress.

Then, the major principal stress

$$= \left(\frac{f_h + f_l}{2}\right) + \sqrt{\left(\frac{f_h - f_l}{2}\right) + f_s^2} \qquad (9.30)$$

and the minor principal stress

$$= \left(\frac{f_h + f_l}{2}\right) - \sqrt{\left(\frac{f_h - f_l}{2}\right)^2 + f_s^2} \qquad (9.31)$$

and the maximum shear stress

$$= \sqrt{\left(\frac{f_h - f_l}{2}\right)^2 + (f_s)^2} \qquad (9.32)$$

SOLVED PROBLEM 9.1

Water in a water main of 100 cm diameter is at a pressure head of 120 cm. The permissible stress on the shell material is 25 N/mm². Find the thickness of the metal shell required for the water main. Take the unit weight of water as 9,810 kN/m³.

Solution:

Given data: Diameter = 100 cm, pressure head = 120 cm, permissible stress = 25 N/mm² and $\gamma = 9{,}810$ kN/m³.

Permissible stress is equal to the hoop stress

i.e., $f_h = 25$ N/mm²

Water pressure in water main

$$= \frac{(9.810 \times 1{,}000) \times 120 \times 1{,}000}{1{,}000^3}$$

$$= 1.177 \text{ N/mm}^2$$

Thickness, $t \geq \frac{pd}{2f_h}$

$$\geq \frac{1.177 \times 100 \times 10}{2 \times 25}$$

$$\geq 23.54 \text{ mm}$$

Design thickness may be 24 mm.

SOLVED PROBLEM 9.2

A cylindrical shell 900 mm long and 200 mm internal diameter having thickness of 8 mm is filled with fluid at atmospheric pressure. If an additional 25×10^3 mm² of fluid is pumped into the cylinder, find the pressure exerted by the fluid on the cylinder, what is the hoop stress induced? $E = 2 \times 10^5$ N/mm², Poisson's ratio, $\mu = 0.3$ and bulk modulus of water = 2.1×10^3 N/mm².

Solution:

Given data: Length = 900 mm, internal diameter = 200 mm, $\delta V = 25 \times 10^3$ mm³, $E = 2 \times 10^5$ N/mm², $\mu = 0.30$ and bulk modulus = 2.1×10^3 N/mm².

Let the internal pressure be p N/mm².

Volumetric strain,
$$\frac{\delta V}{V} = \frac{pd}{2tE}\left(\frac{5}{2} - \frac{2}{m}\right)$$

i.e.,
$$\frac{\delta V}{V} = \frac{p \times 200}{2 \times 8 \times 2 \times 10^5}\left(\frac{5}{2} - 2 \times 0.3\right)$$

$$\frac{\delta V}{V} = 1.1875 \times 10^{-4}\, p$$

i.e.,
$$25 \times 10^3 = 1.1875 \times 10^{-4}\left(\frac{\pi}{4} \times 200^2 \times 900\right) p$$

Solving,
$$p = 7.44 \text{ N/mm}^2$$

Then,
$$f_h = \frac{pd}{2t}$$

$$= \frac{7.44 \times 200}{2 \times 8}$$

$$f_h = 93 \text{ N/mm}^2$$

SOLVED PROBLEM 9.3

A cylindrical shell 1 m diameter and 3 m length is subjected to an internal pressure of 2 N/mm². Calculate the minimum thickness, if the stress should not exceed 50 N/mm². Find the change in diameter and volume of the shell. Poisson's ratio, $1/m = 0.3$ and $E = 200$ kN/mm².

Solution:

Given data: Diameter = 1 m, length = 3 m, $p = 2$ N/mm², allowable stress, $f_a = 50$ N/mm², $m = 0.3$ and $E = 200$ kN/mm².

The allowable stress, $f_a = 50$ N/mm², i.e., the major principal stress (i.e., hoop stress) should not exceed 50 N/mm².

∴ Thickness,
$$t \geq \frac{pd}{2f_a}$$

$$\geq \frac{2 \times 1 \times 1{,}000}{2 \times 50}$$

$$\geq 20 \text{ mm}$$

i.e.,
$$t = 20 \text{ mm}$$

From Equation (9.10) the change in diameter

$$\delta d = \frac{pd^2}{2Et}\left[1 - \frac{1}{2m}\right]$$

$$= \frac{2 \times 1{,}000^2}{2 \times 200 \times 1{,}000 \times 20}\left[1 - \frac{0.3}{2}\right]$$

$$= 0.2125 \text{ mm}$$

From Equation (9.13), the change in length

$$\delta l = \frac{pdl}{2Et}\left(\frac{1}{2} - \frac{1}{m}\right)$$

$$= \frac{2 \times 1{,}000 \times 3{,}000}{2 \times 200 \times 1{,}000 \times 20}\left(\frac{1}{2} - 0.3\right)$$

$$= 0.15 \text{ mm}$$

$$= \frac{\pi}{4}\left(2dl\delta d + \delta l d^2\right)$$

Change in volume $= \frac{\pi}{4}\left[2 \times 1{,}000 \times 3{,}000 \times 0.2125 + 0.15 \times (1{,}000)^2\right]$

$$= 1.12 \times 10^6 \text{ mm}^3$$

SOLVED PROBLEM 9.4

A thin cylindrical shell of internal diameter 80 mm with a wall thickness of 2 mm has rigid end plates and filled with water. The cylinder is subject to an axial pull of 15 kN at the ends and the water pressure has beam observed as 0.006 N/mm^2. Find the poisson's ratio of the material if the modulus of elasticity of the material is 2.0×10^5 N/mm^2 and the bulk modulus of water is 2.0×10^3 N/mm^2.

Solution:

Given data: Diameter = 80 mm, $t = 2$ mm, axial pull = 15 kN, water pressure = 0.06 N/mm^2, bulk modulus of water = 2.0×10^3 N/mm^2 and $E = 2 \times 10^5$ N/mm^2.

For tube

Change is longitudinal stress due to axial load is

$$f_1 = \frac{P - \left(\frac{\pi d^2}{4}\right)p}{\pi dl}$$

$$= \frac{15 \times 1{,}000 - \frac{\pi \times 80^2}{4} \times 0.06}{\pi \times 80 \times 2}$$

$$= 29.3 \text{ N/mm}^2.$$

Change in hoop stress due to axial load is

$$f_h = \frac{pd}{2t}$$
$$= \frac{0.06 \times 80}{2 \times 2}$$
$$= 1.2 \text{ N/mm}^2.$$

The above stress is compressive as p decreases. Let e_h and e_l be the change in strain due to axial load:

$$\therefore \quad e_h = \frac{1}{E}(1.2 + 29.3\mu)$$
$$e_l = \frac{1}{E}(29.3 + 1.2\mu)$$
$$e_v = e_l + 2e_h$$

but

$$= \frac{1}{E}\left[(29.3 + 1.2\mu) - 2(1.2 + 29.3\mu)\right]$$
$$= \frac{1}{E}[27.9 - 57.4\mu]$$

For water

$$e_v = \frac{p}{K} = \frac{0.06}{2 \times 10^3}$$

As there is no escape of water from tube

Volumetric strain of water = Volumetric strain of the tube

i.e.,
$$\frac{1}{E}(27.9 - 57.4\mu) = \frac{0.06}{2 \times 10^3}$$

i.e.,
$$27.9 - 57.4\mu = \frac{0.06 \times 2 \times 10^5}{2 \times 10^3}$$

Solving, Poisson's ratio, $\mu = 0.38$

SOLVED PROBLEM 9.5

A copper cylinder 90 cm long, 40 cm external diameter and wall thickness 6 mm has its both ends closed by rigid blank flanges. It is initially full of oil at atmospheric pressure. Calculate the additional volume of oil which must be pumped into to raise the oil pressure to 5 MN/m^2 above atmospheric pressure. E_{cu} = 100 GN/m^2 and Poisson's ratio $\left(\frac{1}{m}\right) = \frac{1}{3}$. Bulk modulus of oil is 2.6 GN/m^2.

Solution:

Given data: External diameter = 40 cm, length = 90 cm, wall thickness = 6 mm, pressure = 5 MN/m^2 and E_{cu} = 100 GN/m^2.

Internal diameter, $d = 40 - 2 \times 0.6 = 38.8$ cm

Initial volume,
$$V = \frac{\pi}{4} \times d^2 \times l$$
$$= \frac{\pi}{4} \times 38.8^2 \times 90$$
$$= 106,413 \text{ cm}^3$$

Increase in oil pressure, $p = 5$ N/mm^2.
Bulk modulus of oil, $K = 2.6 \times 10^6$ N/mm^2

i.e.,
$$K = \frac{\text{Increase in pressure}}{\text{Decrease in volume/volume}}$$
$$= \frac{dp}{dV/V}$$
$$= V\frac{dp}{dV}$$

Let δV_1 be the increase in volume of cylinder

$$\text{Volumetric strain} = \frac{\delta V_1}{V_1}$$

$$\therefore \quad \frac{\delta V_1}{V_1} = \frac{pd}{2Et}\left[\frac{5}{2} - \frac{1}{m}\right]$$
$$= \frac{5\times10^6\,(38.8)\times10^{-2}}{2\times6\times10^{-3}\times10^9}\left[\frac{5}{2} - 2\times\frac{1}{3}\right]$$
$$= 0.00296$$
$$\therefore \quad \delta V_1 = 0.00296 \times 106{,}413 = 314.98 \text{ cm}^3$$

Let δV_2 be the decrease in volume of oil due to increase of pressure. Because of bulk modulus the volume of the oil will decrease.:

$$\therefore \quad K = \frac{p}{(\delta V_2 / V_2)}$$

i.e.,
$$\delta V_2 = \frac{p}{K} \times V_2$$
$$= \frac{5\times10^6}{2.6\times10^9} \times 106,413$$
$$= 204.65 \text{ cm}^3$$

Resulting additional space created in the cylinder $= \delta V_1 + \delta V_2$
$$= 314.98 + 204.65$$
$$= 519.63 \text{ cm}^3$$

$\therefore$ Additional quantity of oil which must be pumped to raise the oil pressure to 5 MN/m^2 is 519.63 cm^3.

SOLVED PROBLEM 9.6

A cast iron water pipe is 30 cm diameter and 15 mm wall thickness. The pipe is wound closely with a single layer of round steel wire of 5 mm diameter under a tension of 75 N/mm^2. Calculate the initial compressive stress induced in the pipe section. If water under a pressure of 3.5 N/mm^2 is allowed inside the pipe, find the stresses set up in the pipe and steel wire. Take E for cast iron as 0.96×10^5 N/mm^2 and for steel as 2.02×10^5 N/mm^2 and μ as 0.30.

Solution:

Given data: Diameter = 30 cm, thickness = 15 mm, diameter of steel wire = 5 mm, tension = 75 N/mm^2, water pressure = 3.5 N/mm^2, $E_{CI} = 0.96 \times 10^5$ N/mm^2, $E_s = 2.02 \times 10^5$ N/mm^2 and $\mu = 0.30$.

Before filling with water

Let us consider 1 cm length of pipe.

$$\text{Number of turns per cm length of wire} = \frac{\text{Length of pipe}}{\text{Diameter of wire}} = \frac{1}{0.5} = 2$$

Let f_w be the tensile stress in the wire.

Compressive force due to one turn of wire on the cylinder = 2 × (Area for cross section of wire) f_w

$$= 2 \times \frac{\pi}{4} \times 5^2 \times 75$$

$$\text{Total compressive force exerted by wire per cm length of pipe} = 2\left(2 \times \frac{\pi}{4} \times 5^2 \times 75\right)$$

$$= 5{,}890.48 \text{ N.}$$

$$\text{Cross-sectional area of the cylinder over which this force acts} = 2(l \times t) = 2(1 \times 10 \times 15) = 300 \text{ mm}^2.$$

$$\text{Initial compressive stress in the wall of the cylinder} = -\frac{5{,}890.48}{300} = -19.63 \text{ N/mm}^2.$$

Hence, before allowing water in the pipe, the stresses in the wall of pipe and wire are 75 N/mm^2 (tensile).

After filling: Stresses due to fluid pressure alone

Let f_{pf} be stress in pipe due to fluid pressure and f_{wf} be stress in wire due to fluid pressure.

Again considering 1 cm length of pipe, force of fluid pressure causing to burst the pipe along longitudinal section.

$$= pd \times 1 = 3.5 \times 300 \times 1 \times 10 = 10{,}500 \text{ N.} \quad \text{(i)}$$

The force is combinedly resisted by the pipe and the wire

Resisting force of pipe = (Stress in the pipe) × (Area of pipe resisting)

$$\text{Resisting force of pipe} = f_{pf}(2 \times 1 \times t) = f_{pf} \times 2 \times 1 \times 10 \times 15 = 300 f_{pf} \quad \text{(ii)}$$

Resisting force of wire = (Number of turns) × (Area of wire resisting) × (Stress in wire due to fluid pressure)

$$= 2\times\left(2\times\frac{\pi}{4}\times 5^2\right) f_{wf}$$

$$= 78.5 f_{wf} \qquad \text{(iii)}$$

∴ Total resisting force = $300 f_{pf} + 78.5 f_{wf}$

Equating the resisting and longitudinal bursting forces [Equations (i)–(iii)]

$$300 f_{pf} + 78.5 f_{wf} = 10{,}500 \qquad \text{(iv)}$$

The hoop strain in the pipe is equal to the strain in the wire.

$$\text{Hoop straining in pipe} = \frac{\text{(Hoop stress)}}{E_p} - \frac{\text{(Longitudinal stress)}}{mE_p}$$

$$= \frac{f_{pf}}{E_p} - \left(\frac{p\times d}{4t}\right)\frac{1}{mE_p}$$

$$= \frac{1}{E_p}\left(f_{pf} - \frac{3.5\times 300}{4\times 15}\times 0.3\right)$$

$$= \frac{1}{E_p}\left(f_{pf} - 5.25\right) \qquad \text{(v)}$$

$$\text{Strain in the wire} = \frac{f_{wf}}{E_s} \qquad \text{(vi)}$$

Equating the strains [Equations (v) and (vi)]

$$\frac{1}{E_p}\left(f_{pf} - 5.25\right) = \frac{f_{wf}}{E_s}$$

i.e.,

$$f_{wf} = \frac{E_s}{E_p}\left(f_{pf} - 5.25\right) \qquad \text{(vii)}$$

$$= \frac{2.02\times 10^5}{0.96\times 10^5}\left(f_{pf} - 5.25\right)$$

i.e.,

$$f_{wf} = 2.1\left(f_{pf} - 5.25\right)$$

Substituting for f_{wf} in Equation (iv)

$$300 f_{pf} + 78.5 \times 2.1\,(f_{pf} - 5.25) = 10{,}500$$

i.e.,

$$300 f_{pf} + 164.85\, f_{pf} = 10{,}500 + 863.625$$

Solving,

$$f_{pf} = 25.51\ \text{N/mm}^2\ \text{(tensile)}$$

Substituting for f_{pf} in Equation (vii)

$$f_{wf} = 2.1(25.51 - 5.25) = 42.55 \text{ N/mm}^2 \text{ tensile}$$

After filling: Resultant stress in pipe and wire

Resultant stress in pipe = (Initial stress in pipe) + (Stress in pipe due to fluid pressure alone)
= 19.63 N/mm^2 (compressive) + 25.51 N/mm^2 (tensile)
= 5.88 N/mm^2 (tensile)

Resultant stress in wire = (Initial stress in pipe) + (Stress in wire due to fluid pressure alone)
= 75.0 N/mm^2 (tensile) + 42.55 N/mm^2 (tensile)
= 117.6 N/mm^2.

SOLVED PROBLEM 9.7

A thin walled cylinder of internal diameter 1.2 m and length 4 m is subjected to an internal pressure of 1.5 N/mm^2. If the wall thickness of the shell is 10 mm, find the maximum shear stress set up in the shell and the change in the internal volume of the shell. $E = 2.1 \times 10^5$ N/mm^2 and $1/m = 0.28$.

Solution:

Given data: Diameter = 1.2 m, length = 4 m, internal pressure = 1.5 N/mm^2, wall thickness = 10 mm, $E = 2.1 \times 10^5$ N/mm^2 and Poisson's ratio, $1/m = 0.28$

Maximum shear stress, $$f_s = \frac{pd}{8t}$$

$$= \frac{1.5\times1.2\times10^3}{8\times10} = 22.5 \text{ N/mm}^2$$

Volumetric strain, $$\frac{\delta V}{V} = \frac{pd}{2tE}\left[\frac{5}{2} - \frac{2}{m}\right]$$

$$= \frac{1.5\times1.2\times10^3}{2\times10\times2.1\times10^5}\left[\frac{5}{2} - 2\times0.28\right]$$

i.e., $$\frac{\delta V}{V} = 8.314\times10^{-4}$$

i.e., $$\delta V = (8.314\times10^{-4})\, V$$

$$= \left(8.34\times10^{-4}\right)\left(\frac{\pi\times1.200^2}{4}\times4{,}000\right)$$

$$= 3{,}761{,}164.99 \text{ mm}^3.$$

Change in volume of the shell = 3,761,164.99 mm^3.

SOLVED PROBLEM 9.8

A thin cylindrical tube 80 mm internal diameter and 5 mm thick is closed at the ends and is subjected to an internal pressure of 6 N/mm^2. A torque of 20,09,600 kN is also applied in the tube. Find the hoop stress, longitudinal stress, maximum and minimum principle stresses and the maximum shear stress.

Solution:
Given data: Internal diameter = 80 mm, thickness = 5 mm, internal pressure = 6 N/mm² and torque = 20,09,600 kN m.

Hoop stress, $$f_h = \frac{pd}{2t} = \frac{6 \times 80}{2 \times 5} = 48 \text{ N/mm}^2$$

Longitudinal stress, $$f_l = \frac{pd}{4t} = \frac{6 \times 80}{4 \times 5} = 24 \text{ N/mm}^2$$

Stresses f_h and f_l are tensile stresses. Due to torque, shear stress is produced in the tube. The shear stress is assumed to be uniform throughout the thickness because the wall thickness is small.

$$\begin{aligned}\text{Shear force} &= (\text{Shear stress}) \times \text{Area}\\ &= f_s \times (\pi d t)\\ &= f_s \times \pi \times 80 \times 5\\ &= 1{,}256\, f_s\end{aligned}$$

$$\text{Torque, } T = (\text{Shear force}) \times \frac{d}{2}$$

$$20{,}09{,}600 = 1{,}256\ f_s \times \frac{80}{2}$$

Solving $$f_s = 40 \text{ N/mm}^2.$$

9.3 THIN SPHERICAL SHELLS UNDER INTERNAL PRESSURE

A thin spherical shell subjected to an internal pressure p (Figure 9.6) is considered. Let d and t be the internal diameter and thickness of spherical shell, respectively.

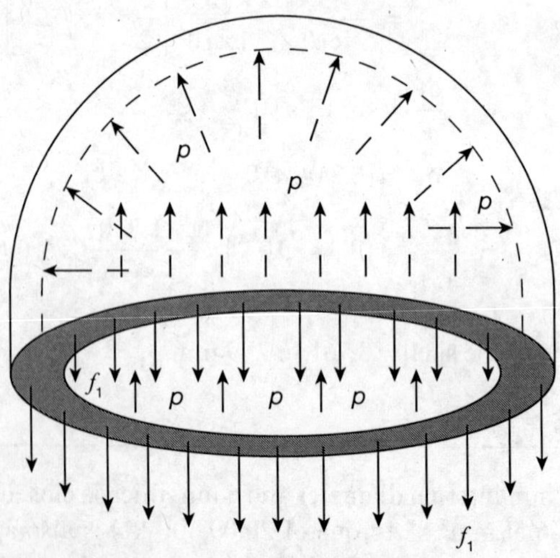

Fig. 9.6 Development of stress in sphere.

The pressure of the fluid tends to split the shell from the other portion of the vessel.

The force P, which has a tendency to split the shell $= p\left(\frac{\pi d^2}{4}\right)$ (9.33)

The resisting force developed due to hoop or tangential stress, $f_h = f_1 \times$ (Area resisting the force)

$$= f_1(\pi dt) \quad (9.34)$$

For equilibrium Equation (9.34) is equal to Equation (9.33)

i.e.,
$$f_1 \pi dt = p\frac{\pi d^2}{4}$$

i.e.,
$$f_1 = \frac{pd}{4t} \quad (9.35)$$

The hoop stress f_1 is tensile in nature. The stress f_1 is the same in every direction tangential to the spherical shell. Thus, the stress acting on a perpendicular plane is also $\frac{pd}{4t}$.

Thus, at any point in the material these principal stresses are

(i) $$f_1 = \frac{pd}{4t} \text{ (tensile)}$$

(ii) $$f_2 = \frac{pd}{4t} \text{ (tensile)}$$

and (iii) Radial stress $= p$ (compressive)

The third principal stress is neglected.

$\therefore$
$$\text{Shear stress} = \frac{1}{2}\left(\frac{pd}{4t}\right) = \frac{pd}{8t} \quad (9.36)$$

$$\text{Hoop strain, } e = \left(\frac{f_1}{E} - \frac{f_2}{mE}\right)$$

i.e.,
$$e = \frac{pd}{4tE}\left(1 - \frac{1}{m}\right) \quad (9.37a)$$

and
$$\text{Strain } e = \frac{\delta d}{d} \quad (9.37b)$$

Now,
$$\text{Volume, } V = \frac{\pi}{6}d^3$$

$\therefore$
$$\delta V = 3\frac{\pi}{6}d^2\delta d = \frac{\pi}{2}d^2\delta d \quad (9.38)$$

$\therefore$
$$\frac{\delta V}{V} = \frac{\frac{\pi}{2}d^2\delta d}{\frac{\pi}{6}d^3}$$

$$= \frac{3\delta d}{d}$$

Substituting for $\frac{\delta d}{d}$ from Equation (9.37b) $\frac{\delta V}{V} = 3e$ (9.39)

$$\text{Volumetric strain, } e_v = \frac{3pd}{4Et}\left(1-\frac{1}{m}\right) \quad (9.40)$$

Considering radial stress also, then volumetric strain is given as

$$(e_v)_r = \frac{3pd}{4Et}\left(1-\frac{1}{m}\right)+\frac{3p}{mE} \quad (9.41)$$

SOLVED PROBLEM 9.9

A spherical shell of 80 mm diameter (internal) has to withstand an internal pressure of 280 N/mm². Calculate the thickness of the shell wall if the maximum permissible stress is 30 N/mm².

Solution:

Given data: Diameter $d = 80$ mm, internal pressure = 280 N/mm² and permissible stress = 30 N/mm².

Thickness of spherical shell, $t \geq \frac{pd}{4f_a}$

$$\geq \frac{280\times 80}{4\times 30}$$

i.e., $t \geq 186.67$

Hence, a thickness of 75 mm is provided.

SOLVED PROBLEM 9.10

A spherical shell of internal diameter and of thickness 10 mm is subjected to an internal pressure of 1.5 MPa. Determine the increase in diameter and volume. $E = 200$ kN/mm² and $\mu = 0.3$.

Solution:

Given data: Diameter = 1 m, thickness = 10 mm, internal pressure = 1.5 MPa, $E = 200$ kN/m² and $\mu = 0.30$.

From Equations (9.37a) and (9.37b) the increase in diameter.

$$e = \frac{\delta d}{d} = \frac{pd}{4Et}\left(1-\frac{1}{m}\right)$$

i.e.,

$$\delta d = \frac{pd^2}{4Et}\left(1-\frac{1}{m}\right)$$

$$= \frac{1.5\times 1{,}000^2}{4\times 200\times 1{,}000\times 10}(1-0.3)$$

i.e., $\delta d = 0.13$ mm

From Equation (9.38) the increase in volume

$$\delta V = \frac{\pi}{2} d^2 \delta d$$

$$= \frac{\pi}{2} \times 1{,}000^2 \times 0.13$$

$$= 2{,}04{,}204 \text{ mm}^3$$

Hence,
Increase in diameter = 0.13 mm
Increase in volume = 2,04,204 mm^3.

SOLVED PROBLEM 9.11

A spherical shell is of 900 mm internal diameter and 6 mm thick. The shell is filled with a liquid and pressure is built in till the volume increases by 180 cc. Find the pressure developed by the fluid in the shell. Take $E = 2.05 \times 10^5$ N/mm^2 and $\mu = 0.28$.

Solution:

Given data: Internal diameter = 900 mm, thickness = 6 mm, increase in volume = 180 cc, $E = 2.05 \times 10^5$ N/mm^2 and $\mu = 0.28$.

$$\text{Volume of shell} = \frac{\pi}{6} d^3 = \frac{\pi}{6} \times 900^3$$

$$= 0.382 \times 10^9 \text{ mm}^3$$

$$\frac{\delta V}{V} = \frac{180 \times 10^3}{0.38 \times 10^9}$$

$$= 4.71 \times 10^{-4}$$

From Equation (9.39)

$$\frac{\delta V}{V} = 3e$$

$\therefore$

$$e = \frac{\delta V}{3V} = \frac{4.71 \times 10^{-4}}{3} = 1.57 \times 10^{-4}$$

But

$$e = \frac{f_1}{E}\left(1 - \frac{1}{m}\right)$$

$\therefore$

$$e = \frac{f_1}{2.05 \times 10^5}(1 - 0.3) = 1.57 \times 10^{-4}$$

$$f_1 = \frac{1.57 \times 10^{-4} \times 2.05 \times 10^5}{0.7} = 45.98 \text{ N/mm}^2$$

But $$f_l = \frac{pd}{4t}$$

$\therefore$ $$p = \frac{4f_l t}{d}$$

$$= \frac{4 \times 45.98 \times 6}{900}$$

$$= 1.23 \text{ N/mm}^2$$

$\therefore$ Pressure developed = 1.23 N/mm².

9.4 EFFICIENCY OF A JOINT

Cylindrical and spherical vessels are made by joining plates to the required shapes. Cylindrical shells like boiler have longitudinal and circumferential joints, whereas spherical shells have only circumferential joints.

To provide a joint, holes are made in the material of the shell for the rivets. Because of the holes the area offering resistance decreases and hence the stress developed in the material also becomes more.

Hence, the hoop stress and longitudinal stresses [Equations (9.3) and (9.6)] have to be modified

$$f_h = \frac{pd}{2t\eta_h} \tag{9.42}$$

and

$$f_l = \frac{pd}{4t\eta_l} \tag{9.43}$$

where η_h and η_l are the efficiency of the hoop joint and the longitudinal joint, respectively.

Similarly, in the case of thick spherical shells the stresses f_1 and f_2 [Equation 9.35] have to be modified as

$$f_1 = f_2 = \frac{pd}{4t\eta} \tag{9.44}$$

where η is the efficiency of the joint.

It has to be noted that in a longitudinal joint the circumferential stress is developed whereas in circumferential joint the longitudinal stress is developed.

SOLVED PROBLEM 9.12

The steam pressure in a boiler is 2.5 N/mm². The thickness of boiler plate is 3.0 cm and the permissible tensile stress is 110 N/mm². Compute the design diameter considering efficiency of longitudinal joint as 86% and that of the hoop joint as 50%.

Solution:

Given data: Pressure, $p = 2.5$ N/mm², thickness = 3.0 cm, tensile stress = 110 N/mm², $\eta_l = 86\%$ and $\eta_h = 50\%$.

From Equation (9.42)

$$f_h = \frac{pd}{2t\eta_l}$$

i.e.,

$$110 = \frac{2.5\,d}{2 \times 3 \times 10 \times 0.86}$$

$$d = \frac{110 \times 2 \times 3 \times 10 \times 0.86}{2.5}$$

$$d = 2{,}270.4 \text{ mm} \approx 2{,}270 \text{ mm} \quad \text{(i)}$$

From Equation (9.43)

$$f_l = \frac{pd}{4t\eta_h}$$

i.e., (ii)

$$110 = \frac{2.5 \times d}{4 \times 3 \times 10 \times 0.50}$$

$$d = \frac{110 \times 4 \times 3 \times 10 \times 0.50}{2.5}$$

i.e.,

$$d = 2{,}640 \text{ mm}$$

The design diameter of the boiler is the minimum value of Equations (i) and (ii). The reason being that if a larger diameter is used the permissible stress will be exceeding the allowable value.

SOLVED PROBLEM 9.13

A boiler shell is to be made of 15 mm thick plate having a limiting tensile stress of 120 N/mm^2. If the efficiency of the longitudinal and circumferential joints are 70% and 30%, respectively, determine,

(i) maximum permissible diameter of the shell for an internal pressure of 2 N/mm^2
and (ii) permissible intensity of internal pressure when the shell diameter is 1.5 m.

Solution:

Given data: Thickness of plate = 15 mm, tensile stress = 120 N/mm^2, η_l = 70% and η_h = 30%

Maximum permissible diameter

The boiler shell should be designed for the limiting tensile stress at 120 N/mm^2. First, the limiting tensile stress is considered as hoop stress and then as longitudinal stress. The minimum of these two diameter shall satisfy the conditions.

Considering circumferential stress as 120 N/mm^2

$$f_h = \frac{pd}{2t \times \eta_l}$$

$$120 = \frac{2 \times d}{2 \times 15 \times 0.70}$$

$$\therefore \quad d = \frac{120 \times 2 \times 15 \times 0.70}{2}$$

$$d = 1{,}260 \text{ mm.}$$

Considering longitudinal stress as 120 N/mm

$$f_l = \frac{pd}{4t \times \eta_h}$$

i.e.,

$$120 = \frac{2 \times d}{4 \times 15 \times 0.30}$$

$$d = \frac{120 \times 4 \times 15 \times 0.30}{2}$$

$$d = 1{,}080 \text{ mm}$$

Hence, the design diameter is 1,100 mm.

Permissible intensity of pressure

Diameter of shell, d = 1.5 m = 1,500 mm. Considering circumferential stress as 120 N/mm²

$$f_h = \frac{pd}{2t\eta_l}$$

$$120 = \frac{p \times 1{,}500}{2 \times 15 \times 0.70}$$

$$\therefore \quad p = \frac{120 \times 2 \times 15 \times 0.70}{1{,}500} = 1.68 \text{ N/mm}^2.$$

Considering longitudinal stress as 120 N/mm²

$$f_l = \frac{pd}{4t\eta_h}$$

$$120 = \frac{p \times 1{,}500}{4 \times 15 \times 0.30}$$

$$\therefore \quad p = \frac{120 \times 4 \times 15 \times 0.30}{1{,}500} = 1.44 \text{ N/mm}^2.$$

The permissible intensity of pressure = 1.44 N/mm².

SOLVED UNIVERSITY QUESTIONS

SOLVED PROBLEM 9.14

A steel cylindrical shell 3 m long which is closed at its ends, had an internal diameter of 1.5 m and a wall thickness of 20 mm. Calculate the circumferential and longitudinal stresses induced and also the change in dimensions of the shell if it is subjected to an internal pressure of 1.0 N/mm². Assume E = 200 kN/mm². Poisson's ratio = 0.3 (Anna Univ., May 2006, ME).

Solution:

Given data: Length of shell, $l = 3$ m, diameter $d = 1.5$ m, internal pressure, $p = 1.0$ N/mm^2 and thickness $t = 20$ mm.

$$\text{Circumferential stress,} \quad f_h = \frac{pd}{2t} = \frac{1 \times 1{,}500}{2 \times 20} = 37.5 \text{ N/mm}^2$$

$$\text{Longitudinal stress,} \quad f_l = \frac{pd}{4t} = \frac{1 \times 1{,}500}{4 \times 20} = 18.75 \text{ N/mm}^2$$

$$\text{Maximum shear stress,} \quad q = \frac{37.5 - 18.75}{2} = 9.375 \text{ N/mm}^2.$$

$$\text{Change in diameter,} \quad \delta d = \frac{pd^2}{2tE}\left[1 - \frac{1}{2m}\right]$$

i.e.,

$$\delta d = \frac{1 \times 1{,}500^2}{2 \times 20 \times 2 \times 10^5}\left[1 - \frac{1}{2} \times 0.3\right]$$

i.e.,

$$\delta d = 0.239 \text{ mm}$$

$$\text{Change in length,} \quad \delta l = \frac{pdl}{2tE}\left[\frac{1}{2} - \frac{1}{m}\right] = \frac{1 \times 1{,}500 \times 3{,}000}{2 \times 20 \times 2 \times 10^5}\left(\frac{1}{2} - 0.3\right)$$

i.e.,

$$\delta l = 0.112 \text{ mm}$$

$$\text{Change in volume,} \quad \delta V = V \times \frac{pd}{2tE}\left[-\frac{5}{2} - \frac{2}{m}\right]$$

$$\text{Volume,} \quad V = \frac{\pi}{4} \times d^2 \times l = \frac{\pi}{4} \times 1{,}500^2 \times 3{,}000 = 53.01 \times 10^8 \text{ mm}^3$$

$$\therefore \quad \delta V = 53.01 \times 10^8 \, \frac{1 \times 1{,}500}{2 \times 20 \times 2 \times 10^5}\left(\frac{5}{2} - 0.3 \times 2\right) = 18{,}88{,}637.15 \text{ mm}^3.$$

Change in volume $= 18{,}88{,}637.15$ mm^3.

SALIENT POINTS

- Fluid, gases or dry materials with or without pressure are stored in specially made vessels with small thickness but adequate enough to sustain the load and pressure.
- If the ratio of the thickness of the wall of the shell to internal diameter is less than $\frac{1}{20}$, it is known as a thin shell.
- In thin walled shells the pressure is uniformly distributed through the thickness of wall and the stress is computed easily.
- If the walls of the shell are thick, the pressures are no longer uniformly distributed and the problem becomes complex.
- Stress acting along the circumference of a cylinder is called the circumferential or hoop stress.
- Stress acting along the longitudinal direction (i.e., along the length of the cylinder) is called longitudinal stress and along the radial direction is called radial stress.
- Hoop strain is given as $e_h = \frac{f_h}{E} - \frac{f_l}{mE}$

where f_h = hoop stress
f_l = longitudinal stress
$1/m$ = Poisson's ratio
E = modulus of elasticity

- In thick cylinders the hoop stress, f_1 and f_2 are given as

$$f_1 = f_2 = \frac{pl}{4t} \text{ (tensile).}$$

Radial stress $= p$ (compressive).

- In thin spherical shells
Volumetric strain $= \frac{3pd}{4Et}\left(1 - \frac{1}{m}\right)$

where p = internal pressure
d = internal diameter
t = thickness of shell.

- Volumetric strain considering radial stress is

Volumetric strain $= \frac{3pd}{4Et}\left(1 - \frac{1}{m}\right) + \frac{3p}{mE}$

- Considering efficiency of joint the hoop and longitudinal stresses are given as

$$f_h = \frac{pd}{2t\eta_l}$$

$$f_l = \frac{pd}{4t\eta_h}$$

where η_l = efficiency of longitudinal joint
η_h = efficiency of hoop joint.

QUESTIONS

1. How are thin and thick cylinders classified? Give example.
2. Derive expressions for hoop and longitudinal stresses in thin cylindrical shells.
3. How do you estimate the change in dimensions of a thin cylinder subjected to an internal pressure? How the volumetric strain computed?
4. Give the design steps of a thin cylindrical shell.
5. How do you estimate the volumetric strain in a thin spherical shell subjected to an internal pressure? How it is changed considering radial stress?
6. Compare (i) the maximum tensile stresses and (ii) the proportional increase in volume of a thin cylinder and a thin spherical shell having the same internal pressure.
7. A cylindrical vessel of 1 m diameter and 2.5 m long is closed with rigid end plates and subjected to an internal pressure of 2.5 N/mm^2. Estimate the thickness of the shell if the maximum principal stress is not to exceed 120 N/mm^2. Also, find the changes in diameter, length and volume of the shell, $E = 2.05 \times 10^5$ N/mm^2 and $\mu = 0.25$.
8. A steel cylinder of 1 m long and 40 cm external diameter with a wall thickness of 8 mm has both ends closed with rigid flat plates. The cylinder is filled with oil of bulk modulus 2.5×10^3 N/mm^2 at atmospheric pressure. If it is intended to raise the oil pressure in the cylinder to a pressure of 5 N/mm^2 above atmospheric pressure, compute the additional volume required. Take $E = 2.0 \times 10^5$ N/mm^2 and $\mu = 1/3$.
9. The internal diameter of a cylindrical shell is 3.2 m and of 8 cm thick and gas is stored. Find the internal pressure of the gas if the tensile stress in the material is not to exceed 75 N/mm^2.
10. A thin cylinder of 40 cm external diameter and wall thickness of 4 mm with 80 cm length is subjected to an internal pressure of 4 N/mm^2. A torque of 12,000 N·m is also applied to the tube. Find the hoop, longitudinal and shear stresses.
11. A spherical shell of 80 cm diameter is under a pressure of 1.5 N/mm^2. Find the thickness of the shell, if the maximum allowable stress is 80 N/mm^2. Take the joint efficiency as 75%.
12. Find the increase in volume of a spherical shell of 90 cm diameter and 10 mm thick when subjected to an internal pressure of 1.5 N/mm^2. Take $E = 2.05 \times 10^5$ N/mm^2 and $\mu = 0.30$.

10

Thick Cylinders and Shells

LEARNING OBJECTIVES

10.1 INTRODUCTION

Thin cylindrical shell analysis has been made based on the assumption that the hoop stress is constant throughout the thickness of the shell and the radial stress is negligible. This assumption is justifiable as long as thickness to internal diameter is less than 1/20.

In a cylinder if the above ratio is greater than 1/20, it is referred to as *thick cylinders.* In thick cylinders, the hoop stress is not constant but varies along the thickness and the radial stress is not negligible. The problem of thick cylinder is, in general, complex in nature.

10.2 LAME'S THEOREM

Lame's theorem gives the solution to thick cylinder problem. The theorem is based on the following assumptions:

(i) Material of the cylinder is homogeneous and isotropic.
(ii) Plane sections of the cylinder perpendicular to the longitudinal axis remain plane under the pressure.

The second assumption implies that the longitudinal strain is same at all points, i.e., the strain is independent of the radius.

Let us consider a thick cylinder subjected to internal and external pressures (Figure 10.1(a)).

Let an annular ring of the cylinder (Figure 10.1(b)) with internal radius r, thickness dr and a length of l be considered.

Let p_r be the internal radial pressure on the annular ring.

$(p_r + \delta p_r)$ be the external radial pressure on the annular ring and f_r be the hoop stress acting on the annular ring.

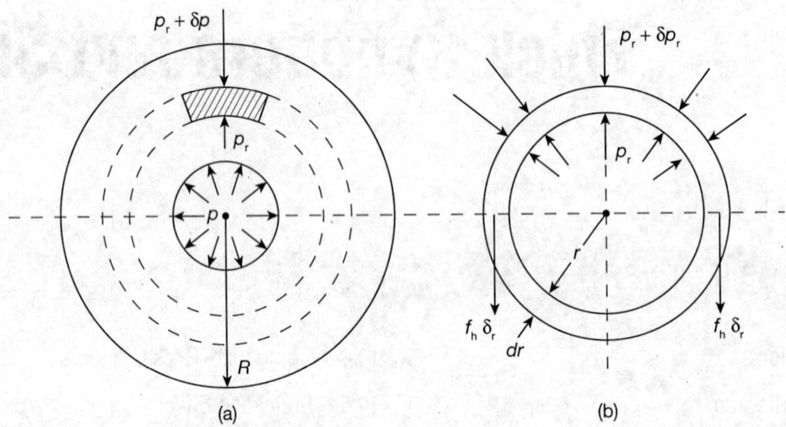

Fig. 10.1 Stress condition in thick cylinder.

Considering the equilibrium of one-half of the ring (similar to thin cylinder)

$$\text{Bursting force} = 2\,p_r(r \times l) - 2(p_r + \delta p_r)(r + \delta r)\,l$$

$$= 2l - p_r\delta_r - r\delta p_r - \delta r\delta p_r$$

$$= -2l\,(p_r\delta r + r\delta p_r) \tag{10.1}$$

After neglecting the products of small quantities

$$\text{Resisting force} = 2f_r\,(\delta r \times l) \tag{10.2}$$

where f_r is the hoop stress at any radius r

For equilibrium the resisting force is equal to bursting force, then

$$2\,f_r(\delta r \times l) = -2l(p_r\delta r \times r\delta p_r)$$

i.e.,

$$f_r = -p_r - r\frac{\delta p_r}{\delta r} \tag{10.3}$$

At the limiting condition the thickness of the annular ring is reduced indefinitely and Equation (10.3) reduces to

$$f_r + p_r + r\frac{dp_r}{dr} = 0 \tag{10.4}$$

From the assumption that the longitudinal strain is independent of r, we have

$$f_r - p_r = 2A \tag{10.5}$$

where A is constant.

Substituting for f_r in Equation (10.4), we have

$$(p_r + 2A) + p_r + \frac{r\,dp_r}{dr} = 0. \tag{10.6a}$$

i.e.,

$$\frac{dp_r}{dr} = \frac{-2(p_r + A)}{r} \tag{10.6b}$$

i.e.,

$$\frac{dp_r}{2(p_r + A)} = \frac{2dr}{r} \tag{10.6c}$$

Integrating,

$$\log_e(p_r + A) = \log_e r^2 + \log_e B \tag{10.7a}$$

where $\log_e B$ is a constant of integration.
i.e.,

$$\log_e (p_r + A) = \log_e \frac{B}{r^2}$$

or

$$p_r = \frac{B}{r_2} - A$$

From Equation (10.5)

$$f_r = p_r + 2A = \frac{B}{r^2} - A + 2A$$

i.e,

$$f_r = \frac{B}{r^2} + A$$

Thus, as per Lame's theorem

$$p_r = \frac{B}{r^2} - A \tag{10.8}$$

$$f_r = \frac{B}{r^2} + A \tag{10.9}$$

Equation (10.8) gives the radial pressure and Equation (10.9) gives the hoop stress at any radius r.

In the above equations, p_r is compressive stress whereas f_r is tensile stress.

The constants A and B in the Lame's equation are obtained after satisfying the boundary conditions.

10.3 THICK CYLINDRICAL SHELLS UNDER INTERNAL PRESSURE

Let R_i and R_e be the internal and external radius of the thick cylinder.

Let p be the internal pressure.

For this case, Lame's theorem conditions are

$$r = R_i \text{ and } p_r = p$$

$$r = R_e \text{ and } p_r = 0.$$

Substituting these values in Equation (10.8)
we have

$$p = \frac{B}{R_i^2} - A \tag{10.10}$$

and

$$0 = \frac{B}{R_e^2} - A \tag{10.11}$$

Solving Equations (10.10) and (10.11), we have

$$A = p\frac{R_i^2}{R_e^2 - R_i^2} \tag{10.12}$$

and

$$B = p\frac{R_i^2 R_e^2}{R_e^2 - R_i^2} \tag{10.13}$$

Maximum hoop stress occurs at $r = R_i$, then Equation(10.9) reduces to $f_r = f_h$.

$$\therefore \quad f_h = p\frac{R_i^2 R_e^2}{\left(R_e^2 - R_i^2\right) R_i^2} + p\frac{R_i^2}{\left(R_e^2 - R_i^2\right)}$$

i.e.,

$$f_h = p\left(\frac{R_e^2 + R_i^2}{R_e^2 - R_i^2}\right) \tag{10.14}$$

10.4 COMPOUND THICK CYLINDERS

It has been shown that the hoop stress is maximum at the inner radius and decreases towards the outer radius. Further, hoop stress is greater than that of internal fluid pressure. Thus, there is a limiting value for the internal fluid pressure as it depends on the hoop stress which in turn depends on the permissible stress of the shell material (Figure 10.2).

To meet the demand of cylinders to withstand high fluid pressures, some methods have to be involved to reduce the hoop stress. One such method was seen in thin cylinders is to wind strong steel wire under tension on the cylinder.

In another method, to reduce the hoop stress the cylinder may be made in two cylinders of different radii of one cylinder being shrink on to the other (after heating). By this process, the inner cylinder is put into compression and the outer one on tension.

Application on an internal fluid pressure causes a tensile hoop stress to be super imposed on the *shrinking stresses* and the resultant stress is the algebraic sum of the two sets.

As a first step, the stresses due to shrinking in each component is calculated from a knowledge of the radial pressure at the common surface. Then, the two cylinders are treated as one and the stresses are computed in the normal way.

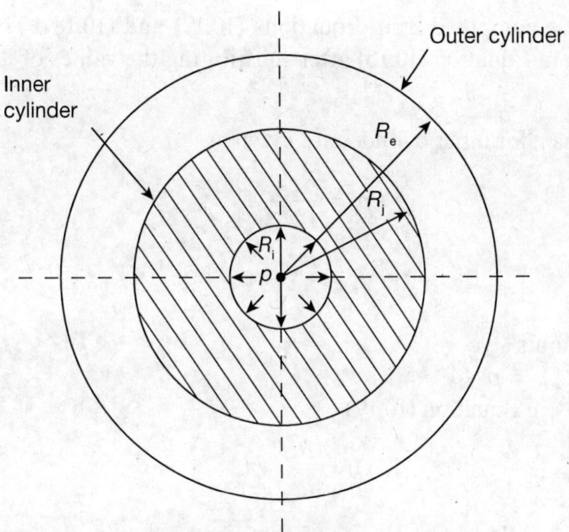

Fig. 10.2 Compound cylinder.

Let the compound cylinder shown in Figure 10.2 be considered.

Let R_o be the outer radius of the compound cylinder

R_j be the radius at the junction of the two cylinders

R_i be the inner radius of the compound cylinder

and p_j be the radial pressure at the junction of the two cylinders.

Now, Lame's equation can be applied under different conditions.

1. Before application of fluid pressure

(a) Outer cylinders

Lame's equations at radius r for outer cylinder are given as

$$p_r = \frac{B_1}{r^2} - A_1 \tag{10.15}$$

$$f_r = \frac{B_1}{r^2} + A_1 \tag{10.16}$$

where A_1 and B_1 are constants.

At $r = R_0$, $p_r = 0$ and $r = R_j$, $p_r = p_j$. Then

$$0 = \frac{B_1}{R_0^2} - A_1 \tag{10.17}$$

$$p_j = \frac{B_1}{R_0^2} + A_1 \tag{10.18}$$

Constants A_1 and B_1 are computed from Equations (10.17) and (10.18). The hoop stress in outer cylinder is determined from Equation (10.16) after substituting the values of A_1 and B_1.

(b) Inner cylinder

Lame's equations at radius r for inner cylinder are given as

$$p_r = \frac{B_2}{r^2} - A_2 \tag{10.19}$$

and

$$f_r = \frac{B_2}{r^2} + A_2 \tag{10.20}$$

where A_2 and B_2 are constants

At $r = R_i$, $p_r = 0$ and $r = R_j$, $p_r = p_j$

Applying these conditions in Equation (10.19)

$$0 = \frac{B_2}{R_i^2} - A_2 \tag{10.21}$$

$$p_j = \frac{B_2}{R_i^2} + A_2 \tag{10.22}$$

Constants A_2 and B_2 are computed from Equations (10.21) and (10.22). Hoop stress in inner cylinder is determined from Equation (10.20) after substituting the values of A_2 and B_2.

2. After application of fluid pressure

Now to compute the stress both the cylinders will be considered as one thick cylinder.

Let p be the internal fluid pressure.

Lame's equation at radius r for the combined cylinder are given as

$$p_r = \frac{\overline{B}}{r^2} - \overline{A} \tag{10.23}$$

$$f_r = \frac{\overline{B}}{r^2} + \overline{A} \tag{10.24}$$

where $\overline{A}$ and $\overline{B}$ are constants

At $r = R_0$, $p_r = 0$ and $r = R_i$, $p_r = p$

Applying these conditions in Equation (10.13)

$$0 = \frac{\overline{B}}{R_0^2} - \overline{A} \tag{10.25}$$

$$p = \frac{\overline{B}}{R_0^2} + \overline{A} \tag{10.26}$$

Constants $\overline{A}$ and $\overline{B}$ are computed from Equations (10.15) and (10.16). Values of $\overline{A}$ and $\overline{B}$ are substituted in Equation (10.14) and the hoop stresses across the section are determined for different boundary conditions. The resultant hoop stresses are the algebric sum of the hoop stresses caused due to shrinking and internal fluid pressure.

10.5 SHRINKAGE EFFECT

To shrink outer tube over the inner tube, it is necessary that the inner diameter of the outer cylinder should be slightly less than the outer diameter of the inner cylinder. This is needed such that after the tubes are filled the required radial pressure at the junction of the tubes may be reached.

Let R_c be the common radius of the tubes at the junction after the shrinkages.

Let $\delta r'$ be the difference between the outer radius of inner tube and R_c.

Let $\delta r''$ be the difference between the inner radius of the outer tube and R_c.

$\therefore$ Original difference of the radius of the two tubes at the junction $= \delta R_c = \delta R_i' + \delta R_i''$

Now for outer tube (using Lame's relation).

$$p_r = \frac{B_1}{r^2} - A_1$$

$$f_r = \frac{B_1}{r^2} + A_1$$

$\therefore$ Circumferential tensile strain for the outer tube at the junction

$$= \frac{\delta R_i''}{R_c} = \frac{1}{E}\left[\left(\frac{B_1}{R_c^2} + A_1\right) + \frac{p'}{m}\right] \tag{10.27}$$

where p' = radial pressure at the junction

$$\frac{1}{m} = \text{Poisson's ratio.}$$

Similarly for the inner tube, circumferential compressive strain for the inner tube at the junction.

$$= \frac{\delta R_i'}{R_c} = \frac{1}{E}\left[\left(\frac{B_2}{R_c^2} + A_2\right) + \frac{p'}{m}\right] \tag{10.28}$$

From Equations (10.27) and (10.28), we have

$$= \frac{\delta R_i'' - \delta R_i'}{R_c} = \frac{\delta R_c}{R_c} = \frac{1}{E}\left[\left(\frac{B_1}{R_c^2} + A_1\right) - \left(\frac{B_2}{R_c^2} + A_2\right)\right] \tag{10.29a}$$

$$\frac{\delta R_c}{R_c} = \frac{\text{Algebraic difference between the hoop stresses in the tubes at the junction}}{E} \tag{10.29b}$$

10.6 THICK SPHERICAL SHELLS

Let a spherical shell of external radius, R_e, and internal radius, R_i, be subjected to an internal fluid pressure p be considered.

Also, an elemental disc of the spherical shell of thickness dr at a radial distance r subtending an angle of $d\theta$ at centre is considered (Figure 10.3).

Let the radius be increased to $(r + u)$ and the thickness to $(dr + du)$ due to the internal fluid pressure. Let e_c and e_r be circumferential strain and radial strain, respectively.

$$\therefore \quad e_c = \frac{(\text{Increased circumference}) - (\text{Original circumference})}{(\text{Original circumference})}$$

$$= \frac{2\pi(r+u) - 2\pi}{2\pi r}$$

i.e.,
$$e_c = \frac{u}{r} \tag{10.30}$$

Similarly

$$e_r = \frac{(\text{Increased thickness of element}) - (\text{Original thickness of element})}{(\text{Original thickness of element})}$$

i.e.,
$$= \frac{(dr+du) - dr}{dr}$$

$$e_r = \frac{du}{dr} \tag{10.31}$$

From Equation (10.30)

$$u = re_c$$

$$\therefore \quad e_r = \frac{du}{dr} = e_c + r\frac{de_c}{dr} \tag{10.32}$$

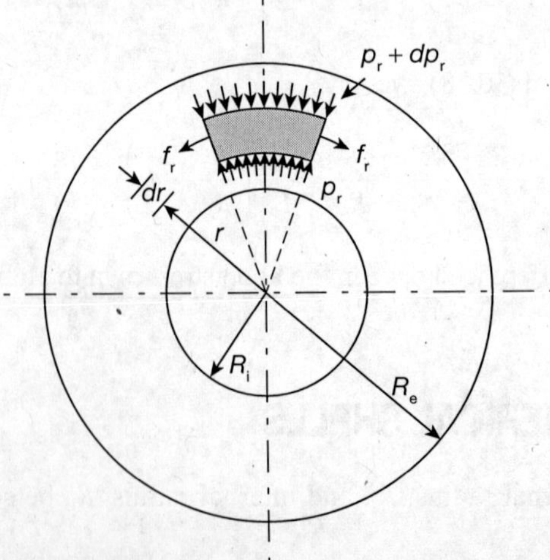

Fig. 10.3 Spherical shell subjected to internal pressure.

Let an elemental spherical shell of radius r and thickness dr (Figure 10.3) be considered.

Let the radial pressure be p_r and $(p_r + dp_r)$ at radius r and $(r + dr)$, respectively.

Let f_c be the circumferential tensile stress which is equal in all directions.

Forces acting on the half of the elemental spherical shell are:

(i) an upward force of $\pi r p_r$ due to internal radial pressure, p_r.

(ii) a downward force of $\pi(r + dr)^2 (p_r + dp_r)$ due to radial pressure $(p_r + dp_r)$

(iii) a downward resisting force $f_c\,(2\pi r dr)$

Considering the equilibrium of the half of the elementary spherical shell by equating the forces, we have

$$\pi r^2 p_r = \pi(r + dr)^2 (p_r + dp_r) + 2\pi r dr f_c \tag{10.33}$$

Neglecting squares and products of dr and dp and reducing we have (Figure 10.4)

$$2f_c = 2p_r - r\frac{dp_r}{dr}$$

Dividing both sides by rdr

$$f_c = -p_r - \frac{r}{2}\frac{dp_r}{dr} \tag{10.34}$$

Differentiating Equation (10.34) with respect to r, we get

$$\frac{df_c}{dr} = \frac{d}{dr}(-p_r) - \frac{1}{2}\frac{d}{dr}\left(r \cdot \frac{dp_r}{dr}\right)$$

i.e.,

$$\frac{df_c}{dr} = \frac{-dp_r}{dr} - \frac{1}{2}\left(\frac{rd^2 p_r}{dr^2} + \frac{dp_r}{dr}\right) \tag{10.35}$$

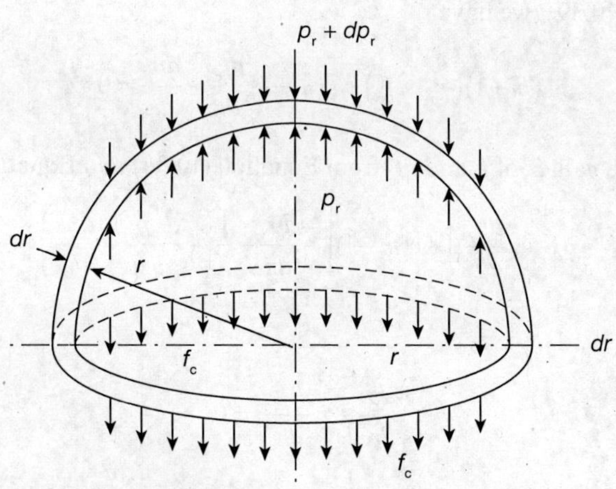

Fig. 10.4 Elemental spherical shell.

At any point in the elementary spherical shell, there are three principal stresses:

(i) the radial pressure p_r is compressive
(ii) the circumferential stress f_c is tensile and
(iii) stress acting on a place at right angles to the plane of f_c is also of magnitude f_c and tensile.

Now the radial strain,

$$e_r = \frac{p_r}{E} + \frac{f_c}{mE} + \frac{f_c}{mE}$$

i.e.,

$$e_r = \frac{p_r}{E} + \frac{2f_c}{mE} \quad \text{(compressive)}$$

$$e_r = -\left(\frac{p_r}{E} + \frac{2f_c}{mE}\right) \quad \text{(tensile)} \tag{10.36}$$

and

Circumferential strain

$$e_c = \frac{f_c}{E} - \frac{f_c}{mE} + \frac{p_r}{mE} \tag{10.37}$$

i.e.,

$$e_c = \frac{1}{E} + \left[f_c\left(\frac{m-1}{m}\right) + \frac{p_r}{m}\right] \tag{10.38}$$

Substituting the values of e_r and e_c from Equations (10.35) and (10.36) in Equation (10.30), we have

$$-\left(\frac{p_r}{E} + \frac{2f_c}{mE}\right) = \frac{1}{E}\left[f_c\left(\frac{m-1}{m}\right) + \frac{p_r}{m}\right] + \frac{rd}{dr}\left[\frac{1}{E}\left\{f_c\left(\frac{m-1}{m}\right) + \frac{p_r}{m}\right\}\right] \tag{10.39}$$

Reducing Equation (10.39), we have

$$(m+1)(p_r + f_c) + r(m-1)\frac{df_c}{dr} + \frac{dp_r}{dr} = 0 \tag{10.40}$$

Now, substituting the values of f_c and df_c from Equation (10.34) and Equation (10.35), we get

$$(m+1)\left(p_r - p_r - r_2\frac{dp_r}{dr}\right) + r(m-1)\left[-\frac{dp_r}{dr} - \frac{1}{2}\left(r\frac{d^2p_r}{dr^2} + \frac{dp_r}{dr}\right)\right] + r\frac{df_c}{dr} = 0 \tag{10.41}$$

Reducing Equation (10.41), we get

$$\frac{4dp_r}{dr} + r\frac{d^2p_r}{dr^2} = 0 \tag{10.42}$$

Substituting $\frac{dp_r}{dr} = Z$ in Equation (10.42) we get

$$4Z + \frac{r}{dr}\left(\frac{dpr}{dr}\right) = 0$$

i.e.,
$$4Z + r\frac{dZ}{dr} = 0 \tag{10.43}$$

or
$$\frac{dZ}{Z} = -4\frac{dr}{r}$$

Integrating Equation (10.43), we get

$$\log_e Z = -4\log_e r + \log_e C_1 \tag{10.44}$$

where C_1 is the constant of integration.

Equation (10.44) can be rewritten as

i.e.,
$$\log_e Z = \log_e r^{-4+\log_e C_1} \tag{10.45}$$

i.e.,
$$\log_e Z = \log_e\left(\frac{C_1}{r^4}\right) \tag{10.46}$$

$$Z = \frac{C_1}{r^4}$$

Substituting $Z = \frac{dp_r}{d_r}$, we have

$$\frac{dp_r}{dr} = \frac{C_1}{r^4}$$

i.e.,
$$dp_r = \frac{C_1}{r^4}dr \tag{10.47}$$

Integrating Equation (10.47) we get

$$p_r = \frac{C_1}{3r^3} + C_2 \tag{10.48}$$

where C_2 is constant of integration.

Substituting for p_r in Equation (10.44), we have

$$f_c = -\left(-\frac{C_1}{3r^3} + C_2\right) - \frac{r}{2}\frac{dp_r}{dr} \tag{10.49}$$

Substituting for $\frac{dp_r}{dr} = \frac{C_1}{r^4}$, we have

$$f_c = -\frac{C_1}{6r^3} - C_2 \tag{10.50}$$

Now, substituting $C_1 = -6b$ and $C_2 = -a$ in Equations (10.46) and (10.48), we have

$$p_r = -\left(\frac{-6b}{3r^3}\right) + (-a)$$

i.e.,
$$p_r = \frac{2b}{r^3} - a \tag{10.51}$$

and
$$f_c = -\frac{(-6b)}{6r^3} - (-a)$$

i.e., $$f_c = \frac{b}{r^3} + a \tag{10.52}$$

Constants a and b are obtained from initial given conditions.

Considering the initial conditions, $p_r = 0$ at $r = R_i$ and $p_r = p$ at $r = R_e$ and substituting in Equation (10.51) we have

$$0 = \frac{2b}{R_i^3} - a \tag{10.53}$$

Solving

$$p = \frac{2b}{R_e^3} - a \tag{10.54}$$

Solving for a and b, we have

$$a = \frac{pR_e^3}{R_i^3 - R_e^3} \tag{10.55}$$

and $$b = \frac{pR_1^3 R_e^3}{2(R_i^3 - R_e^3)} \tag{10.56}$$

SOLVED PROBLEM 10.1

A thick cylinder of 10 cm inner diameter and 20 cm outer diameter is subjected to an internal fluid pressure of 60 MPa. Calculate the hoop stress at the inside and outside surfaces and also plot the stress distribution curves.

Solution:

Given data: Inner diameter = 10 cm, outer diameter = 20 cm and internal pressure = 60 MPa.

From Lame's theorem,

$$p_r = \frac{B}{r^2} - A \tag{i}$$

$$f_r = \frac{B}{r^2} + A \tag{ii}$$

At $r = 100$ mm, $p = 60$ N/mm^2

$r = 200$ mm, $p = 0$

Substituting in Equations (i) and (ii)

$$60 = \frac{B}{100^2} - A \tag{iii}$$

and $$0 = \frac{B}{200^2} + A \tag{iv}$$

Substituting for A in Equation (iii) from Equation (iv)

$$60 = \frac{B}{100^2} - \frac{B}{200^2}$$

i.e.,

$$B = \frac{60 \times 200^2 \times 100^2}{200^2 - 100^2} = 800{,}000$$

$\therefore$

$$A = \frac{B}{200^2} = \frac{800{,}000}{200^2} = 20$$

At $r = 100$ mm

$$p_{\mathrm{r}} = \frac{800{,}000}{100^2} - 20 = 60 \text{ N/mm}^2$$

Here, $f_{\mathrm{r}} = f_{\mathrm{h}}$ = hoop stress. Then

$$f_{\mathrm{h}} = \frac{800{,}000}{200^2} - 20 = 100 \text{ N/mm}^2$$

At $r = 200$ mm, $p_{\mathrm{r}} = 0$ and $f_{\mathrm{h}} = \dfrac{800{,}000}{200^2} + 20 = 40 \text{ N/mm}^2$

The stress distribution is shown in Figure 10.5.

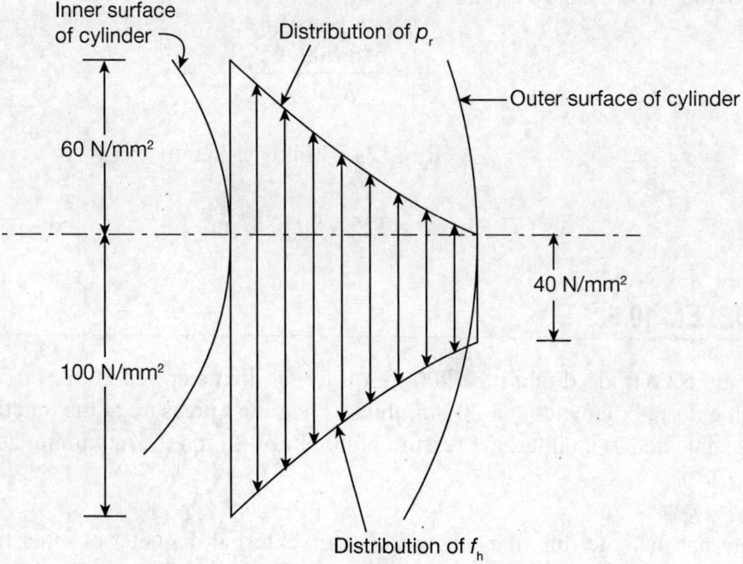

Fig. 10.5

SOLVED PROBLEM 10.2

Find the thickness of a thick metal cylinder with internal diameter 160 mm to withstand an internal pressure of 60 N/mm². The maximum hoop stress in the section is not to exceed 145 N/mm².

Solution:

Given data: Internal diameter = 160 mm, internal pressure = 60 N/mm² and hoop stress = 145 N/mm².

At $r = 20$ mm, $p_r = 60\ \text{N/mm}^2$ and $f_r = f_h = 145\ \text{N/mm}^2$.

Applying Lame's theorem

$$60 = \frac{B}{80^2} - A \quad \text{(i)}$$

$$145 = \frac{B}{80^2} + A \quad \text{(ii)}$$

Substituting $\frac{B}{80^2}$ from Equation (i) in Equation (ii)

$$145 = 60 + A + A$$

$$\therefore \quad A = \frac{145 - 60}{2} = 42.5$$

$$\therefore \quad B = (145 - 42.5) \times 80^2 = 656{,}000$$

At $r = R_e$, $p_r = 0$ then from Lame's Equation

$$0 = \frac{656{,}000}{R_e^2} - 42.5$$

Solving $\quad R_e = 124.24\ \text{mm} \simeq 125\ \text{mm}$

$\therefore$ Thickness = 125 – 80 = 45 mm.

SOLVED PROBLEM 10.3

A compound tube is composed of a tube 200 mm internal diameter and 20 mm thick shrunk on a tube of 200 mm external diameter and 20 mm thick. The radial pressure at the junction 6.4 N/mm². The cylinder is subjected to an internal pressure of 80 N/mm². Find the maximum hoop stress.

Solution:

Given data: Internal diameter of outer tube = 200 mm, external diameter of inner tube = 200 mm, thickness of inner tube = 20 mm, thickness of external tube = 20 mm, radial pressure = 6.4 N/mm² and internal pressure 80 N/mm².

Outer radius, $R_0 = 20 + \frac{200}{2} = 120$ mm

Radius at junction, $R_j = \frac{200}{2} = 100$ mm

Inner radius, $R_i = \frac{200}{2} - 20 = 80$ mm.

Radial pressure at junction, $p_r = 6.4$ N/mm^2

1. Before application of fluid pressure

(a) Outer cylinder

From Equations (10.17) and (10.18)

$$0 = \frac{B_1}{120^2} - A_1 \quad \text{(i)}$$

$$6.4 = \frac{B_1}{100^2} + A_1 \quad \text{(ii)}$$

Substituting for A_1 from Equation (i) in Equation (ii)

$$6.4 = \frac{B_1}{100^2} - \frac{B_1}{120^2} = \frac{B_1(120^2 - 100^2)}{100^2 \times 120^2}$$

Solving $B_1 = 209{,}454.6$

$$\therefore \quad A = \frac{B_1}{R_e^2} = \frac{209{,}454.6}{120^2} = 14.554$$

Then, from Equation (10.16)

$$f_r = \frac{B_1}{r^2} + A_1 = \frac{209{,}454.6}{r^2} + 14.554$$

The above equation gives the hoop stress in the outer cylinder due to shirking

$$(f_r)_{120} = \frac{209{,}454.6}{120^2} + 14.554 = 29.09 \text{ N/mm}^2 \text{ (tensile)}$$

$$(f_r)_{100} = \frac{209{,}454.6}{100^2} + 14.554 = 35.50 \text{ N/mm}^2 \text{ (tensile)}$$

(b) Inner cylinder

From Equations (10.21) and (10.22)

$$0 = \frac{B_2}{R_0^2} - A_2 = \frac{B_2}{80^2} - A_2 \quad \text{(iii)}$$

and $$6.4 = \frac{B_2}{R_j^2} - A_2 = \frac{B_2}{100^2} - A_2 \qquad \text{(iv)}$$

Substituting for A_2 from Equation (iii) to Equation (iv)

$$6.4 = \frac{B_2}{100^2} - \frac{B_2}{80^2}$$

Solving $$B_2 = -113{,}777.8$$

and $$A_2 = \frac{B_2}{80^2} = -\frac{113{,}777.8}{80^2} = -17.78$$

Using

$$f_r = f_h = \frac{B_2}{r^2} + A_2 = -\frac{113{,}777.8}{80^2} - 17.78$$

The above equation gives the hoop stress in the inner cylinder.

i.e., $$(f_h)_{100} = -\frac{113{,}777.8}{100^2} - 17.78 = 29.16 \text{ N/mm}^2 \quad \text{(compressive)}$$

$$(f_h)_{80} = -\frac{113{,}777.8}{80^2} - 17.78 = 35.56 \text{ N/mm}^2 \quad \text{(compressive)}$$

2. After application of fluid pressure

From Equations (10.46) and (10.47)

$$0 = \frac{\overline{B}}{R_0^2} - \overline{A} = \frac{\overline{B}}{120^2} - \overline{A} \qquad \text{(v)}$$

$$p = \frac{\overline{B}}{R_i^2} - \overline{A}$$

i.e., $$80 = \frac{\overline{B}}{80^2} - \overline{A} \qquad \text{(vi)}$$

Substituting for $\overline{A}$ from Equation (v) is Equation (vi)

$$80 = \frac{\overline{B}}{80^2} - \frac{\overline{B}}{120^2} = \frac{\overline{B}(120^2 - 80^2)}{80^2 \times 120^2}$$

i.e., $$\overline{B} = \frac{80 \times 80^2 \times 120^2}{8{,}000} = 921{,}600$$

$\therefore$ $$\overline{A} = \frac{\overline{B}}{120^2} = \frac{921{,}600}{120^2} = 64$$

From Equation (10.25), the hoop stress due to internal fluid pressure can be obtained as

$$(f_h)_{120} = \frac{921{,}600}{120^2} + 64 = 128 \text{ N/mm}^2 \quad \text{(tensile)}$$

$$(f_h)_{100} = \frac{921{,}600}{100^2} + 64 = 156.16 \text{ N/mm}^2 \quad \text{(tensile)}$$

$$(f_h)_{80} = \frac{921{,}600}{80^2} + 64 = 208.0 \text{ N/mm}^2 \quad \text{(tensile)}$$

The resultant hoop stress will be the algebraic sum of the stresses due to shrinking and those due to internal fluid pressure.

(a) Outer cylinder

$$(f_h)_{120} = [(f_h)_{120} \text{ due to shrinking}] + [(f_h)_{120} \text{ due to internal fluid pressure}]$$
$$= 29.09 + 128 = 157.09 \text{ N/mm}^2 \text{ (tensile)}$$
$$(f_h)_{100} = [(f_h)_{100} \text{ due to shrinking}] + [(f_h)_{100} \text{ due to internal fluid pressure}]$$
$$= 35.50 + 156.16 = 191.66 \text{ N/mm}^2 \text{ (tensile)}$$

(b) Inner cylinder

$$(f_h)_{100} = [(f_h)_{100} \text{ due to shrinkage}] + [(f_h)_{100} \text{ due to internal fluid pressure}]$$
$$= -29.16 + 156.16 = 127.0 \text{ N/mm}^2. \text{ (tensile)}$$
$$(f_h)_{80} = [(f_h)_{80} \text{ due to shrinkage}] + [(f_h)_{80} \text{ due to internal fluid pressure}]$$
$$= -35.56 + 208 = 172.44 \text{ N/mm}^2 \text{ (tensile)}$$

Hence, the maximum hoop stress is 191.66 N/mm^2 (tensile)

SOLVED PROBLEM 10.4

A thick spherical shell of 400 mm internal diameter is subjected to an internal fluid pressure of 1.5 N/mm^2. If the permissible tensile stress is 3 N/mm^2, find the thickness of the shell and the minimum value of hoop stress.

Solution:

Given data: Internal diameter = 400 mm, internal fluid pressure = 1.5 N/mm^2, permissible tensile stress = 3 N/mm^2 and internal radius of shell, $R_i = \frac{400}{2} = 200$ mm

Radial pressure and hoop stress at any radius of thick spherical shell are given by

$$p_i = \frac{2b}{x^3} - a$$

$$f_i = \frac{b}{x^3} + a$$

Hoop stress will be maximum at the inner radius

At $$x = \frac{200}{2} = 100 \text{ mm}, p_i = 1.5 \text{ N/mm}^2$$

$$1.5 = \frac{2b}{100^3} - a = \frac{2b}{1,000,000} - a \qquad \text{(i)}$$

At $$r = \frac{200}{2} = 100 \text{ mm}; f_i = 3 \text{ N/mm}^2$$

$$3 = \frac{b}{100^3} + a = \frac{b}{1,000,000} + a \qquad \text{(ii)}$$

Adding (i) and (ii)

$$4.5 = \frac{3b}{1,000,000}$$

∴ $$b = 1,500,000$$

Substituting the value of b in Equation (ii), we have

$$3 = \frac{b}{1,000,000} + a$$

∴ $$3 = \frac{1,500,000}{1,000,000} + a$$

$$a = 1.5$$

Substituting the values of a and b in Equation (i)

$$p_i = \frac{2 \times 1,500,000}{r^3} - 1.5$$

Let R_e be the external radius of the shell

At the outer surface, i.e., at $r = R_e$, $p_x = 0$

$$0 = \frac{2 \times 1,500,000}{R_i^3} - 1.5$$

Solving, $$R_i = 125.38 \text{ mm}$$

∴ Thickness of shell, $$t = R_e - R_i$$

$$= 125.38 - 100.00$$

$$= 25.38 \text{ mm}$$

The hoop stress will be minimum at the external radius, i.e., at $R_e = 125.38$ mm.

i.e., $$f_h = \frac{1,500,000}{(125.38)^2} + 1.5$$

$$= 2.20 \text{ N/mm}^2.$$

SOLVED PROBLEM 10.5

A cylinder has an internal radius of 200 mm and external radius of 300 mm. Permissible stress for the material is 15.5 N/mm^2. If the cylinder is subjected to an external pressure of 4 N/mm^2, find the internal pressure that can be applied.

Solution:

Given data: External radius = 300 mm, permissible stress = 15.5 N/mm^2 and external pressure = 4 N/mm^2.

At $r = 200$ mm; $f_r = 15.5$ N/mm^2.

Applying Lame's theorem

$$15.5 = \frac{B}{(200)^2} + A \quad \text{(i)}$$

At $r = 300$ mm, $p_r = p_e$

$$\therefore \quad 4 = \frac{B}{300^2} - A \quad \text{(ii)}$$

Adding Equations (i) and (ii)

$$\therefore \quad 19.5 = \frac{B}{(200)^2} + \frac{B}{(300)^2}$$

$$B = 540{,}000$$

and

$$A = \frac{540{,}000}{(300)^2} - 4 = 2$$

At $x = 200$ mm, the internal pressure

$$(p)_{200} = \frac{B}{x^2} - A$$

$$= \frac{540{,}000}{(200)^2} - 2$$

$$= 11.5 \text{ N/mm}^2.$$

SALIENT POINTS

- In a cylinder, if the ratio of the thickness and internal diameter is greater than 1/20, then the cylinder is referred to as thick cylinder.
- In thick cylinder the hoop stress is not constant but varies along the thickness and the radial stress is not negligible.
- Lame's theorem is based on the following assumptions:
  (i) the material of the cylinder is homogeneous and isotropic and
  (ii) the plane sections of the cylinder perpendicular to the longitudinal axis remain plane under the pressure.

- Hoop stress is maximum at the inner radius and decreases towards the outer radius. Further hoop stress is greater than that of internal fluid pressure.
- One method of withstanding high fluid pressures in cylinders is to wind strong steel wire under tension on the cylinder. Another method is that the cylinders may be made of two cylinders of different radii one cylinder being shrunk on to the other. Such cylinders are called as compound thick cylinders.
- In compound thick cylinders the inner cylinder is put into compression and the outer one in tension.
- Application of an internal fluid pressure in a compound thick cylinder causes a tensile hoop stress to be super imposed on the shrinking stresses and the resultant stress is the algebraic sum of the two sets.

QUESTIONS

1. Write Lame's equation for stress in thick cylinder. How it is used in the analyses of stresses in thick cylinder?
2. How are thin and thick cylinders classified?
3. Discuss the methods of reducing the internal hoop stress in thick cylinders.
4. Where will hoop stress be maximum in a thick cylinder? Why?
5. What is a compound cylinder? State the advantages of a compound cylinder.
6. A thick cylinder 180 mm internal diameter is subjected to an internal pressure of 3 N/mm^2. Determine the thickness required if the permissible stress is 22.5 N/mm^2.
7. The allowable maximum stress in a thick cylinder of radii 10 cm and 15 cm is 25 N/mm^2. If the external pressure is zero what internal pressure can be applied?
8. A compound tube 12 cm internal diameter and 24 cm external diameter is made by shrinking one tube on to another. After cooling a radial stress of 20 N/mm^2 is produced at the common surface which has a diameter of 18 cm. If the compound tube is subjected to an internal pressure of 80 N/mm^2, calculate the maximum hoop stress developed.

11

Columns and Struts

LEARNING OBJECTIVES

11.1 INTRODUCTION

A strut is a structural member which is subjected to an axial compressive force. A strut may be horizontal, vertical or inclined with any end fixity condition, e.g., connecting rods, piston rods, etc. But a vertical strut is called a column, e.g., a vertical pillar between a roof and a floor in a building.

Sometimes, a compression member like a column may be subjected to an axial load which may not pass through the geometric axis of the member. In that case, the load induces bending and direct stresses. Such members have to be designed based on bending and direct stresses.

11.2 DIRECT AND BENDING STRESSES

Let a rectangular column subjected to a compressive force p acting along the axis of the column (Figure 11.1(a)) be considered. This force will cause a direct stress whose intensity will be distributed uniformly across the cross section of the column. That is

$$p_o = \frac{P}{A} \tag{11.1}$$

where p_o = the intensity of direct stress
A = area of cross section

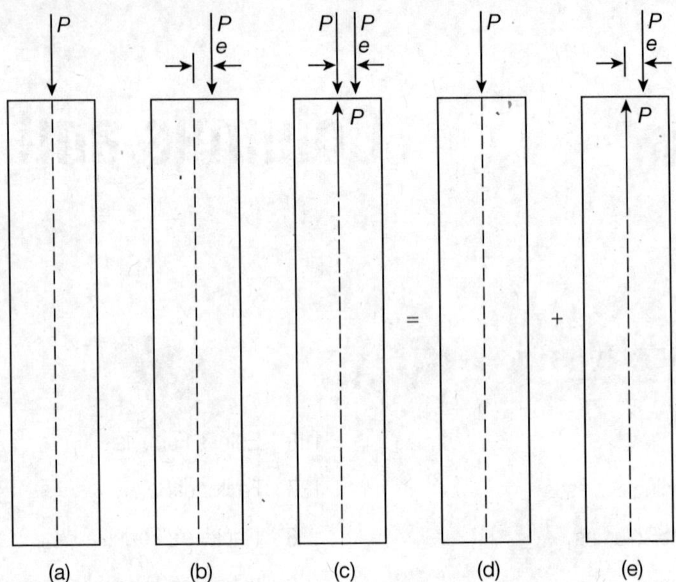

Fig. 11.1 Columns under different loading conditions.

Let a similar column be subjected to a compressive force P whose line of action is at a distance e from the axis of the column (Figure 11.1(b)) be considered. Here, e is known as the *eccentricity* of the force. Under this condition, the eccentric force will cause direct stress and bending stress.

Let two equal and opposite force P_s be applied (Figure 11.1(c)) to the eccentrically loaded column. Now, there are three forces, acting in the column which are represented in Figure 11.1(d) and Figure 11.1(e).

In Figure 11.1(d), the compressive force is acting along the axis of the column and hence this force causes a direct stress, p_o

$$p_o = \frac{P}{A}$$

In Figure 11.1(e), the column is subjected to a couple with a moment of $M = Pe$. This couple causes bending stresses, i.e., longitudinal tensile and compressive stresses.

Fibre stress, p_b, at any distance y from the neutral axis is given by

$$p_b = \frac{My}{I_{xx}} = \frac{Pey}{I_{xx}} \tag{11.2}$$

where I_{xx} is the moment of inertia along $x - x$ axis.

Hence, the resultant stress at any section is given as

Resultant stress = (Direct stress) + (Bending stress)

i.e.,
$$f = \frac{P}{A} \pm \frac{Pey}{I_{xx}} \tag{11.3}$$

i.e.,
$$f = p_o \pm p_b \tag{11.4}$$

Substituting $Z_x = \frac{I_{xx}}{y}$ and $M = P_e$ in Equation (11.3)

$$f = \frac{P}{A} \pm \frac{M}{Z_x} \tag{11.5}$$

Thus, the maximum and minimum resultant stresses are

$$f_{max} = p_o + p_b = \frac{P}{A} + \frac{M}{Z_x} \tag{11.6}$$

$$f_{min} = p_o - p_b = \frac{P}{A} - \frac{M}{Z_x} \tag{11.7}$$

Depending on the magnitudes of p_o and p_b, three stress conditions may exist, i.e.,

$$p_o > p_b$$
$$p_o < p_b$$
$$p_o = p_b$$

If $p_o > p_b$, the stress throughout the section will be of the same sign.

If $p_o < p_b$, the stress will change sign, being partly tensile and partly compressive across the section.

If $p_o = p_b$, the stress will be zero at one side and $2p_o$ at the other end.

Pressure distributions for these three cases are shown in Figure 11.2.

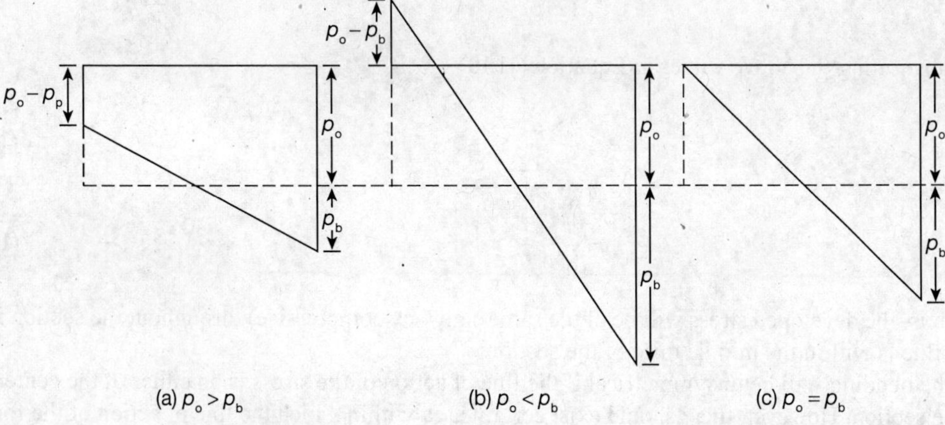

Fig. 11.2 Pressure distribution for different stress conditions.

Development of tensile stress in the section (under the condition $p_o < p_b$) is not desirable in certain materials which are weak in tension such as masonry, concrete, etc. This situation can be avoided by fixing a limit for the eccentricity, *e*.

Thus, in order that the stress not to change from compressive to tensile the condition needed is

$$p_o \geq p_b \geq \frac{M}{Z_{xx}} \tag{11.8}$$

i.e.,

$$\frac{P}{A} \geq \frac{Pe}{A} \geq \frac{Pe}{AI_{xx}/y}$$

i.e.,

$$\frac{P}{A} \geq \frac{Pe}{A \times Ak^2/d/2} \quad (\because I = Ak^2 \text{ and } y = d/2)$$

$$\frac{P}{A} \geq \frac{Ped}{2Ak^2} \tag{11.9}$$

or Eccentricity,

$$e \leq \frac{2k^2}{d} \tag{11.10}$$

where *k* is the radius of gyration of the section with reference to the neutral axis and *d* is the depth of the section.

Equation (11.10) is the required condition to avoid tension in a section.

For example, in a rectangular section of width *b* and depth *d*

$$I = \frac{1}{12}bd^2$$

and

$$A = bd$$

$$k^2 = \frac{d^2}{12}$$

Substituting the above values in Equation (11.10)

$$e \leq \frac{2 \times d^2}{d \times 12} \leq \frac{d}{6} \tag{11.11}$$

i.e.,

$$e_{max} = \frac{d}{6}. \tag{11.12}$$

Thus, the developed stress will be of the same sign (say compressive) throughout the section if the load line is within the middle third of the section.

The bending will be unsymmetrical if the line of action of the stress is on either of the center line of the section. However, there should exist certain area within which the line of action of the force *P* must cut the cross section. This area is called a *core* or *kernel* of the section.

11.3 CLASSIFICATION OF COLUMNS

The resistance of any member to bending depends on the flexural rigidity (i.e., EI) and moment of inertia ($I = Ak^2$). Thus, for a given material, the load per unit area the member can withstand is related to k. For a given cross section there will be two principal moments of inertia. Considering the length of the member and the least moment of inertia or the radius of gyration, a ratio called *slenderness ratio* is defined as

$$s_r = \frac{\text{Length of the member}}{\text{Least radius of gyration}} = \frac{l}{k} \tag{11.13}$$

The numerical value of slenderness ratio classifies a column as *short column* or *long column.* Mild steel columns with slenderness ratio more than 80 are called long columns and those with slenderness ratio less than 80 are known as short columns. However, this limiting ratio may vary depending on the material properties.

11.4 FAILURE OF COLUMNS

Based on the type of loading and the slenderness ratio, a column may fail due to any one of the following stresses:

(i) Direct compressive stress
(ii) Bending stress
(iii) Combined direct and bending stresses.

Let a short column with area of cross section A be loaded with a compressive load P. Then, the compressive stress induced in the column is given as

$$p = \frac{P}{A} \tag{11.14}$$

If the compressive force is increased gradually, a stage will reach when the column will be on the point of failure by crushing. The corresponding load at the time of failure is called the *crushing load* P_c and the induced stress is called *crushing stress* f_c Then

$$f_c = \frac{P_c}{A} \tag{11.15}$$

All short columns fail under crushing.

Let a long column with area of cross section A and of length l be subjected to a compressive load P (Figure 11.3). Long columns do not fail by crushing alone but also by bending (also known as buckling). The load at which a column just buckles is known as *buckling load* or *crippling load.*

For a long column, the buckling load is less than the crushing load. Further, for long columns the buckling load is low compare to short columns.

Let P_{cb} be the load at which the column has buckled. The stress due to direct load is given as

$$p_o = \frac{P_{cb}}{A} \tag{11.16}$$

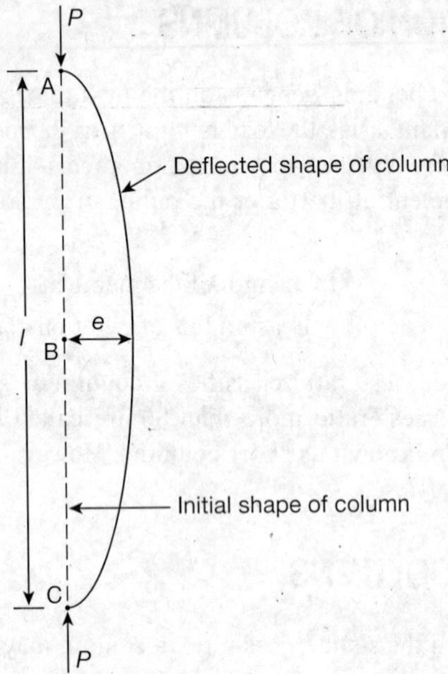

Fig. 11.3 Initial and deflected shapes of long column.

and the stress due to bending at the centre of the column is

$$p_b = \frac{P_{cb} \times e}{Z_b} \tag{11.17}$$

where e = maximum bending of the column at the centre
Z_b = section modulus about the axis of bending.

Resultant stresses at the mid-section are given as

$$f_{max} = p_o + p_b \tag{11.18}$$

$$f_{min} = p_o - p_b \tag{11.19}$$

The column will fail when maximum stress $(p_o + p_b)$ is more than the crushing stress f_c.

In the case of long columns, the buckling stress will be predominating and the compressive stress is negligible. Hence, very long columns are subjected to buckling stresses only.

11.5 END CONDITIONS FOR LONG COLUMNS

In practice, the failure of long columns have been observed essentially due to buckling. The following four end conditions have to be adopted in practice (Figure 11.4).

(i) Both ends are hinged or pinned
(ii) One end is free and the other is fixed
(iii) Both ends are fixed
(iv) One end is fixed and the other end is pinned

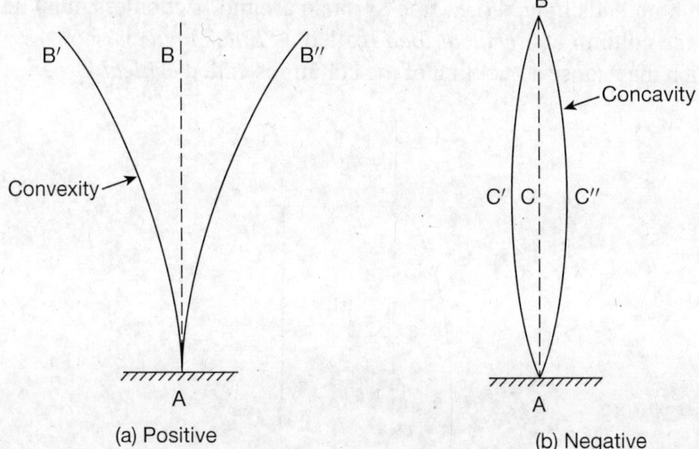

Fig. 11.4 Sign conventions.

The deflection is zero for hinged end.
The deflection and slope are zero for fixed end.
The following sign conventions for the bending of the columns will be adopted.

(i) A moment causing the column to bend with its *convexity* towards its initial centre line (Figure 11.4(a)), i.e., positions AB′ or AB″ are considered positive.

(ii) A moment causing the column to bend with its *concavity* towards its initial centre line (Figure 11.4(b)), i.e., positions AC′B or AC″B are considered negative.

11.6 EULER'S FORMULA

Euler postulated a theory for columns based on the following assumptions:

(i) Column is very long in proportion to its cross sectional dimensions
(ii) Column is initially straight and the compressive load is applied axially
(iii) Material of the column is elastic, homogeneous and isotropic
(iv) Effect of direct stress is very small in comparison with bending stress.
(v) Column shall fail by buckling alone.
(vi) Effect of self-weight of column is negligible

The behavior of a column to resist buckling is very much influenced by the end condition and the flexural rigidity of the column. In turn, the flexural rigidity of the column depends on the minimum

I and at the critical load it buckles in that plane about which I is minimum. The expression for crippling load for various end conditions are discussed below.

11.6.1 Column with Both Ends Hinged

Let a column with both ends hinged (i.e., free to rotate around frictionless pins) be considered. The buckled shape of the column at a *critical load* (called as *Euler load*) is shown in Figure 11.5. The smallest force which may cause a buckling of the column is called *critical force*.

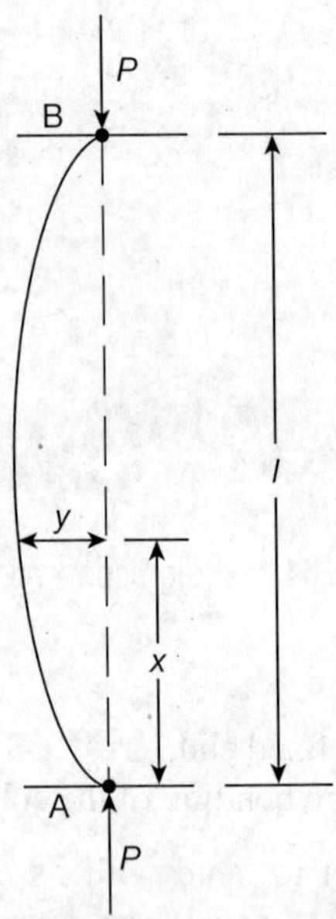

Fig. 11.5 Column with both ends hinged.

Let the deflection of column of length l be y from the centre line at a distance x from the bottom hinge A.

Any section at a distance x from the end A is considered. Let y be the lateral deflection.

The moment, M, due to the crippling load at the section is

$$M = -P \times y \text{ (−ve for concave bending)}$$

But,
$$\text{moment} = EI\frac{d^2y}{dx^2}$$

Hence
$$EI\frac{d^2y}{dx^2} = M = -P \times y$$

i.e.,
$$EI\frac{d^2y}{dx^2} + Py = 0 \tag{11.20a}$$

Equation (11.20a) may be rearranged as

$$\frac{d^2y}{dx^2} + \frac{P}{EI}y = 0 \tag{11.20b}$$

Equation (11.20b) is a standard differential equation whose solution is

$$y = a_1 \cos\left(x\sqrt{\frac{P}{EI}}\right) + a_2 \sin\left(x\sqrt{\frac{P}{EI}}\right) \tag{11.21}$$

where a_1 and a_2 are constants of integration which may be determined after applying the appropriate end conditions.

(i) At A, $x = 0$ and $y = 0$
Substituting these values in Equation (11.21)

$$0 = a_1 \cos 0° + a_2 \sin 0° = a_1$$

i.e.,
$$a_1 = 0$$

(ii) At B, $x = l$ and $y = 0$
Substituting these values in Equation (11.21)

$$y = a_1 \cos\left(l \times \sqrt{P/EI}\right) + a_2 \sin\left(l \times \sqrt{P/EI}\right)$$

$$0 = 0 + a_2 \sin\left(l \times \sqrt{P/EI}\right)$$

From the above equation, either $a_2 = 0$ or $\sin\left(l \times \sqrt{P/EI}\right) = 0$. As $a_1 = 0$ and if $a_2 = 0$, from Equation (11.21), we get $y = 0$. This means that the bending of the column will be zero or the column will not bend at all which is not true.

$\therefore$
$$\sin\left(l \times \sqrt{\frac{P}{EI}}\right) = 0$$

i.e.,
$$\sin\left(l \times \sqrt{\frac{P}{EI}}\right) = \sin 0 \text{ or } \sin \pi \text{ or } \sin 2\pi.$$

Taking the least significant value

i.e.,
$$l\sqrt{\frac{P}{EI}} = \pi$$

or
$$P = \frac{\pi^2 EI}{l} \tag{11.22}$$

11.6.2 Column with One End Fixed and Other End Free

A column AB of the length l_1 fixed at the end A and free at the end B is considered (Figure 11.6)

Let y be the deflection of the column of length l, at any distance x from the bottom end A. Let y_m be the maximum deflection at the free end B.

Moment at the section

$$M_x = P(y_m - y) \quad (11.23a)$$

$$\therefore \quad EI\frac{d^2y}{dx^2} = M_x = P(y_m - y) \quad (11.23b)$$

i.e.,

$$\frac{d^2y}{dx^2} + \frac{Py}{EI} = \frac{Py_m}{EI} \quad (11.24)$$

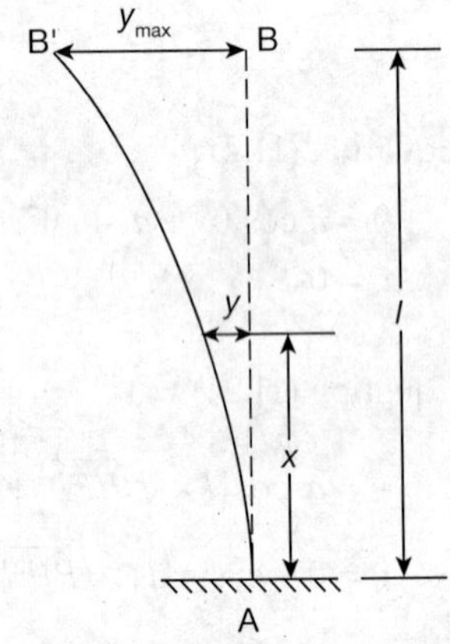

Fig. 11.6 Column with one end fixed and the other end free.

Solution of Equation (11.24) yields

$$y = a_1 \cos\left(x\sqrt{\frac{P}{EI}}\right) + a_2 \sin\left(x\sqrt{\frac{P}{EI}}\right) + y_m \quad (11.25)$$

where a_1 and a_2 are constants of integration and determined after applying appropriate end conditions.

(i) At A, $x = 0$ and $y = 0$.

Substituting these values in Equation (11.25) we get

$$a_1 = -y_m$$

Differentiating Equation (11.25), we have

$$\frac{dy}{dx} = -a_1\sqrt{\frac{P}{EI}}\sin\left(x\sqrt{\frac{P}{EI}}\right) + a_2\sqrt{\frac{P}{EI}}\cos\left(x\sqrt{\frac{P}{EI}}\right)$$

(ii) At A, $x = 0$ and $\frac{dy}{dx} = 0$

Substituting for $\frac{dy}{dx}$, and x in the above equation, we have

$$0 = -a_1\sqrt{\frac{P}{EI}}\sin 0° + a_2\sqrt{\frac{P}{EI}}\cos 0°$$

i.e.,

$$= a_2\sqrt{\frac{P}{EI}}$$

$$a_2\sqrt{\frac{P}{EI}} = 0$$

From the above equation either $a_2 = 0$ or $\sqrt{\frac{P}{EI}} = 0$

But for the crippling load P, the value $\sqrt{\frac{P}{EI}}$ cannot be zero.

$\therefore$

$$a_2 = 0$$

Substituting these values a_1 and a_2 in Equation (11.25)

$$y = -y_m \cos\left(x\sqrt{\frac{P}{EI}}\right) + y_m$$

(iii) At the free end B, $x = l$ and $y = y_m$.

$$y_m = -y_m \cos\left(l\sqrt{\frac{P}{EI}}\right) + y_m$$

i.e.,

$$0 = -y_m \cos\left(l\sqrt{\frac{P}{EI}}\right)$$

i.e.,

$$y_m \cos\left(l\sqrt{\frac{P}{EI}}\right) = 0$$

As y_m cannot be zero,

i.e.,

$$\cos\left(l\sqrt{\frac{P}{EI}}\right) = 0 = \cos\frac{\pi}{2} \text{ or } \cos\frac{3\pi}{2}$$

$$l\sqrt{\frac{P}{EI}} = \frac{\pi}{2} \text{ or } \frac{3\pi}{2}$$

Taking the least significant value

$$l\sqrt{\frac{P}{EI}} = \frac{\pi}{2}$$

$$P = \frac{\pi^2 EI}{4l^2} \tag{11.26}$$

11.6.3 Column with Both Ends Fixed

Let a column whose both ends be fixed direction wise and position wise. The deformed shape of the column of length l is shown in Figure 11.7. From either ends, the column is convex for a lengths of $l/2_f$ and the middle half is concave.

A section on the column at a distance x from the end A and the deflection y be considered. Both the ends of the column are fixed and carries a crippling load, so there will be some fixed moments at A and B.

Let M be the fixed end moment at A and B

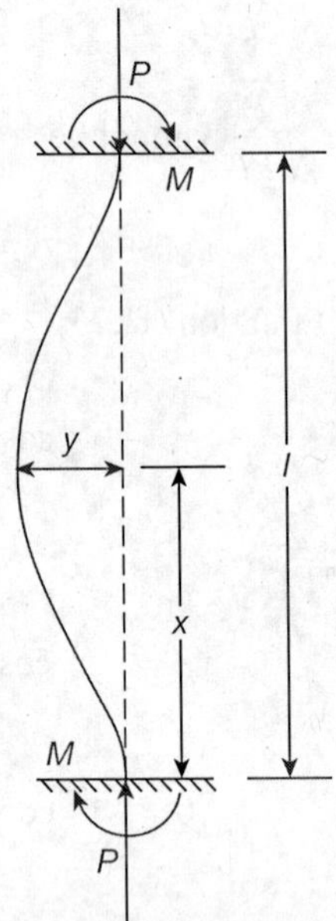

Fig. 11.7 Column with both ends fixed.

∴ Moment at this section $= M - P_y$

i.e.,

$$EI\frac{d^2y}{dx^2} = M - Py \tag{11.27}$$

Rearranging,

$$\frac{d^2y}{dx^2} + \frac{P}{EI}y = \frac{M}{EI} \tag{11.28}$$

Solution of Equation (11.28) is given as

$$y = a_1 \cos\left(x\sqrt{\frac{P}{EI}}\right) + a_2 \sin\left(x\sqrt{\frac{P}{EI}}\right) + \frac{M}{P} \tag{11.29}$$

Differentiating Equation (11.29), we have

$$\frac{dy}{dx} = -a_1\sqrt{\frac{P}{EI}}\sin\left(x\sqrt{\frac{P}{EI}}\right) + a_2\sqrt{\frac{P}{EI}}\cos\left(x\sqrt{\frac{P}{EI}}\right) + 0$$

(i) At the end A, $x = 0$ and $y = 0$

Substituting these values in Equation (11.29)

$$0 = a_1 \times 1 + a_2 \times 0 + \frac{M}{P}$$

$$= a_1 + \frac{M}{P}$$

i.e.,

$$a_1 = -\frac{M}{P}$$

(ii) Also, at end A, $x = 0$ and $\frac{dy}{dx} = 0$

∴

$$0 = -a_1 \times 0 + a_2 \times 1 \times \sqrt{\frac{P}{EI}}$$

$$= a_2\sqrt{\frac{P}{EI}}$$

From the above equation, either $a_2 = 0$ or $\sqrt{\frac{P}{EI}} = 0$. But, for a given load, P, $\sqrt{\frac{P}{EI}}$ cannot be zero.

∴

$$a_2 = 0$$

Substituting $a_1 = \frac{M}{P}$ and $a_2 = 0$ in Equation (11.29)

$$y = -\frac{M}{P}\cos\left(x\sqrt{\frac{P}{EI}}\right) + 0 + \frac{M}{P}$$

(iii) At the end B, $x = l$ and $y = 0$

Substituting these values in the above equation

$$0 = -\frac{M}{P}\cos\left(l \times \sqrt{\frac{P}{EI}}\right) + \frac{M}{P}$$

∴

$$\frac{M}{P}\cos\left(l\sqrt{\frac{P}{EI}}\right) = \frac{M}{P}$$

i.e.,

$$\cos\left(l\sqrt{\frac{P}{EI}}\right) = 1.$$

i.e., $$\cos\left(l\sqrt{\frac{P}{EI}}\right) = \cos 0° \text{ or } \cos 2\pi \text{ or } \cos 4\pi$$

i.e., $$l\sqrt{\frac{P}{EI}} = 0° \text{ or } 2\pi \text{ or } 4\pi$$

Taking the least significant value

$$l\sqrt{\frac{P}{EI}} = 2\pi$$

Squaring both sides and reducing

$$P = \frac{4\pi^2 EI}{l^2} \tag{11.30}$$

11.6.4 Column with One End Fixed and Other End Hinged

Let a column of length l be considered with bottom end fixed in direction and position and top end hinged. The deflected shape of the column is shown in Figure 11.8.

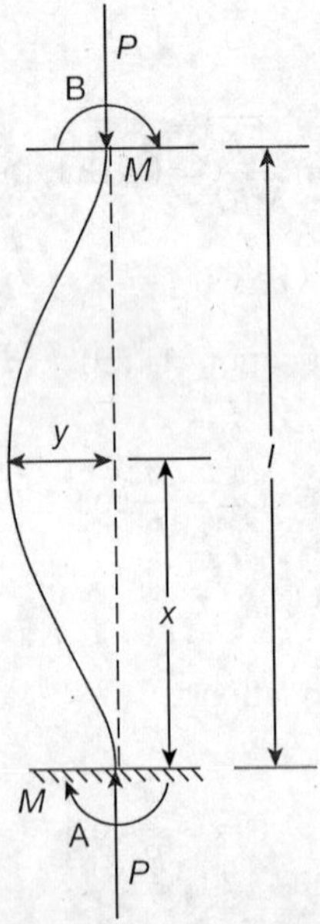

Fig. 11.8 Column with one end fixed and other end hinged.

Since the end B's direction is fixed, the bending moment M induced at A involves the presence of a force R at right angles to AB to maintain equilibrium. Hence, there is a moment at the section which is equal to moment due to load at B plus moment due to reaction at B. Let R be the horizontal reaction at B which be equal to $-P_y + R(l-x)$.

Equating the moments, we have

$$EI\frac{d^2y}{dx^2} = -P_y + R(l-x)$$

i.e.,

$$\frac{d^2y}{dx^2} + \frac{P}{EI}y = \frac{R}{EI}(l-x) \tag{11.31}$$

Solution of the above differential equation is

$$y = a_1\cos\left(x\sqrt{\frac{P}{EI}}\right) + a_2\sin\left(x\sqrt{\frac{P}{EI}}\right) + \frac{R}{P}(l-x) \tag{11.32}$$

Differentiating the above equation

$$\frac{dy}{dx} = -a_1\sqrt{\frac{P}{EI}}\sin\left(x\sqrt{\frac{P}{EI}}\right) + a_2\sqrt{\frac{P}{EI}}\cos\left(x\sqrt{\frac{P}{EI}}\right) - \frac{R}{P} \tag{11.33}$$

(i) At the end A, $x = 0$ and $y = 0$

Substituting these values in Equation (11.32)

$$a_1 = \frac{Rl}{P}$$

(ii) Also at the end A, $x = 0$ and $\frac{dy}{dx} = 0$

Substituting these values in Equation (11.33)

$$0 = a_2\sqrt{\frac{P}{EI}} - \frac{R}{P}$$

$$a_2 = \frac{R}{P}\sqrt{\frac{EI}{P}}.$$

Substituting the values of a_1 and a_2 in Equation (11.32)

$$y = -\frac{Rl}{P}\cos\left(x\sqrt{\frac{P}{EI}}\right) + \frac{R}{P}\sqrt{\frac{EI}{P}}\sin\left(x\sqrt{\frac{P}{EI}}\right) + \frac{R}{P}(l-x)$$

(iii) At the end B, $x = l$ and $y = 0$

$$0 = -\frac{Rl}{P}\cos\left(l\sqrt{\frac{P}{EI}}\right) + \frac{R}{P}\sqrt{\frac{EI}{P}}\sin\left(l\sqrt{\frac{P}{EI}}\right) + \frac{R}{P}(l-l)$$

or

$$\frac{R}{P}\sqrt{\frac{EI}{P}}\sin\left(l\sqrt{\frac{P}{EI}}\right) = \frac{R}{P}l\cos\left(l\sqrt{\frac{P}{EI}}\right)$$

or
$$\sin\left(l\sqrt{\frac{P}{EI}}\right) = \frac{R}{P}l \times \frac{P}{R} \times \sqrt{\frac{P}{EI}}\cos\left(l\sqrt{\frac{P}{EI}}\right)$$

$$= l\sqrt{\frac{P}{EI}}\cos\left(l\sqrt{\frac{P}{EI}}\right)$$

or
$$\tan\left(l\sqrt{\frac{P}{EI}}\right) = l\sqrt{\frac{P}{EI}}$$

The solution to the above equation is $l\sqrt{\frac{P}{EI}} = 4.5$ radians

Squaring both sides

$$\therefore \quad \frac{l^2 P}{EI} = (4.5)^2 = 20.25$$

$$P = 20.25\frac{EI}{l^2}$$

Approximately $20.25 = 2\pi^2$, then

$$P = \frac{2\pi^2 EI}{l^2}. \qquad (11.34)$$

11.6.5 Limitations of Euler's Formula

The assumption of initial perfect straightness and axial transfer of load may not be fully realized in practice as there may be some initial curvature or distortion in the column.

It is to be noted that Euler's theory does not consider any strength property of the material. But, the only property involved is the elastic modulus, then the crippling stress is given as

$$\text{Crippling stress} = \frac{P_E}{A} = \frac{\pi^2 EI}{Al^2}$$

$$= \frac{\pi^2 EAk^2}{Al^2}$$

$$= \frac{\pi^2 E}{\left(\frac{l}{k}\right)^2} \qquad (11.35)$$

for the standard case where P_E is the Euler's crippling load. The above expression shows that it depends on modulus of elasticity and slenderness ratio. The modulus of elasticity of steel does not vary much, so, the critical stress forms a function of slenderness ratio only. Hence, Euler's theory is applicable only if l/k is greater than 80 for steel column hinged at both the ends. Further, it does not consider the direct stress and will be applicable for long columns only.

11.6.6 Equivalent Length

For four different end conditions, Euler's crippling load were given in the preceeding sections. Now, Euler's crippling load can be expressed in a generalized form as

$$P_E = \frac{\pi^2 EI}{Cl^2} \tag{11.36}$$

where C is a constant which is a function of end conditions.

Column hinged at both ends (Equation 11.5) is considered as a standard case and for which C is taken as 1. Accordingly, C for other end conditions are:

$C = 4$ one end fixed and other end free
$= \frac{1}{4}$ both ends fixed
$= \frac{1}{2}$ one end fixed and other hinged.

Instead of using the actual lengths of column, length of the column, which is equivalent to that of a pin jointed column may be considered.

Thus

$$P_E = \frac{\pi^2 EI}{l_e^2} \tag{11.37}$$

where l_e = equivalent length

$= l\sqrt{C}$

The above relationship is applicable for all end conditions. The equivalent length of different columns are given in Table 11.1.

As the compression members in actual practice are not truly hinged or perfectly fixed, the concept of equivalent length has more practical significance.

Table 11.1 Values of l_e

End conditions	Both ends hinged	One end fixed and other free	Both ends fixed	One end fixed and other hinged
Equivalent length l_e	l	$2l$	$\frac{l}{2}$	$\frac{l}{\sqrt{2}}$

11.6.7 Eccentrically Loaded Long Column

A circular long column of length l, with both ends hinged and subjected to an eccentric load of p with an eccentricity, e, is considered. (Figure 11.9).

Bending moment by the centre load $= -Py$

$$\therefore \quad EI\frac{d^2y}{dx^2} = -Py$$

i.e.,

$$\frac{d^2y}{dx^2} + \frac{P}{EI}y = 0 \tag{11.38}$$

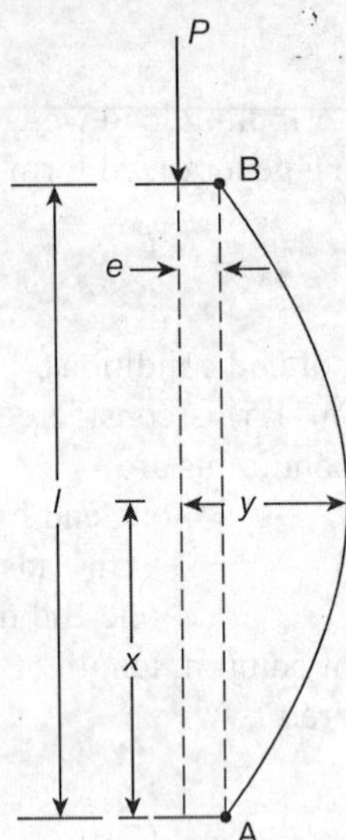

Fig. 11.9 Eccentrically loaded long column.

The solution to Equation (11.38) is

$$y = a_1 \sin\left(x\sqrt{\frac{P}{EI}}\right) + a_2 \cos\left(x\sqrt{\frac{P}{EI}}\right) \tag{11.39}$$

(i) At, A, $x = 0$ and $y = e$

$\therefore \quad a_2 = e.$

Differentiating Equation (11.39)

$$\frac{dy}{dx} = a_1\sqrt{\frac{P}{EI}} \cos\left(x\sqrt{\frac{P}{EI}}\right) - a_2 \sin\left(x\sqrt{\frac{P}{EI}}\right) \tag{11.40}$$

(iv) At the mid-height of the column $\dfrac{dy}{dx} = 0$ and $x = \dfrac{l}{2}$

i.e.,

$$0 = a_1\sqrt{\frac{P}{EI}} \cos\left(\frac{l}{2}\sqrt{\frac{P}{EI}}\right) - e\sqrt{\frac{P}{EI}} \sin\left(\frac{l}{2}\sqrt{\frac{P}{EI}}\right)$$

or

$$a_1 = e\frac{\sin\left(\dfrac{l}{2}\sqrt{\dfrac{P}{EI}}\right)}{\cos\left(\dfrac{l}{2}\sqrt{\dfrac{P}{EI}}\right)}$$

Substituting for a_1 and a_2 in Equation (11.40)

$$y = e\left[\frac{\sin\left(\frac{l}{2}\sqrt{\frac{P}{EI}}\right)}{\cos\left(\frac{l}{2}\sqrt{\frac{P}{EI}}\right)}\sin\left(x\sqrt{\frac{P}{EI}}\right) + \cos\left(x\sqrt{\frac{P}{EI}}\right)\right]$$

(ii) At $x = \frac{l}{2}$ and $y = y_{max}$

$$y_{max} = e\left[\frac{\sin\left(\frac{l}{2}\sqrt{\frac{P}{EI}}\right)}{\cos\left(\frac{l}{2}\sqrt{\frac{P}{EI}}\right)} + \cos\left(\frac{l}{2}\sqrt{\frac{P}{EI}}\right)\right] = e\sec\left(\frac{l}{2}\sqrt{\frac{P}{EI}}\right)$$

Maximum BM occurs at $x = \frac{l}{2}$ where y is maximum

$$\therefore \qquad M_{max} = Py_{max} = Pe\sec\left(\frac{l}{2}\sqrt{\frac{P}{EI}}\right)$$

Maximum compressive stress is obtained from

$$f_{max} = \frac{P}{A} + \frac{M}{Z}$$

i.e.,

$$f_{max} = \frac{P}{A} + \frac{Pe\ \sec\left(\frac{l}{2}\sqrt{\frac{P}{EI}}\right)}{Z}$$

i.e.,

$$f_{max} = \frac{P}{A}\cdot\left[1 + \frac{ey_c}{k^2}\sec\left(\frac{l}{2}\sqrt{\frac{P}{EA}}\right)\right] \qquad \left(\because\ \bar{Z} = \frac{Ak_e^2}{y_c}\right) \qquad (11.41)$$

where y_c is the deflection.

As the above equation contains a secant term it is known as the *secant formula for column.*

SOLVED PROBLEM 11.1

A solid square bar of 2 m long and 6 cm side is used as a short column and is subjected to compressive load of 50 kN acting 0.5 cm eccentricity from the central axis. Determine the maximum and minimum stresses.

Solution:

Given data: Side of rod = 6 cm, length = 2 m, compressive load =50 kN, e = 0.5 cm and area, $A = 6 \times 10 \times 6 \times 10 = 3{,}600$ mm^2.

Section modulus, $Z = \frac{1}{6}bd^2$

$$Z = \frac{1}{6} \times 60 \times (60)^2 = 36,000 \text{ mm}^3$$

$$p_o = \frac{P}{A} = \frac{50,000}{3,600} = 1.38 \text{ N/mm}^2$$

$$p_b = \frac{M}{Z} = \frac{Pe}{Z}$$

$$= \frac{50,000 \times 0.5 \times 10}{36,000}$$

$$= 0.6944 \text{ N/mm}^2$$

$$f_{max} = p_o + p_b = 1.38 + 0.6944 = 2.07 \text{ N/mm}^2$$

$$f_{min} = p_o - p_b = 1.38 - 0.694 = 0.686 \text{ N/mm}^2$$

SOLVED PROBLEM 11.2

A cast iron column is of hollow circular section with 30 cm external diameter and 6 cm thick. A vertical compressive load of 20 kN acts at an eccentricity of 9 cm from the axis. Determine the maximum and minimum stresses considering the column as a short column.

Solution:

Given data: External diameter = 30 cm, thickness = 6 cm, compressive load = 20 kN, e = 9 cm and internal diameter = 30 – (2 × 6) = 18 cm.

Area, $A = \frac{\pi}{4}\left(300^2 - 180^2\right) = 45,216 \text{ mm}^2$

$$Z = \frac{I_{xx}}{y} = \frac{\pi}{64}\left(\frac{300^4 - 180^4}{150}\right)$$

$$= 2.31 \times 10^6 \text{ mm}^3$$

$$p_o = \frac{P}{A} = \frac{20 \times 1,000}{45,216} = 0.44 \text{ N/mm}^2$$

$$p_b = \frac{Pe}{Z} = \frac{20 \times 1,000 \times 90}{2.31 \times 10^6}$$

$$= 0.77 \text{ N/mm}^2$$

$$f_{max} = p_o + p_b = 0.44 + 0.77 = 1.21 \text{ N/mm}^2$$

$$f_{min} = p_o + p_b = 0.44 - 0.77 = 0.33 \text{ N/mm}^2.$$

SOLVED PROBLEM 11.3

Determine Euler's buckling load for a column of *I* section with flanges 300 mm × 10 mm and web 40 mm × 8 mm. The length of the column is 6 m. One end is fixed and other end is hinged. $E = 200$ kN/mm².

Solution:

Given data: Flanges = 300 mm × 10 mm, web = 40 mm × 8 mm, length, $l = 6$ m and $E = 200$ kN/mm².

Referring to *I*-section given in Figure 11.10.

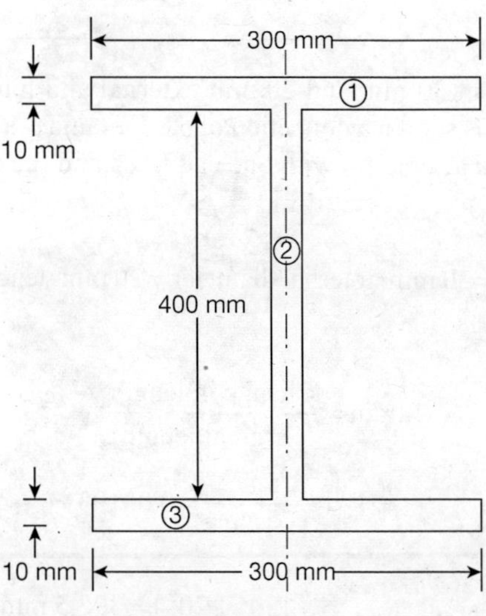

Fig. 11.10

Area of element (1) = 300 × 10 = 3,000 mm²

Area of element (2) = 400 × 8 = 3,200 mm²

Area of element (3) = 300 × 10 = 3,000 mm²

$$I_{xx} = \frac{1}{12} \times 8 \times 400^3 + 2\left[\frac{1}{2} \times 300 \times 10^3\right] + 3,000\left(10 + \frac{400}{2}\right)^2$$

$$= 3.07 \times 10^8 \text{ mm}^4$$

$$I_{yy} = \frac{1}{12} \times 400 \times 8^3 + 2\left[\frac{1}{2} \times 10 \times 300^3\right]$$

$$= 2.25 \times 10^7 \text{ mm}^4.$$

$I_{yy} < I_{xx}$, therefore the column will tend to buckle in *y*-direction and the least moment of inertia has to be used in the Euler's formula.

For the condition of one end fixed and the other end hinged the Euler's buckling load is

$$P_E = \frac{2\pi^2 EI}{l^2}$$

$$= \frac{2 \times \pi^2 \times 200 \times 2.25 \times 10^7}{(6 \times 1,000)^2}$$

$$= 2,464.5 \text{ kN.}$$

∴ Euler's buckling load = 2,464.5 kN.

SOLVED PROBLEM 11.4

A hollow alloy tube 3 m long of 30 mm and 20 mm external and internal diameter respectively, is subjected to a tension text. It has shown a deflection of 5.8 mm under a tensile load of 50 kN. Find the bucking load if the tube is used as column with one end fixed and the other end free.

Solution:

Given data: External diameter = 30 mm, internal diameter = 20 mm, length, $l = 3$ m, deflection = 5.8 mm and tensile load = 50 kN.

$$\text{Strain} = \frac{\text{Change in length}}{\text{Original length}}$$

$$= \frac{5.8}{3 \times 1,000} = 0.00193$$

Area of cross section $= \frac{\pi}{4}\left(30^2 - 20^2\right) = 392.5 \text{ mm}^2$

$$\text{Stress} = \frac{P}{A} = \frac{50 \times 1,000}{392.5} = 127.4 \text{ N/mm}^2$$

$$E = \frac{\text{Stress}}{\text{Strain}} = \frac{127.4}{0.00193} = 66,000 \text{ N/mm}^2$$

$$I = \frac{\pi}{64}\left(30^4 - 20^4\right) = 31,890 \text{ mm}^4$$

Equivalent length, $l_e = 2l = 2 \times 3 \times 1,000 = 6,000$ mm

Euler's buckling load, $P_E = \frac{\pi^2 EI}{l_e^2}$

$$= \frac{\pi^2 \times 66,000 \times 31,890}{6,000^2}$$

∴ $P_E = 576.4$ N

SOLVED PROBLEM 11.5

A straight cylindrical bar is 15 mm diameter and 1.2 m long. It is freely supported at its two ends in a horizontal position and loaded at the centre with a concentrated load of 900 N when the central deflection is found to be 5 mm. If the bar is placed vertical and loaded along axis, with both ends hinged what is the buckling load?

Solution:

Given data: Diameter = 15 mm, length = 1.2 m, concentrated load = 900 N and deflection = 5 mm.

For a simply supported beam subjected to a concentrated load at the centre, the central deflection

$$y_c = \frac{Wl^3}{48EI}$$

Hence,

$$EI = \frac{Wl^3}{48y_c} = \frac{900\times(1.2\times1,000)^3}{48\times5} = 0.648\times10^{10}\ \text{N/m}$$

Buckling load,

$$P_E = \frac{\pi^2 EI}{l^2}$$

$$= \frac{\pi^2\times0.4\times10^{10}}{(1.2\times1,000)^2}$$

$$= 27,415.56\ \text{N}$$

Hence, buckling load, $P_E = 27.42$ kN.

SOLVED PROBLEM 11.6

Two columns, one hollow and the other solid, have same cross-sectional area, same length and made of same material and have same end conditions. The internal diameter of hollow column is half of its external diameter. Find the ratio of crippling strengths of two columns.

Solution:

Given data: Hollow and solid = same cross section, hollow and solid = same length, same material and same end conditions, internal diameter of hollow column = $\frac{1}{2}$ of its external diameter.

Let D_h be the external diameter of the hollow column and D_s be the diameter of solid column.

Then, $D_h/2$ is the internal diameter of the hollow column.

$\therefore$ Area of solid shaft = Area of hollow shaft

$$\frac{\pi}{4}D_s^2 = \frac{\pi}{4}\left[D_h^2 - \left(\frac{D_h^2}{2}\right)^2\right]$$

i.e., $$D_s^2 = \frac{3}{4} D_h^2$$

$\therefore$ $$D_s = \frac{\sqrt{3}}{2} D_h$$

Least moment of inertia of hollow column $$I_h = \frac{\pi}{64}\left[D_h^4 - \left(\frac{D_h}{2}\right)^4\right]$$

$$= \frac{\pi}{64} \times \frac{15}{16} D_h^4$$

Least moment of inertia of solid column $$I_s = \frac{\pi}{64} D_s^4$$

Generalized crippling load is given as $$P_E = \frac{\pi^2 EI}{Cl^2}$$

As the material, length and end conditions are same for the columns, then E, l and C are same for both columns. Then

i.e., $$\frac{(P_E)_h}{I_h} = \frac{(P_E)_s}{I_s}$$

$$\frac{(P_E)_h}{(P_E)_s} = \frac{I_h}{I_s}$$

$$\frac{(P_E)_h}{(P_E)_s} = \frac{I_h}{I_s}$$

$$= \frac{\frac{\pi}{64} \times \frac{15}{16} \times D_h^4}{\frac{\pi}{64} \times D_s^4}$$

$$= \frac{15}{16} \frac{D_h^4}{\left(\frac{\sqrt{3D_h^2}}{2}\right)^4}$$

$$= \frac{15}{16} \times \left(\frac{2}{\sqrt{3}}\right)^4$$

$$= 1.667$$

Hence, the ratio of crippling strength of the hollow circular column is 1.667 times of the solid circular column.

SOLVED PROBLEM 11.7

A mild steel long column hinged at both ends is of hollow circular section with 100 mm and 80 mm as external and internal diameters. The column is 2.5 m long and loaded eccentrically by a compressive load of 70 kN. The eccentricity is 10 mm from the geometric axis. Calculate the maximum and minimum intensities of the stresses. Take $E = 2.0 \times 10^3$ N/mm^2.

Solution:

Given data: External diameter = 100 mm, e = 10 mm, internal diameter = 800 mm, length = 2.5 m and $E = 2.0 \times 10^3$ N/mm^2

Area of column

$$A = \frac{\pi}{4}\left(100^2 - 80^2\right) = 2{,}826 \text{ mm}^2$$

$$I = \frac{\pi}{64}\left(100^4 - 80^4\right) = 2{,}896{,}650 \text{ mm}^4$$

$$k^2 = \frac{I}{A} = \frac{2{,}896{,}650}{2{,}826} = 1{,}025 \text{ mm}^2$$

$$Z = \frac{I}{y} = \frac{2{,}896{,}650}{50} = 57{,}933 \text{ mm}^3$$

$$M = Pe\sec\left[\frac{l}{2}\sqrt{\frac{P}{EI}}\right]$$

$$= 70 \times 1{,}000 \times 10 \times \sec\left[\frac{2.5 \times 1{,}000}{2}\sqrt{\frac{70 \times 1{,}000}{2.0 \times 10^5 \times 2{,}896{,}650}}\right]$$

$$= 70{,}000 \times 10 \sec(0.435)$$

$$= 700{,}021 \text{ N} \cdot \text{mm}$$

$$p_b = \frac{M}{Z} = \frac{700{,}021}{57{,}933} = 12.08 \text{ N/mm}^2$$

$$p_o = \frac{70{,}000}{2{,}826} = 24.77 \text{ N/mm}^2$$

$$f_{max} = p_o + p_b = 24.77 + 12.08 = 36.85 \text{ N/mm}^2$$

$$f_{min} = p_o - p_b = 24.77 - 12.08 = 12.69 \text{ N/mm}^2.$$

SOLVED PROBLEM 11.8

A cast iron hollow circular column is of 300 mm external diameter and 260 mm internal diameter 2.4 m long. It is fixed, at its both ends and subjected to an eccentric load of 175 kN. Determine the maximum eccentricity, in order that there is no tension anywhere in the section. Take $E = 0.94 \times 10^5$ N/mm^2.

Solution:

Given data: External diameter = 300 mm, length = 2.4 m, internal diameter = 260 mm, eccentric load = 175 kN and $E = 0.94 \times 10^5$ N/mm^2.

Area of hollow column $= \frac{\pi}{4}\left(300^2 - 260^2\right)$

$= 17{,}592 \text{ mm}^2$

$$I = \frac{\pi}{64}\left(300^4 - 260^4\right)$$

$$= 1{,}732 \times 10^5 \text{ mm}^4$$

$$Z = \frac{1{,}732 \times 10^5}{300/2}$$

$$= 1{,}155{,}268 \text{ mm}^3$$

For column with fixed ends, $l_e = l/2 = 4/2 = 2$ m

Total stress, f = (Direct stress) + (Bending stress)

$$= \frac{P}{A} + Pe\sec\left(\frac{l_e}{2}\sqrt{\frac{P}{EI}}\right)$$

Now, $\frac{l_e}{2}\sqrt{\frac{P}{EI}} = \frac{2{,}000}{2}\sqrt{\frac{175 \times 10^3}{0.94 \times 10^5 \times 1{,}732 \times 10^5}}$

$= 0.1060$ radian

$= 6.07°$

In order that there is no tension anywhere in the section, the direct stress should be equal to the bending stress.

i.e.,

$$\frac{P}{A} = \frac{Pe\sec\left(\frac{l_e}{2}\sqrt{\frac{P}{EI}}\right)}{Z}$$

i.e.,

$$\frac{175 \times 10^3}{17{,}592} = \frac{175 \times e \times \sec 6.07°}{1{,}155{,}208}$$

Solving, $e = 64.60$ mm

SOLVED PROBLEM 11.9

A straight bar of steel 2.5 m long of 30 mm × 16 mm is used as strut with both ends hinged. Assuming Euler's formula for pinned ends to apply, estimate the maximum central deflection before the material attains its yield. Strength 290 N/mm^2. Take $E = 2 \times 10^5$ N/mm^2.

Solution:

Given data: Cross section = 30 mm × 16 mm, l = 2.5 m, yield strength = 290 N/mm² and $E = 2 \times 10^5$ N/mm².

Moment of inertia,
$$I_{min} = \frac{bd^3}{12}$$
$$= \frac{30 \times 16^3}{12} = 10,240 \text{ mm}^4$$

For both ends hinged, Euler's buckling load is given as

$$P_e = \frac{\pi^2 EI}{l^2}$$
$$= \frac{\pi^2 \times 2 \times 10^5 \times 10,240}{(2,500)^2}$$
$$= 3,680.67 \text{ N.}$$

Maximum deflection, y, occurs at half the height:

$$f_{max} = \frac{P}{A} + \frac{M}{Z}$$
$$= \frac{P_e}{A} + \frac{P_e \cdot e}{Z}$$
$$290 = \frac{3,684.64}{30 \times 16} + \frac{3,684.67 \times e}{\frac{30 \times 16^2}{6}}$$

Solving $e = 98.01$ mm

∴ Central deflections $= 98.01$ mm

SOLVED PROBLEM 11.10

Find the shortest length of a steel column with hinged ends having a cross-sectional area of 40 mm × 80 mm for which the elastic Euler's formula is applicable. Take E = 200 GPa and yield strength be 250 MPa.

Solution:

Given data: Cross section = 40 mm × 80 mm, E = 200 GPa and yield strength = 250 MPa.

For the shortest length of column, the elastic load and Euler's buckling load should be equal.

$$f_c \times A = \frac{\pi^2 EI}{l_e^2}$$

i.e.,
$$l_e = \sqrt{\frac{\pi^2 EI}{f_c A}}$$

Now, $$I = \frac{1}{12} \times bd^3 = \frac{1}{12} \times 80 \times 40^3$$
$$= 426,666 \text{ mm}^4$$

$\therefore$ Minimum length, $$l_e = \sqrt{\frac{\pi^2 \times 200 \times 10^9 \times 126,666}{250 \times 10^6 \times 40 \times 30}}$$

$$l_e = 912.923 \text{ mm}$$

11.7 FACTOR OF SAFETY

Factor of safety (FS) with reference to load is defined as the ratio of critical load to safe load.

$\therefore$ $$\text{FS} = \frac{\text{Critical load}}{\text{Safe load}}$$

SOLVED PROBLEM 11.11

A solid round bar 4.5 m long and 60 mm in diameter was found to extend 4.6 mm under a tensile load of 60 kN. This bar is used as a strut with both ends fixed. Determine the buckling load for the bar and also the safe load taking factor of safety as 4.

Solution:

Given data: Diameter of the bar = 60 mm, length = 4.5 m, tensile load = 60 kN, FS = 4 and deflection = 4.6 mm.

Area of the bar, $$A = \frac{\pi}{4} \times 60^2 = 2,827 \text{ mm}^2$$

Young's modulus, $$E = \frac{\text{Tensile stress}}{\text{Tensile strain}} = \frac{P/A}{\delta l/l}$$
$$= \frac{(60 \times 1,000)/2,827}{4.6/4,500}$$
$$= 2.076 \times 10^4 \text{ N/mm}^2$$

The strut is fixed at both the ends.

$\therefore$ Effective length, $$l_e = 2l = 2 \times 4.5 = 9.0 \text{ m}$$

Buckling load, $$P = \frac{4\pi^2 EI}{l^2}$$

$$= \frac{4 \times \pi^2 \times 2.076 \times 10^4 \times \frac{\pi}{64} \times 60^4}{(4,500)^2}$$

$$= 25,747 \text{ N}$$

$$FS = \frac{\text{Critical load}}{\text{Safe load}}$$

Here, the critical load is the crippling load

$$\therefore \qquad \text{Safe load} = \frac{\text{Crippling load}}{\text{FS}}$$

$$= \frac{25,747}{4} = 6,436 \text{ N}$$

SOLVED PROBLEM 11.12

A pinned-ends circular column of length 3 m is subjected to a compressive stress of 40 MPa. Using a factor of safety of 2, find the diameter of the column to safely support a load of 100 kN. $E = 2 \times 10^5$ N/mm^2. Use Euler's formula.

Solution:

Given data: Length = 3 m, compressive stress = 40 MPa, FS = 2, safe load = 100 kN, and $E = 2 \times 10^5$ N/mm^2.

The given length is taken as the effective length

i.e., $\quad l_e = l = 3$ m

Column has pinned ends and the critical load $= (\text{Safe load}) \times (\text{Factor of safety})$

$$= 100 \times 10^3 \times 2 = 20,000 \text{ N}$$

Using Euler's formula,

$$P_c = \frac{\pi^2 EI}{l_e^2} = \frac{\pi^2 \times 2 \times 10^5 \times I}{(3,000)^2}$$

Rearranging and solving, $\quad I = 911,890$ mm^4

Let d be the diameter of the column. Then

$$I = \frac{\pi}{64} d^4$$

i.e.,

$$911,890 = \frac{\pi}{64} d^4$$

Solving, $\quad d = 65.65$ mm

Normal stress produced in the column,

$$p = \frac{100 \times 10^3}{\frac{\pi}{4} \times (65.65)^2}$$

$$= 23.66 \text{ MPa.}$$

which is less than the 40 Pa.

Hence, the diameter of 65.65 is safe.

11.8 RANKINE'S FORMULA

It has been shown that Euler's formula is valid for long column having l/k ratio greater than a certain value for a particular material. Euler's formula does not give a reliable result for short column and length of column intermediate between very long to short.

An empirical formula has been proposed by Rankine for columns of all lengths. The proposed formula is

$$\frac{1}{P} = \frac{1}{P_c} + \frac{1}{P_E} \tag{11.42}$$

Where

P = crippling load

P_c = crushing load $= f_c \times A$

P_E = crippling load by Euler's formula

$$= \frac{\pi^2 EI}{l_e^2}$$

f_c = ultimate crushing stress

A = area of cross section

In Equation (11.42), $\frac{1}{P_c}$ is constant for a given column of a particular material.

For short columns, P_E is very large and hence $\frac{1}{P_E}$ is small and hence the crippling load P is approximately equal to P_c.

For long columns, P_E is very small and hence $\frac{1}{P_E}$ is large and hence the crippling load P is approximately equal to P_E.

The above relation can be rearranged as

$$P = \frac{P_c P_E}{P_E + P_c} = \frac{P_c}{1 + \frac{P_c}{P_E}} = \frac{f_c A}{1 + \frac{f_c A l_e^2}{\pi^2 EI}}$$

i.e.,

$$P = \frac{f_c A}{1 + \left(\frac{f_c}{\pi^2 E}\right)\left(\frac{l_e}{k}\right)^2} \qquad (\because I = Ak^2)$$

i.e.,

$$P = \frac{f_c A}{1 + a\left(\frac{l_e}{k}\right)^2} \tag{11.43}$$

where $a = \frac{f_c}{\pi^2 E}$ is Rankine's constant for the material which is determined experimentally.

Values of f_c and a for materials most commonly used for column and struts are given in Table 11.2

Table 11.2 Values of f_c and a

Sl.No	Material	f_c, N (mm²)	a for hinged ends
1	Wrought iron	255	1/9,000
2	Cast iron	567	1/1,600
3	Mild steel	330	1/7,500
4	Timber	50	1/750

Rankine's formula for the standard case of two-hinged condition is sometimes known as Rankine-Gordon formula.

Rankine's formula is modified to account for different end conditions other than the standard case as

$$P = \frac{f_c A}{1 + a\left(\frac{l_e}{k}\right)^2} \tag{11.44}$$

where a is taken from Table 11.2 and l_e from Table 11.1.

11.9 JOHNSON'S PARABOLIC FORMULA

The critical load P is given by Johnson as

$$P = f_c \times A - b\left(-\frac{l_e}{k}\right)^2 A \tag{11.45}$$

where f_c = compressive yield stress.

b = constant whose value depends on the material of the column.

$$= \frac{f_c}{4\pi^2 E^2}$$

$\frac{l_e}{k}$ = slenderness ratio

and A = cross-sectional area of the column.

SOLVED PROBLEM 11.13

A cast iron column 10 cm external diameter and 6 m internal diameter is 4 m long. Calculate the safe load using Rankine's formula when (a) both ends are hinged (b) both ends are fixed; $f_c = 600$ N/mm² and $a = 1/1,600$. Adopt a factor of safety of three.

Solution:

Given data: External diameter = 10 cm, internal diameter = 4 cm, length = 4 m, f_c = 600 N/mm², $a = 1/1{,}600$ and FS = 3.0

Moment of inertia,
$$I = \frac{\pi}{64}\left(100^4 - 60^4\right)$$
$$= 4{,}270{,}400 \text{ mm}^4$$

Area of cross section,
$$A = \frac{\pi}{4}\left(100^2 - 60^2\right)$$
$$= 5{,}024 \text{ mm}^2$$

$$k = \sqrt{\frac{I}{A}} = \sqrt{\frac{4{,}270{,}400}{5{,}024}}$$
$$= 29.16 \text{ mm}$$

From modified Rankine's formula

$$P = \frac{f_c A}{1 + a\left(\frac{l_e}{k}\right)^2}$$

(i) Both ends hinged

From Table 11.1,
$$l_e = l = 4 \times 1{,}000 = 4{,}000 \text{ mm.}$$

$$P = \frac{600 \times 5{,}024}{1 + \frac{1}{6{,}000}\left(\frac{4{,}000}{29.16}\right)^2} \times \frac{1}{1{,}000}$$

$$= 728 \text{ kN}$$

$$\therefore \quad \text{Safe load} = \frac{728}{3} = 243 \text{ kN}$$

(ii) Both ends fixed

From Table 11.1,
$$l_e = \frac{l}{2} = \frac{4{,}000}{2} = 2{,}000 \text{ mm}$$

$$\therefore \quad P = \frac{600 \times 5{,}024}{1 + \frac{1}{6{,}000}\left(\frac{2{,}000}{29.16}\right)} \times \frac{1}{1{,}000}$$

$$= 1{,}693.5 \text{ kN}$$

$$\therefore \quad \text{Safe load } \frac{1{,}693.5}{3} = 564.5 \text{ kN}$$

SOLVED PROBLEM 11.14

A strut of T-section has a flange of 120 mm × 30 mm and web of 120 mm × 30 mm. It is 4 m long and hinged at both ends. Calculate the safe load the column can carry by Rankine's formula. Use a safety factor of 3; a = 1/ 7,500 and f_c = 330 N/mm^2.

Solution:

Given data: Flange = 120 mm × 30 mm, web = 120 mm × 30 mm, length = 4 m, FS = 3, a = 1/7,500 and f_c = 330 N/mm^2.

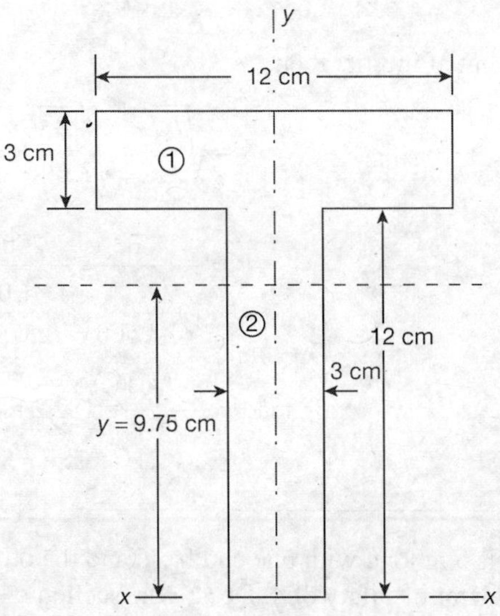

Fig. 11.11

The centre of gravity lies as the y -y axis at $\bar{y}$ distance from the base $x-x$.

Area of elements ① = 120 × 30 = 3,600 mm^2 = 36 cm^2

Area of elements ② = 120 × 30 = 3,600 mm^2 = 36 cm^2

Total area 7,200 mm^2 = 72 cm^2.

$$\bar{y} = \frac{3,600(120+15)+3,600(60)}{7,200}$$

$$\bar{y} = 9.75 \text{ cm}$$

$$I_{xx} = I_{cG} + Ah^2$$

I_{xx} = [M·I due to the area of element ①] + [M·I due to the area of element ②]

$$= \left[\frac{1}{12}\times 12\times 3^3 + 36(3+12-9.75)^2\right] + \left[\frac{1}{12}\times 3\times 12^3 + 36(9.75-60)^2\right]$$

$$= 1{,}019.25 + 938.25 = 1{,}957.5 \text{ cm}^4$$

$$I_{yy} = \left[\frac{1}{12}\times 3\times 12^3 + \frac{1}{2}\times 12\times 3^3\right] = 459 \text{ cm}^4.$$

Rankine's crippling load, $$P = \frac{f_c A}{1 + a\left(\frac{l_e}{k}\right)^2}$$

Considering the least moment of inertia, then,

$$k = \sqrt{\frac{I_{yy}}{A}} = \sqrt{\frac{459\times 10^4}{3{,}600 + 3{,}600}} = 25.25 \text{ mm}$$

$$\therefore \quad P = \frac{330\times 7{,}200}{1 + \frac{1}{7{,}500}\left(\frac{4\times 1{,}000}{25.25}\right)^2} = 546{,}709.6 \text{ N}$$

$$\therefore \quad \text{Safe load} = \frac{546.71}{3} = 182.24 \text{ kN}$$

SOLVED PROBLEM 11.15

A hollow cylindrical column is 5 m long with one end fixed and the other end hinged. It has to carry a safe load of 200 kN with a factor of safety of 4. Design the section taking the inner diameter as 0.8 times the outer. The crushing strength for the material is 550 MPa and Rankine's constant is 1/1,600.

Solution:

Given data: Length = 5 m, safe load = 200 kN, FS = 4, $d = 0.8\ D$, crushing strength = 550 MPa and Rankine's constant = 1/1,600.

From modified Rankine's formula $$p = \frac{f_c^A}{1 + a\left(\frac{l_e}{k}\right)^2}$$

For one end fixed and the other end hinged $$l_e = \frac{l}{\sqrt{2}} = \frac{5\times 1{,}000}{\sqrt{2}}$$

$$= 3{,}536.07 \text{ mm.}$$

Area $$A = \frac{\pi}{4}\left[D^2 - (0.8D)^2\right]$$

$$= 0.283D^2$$

Moment of inertia, $I = \frac{\pi}{64}\left[D^4 - (0.8D)^4\right]$

$$= 0.0290\ D^4$$

$$k = \sqrt{\frac{I}{A}} = \sqrt{\frac{0.290D^4}{0.28D^4}} = 0.32D$$

$$\text{Factor of safety} = \frac{\text{Crippling load}}{\text{Safe load}}$$

∴ $\text{Crippling load} = 200 \times 4 = 800 \text{ kN}$

$$800 \times 1,000 = \frac{550 \times 0.283D^2}{1 + \frac{1}{1,000}\left(\frac{3,536.07}{0.32D}\right)^2}$$

Rearranging $800,000\ D^2 + 76,317.1 \times 800,000 = 155.65\ D^4$

i.e., $D^4 - 5,139.7\ D^2 - 392,226,148.4 = 0$

This is a quadratic equation in D^2.

$$D^2 = \frac{+5,139.7 \pm \sqrt{5,139.7^2 + 4 \times 1 \times 392,226,148.4}}{2}$$

∴ $$= \frac{5,139.7 \pm 39,941.67}{2}$$

$$= 22,540.6 \text{ mm}^2$$

$$D = 150.13 \text{ mm}$$

∴ External diameter = 160 mm

∴ Internal diameter = 0.8 × 160 = 124 mm

SOLVED PROBLEM 11.16

A column is made up of two rolled steel joints of *I* section 160 mm × 80 mm × 10 mm thick with plate 200 mm × 10 mm riveted with flanges one each on the top and on the bottom. The edges of the plates being flush with the outside edges joints flanges. Determine, the safe load by Rankine's formula the column is 4 m length with both ends hinged, can carry with factor of safety 3. Take $a = 1/7,500$ and $f_c = 320 \text{ MN/m}^2$.

Solution:

Given data: I sections = 160 mm × 80 mm × 10 mm, plates 200 mm × 10 mm, length = 4 m, FS = 3, $a = 1/7,500$ and $f_c = 320 \text{ MN/m}^2$.

Area of cross section of the column,

$$A = 2[80 \times 10 \times 20 + 140 \times 10] + 2 \times 200 \times 2$$

$$A = 10,000 \text{ mm}^2$$

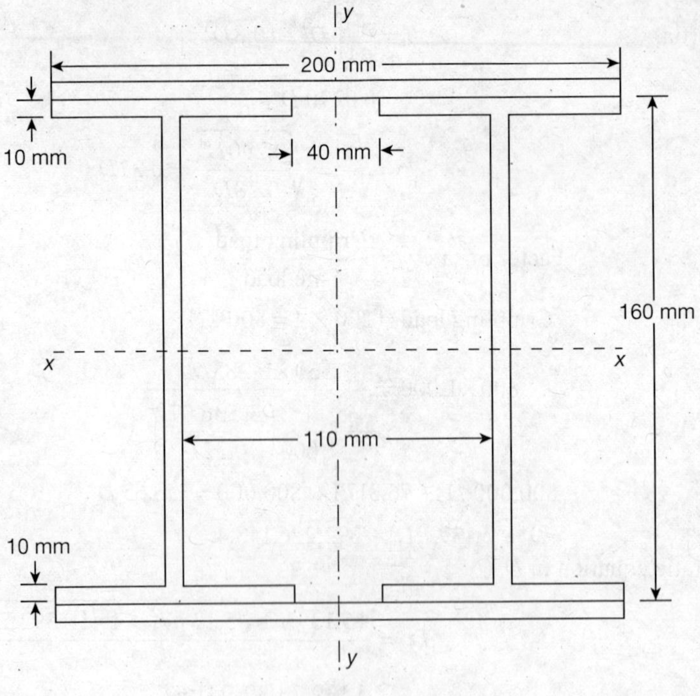

Fig. 11.12

As the section is symmetrical, the centre of gravity will lie at the point of intersection of two axes of symmetry.

i.e., $$I_{xx} = 4\left[\frac{8\times1^3}{12} + 8\times1\times(8-0.5)^2\right] + 2\left[\frac{1\times14^3}{12}\right] + 2\left[\frac{20\times1^3}{12} + 20\times1\times(8+0.5)^2\right]$$

$= 5,153.2\ \text{cm}^4$

$= 51,53,200\ \text{mm}^4$.

I_{yy} for one I section is

$$= 2\left[\frac{1\times8^3}{12}\right] + \frac{14\times1^3}{12}$$

$= 86.49\ \text{cm}^4$

$= 864,900\ \text{mm}^4$

Moment of inertia for the whole section about $y - y$ axis

$$I_{yy} = 2\left[8,649 + 30\times6^2\right] + 2\times\frac{1\times20^3}{12}$$

$= 3,666.31\ \text{cm}^4$.

As $I_{yy} < I_{xx}$, the column will tend to buckle in the $y - y$ direction

As the end condition is both ends hinged

$$l_e = l = 4 \text{ m}$$

$$k^2 = \frac{I}{A} = \frac{3,666.31\times10^4}{10,000} = 3,666 \text{ mm}^2$$

Using Rankine's formula,

$$P = \frac{f_c A}{1 + a(l_e(k))^2}$$

$$= \frac{320\times10^6\times100\times10^{-4}}{1+\dfrac{1}{7,500}\left(\dfrac{4^2}{3,666}\right)\times10^{-4}}$$

$$= 2,022 \times 10^3 \text{ N}$$

Safe load $= \dfrac{2,022\times10^3}{3}$

$$= 674,284 \text{ N.}$$

$$= 674.284 \text{ kN.}$$

SOLVED PROBLEM 11.17

Compare the crippling loads given by Rankine's and Euler's formula for a tubular column 225 cm long having outer and inner diameters of 37.5 mm and 32.5 mm, respectively loaded through pin joints at both ends; $f_c = 315$ MPa, $a = \dfrac{1}{7,500}$ and $E = 2 \times 10^5$ MPa.

Solution:

Given data: Length = 225 cm, outer diameter = 37.5 mm, inner diameter = 32.5 mm, $f_c = 315$ MPa, $a = \dfrac{1}{7,500}$ and $E = 2 \times 10^5$ MPa.

Moment of inertia, $I = \dfrac{\pi}{64}(37.5^4 - 32.5^4)$

$$= 4.23 \times 10^4 \text{ mm}^4$$

Area, $A = \dfrac{\pi}{4}(37.5^2 - 32.5^2)$

$$= 274.75 \text{ mm}^2$$

(ii) Euler's formula

$$k = \sqrt{\frac{I}{A}} = \sqrt{\frac{4.23\times10^4}{274.71}}$$

$$= 12.40 \text{ mm}$$

For pin joints, Euler's load is given as

$$P_E = \frac{\pi^2 EI}{l_e^2}$$

$$= \frac{\pi^2 \times 2\times 10^5 \times 4.23\times 10^4}{(2,250)^2} \times \frac{1}{1,000}$$

$$= 16.48 \text{ kN}$$

(ii) Rankine's formula

For pin joints, the load as per Rankine's formula is given as

$$P = \frac{f_c A}{1 + a\left(\frac{l_e}{k}\right)^2}$$

$$= \frac{315 \times 274.75}{1+\frac{1}{7,500}\left(\frac{2.250}{12.40}\right)^2} \times \frac{1}{1,000}$$

$$= 14.97 \text{ kN}$$

SOLVED PROBLEM 11.18

A 1.2 m long column has a circular cross section of 45 mm diameter one of the ends of the column is fixed in direction and position and other end is free. Taking factor of safety as 3, calculate the safe load.

(i) Rankine's formula, take yield stress = 560 N/mm² and $a = 1/1,600$

(ii) Euler's formula, E for cast iron = 1.2×10^5 N/mm.

Solution:

Given data: Length = 1.2 m, diameter = 45 mm, $FS = 3.0$, yield stress = 560 N/mm², $a = 1/1,600$, $E = 1.2 \times 10^5$ N/mm.

Area of the column, $A = \frac{\pi}{4} \times 45^2 = 1,589.62 \text{ mm}^2$ and

Moment of inertia, $I = \frac{\pi}{64} \times 45^4 = 20.12 \times 10^4 \text{ mm}^4$

$$\text{Least radius of gyration} = \sqrt{\frac{I}{A}}$$

$$= \sqrt{\frac{20.12\times 10^4}{1,589.62}}$$

$$= 11.25 \text{ mm}$$

For one end fixed and other end free, the effective length, $l_e = 2l$.

i.e., $$l_e = 2 \times 1{,}200 = 2{,}400 \text{ mm}$$

(i) Rankine's formula

$$P = \frac{f_c A}{1 + a\left(\frac{l_e}{k}\right)^2}$$

$$= \frac{560 \times 1{,}589.62}{1 + \frac{1}{1{,}600}\left(\frac{2{,}400}{11.25}\right)^2}$$

$$= 30{,}237 \text{ N}$$

$$\text{Safe load} = \frac{30{,}237}{3} = 10{,}079 \text{ N} = 10.079 \text{ kN}$$

(ii) Euler's formula

$$P_E = \frac{\pi^2 EI}{l_e^2}$$

$$= \frac{\pi^2 \times 1.2 \times 10^5 \times 20.12 \times 10^4}{(2{,}400)^2}$$

$$= 41{,}328 \text{ N}$$

$$\text{Safe load} = \frac{41{,}328}{3} = 13{,}776 \text{ N}$$

$$= 13.77 \text{ kN.}$$

SOLVED UNIVERSITY QUESTIONS

SOLVED PROBLEM 11.19

Find the Euler's critical load for a hollow cylindrical cast iron column 150 mm external diameter, 20 mm wall thickness if it is 6 m long with hinges at both ends. Assume Young's modulus of cast iron as 80 kN/mm². Compare this load with given by Rankine's formula using Rankine's constant $a = 1/1{,}600$ and $f_c = 567$ N/mm² (Anna Univ., May 2006, EEE).

Solution:

Given data: External diameter = 150 mm, thickness = 20 mm, length = 6 m, $E = 80$ kN/mm², $a = 1/1{,}600$ and $f_c = 567$ N/mm².

Area of column, $A = \frac{\pi}{4}\left(D_0^2 - D_i^2\right)$

$$= \frac{\pi}{4}\left(150^2 - 110^2\right) = 8{,}167.14 \text{ m}^2$$

Moment of inertia, $I = \frac{\pi}{64}\left(D_0^4 - D_i^4\right)$

$$= \frac{\pi}{64}\left(150^4 - 110^4\right) = 17.66 \times 10^6 \text{ mm}^4$$

Least radius of gyration, $k = \sqrt{\frac{I}{A}}$

$$= \sqrt{\frac{17.66 \times 10^6}{8{,}167.14}} = 46.55 \text{ mm}$$

(i) Euler's formula

For both ends hinged condition, effective lengths, $l_e = l$ (actual lengths)

∴ Crippling load, $P_E = \frac{\pi^2 EI}{l_e^2}$

$$= \frac{\pi^2 \times 80 \times 10^3 \times 17.66 \times 10^6}{(6{,}000)^2}$$

$$= 387.33 \times 10^3 \text{ N}$$

(ii) Rankine's formula

Crippling load, $P_R = \frac{f_c A}{1 + a\left(\frac{l_e}{k}\right)^2}$

$$= \frac{567 \times 8{,}168.14}{1 + \frac{1}{1{,}600}\left(\frac{6{,}000}{46.49}\right)^2}$$

$$= 405.80 \times 10^3 \text{ N}$$

(iii) Comparison

$$\frac{\text{Crippling load by Euler's formula}}{\text{Crippling load by Rankime's formula}} = \frac{P_E}{P_R}$$

$$= \frac{387.3 \times 10^3}{405.8 \times 10^3}$$

$$= 0.954$$

SOLVED PROBLEM 11.20

A hollow cylindrical column of 150 mm external diameter and 15 mm thick, 3 m long is hinged at one end and fixed at the other end. Find the ratio of Euler's and Rankine's critical load. $E = 8 \times 10^4$ N/mm², $f_c = 550$ N/mm² and Rankine's constant = 1/1,600 (Anna Univ., Nov. 2008, ME).

Solution:

Given data: Length = 3 m, external diameter = 150 mm, thickness = 15 mm, $E = 8 \times 10^4$ N/mm², $f_c = 550$ N/mm² and $a = 1/1{,}600$.

Area,
$$A = \frac{\pi}{4}(150^2 - 120^2)$$
$$= 6{,}361.725 \text{ mm}^2$$

Moment of inertia,
$$I = \frac{\pi}{64}(D_e^4 - D_i^4)$$
$$= \frac{\pi}{64}(150^4 - 120^4)$$
$$= 1.467 \times 10^7 \text{ mm}^4$$

(i) Euler's formula

Euler's critical load,
$$P_E = \frac{\pi^2 EI}{l_e^2}$$

For one end hinged and other fixed condition, the effective length,
$$l_e = l\sqrt{2} = \frac{3{,}000}{\sqrt{2}} = 2{,}121.32 \text{ mm}$$

$$\therefore \quad P_E = \frac{\pi^2 \times 8 \times 10^4 \times 1.467 \times 10^7}{(2{,}121.32)^2}$$
$$P_E = 2.574 \times 10^6 \text{ N}$$

(ii) Rankine's formula

Rankine's critical load,
$$P_R = \frac{f_c A}{l + a\left(\frac{l_e}{k}\right)^2}$$

Now,
$$k = \sqrt{\frac{I}{A}} = \sqrt{\frac{1.467 \times 10^7}{6{,}361.72}} = 48.02$$

$$P_R = \frac{500 \times 6{,}361.72}{1 + \frac{1}{1{,}600}\left(\frac{2{,}121.32}{48.02}\right)^2}$$
$$P_R = 1.57 \times 10^6 \text{ N}$$

(iii) Ratio

$$\text{Ratio of } \frac{P_E}{P_R} = \frac{2.57\times10^6}{1.57\times10^6} = 1.63$$

SALIENT POINTS

- A strut is a structural member which is subjected to an axial compressive force. A strut may be horizontal, vertical or inclined with any end fixity condition.
- A vertical strut is called column, e.g. a pillar between a roof and a floor.
- A column subjected to a compressive force, *P*, whose line of action is at a distance, *e*, from the axis of the column is called eccentrically loaded column and *e* is called the eccentricity.
- Slenderness ratio, s_r, is the ratio of length of a member and its least radius of gyration.
- Numerical value of slenderness ratio classifies a column as short column or long column. Mild steel columns with slenderness ratio more than 80 are called long columns and those with slenderness ratio less than 80 are called short columns. This limiting ratio may vary depending on the material properties.
- Based on the type of loading and the slenderness ratio, a column may fail due to any one of the following stresses: (i) direct compressive stress, (ii) bending stress or (iii) combined direct and bending stresses.
- The load at the time of failure is called the crushing load, P_c and the induced stress is called crushing stress, f_c. All short columns fail under crushing.
- The load at which a column just buckles is known as buckling load or critical load or crippling load. For long column the buckling load is less than the crushing load.
- The four end conditions of column are:

  (i) Both ends are hinged or pinned
  (ii) One end is free and the other end is fixed.
  (iii) Both ends are fixed
  (iv) One end is fixed and the other end is pinned

- The following sign conventions for the bending of the columns will be adopted:

  (i) A moment causing the column to bend with its convexity towards its initial central line is considered positive.
  (ii) A moment causing the column to bend with its concavity towards its initial line is considered negative.

- Euler's Formula is based on the following assumptions.

  (i) Column is very long in proportion to its cross-sectional dimension.
  (ii) Column is initially straight and the compressive load is applied axially.
  (iii) Material of the column is elastic, homogeneous and isotropic
  (iv) Effect of direct stress is very small in comparison with bending stress.
  (v) Column shall fail by buckling alone.
  (vi) Effect of self-weight of column is negligible.

- Euler's crippling load is expressed in a generalized form as $P_E = \dfrac{\pi^2 EI}{Cl^2}$

  where C is a constant which is a function of end conditions, thus, $P_E = \dfrac{\pi^2 EI}{l_e^2}$

  where l_e = equivalent length

$$l_e = l\sqrt{C}$$

- Euler's formula is valid for long columns having l_e/k ratio greater than a certain value for a particular material. Euler's formula does not give a reliable result for short column and length of column intermediate beteween very long to short.
- Rankine's formula is applicable for all lengths which is given as

$$\frac{1}{P} = \frac{1}{P_C} + \frac{1}{P_E}$$

  where P = Crippling load
  P_c = Crushing load
  P_E = Crippling load for Euler's formulae

- Critical load P is given by Johnson as $P = f_C \times A - b\left(\dfrac{l_e}{k}\right)^2 A$

  where f_c = Compressive yield stress
  b = Constant whose value depends upon the material of the column
  $\dfrac{l_e}{k}$ = Slenderness ratio
  A = Cross-sectional area of column.
- The ratio of the critical load to the safe load is known as factor of safety.
- Maximum compressive stress f_{max} for an eccentrically loaded long column is given as

$$f_{max} = \frac{P}{A}\left[1 + \frac{ey_c}{k^2}\sec\left(\frac{l}{2}\sqrt{\frac{P}{EI}}\right)\right]$$

  where y_c = central deflection.
  As the above equation contains a secant term, it is known as the secant formula for columns.

QUESTIONS

1. Distinguish between a long and short column. How slenderness ratio play a role in deciding the kind of a column?
2. What are end conditions of a column? How do they affect the performance of a column?
3. What is the effective length of a column? How the end conditions affect the effective length?
4. State the assumptions made in Euler's formula. Give generalized Euler's equation for different conditions.

5. What are the limitations of Euler's formula? Under what conditions Eurler's formula is reliable?
6. How Rankine's formula gains importance over Euler's formula? Compare their merits.
7. Give Johnson's parabolic formula. In which way it is differenct from Rankine's formula?
8. An eccentric point laod W is carried by a short column of external diameter D and internal diameter d. Prove that for the column not to develop any tension the max eccentricity can be $\dfrac{(D^2+d^2)}{8D}$.
9. A solid circular column of 240 mm diameter carries a load of 30 kN at an eccentricity of 30 mm from the vertical axis. Calculate the maximum and minimum stresses at the section.
10. A straight cylindrical bar is of 15 mm diameter and 1.2 m long. It is freely supported at its ends in a horizontal position and loaded at the centre with a concentrated load of 90 N when the central deflection is found to be 5 mm. If the bar is placed vertically and loaded along its axis, what load would cause it to buckle? The ends of the bar are pinned.
11. Calculate the Euler's critical load for a strut of T-section, the flange width being 10 cm, overall depth 8 cm and both flange and stem 1 cm thick. The strut is 3 m long and is built in at both ends. Take $E = 2 \times 10^5$ N/mm^2.
12. A hollow C.I. column ($d_o = 200$ mm and $d_i = 160$ mm) is 4.5 m. If the allowable maximum compressive stress is 110 N/mm^2, calculate the safe load by Rankine's formula using $FS = 4$, $a = 1/1{,}600$ and both the ends are hinged.
13. A hollow column with fixed ends, supports an axial load of 1,000 kN. The column is 5 m long and has an external diameter of 250 mm. Find the thickness of metal required using Rankine's formula. Working stress is 80 N/mm^2. Rankine's constant is 1/6,000.
14. Find the Euler's critical load for a hollow cylindrical cast iron column 15 cm external diameter and 2 cm wall thickness, if it is 6 m long with both ends hinged. Compare this load with that obtained by Rankine's formula using the following constant: $f_c = 620$ MPa, $a = 1/1{,}600$ and $E = 80$ MN/mm^2.

12

Analysis of Framed Structures

LEARNING OBJECTIVES

12.1 Introduction
12.2 Types of Frames
12.3 Analysis of Perfect Frames
12.4 Method of Joints or Method of Resolution
12.5 Method of Sections or Method of Moments

12.1 INTRODUCTION

A frame is a structure made up of several steel bars which are rivetted or welded together. These are made up of angle irons or channel sections and are called members of the frame or framed structure. Although the members are welded or rivetted together at their joints, they are considered as hinged or pin-jointed for the purpose of calculations. Determination of forces in a frame is needed in many engineering structures. The forces are determined based on the application of the principles of either statics or graphics.

12.2 TYPES OF FRAMES

The simplest frame is a triangle which has three members and three joints (pin-jointed).

Although there are many types of frames, which can be thought of, the frames may be grouped under two classes viz.,

(i) Perfect frames
(ii) Imperfect frames

12.2.1 Perfect Frames

A perfect frame may be defined as that one which is made up of members just sufficient to keep the frame in equilibrium, when loaded, without any change in the shape.

The simplest example of a perfect frame is a triangle. It is to be noted that the shape will not be distorted when the structure is loaded (Figure 12.1(a)). Thus, for three jointed frame, there should be three members to prevent distortion. If it is needed to increase a joint to this triangular frame, ABC, two members AD and BD are to be added (Figure 12.1(b)). Thus, to keep a frame as a perfect frame two additional members at one joint have to be added (Figure 12.1(c)).

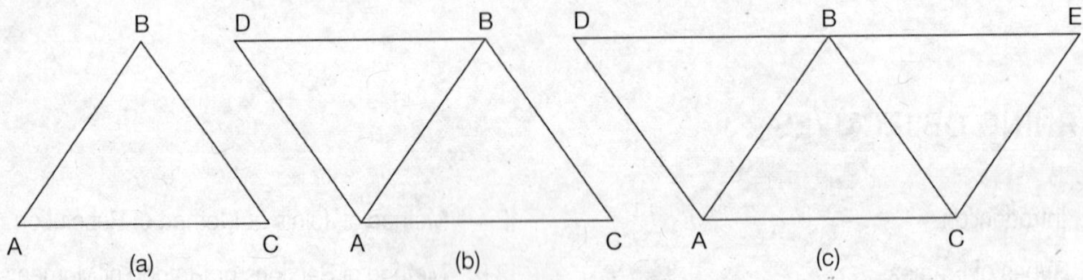

Fig. 12.1 Examples of perfect frames.

A relationship for a perfect frame can be obtained in terms of joints and members. Let n be the number of members and j be the number of joints. In Figure 12.2(a) let the three members AB, BC and AC and three joints A, B and C are removed. Now, we have $(n - 3)$ members and $(j - 3)$ joints in the frame (Figure 12.2(b)). It can be noted that for $(j - 3)$ joints, there are $2(j - 3)$ members, i.e., 3 joints and six members are remaining.

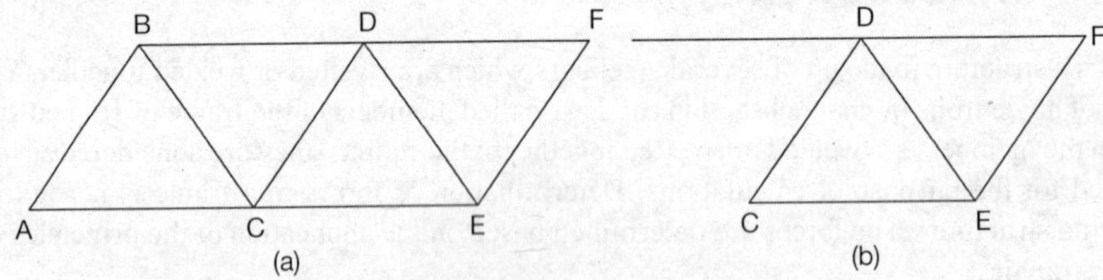

Fig. 12.2 Requirement of number of members.

i.e., $$n - 3 = 2(j - 3)$$

i.e., $$n = (2j - 3)$$

So for a stable frame the minimum number of members should be equal to two times the number of joints minimum three.

12.2.2 Imperfect Frame

Imperfect frame is one in which the number of members are more or less than $(2j - 3)$. The imperfect frame may be further classified as (i) deficient frame and (ii) redundant frame.

1. Deficient Frame

Let us consider a square frame which has four members and four joints (Figure 12.3(a)). As per the relationship we should have $(2 \times 4) - 3 = 5$ members for the frame to be stable. As only four members are present it is not a stable frame. So, if a frame has less number of members than that required by the relation $n = (2j - 3)$, then the frame is known as deficient frame.

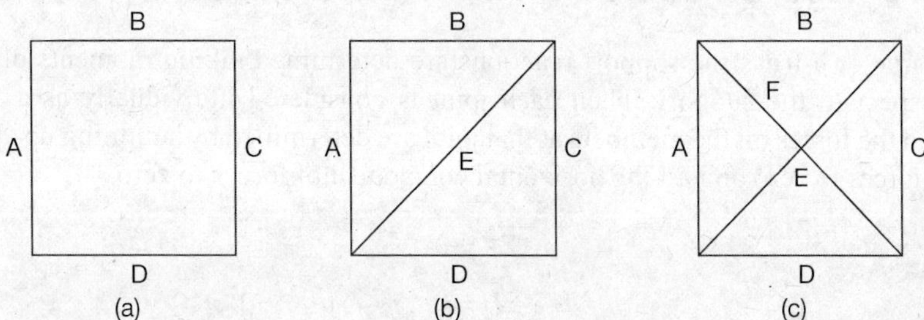

Fig. 12.3 Deficient and redundant frames.

By adding one more member, E, the square frame satisfies the relation (i.e., $2 \times 4 - 3 = 5$) and hence forms a perfect frame (Figure 12.3(b)).

2. Redundant Frame

If another additional diagonal member is added (Figure 12.3(c)) the frame still remains stable. But it has more members than required, i.e., 6 members instead of 5 members. Such a frame is called a redundant frame. Thus, a redundant frame is an imperfect frame, in which the number of members are more than $(2j - 3)$.

Thus, it can be summarized that:

if $n = (2j - 3)$ then the frame is a perfect frame
if $n < (2j - 3)$ then the frame is a deficient frame
if $n > (2j - 3)$ then the frame is a redundant frame.

Perfect frames are stable, whereas deficient frames are unstable. Hence, perfect frames can be analyzed using equilibrium conditions. Redundant frames, although stable, cannot be fully analyzed by using equilibrium conditions.

12.3 ANALYSIS OF PERFECT FRAMES

Following are the assumptions made in finding the forces in the members of a frame.

(i) The frame is a perfect frame
(ii) The frame is loaded only at the joints

(iii) All the members of the frame are pin-jointed
(iv) Self-weight of the members is neglected.

Forces in various members of a perfect frame may be found out either by analytical method or graphical method. Here, the discussion is confined to only analytical method.

A perfect frame can be analytically analyzed by the following methods:

(i) Method of joints or method of resolution
(ii) Method of sections or method of moments

12.4 METHOD OF JOINTS OR METHOD OF RESOLUTION

For a given frame or a truss the support reactions are determined taking moments of the external forces with respect to the support. Then each joint is considered individually as a free body in equilibrium and the forces on the members at that joint are determined by summing up all the vertical component of forces to zero and all the horizontal component of forces to zero.

i.e.,
$$\Sigma V = 0$$
and
$$\Sigma H = 0$$

Joints should be selected such that forces for only two members are unknown in that joint. The force is said to be tensile if it pulls the joint to which it is connected.

This method is explained in detail in the solved problems given below.

SOLVED PROBLEM 12.1

Determine the forces in the members of the truss shown in Figure 12.4.

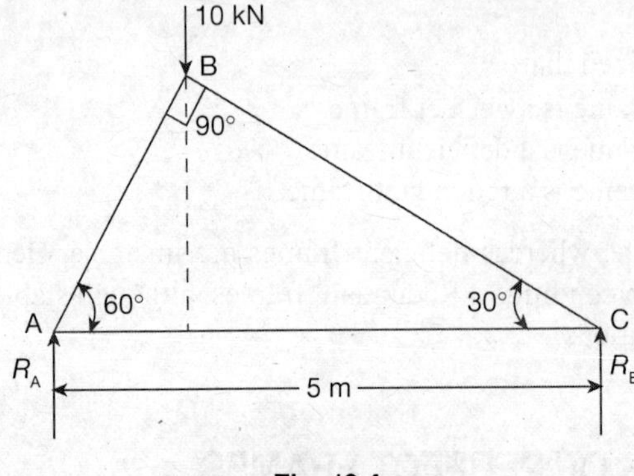

Fig. 12.4

Solution:

First support reactions are found.

Forces in members are found by analyzing the equilibrium of forces at each joint.

(i) Support reactions

From Δ ABC, AB = AC $\times$ cos 60° = 5 $\times$ 0.5 = 2.5 m

From Δ ABD, AD = AB $\times$ cos 60° = 2.5 $\times$ 0.5 = 1.25 m

Taking moments of forces with respect to A and equating

$$R_C \times 5 = 10 \times 1.25$$

$$R_C = 2.5 \text{ kN}$$

$$\therefore \quad R_A = 10 - 2.5 = 7.5 \text{ kN.}$$

(ii) Joint A

Let F_{AB} be the force in the member AB

F_{AC} be the force in the member AC

Let F_{AB} act towards the joint A and force F_{AC} act away from the joint A.

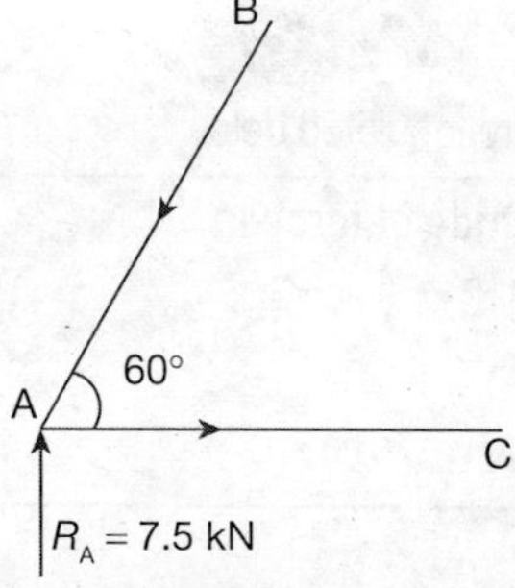

Fig. 12.5

Resolving the forces vertically and equating

$$R_A = F_{AB} \sin 60°$$

$$\therefore \quad F_{AB} = \frac{R_A}{\sin 60°} = \frac{7.5}{0.866}$$

$$= 8.66 \text{ kN (compressive)}$$

Resolving the forces horizontally and equating

$$F_{AC} = F_{AB} \sin 60°.$$

$$= 8.66 \times 0.5 = 4.33 \text{ kN (tensile)}$$

Since the values are positive, the assumed directions of force are correct.

(iii) Joint C

Let F_{BC} be the force in the member BC and F_{AC} be the force in the member AC

The force F_{AC} has already been found out and acts away from the joint C.

Let the force F_{BC} act towards the joint C.

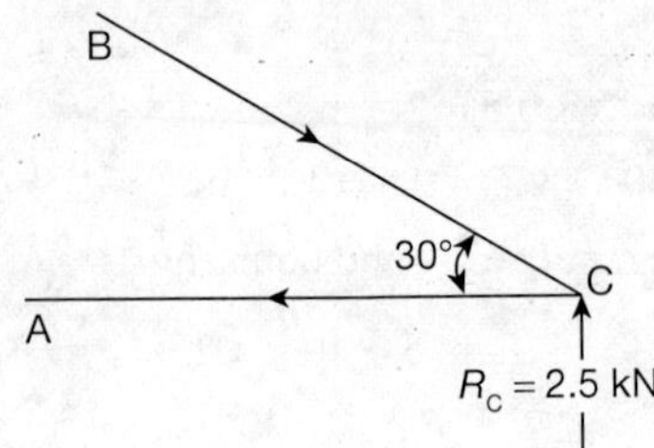

Fig. 12.6

Resolving the forces vertically and equating, $R_C = F_{BC}$ sin 30°

$$\therefore \quad F_{BC} = \frac{R_C}{\sin 30°} = \frac{2.5}{0.5} = 5 \text{ kN (compressive)}$$

(iv) Summary

Forces in members are tabulated below

Member	Magnitude of force in kN	Nature of force
AB	8.66	Compressive
AC	4.33	Tensile
BC	5.00	Compressive

SOLVED PROBLEM 12.2

Figure 12.7 shows a Warner girder consisting of members of 5 m length. The truss is freely supported at its end points. The girder is loaded with external loads. Determine the forces in the members and their nature.

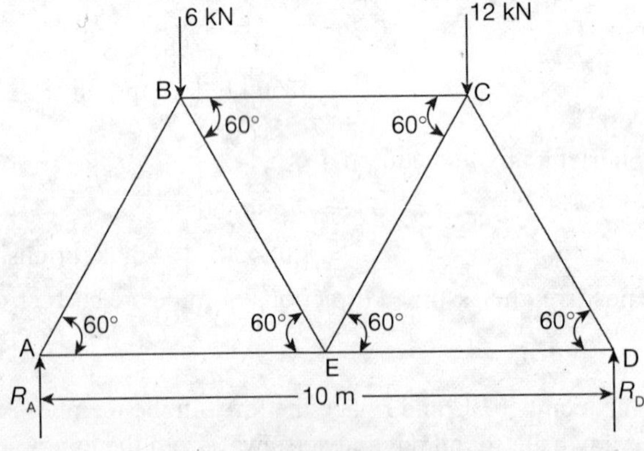

Fig. 12.7

Solution:

(i) Support reactions

Taking moments of forces about A and equating

$$R_D \times 10 = (12 \times 7.5) + (6 \times 2.5) = 105$$

$$\therefore \quad R_D = 10.5 \text{ kN}$$

$$R_A = (6 + 12) - 10.5 = 7.5 \text{ kN}$$

(ii) Joint A

Let F_{AB} be the force in the member AB and act towards joint A

F_{AE} be the force in the member AE and act away from joint A.

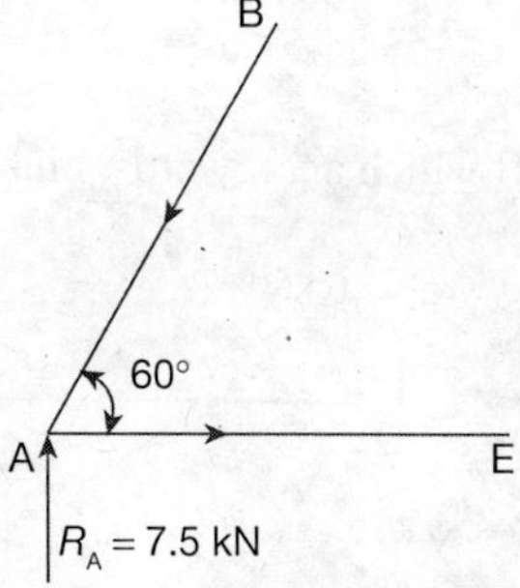

Fig. 12.8

Resolving the forces vertically and equating

$$R_A = F_{AB} \sin 60°$$

$$F_{AB} = \frac{7.5}{\sin 60°} = 8.66 \text{ kN (compressive)}$$

Resolving the forces horizontally and equating, $F_{AE} = F_{AB} \cos 60° = 8.66 \times 0.5 = 4.33$ kN (tensile)

Since the values are positive, the assumed nature of force is correct.

(iii) Joint D

Let F_{CD} be the force in the member CD and act towards joint D

and F_{DE} be the force in the member DE and act away from the Joint D.

Resolving the forces vertically and equating

$$R_D = F_{CD} \sin 60°$$

$$\therefore \quad F_{CD} = \frac{R_D}{\sin 60°} = \frac{10.5}{0.866} = 12.12 \text{ kN (compressive)}$$

Resolving the forces horizontally and equating, $F_{DE} = F_{CD} \cos 60°$

$$= 12.12 \times 0.5 = 6.06 \text{ kN (tensile)}$$

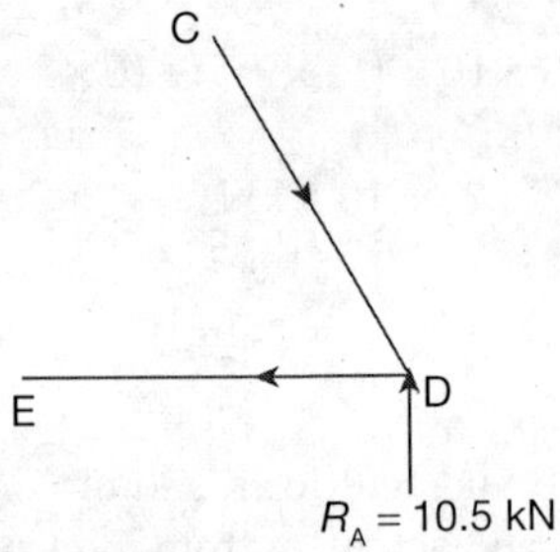

Fig. 12.9

(iv) Joint B

Let F_{AB} be the force in the member AB which act towards joint A (as already found)

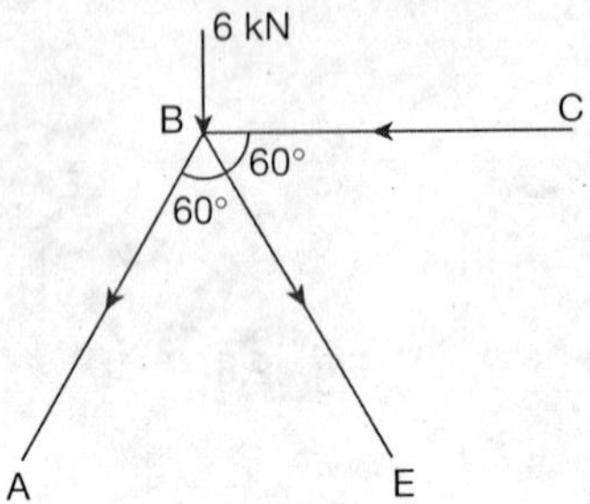

Fig. 12.10

Let force F_{BC} act towards the joint B and force F_{BE} act away from the joint B.

Resolving the forces vertically and equating, $F_{BE} \sin 60° + 6 = F_{AB} \sin 60°$

$$\therefore \quad F_{BE} = \frac{F_{AB} \sin 60° - 6}{\sin 60°}$$

$$= \frac{(8.66 \times 0.866) - 6}{0.866} = 1.73 \text{ kN (tensile)}$$

Resolving the forces horizontally and equating

$$F_{BC} = F_{AB} \cos 60° + F_{BE} \cos 60°$$

$$= (8.66 \times 0.5) + (1.73 \times 0.5)$$

$$= 5.195 \text{ N (compressive)}$$

(v) Joint C

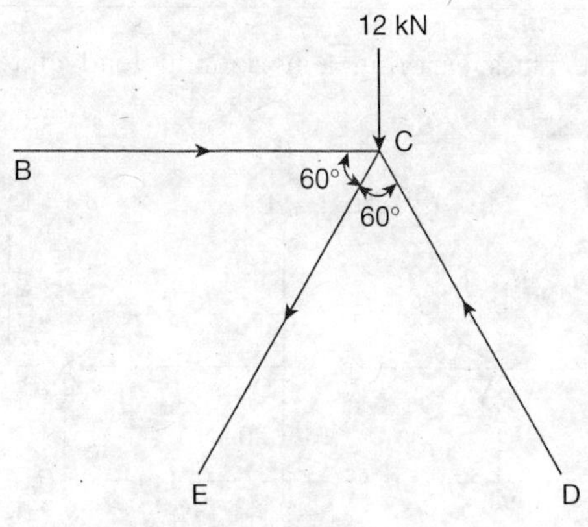

Fig. 12.11

Let F_{BC} be the force in member BC and act as compressive

F_{CD} be the force in the member CD and as compressive

F_{CE} be the force in the member CE and assumed to act away from joint C.

Resolving the forces vertically and equating

$$F_{CE} \sin 60° + 12 = F_{CD} \sin 60°$$

i.e.,

$$F_{CE} = \frac{F_{CD} \sin 60° - 12}{\sin 60°}$$

$$F_{CE} = 1.736 \text{ kN}$$

As the value of F_{CE} is negative, the assumed nature of the force is not correct. Thus, the force is compressive.

i.e.,

$$F_{CE} = 1.736 \text{ (compressive)}$$

(vi) Summary

Forces in members are tabulated below

Member	Magnitude of force in kN	Nature of force
AB	8.66	Compressive
AE	4.33	Tensile
BC	5.195	Compressive
BE	1.730	Tensile
CD	12.12	Compressive
CE	1.736	Compressive
DE	6.06	Tensile

SOLVED PROBLEM 12.3

Determine the forces in the truss carrying a horizontal load and vertical load as shown in Figure 12.12.

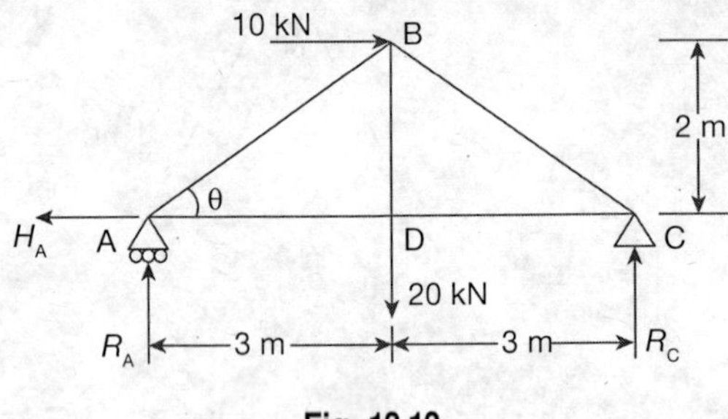

Fig. 12.12

Solution:

(i) Support reactions

Here, the truss has a hinged support and a roller support. The reaction at the roller support will be perpendicular to the roller plane and at the hinged support. There are two reactions, one is a perpendicular reaction and the other is a horizontal reaction.

The horizontal reaction is determined by adding algebraically all the horizontal forces.

Taking moment of all the forces with respect to A and equating

$$R_C \times 6 = (20 \times 3) + (10 \times 2) = 80$$

$$\therefore \quad R_C = \frac{80}{6} = 13.33 \text{ kN}(\uparrow)$$

$$\therefore \quad R_A = 20 - 13.33 = 6.67 \text{ kN}(\uparrow)$$

Taking sum of all horizontal forces,

$$H_A = 10 \text{ kN } (\leftarrow)$$

(ii) Joint A

Length of member, $AB = \sqrt{2^2 + 3^2} = 3.6 \text{ m}$

$$\cos\theta = \frac{AD}{AB} = \frac{3}{3.6} = 0.83$$

$$\sin\theta = \frac{AD}{AB} = \frac{2}{3.6} = 0.56$$

Let F_{AB} be the force in member AB and assumed to act towards the joint A

F_{AD} be the force in member AD and assumed to act away from the joint A

Resolving the forces vertically

$$R_A = F_{AB} \sin\theta$$

$$\therefore \quad F_{AB} = \frac{R_A}{\sin\theta} = \frac{6.67}{0.56} = 11.91 \text{ kN (compressive)}$$

Resolving the forces horizontally

$$\begin{aligned} F_{AD} &= H_A + F_{AB} \cos\theta \\ &= 10 + (11.91 \times 0.83) \\ &= 19.89 \text{ kN (tensile)} \end{aligned}$$

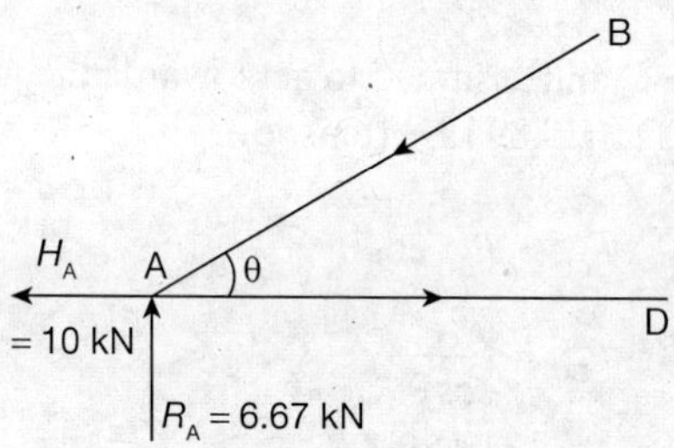

Fig. 12.13

(iii) Joint D

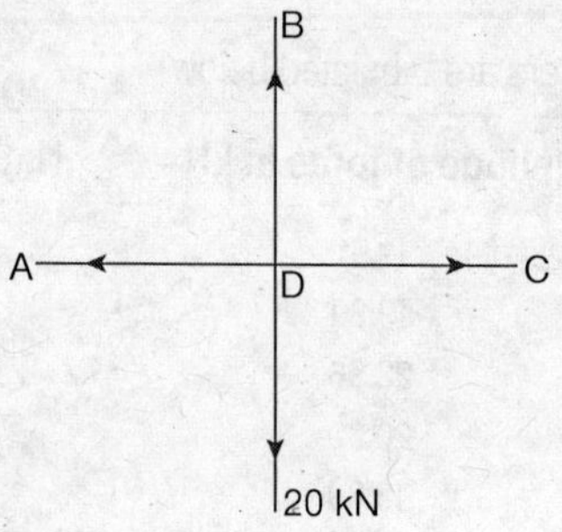

Fig. 12.14

Let F_{AD} be the force in member AD (=19.89 kN) that is tensile
F_{BD} be the force in member BD and assumed to act away from joint D
F_{CD} be the force in member CD and assumed to act away from joint D
Resolving the forces vertically

$$F_{BD} = 20 \text{ kN (tensile)}$$

Resolving the forces horizontally

$$F_{AD} = F_{CD} = 19.89 \text{ kN (tensile)}$$

(iv) Joint C

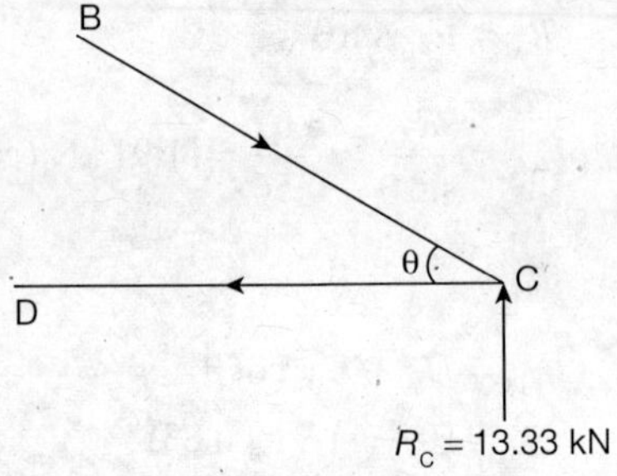

Fig. 12.15

Let F_{BC} be the force in member BC and assumed to act towards the joint C

F_{CD} be the force in member CD (=19.89 kN) (tensile)

Resolving forces horizontally

$$F_{CD} = F_{BC}\cos\theta$$

$$F_{BC} = \frac{F_{CD}}{\cos\theta} = \frac{19.89}{0.83} = 23.96\text{ kN}(\text{compressive})$$

(v) Summary

Forces in members are tabulated below

Member	Magnitude of force in kN	Nature of force
AB	11.91	Compressive
AD	19.89	Tensile
BC	23.96	Compressive
BD	20.00	Tensile
CD	19.89	Tensile

SOLVED PROBLEM 12.4

Find out the forces in every member of the truss shown in Figure 12.16.

Solution:

Considering $\Delta\, ADB,\ \tan 30° = \dfrac{\text{AB}}{\text{AD}}$

$$\therefore \quad \text{AD} = \frac{2.5}{\tan 30°} = 4.33\text{ m}$$

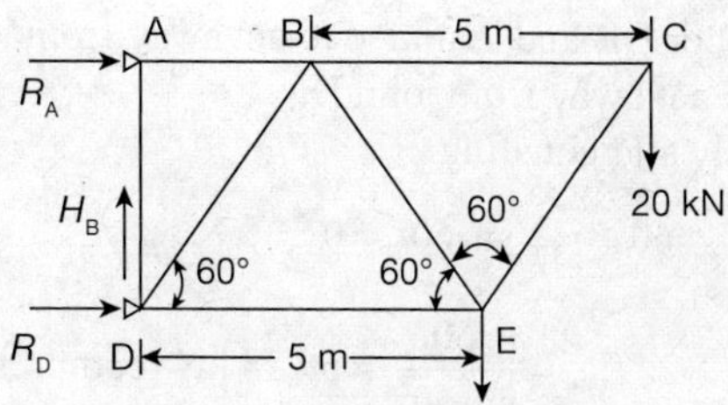

Fig. 12.16

(i) Support reactions

The hinged support at D has two reaction components one in the vertical direction and another in the horizontal direction

$$R_A + R_D = 0 \text{ and } H_D = 40 + 20 = 60 \text{ kN}$$

Taking moments of forces about A and equating clockwise moments and anticlockwise moments

$$20 \times (5 + 2.5) + 40 \times 5 = R_D \times 4.33$$

i.e.,

$$R_D = 80.8 \text{ kN}$$

$$R_A = -80.83 \text{ kN}$$

Reaction R_A is acting opposite to the selected direction.

(ii) Joint A

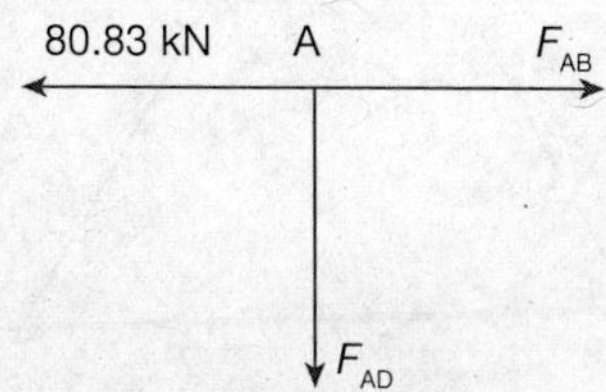

Fig. 12.17

Let F_{AB} be the force in the member AB and assumed to act away from Joint A

F_{AD} be the force in the member AD and will be zero

Hence, $F_{AB} = 80.83$ kN

(iii) Joint D

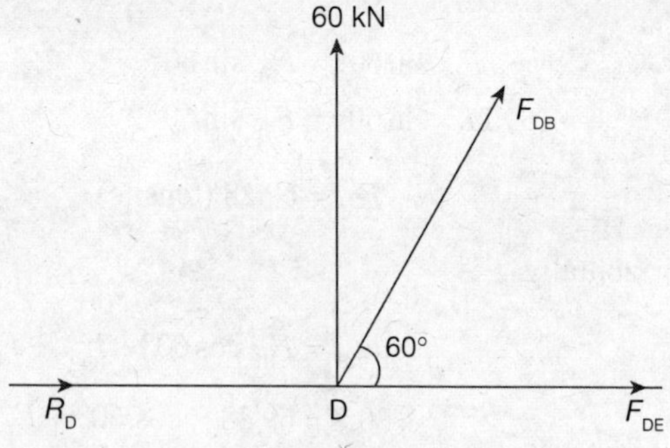

Fig. 12.18

Let F_{DB} be the force in the member DB and assumed to act away from joint D and F_{DE} be the force in the member DE and assumed to act away from joint D.

Resolving the forces vertically and equating

$$60 + F_{DB} \sin 60° = 0$$

$$\therefore \quad F_{DB} = \frac{60}{\sin 60°} = 69.28 \text{ kN (compressive)}$$

Resolving the forces horizontally and equating $F_{DE} + 80.8 + F_{DB} \cos 60° = 0$

i.e., $\quad F_{DE} = -46.16$ kN

The assumption is wrong and F_{DE} is compressive

(iv) Joint B

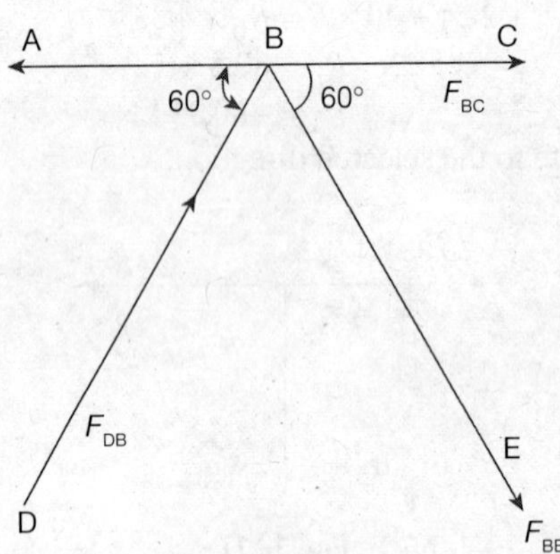

Fig. 12.19

Let F_{BE} be the force in the member BE and assumed to act away from joint B

F_{BC} be the force in the member BC and assumed to act away from joint B

Resolving the forces vertically and equating

$$F_{DB} \sin 60° = F_{BE} \sin 60°$$

$$69.28 \times \sin 60° = F_{BE} \sin 60°$$

$$\therefore \quad F_{BE} = 69.28 \text{ (tensile)}$$

Resolving the forces horizontally,

$$F_{AB} = F_{BE} \cos 60° + F_{BC} + F_{DB} \cos 60°$$

$$80.8 = 69.28 \times \cos 60° + F_{BC} + 69.28 \times \cos 60°$$

Solving, $\quad F_{BC} = 11.52$ (tensile)

(v) Joint E

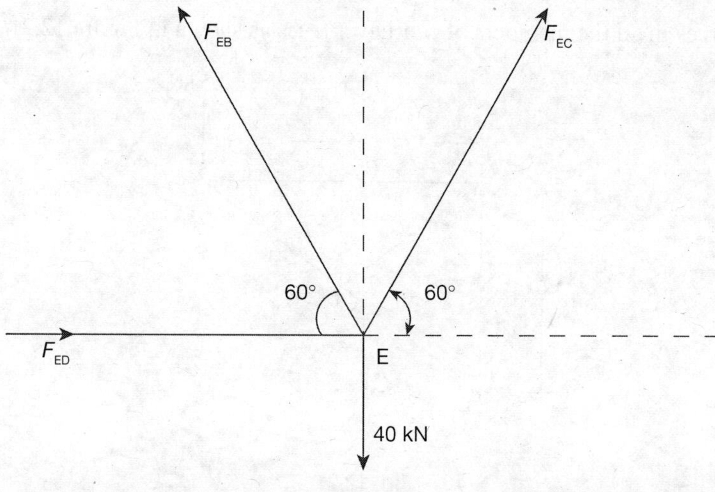

Fig. 12.20

Direction of forces are assumed in resolving the forces horizontally and equating

$$F_{EC}\cos 60° + F_{ED} - F_{EB}\cos 60° = 0$$

i.e.,

$$F_{EC}\cos 60° = 69.26 \times \cos 60° - 46.16$$

$$F_{EC} = -23.04 \text{ kN}$$

The assumption is not correct, hence F_{EC} is compressive

$$F_{EC} = 23.04 \text{ kN (compressive)}$$

(vi) Summary

Forces in members are summarized and given below.

Member	Magnitude of force in kN	Nature of force
AB	80.83	Tensile
DB	69.28	Compressive
DE	46.16	Compressive
BE	69.28	Tensile
BC	11.52	Tensile
EC	23.04	Compressive

SOLVED PROBLEM 12.5

Determine the forces in all the members of cantilever truss as shown in Figure 12.21.

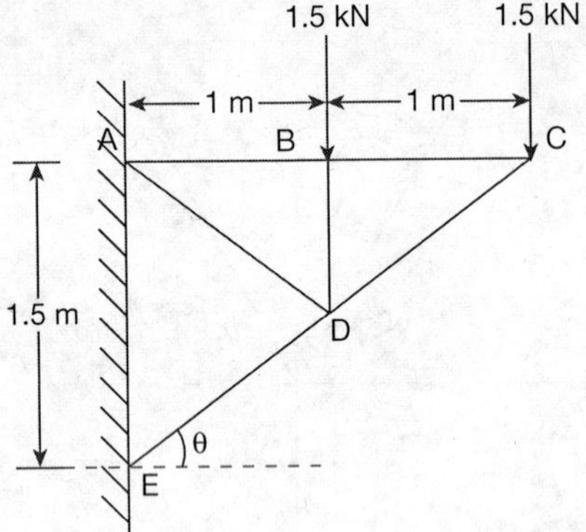

Fig. 12.21

Solution:

From ΔACE,

$$EC = \sqrt{1.5^2 + 2^2} = 2.5 \text{ m}$$

$$\cos\theta = \frac{AC}{CE} = \frac{2}{2.5} = 0.80$$

$$\sin\theta = \frac{AE}{CE} = \frac{1.5}{2.5} = 0.60$$

In the case of cantilever trusses, the forces in the members can be obtained by starting the calculations from the free end of the cantilever.

(i) Joint C

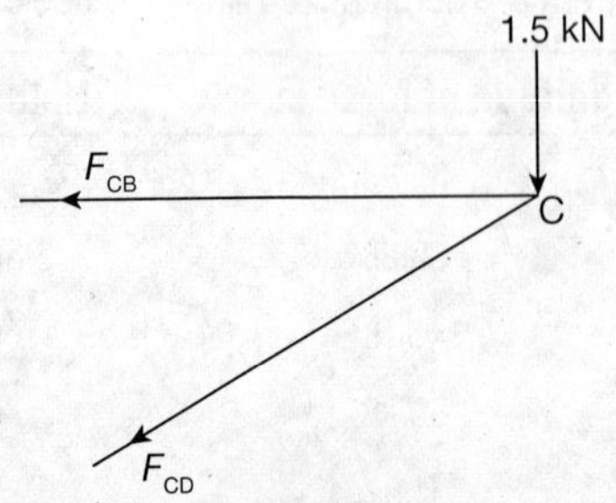

Fig. 12.22

Let F_{CB} and F_{CD} be the forces in members CB and CD and the forces are assumed to act as shown in Figure 12.22.

Resolving the forces vertically and equating

$$-1.5 - F_{CD} \sin\theta = 0$$

$$\text{i.e., } -F_{CD} \times 0.6 = 1.5$$

$$\therefore \quad F_{CD} = -2.5 \text{ kN.}$$

The assumption is wrong and hence $F_{CD} = 2.5$ kN (compressive)

Resolving the forces horizontally and equating

$$-F_{CB} - F_{CD}\cos\theta = 0$$

$$-F_{CB} - (-2.5) \times 0.8 = 0$$

$$\therefore \quad F_{CB} = 2 \text{ kN (tensile)}$$

(ii) Joint B

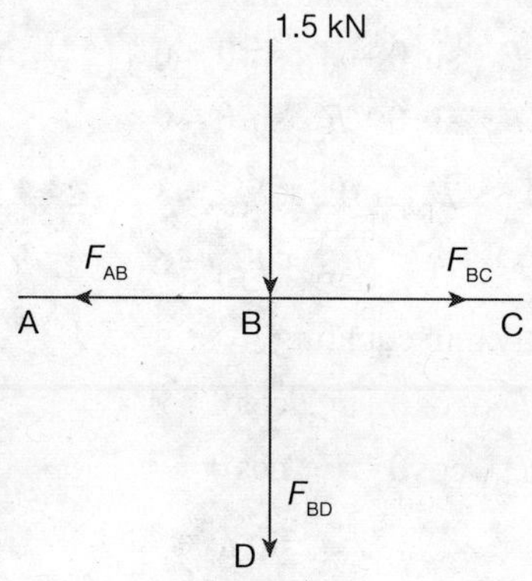

Fig. 12.23

Resolving the forces vertically and equating

$$-1.5 - F_{BD} = 0$$

$$F_{BD} = -1.5 \text{ kN}$$

The assumed direction is wrong, hence

$$F_{BD} = 1.5 \text{ kN (compressive)}$$

Resolving the forces horizontally and equating

$$-F_{BA} + F_{BC} = 0$$

$$\therefore \quad F_{BA} = 2 \text{ kN}$$

$$F_{BA} = F_{CB} = 2 \text{ kN (tensile)}$$

(iii) Joint D

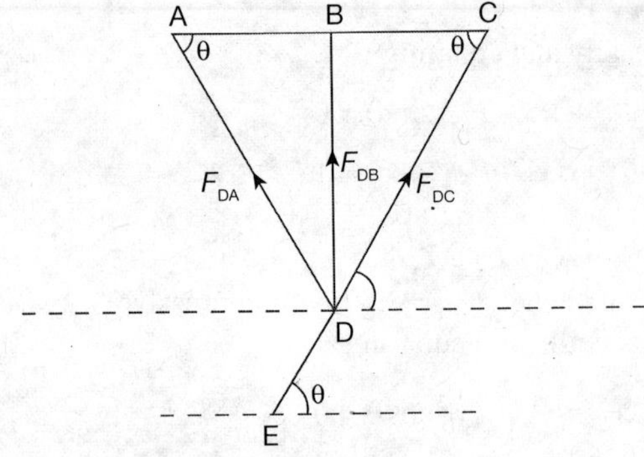

Fig. 12.24

Forces in members CD and BD have already been calculated as 1.5 kN and 2.5 kN.

Now, resolving the forces vertically and equating

$$F_{BD} + F_{CD}\sin\theta + F_{DA}\sin\theta + F_{ED}\sin\theta = 0$$

$$-1.5 + (-2.5)\times(0.6) + F_{DA}\sin\theta + F_{ED}\sin\theta = 0$$

i.e.,
$$0.6\times F_{DA} + 0.6\times F_{ED} = 3$$

i.e.,
$$F_{DA} + F_{ED} = S \qquad \text{(i)}$$

Resolving the forces horizontally and equating

$$F_{DC}\cos\theta - F_{DA}\cos\theta + F_{ED}\cos\theta = 0$$

$$-2.5\times 0.8 - F_{DA}\cos\theta + F_{ED}\cos\theta = 0$$

i.e,
$$F_{ED} - F_{DA} = 2.5 \text{ kN} \qquad \text{(ii)}$$

Solving, Equations (i) and (ii)

$$F_{ED} = 3.75 \text{ kN (compressive)}$$
$$F_{DA} = 1.25 \text{ kN (tensile)}$$

(iv) Summary

Forces in the members are tabulated as under

Member	Magnitude of force in kN	Nature of force
CD	2.5	Compressive
CB	2.0	Tensile
BD	1.5	Compressive
BA	2.0	Tensile
DA	1.25	Tensile
ED	3.75	Compressive

SOLVED PROBLEM 12.6

Determine the forces in the members of the truss shown in Figure 12.25

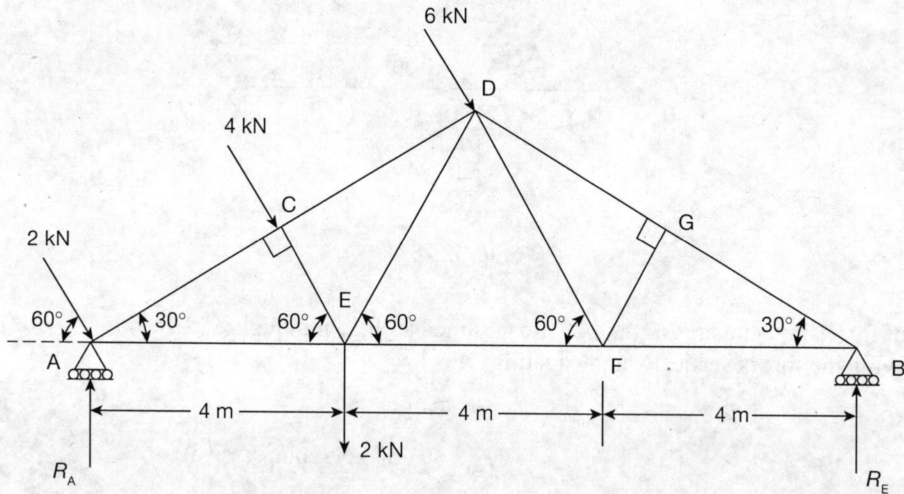

Fig. 12.25

Solution:

In the Δ ACE,

$$\cos 30° = \frac{AC}{4}$$

i.e.,

$$AC = 3.464 \text{ m}$$

and

$$\text{length } AD = 2 \times AC = 6.928 \text{ m}$$

(i) Support reactions

The truss is supported on rollers at B and hence R_B will be vertical (normal to the roller base). The truss is hinged at A and hence the support reactions at A will consist of a horizontal reaction H_A and a vertical reaction R_A.

Taking moments about A

$$R_B \times 12 = 4 \times AC + 6 \times AD + 1 \times AE$$

$$= 4 \times 3.464 + 6 \times 6.928 + 1 \times 4$$

$$R_B = 4.834 \text{ kN.}$$

Vertical components of all the inclined loads = (2 + 4 + 6) × cos 60° = 6 kN

$$\therefore \quad R_A = 11.39 - 6.834 = 6.556 \text{ kN}$$

Sum of all horizontal components = 6 kN

Hence,

$$H_A = 6 \text{ kN.}$$

(ii) Joint A

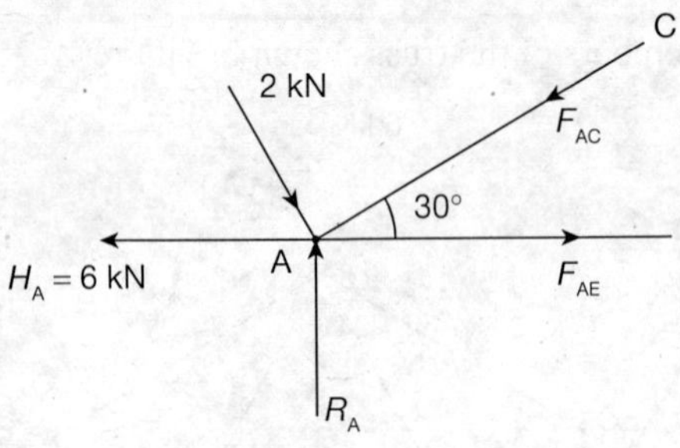

Fig. 12.26

Nature of forces in members AC and AE are assumed as shown in Figure 12.26.

Resolving the forces vertically and equating

$$F_{AC} \times \sin 30° + 2 \times \sin 60° = 6.556$$

i.e., $$F_{AC} = 9.664 \text{ kN (compressive)}$$

Resolving the forces horizontally and equating

$$F_{AE} + 2 \cos 60° = 6 + F_{AC} \cos 30°$$

i.e., $$F_{AE} = 6 + 9.664 \times 0.860 - 2 \times 0.5$$

$$= 13.36 \text{ kN (tensile)}$$

(iii) Joint C

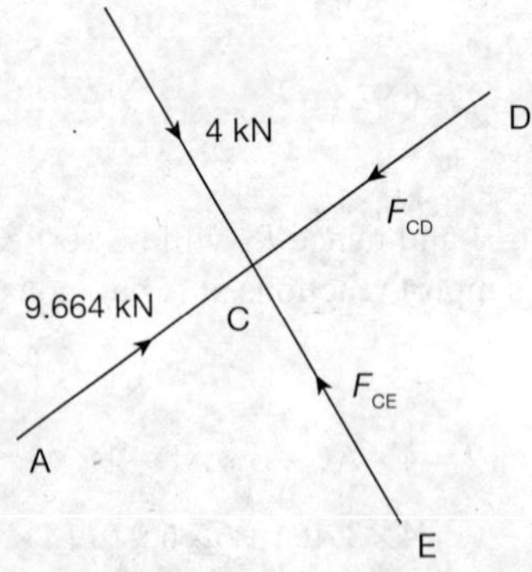

Fig. 12.27

From Figure 12.27 it can be assummed as

$$F_{CD} = F_{AC} = 9.664 \text{ kN (compressive)}$$

$$F_{CE} = 4 \text{ kN (compressive)}$$

(iv) Joint E

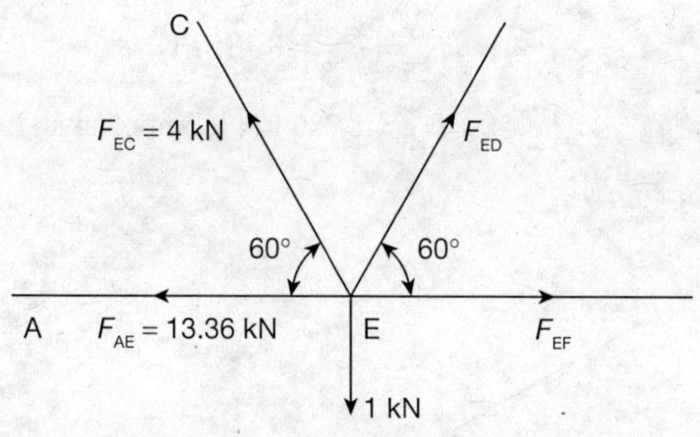

Fig. 12.28

Resolving the forces vertically and equating

$$1 + 4\sin 60° = F_{ED} + \sin 60°$$

Solving, $F_{ED} = 5.155$ kN (tensile)

Resolving the forces horizontally and equating

$$13.364 - 4 \times \cos 60° - F_{ED} \cos 60° - F_{EF} = 0$$

$$13.364 - 4 \times 0.5 - 5.155 \times 0.5 - F_{EF} = 0$$

$\therefore$ $F_{EF} = 8.786$ kN (tensile)

(v) Joint G

At this joint G, two forces F_{BG} and F_{DG} are in the same straight line and hence the third force, i.e., F_{GF} should be zero.

$\therefore$ $F_{GF} = 0$

(vi) Joint F

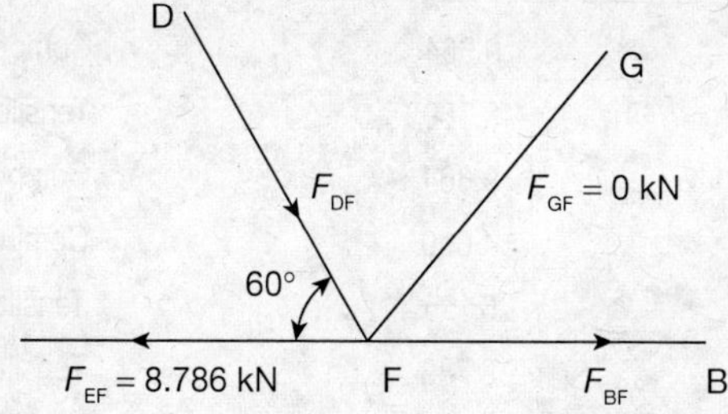

Fig. 12.29

Resolving forces vertically and equating $F_{DF} = 0$

Hence the assumption of F_{DF} as compressive force is wrong.

Resolving horizontally and equating

$$F_{EF} = F_{FB} = 8.786 \text{ kN.}$$

$$\therefore \quad F_{FB} = 8.786 \text{ kN (compressive)}$$

(vii) Joint B

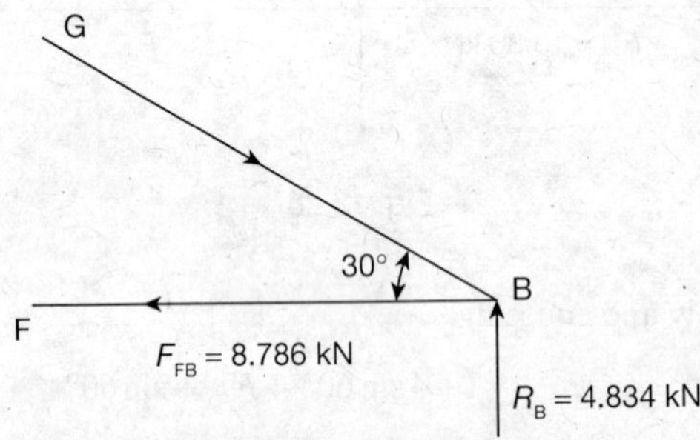

Fig. 12.30

Resolving the forces vertically $\quad F_{BG} \sin 30° = 4.834$ kN

i.e., $\quad F_{BG} = 9.668$ kN (compressive)

and $\quad F_{GD} = F_{BG} = 9.668$ kN (compressive)

(viii) Summary

Forces in each member of the truss are tabulated below

Member	Magnitude of force in kN	Nature of force
AC	9.664	Compressive
AE	13.360	Tensile
CD	9.664	Compressive
CE	4.000	Compressive
ED	5.155	Tensile
EF	8.786	Tensile
GF	0	Nil
FB	8.786	Compressive
BG	9.668	Compressive
GD	9.668	Compressive

SOLVED PROBLEM 12.7

A truss of span 12 m is as shown in Figure 12.31. Determine the forces in members.

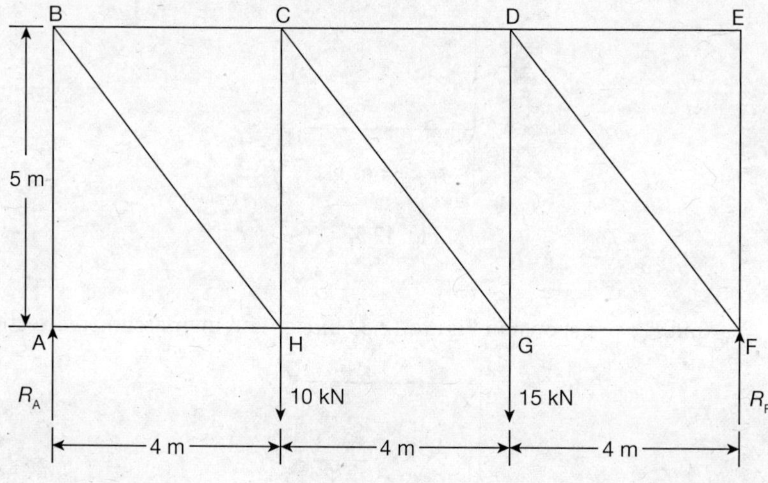

Fig. 12.31

Solution:

$$\cos\theta = \frac{AB}{BH} = \frac{5}{6.4} = 0.78 \qquad \left(\because BH = \sqrt{5^2 + 4^2} = 6.4\right)$$

$$\sin\theta = \frac{4}{6.4} = 0.625$$

(i) Support reactions

Taking moments of all the forces about A and equating

$$R_F \times 12 = (15 \times 8) + 10 \times 4$$

$$\therefore \quad R_F = \frac{160}{12} = 13.33 \text{ kN}$$

$$R_A = 10 + 15 - 13.33 = 11.67 \text{ kN}$$

(ii) Joint A

Before considering joint A, it is necessary to identify members which do not have any force.

As a general rule if three forces act at a joint and two of them are along the same straight line then for equilibrium of the joint, the third force should be equal to zero.

At joint A, there are two members and one reaction. F_{AB} and R_A are in a straight line.

So, F_{AH} is perpendicular and will have zero force.

At point F, a similar condition exists then F_{FG} is zero.

At joint D, F_{CE} and F_{DE} are in a straight line. F_{DG} is perpendicular and is zero.

The forces F_{AB} and F_{AH} are the two forces out of which $F_{AH} = 0$ and the assumed direction of F_{AB} is shown in Figure 12.32.

Now, $F_{AB} = 11.67$ kN (compressive)

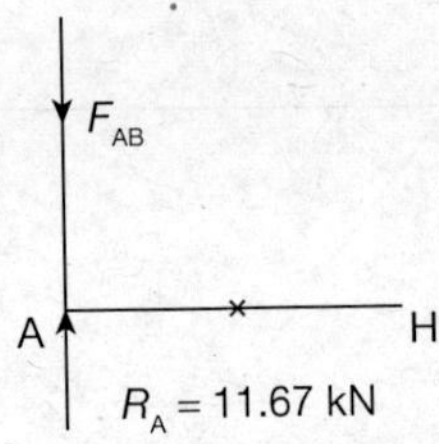

Fig. 12.32

(iii) Joint B

The forces F_{AB}, F_{BC} and F_{BH} are shown in Figure 12.33 along with their assumed direction of action.

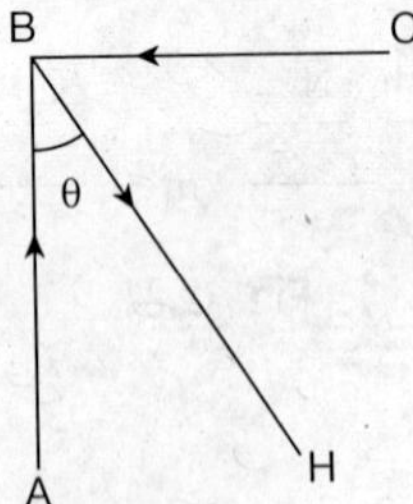

Fig. 12.33

$$F_{AB} = 11.67 \text{ kN (compressive)}$$

Resolving the forces vertically and equating

$$F_{AB} = F_{BH} \cos\theta$$

$$\therefore \quad F_{BH} = \frac{F_{AB}}{\cos\theta} = \frac{11.67}{0.78} = 14.96 \text{ kN (tensile)}$$

Resolving the forces horizontally and equating.

$$F_{BC} = F_{BH} \sin\theta = 14.96 \times 0.625$$

$$F_{BC} = 9.35 \text{ kN (compressive)}$$

(iv) Joint H

Forces F_{CH} and F_{GH} be the forces in the members *CH* and *HG* and the assumed directions are shown is Figure 12.34.

Resolving the forces vertically

$$F_{CH} + 10 = F_{BH} \cos\theta$$

$$F_{CH} = F_{BH} \cos\theta - 10 = (14.96 \times 0.78) - 10$$

$$= 1.67 \text{ kN (compressive)}$$

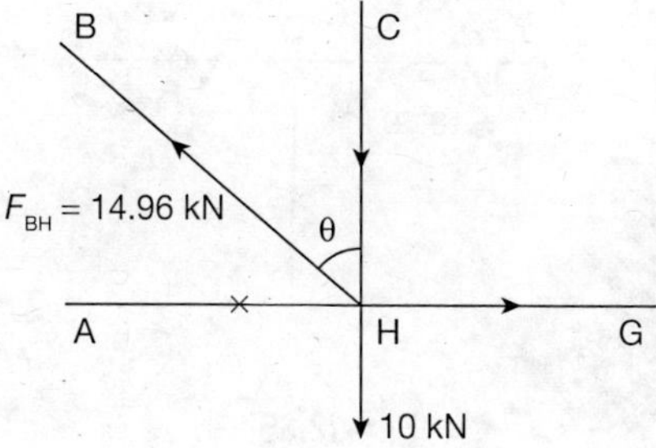

Fig. 12.34

Resolving the forces horizontally

$$\begin{aligned} F_{GH} &= F_{BH} \sin\theta \\ &= 14.96 \times 0.625 \\ &= 9.35 \text{ kN (compressive)} \end{aligned}$$

(v) Joint C

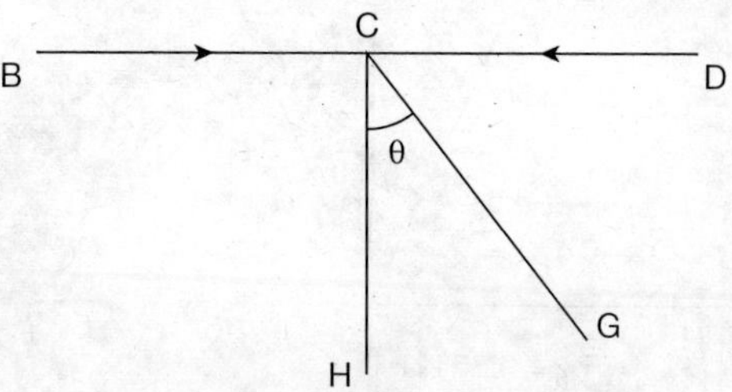

Fig. 12.35

The forces F_{BC} and F_{CD} and the assumed directions are shown in Figure 12.35.
Resolving the forces vertically

$$F_{CH} = F_{CG} \cos\theta$$

$$F_{CG} = \frac{F_{CH}}{\cos\theta} = \frac{1.67}{0.78} = 2.14 \text{ kN (tensile)}$$

Resolving the forces horizontally

$$F_{BC} + F_{CG} \sin\theta = F_{CD}$$

$$F_{CD} = 9.35 + (2.14 \times 0.625)$$

$$F_{CD} = 10.69 \text{ kN (compressive)}$$

(vi) Joint D

F_{CD} and F_{DE} are the forces in members CD and DE and the assumed directions are shown in Figure 12.36.

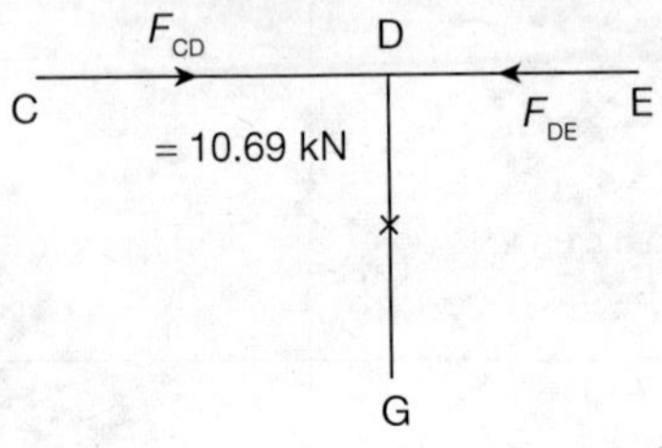

Fig. 12.36

As already discussed the force in DG is zero. Therefore, for equilibrium condition for the joint D,

$$F_{CD} = F_{DE} = 10.69 \text{ kN (compressive)}$$

(vii) Joint G

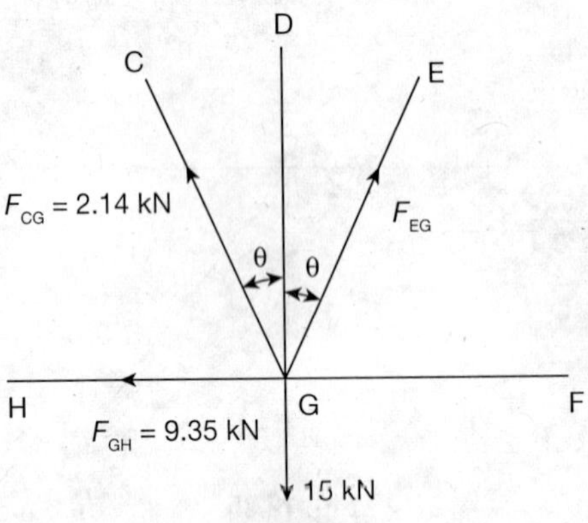

Fig. 12.37

F_{CG} and F_{EG} are the forces assumed on the members CG and EG along with their directions.
Resolving the forces vertically

$$F_{CG} \cos\theta + F_{EG} \cos\theta = 15$$

$$F_{EG} = \frac{15 - F_{CG} \cos\theta}{\cos\theta}$$

$$= \frac{15 - (2.14 \times 0.78)}{0.78}$$

$$= 17.09 \text{ kN (tensile)}$$

(viii) Joint F

F_{EF} is the assumed direction which is shown in Figure 12.38.

$$F_{EF} = R_F = 13.33 \text{ kN (compressive)}$$

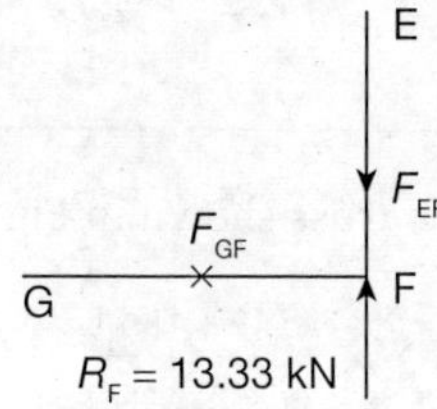

Fig. 12.38

(ix) Summary

Forces in the members are tabulated below.

Member	Magnitude of force in kN	Nature of force
AB	11.67	Compressive
AH	0	Nil
BC	9.35	Compressive
BH	14.96	Tensile
CD	10.69	Compressive
CG	2.14	Tensile
CH	1.67	Compressive
DE	10.69	Compressive
DG	0	Nil
EF	13.33	Compressive
EG	17.09	Tensile
FG	0	Nil
GH	9.35	Tensile

12.5 METHOD OF SECTIONS OR METHOD OF MOMENTS

When forces in a few members of a truss are to be determined then this method is the most simple one. This method is easy since we do not need the solutions from other joints. Here, we pass a section line passing through the members in which the forces are to be determined. The section line should be such that it does not cut more than three members in which the forces are unknown. The truss on one side of the section line is treated as a free body in equilibrium under the action of external forces. The unknown forces are then determined using the equilibrium equations, $\Sigma M = 0$, $\Sigma F_x = 0$ and

$\Sigma F_y = 0$. When we get a negative value of force in a member then the assumed direction is not correct and it has to be reversed.

SOLVED PROBLEM 12.8

Determine the forces in the member of the truss shown in Figure 12.39 by the method of moments.

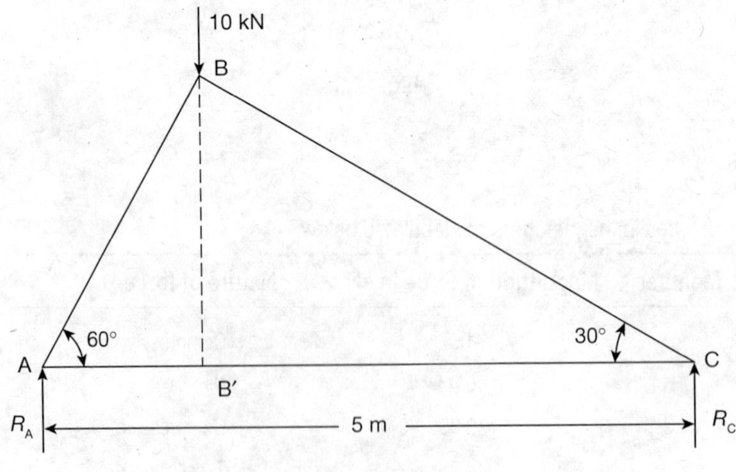

Fig. 12.39

Solution:

Length of members are calculated as:

$$AB = AC\cos 60° = 5 \times 0.5 = 2.5 \text{ m}$$

$$BC = AC\cos 60° = 5 \times 0.866 = 4.33 \text{ m}$$

$$BB' = AB\cos 60° = 2.5 \times 0.866 = 2.165 \text{ m}$$

$$AB' = AB\cos 60° = 2.5 \times 0.5 = 1.25 \text{ m}$$

$$B'C = AC\cos 60° = 5 - 1.25 = 3.75 \text{ m}$$

(i) Support reactions

Support reactions have already been worked out in Problem 12.1 as

$$R_A = 75 \text{ kN and } R_c = 2.5 \text{ kN}$$

(ii) Forces in members

A section (1) – (1) cutting through AB and AC splitting the truss into two is considered.

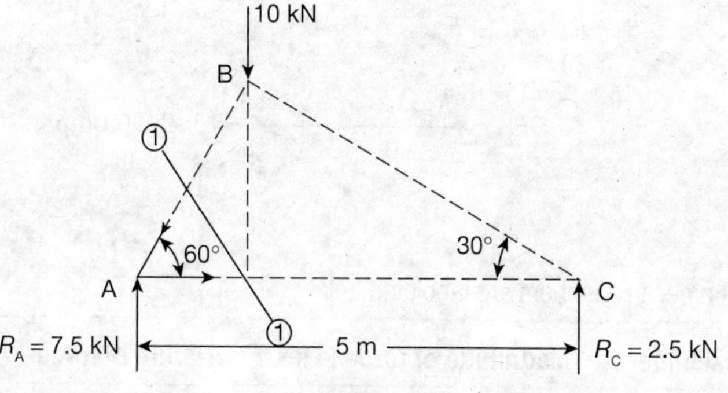

Fig. 12.40

Considering the left part of the truss, taking moments of the forces acting on the left part of the truss with respect to C,

$$F_{AB} \times BC = R_A \times AC$$

i.e.,
$$F_{AB} = \frac{R_A \times AC}{BC} = \frac{7.5 \times 5}{4.33} = 8.66 \text{ kN (compressive)}$$

Taking moments with respect to B

$$-(R_A \times AB') + (F_{AC} \times BB') = 0$$
$$-(7.5 \times 1.25) + F_{AC} \times 2.163 = 0$$

∴
$$F_{AC} = \frac{7.5 \times 1.25}{2.163} = 4.33 \text{ (tensile)}$$

Let another section (2) – (2) cutting through AC and BC to split the truss into two parts be considered.

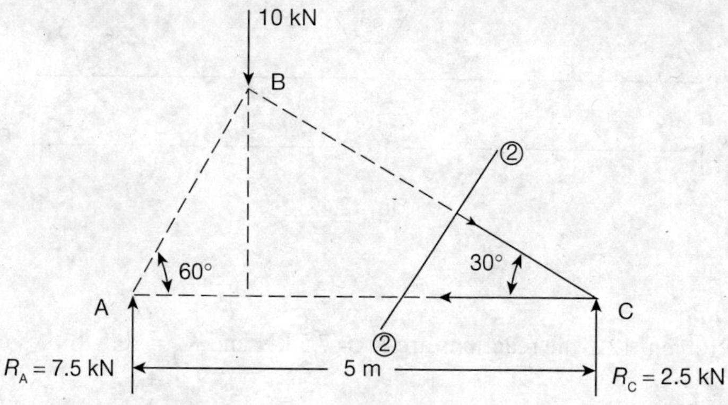

Fig. 12.41

$$R_c \times AC - F_{BC} \times AB = 0$$

$$\therefore \qquad F_{BC} = \frac{R_c \times AC}{AB} = \frac{2.5 \times 5}{2.5} = 5 \text{ kN (compressive)}$$

(iii) Summary

Forces in members are tabulated below.

Member	Magnitude of force in kN	Nature of force
AB	8.66	Compressive
AC	4.33	Tensile
BC	5.0	Compressive

SOLVED PROBLEM 12.9

Figure 12.42 shows a Warren girder consisting of members of 5 m length. The truss is freely supported at its end points. The girder is loaded with external loads. Determine the forces in the member and their nature.

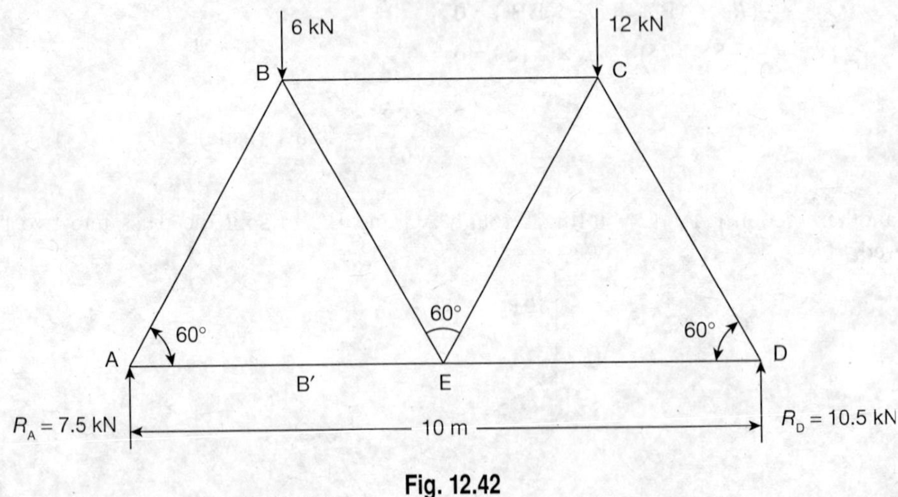

Fig. 12.42

(i) Support reactions

As discussed in Problem 12.2 the reactions are $R_A = 7.5$ kN and $R_D = 10.5$ kN

(ii) Forces in members

Length of each member = 5 m

$$AB' = B'E = EC' = C'D = \frac{5}{2} = 2.5 \text{ m}$$

$$BB' = CC' = DD' = EE' = AA' = 5 \times \sin 60^\circ$$

$$= 5 \times 0.866 = 4.33 \text{ mm.}$$

A section (1) – (1) cutting through the left portion of the truss taking moments of the force acting left of the section with reference to joint E.

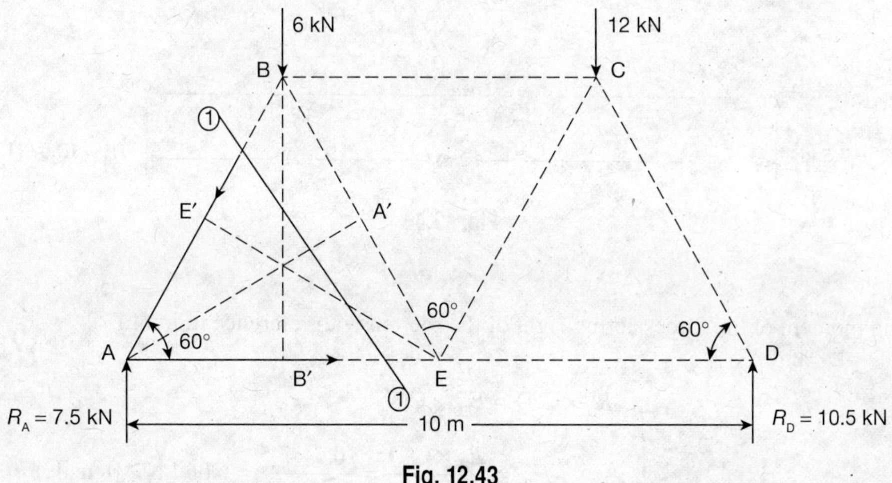

Fig. 12.43

Taking moments about *E*

$$F_{AB} \times EE' = R_A \times AE$$

$$F_{AB} = \frac{R_A \times AE}{EE'} = \frac{7.5 \times 5}{4.33} = 8.66 \text{ kN} \quad \text{(compressive)}$$

Taking moments about the joint *B*

$$R_A \times AB' = F_{AE} \times BB'$$

$$F_{AE} = \frac{R_A \times AB'}{BB'} = \frac{7.5 \times 2.5}{4.33} = 4.33 \text{ kN} \quad \text{(tensile)}$$

Assuming a section (2)–(2) cutting through the members CD and DE is considered (Figure 12.44). The right portion of the truss is considered.
Taking moments of the forces about *E*.

$$F_{CD} \times EE'' = R_D \times DE$$

$$F_{CD} = \frac{R_D \times DE}{EE''} = \frac{10.5 \times 5}{4.33} = 12.12 \text{ kN} \quad \text{(tensile)}$$

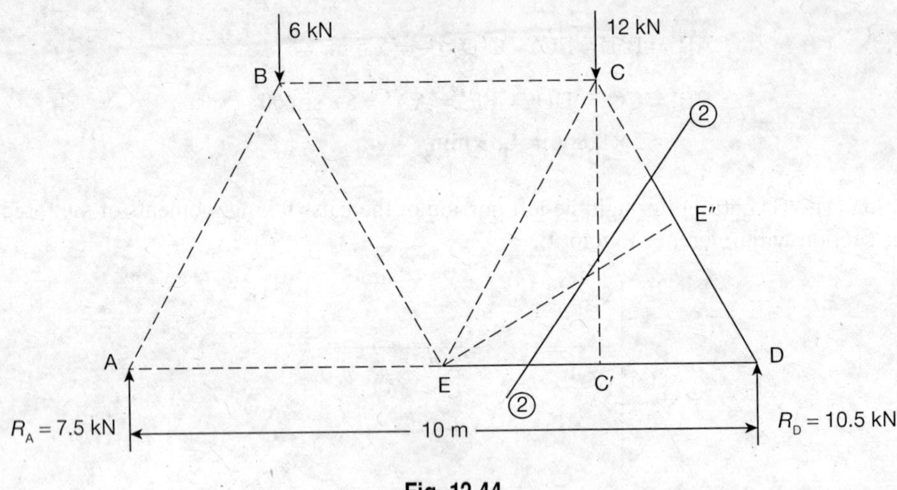

Fig. 12.44

Taking moment of the forces acting right of the section with reference to joint E

$$R_D \times C'D = F_{ED} \times CC'$$

$$\therefore \quad F_{ED} = \frac{R_D \times C'D}{CC'} = \frac{10.5 \times 2.5}{4.33} = 6.06 \text{ kN} \quad \text{(tensile)}$$

A section (3) – (3) cutting through the members AE, BE and BC is considered (Figure 12.45). The left portion of the truss is considered.
Taking moments of the forces acting left of the section with reference to joint E.

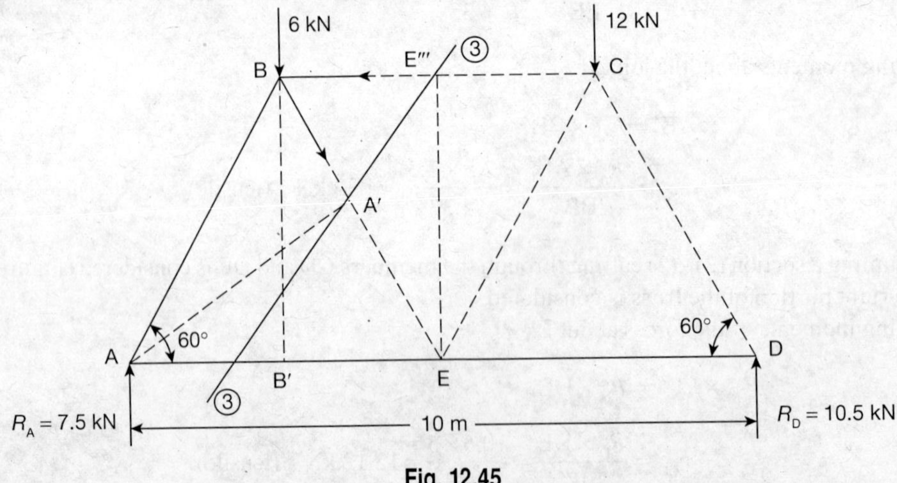

Fig. 12.45

$$R_A \times AE = (6 \times B'E) + (F_{BC} \times EE''')$$

$$\therefore \quad F_{BC} = \frac{(R_A \times AE) - (6 \times B'E)}{EE'''}$$

$$= \frac{(7.5 \times 5) - (6 \times 2.5)}{4.33}$$

$$= 5.2 \text{ kN (compressive)}$$

Taking moments of forces with reference to joint A

$$(6 \times AB') + (F_{BE} \times AA') = (F_{BC} \times BB')$$

$$F_{BE} = \frac{(F_{BC} \times BB') - (6 \times AB')}{AA'}$$

$$= \frac{(5.2 \times 4.33) - (6 \times 2.5)}{4.33}$$

$$= 1.75 \text{ kN (tensile)}$$

A section (4)–(4) cutting through the members BC, CD and CE is considered (Figure 12.46). The right portion of the-truss is considered.

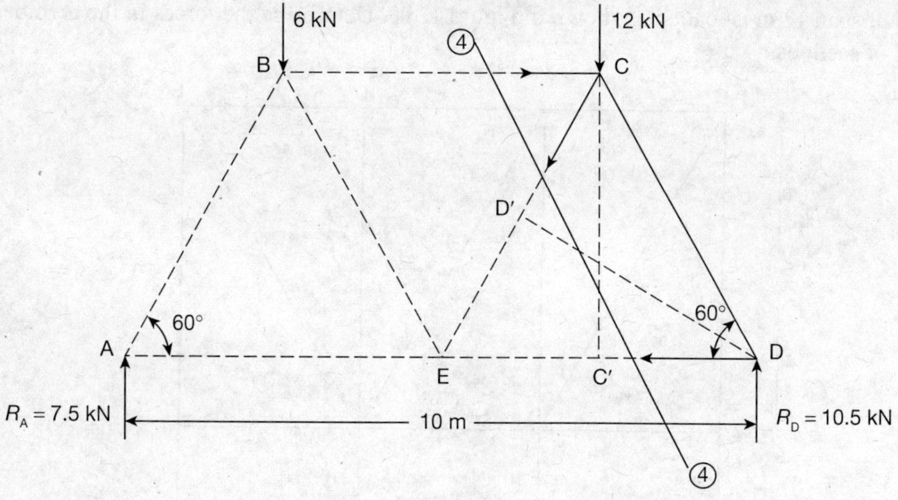

Fig. 12.46

Taking moments of the forces acting right of the section with reference to joint D.

$$(12\times C'D)+(F_{CE}\times DD')=(F_{BC}\times CC')$$

$$F_{CE}=\frac{(F_{BC}\times CC')-(12\times C'D)}{DD'}$$

$$=\frac{(5.2\times 4.33)-(12\times 2.5)}{4.33}$$

$$=1.73\text{ kN (tensile)}$$

(iii) Summary

Forces in the members of the truss are tabulated below.

Member	Magnitude of force in kN	Nature of force
AB	8.66	Compressive
AE	4.33	Tensile
BC	5.20	Compressive
BE	1.74	Tensile
CD	12.12	Compressive
CE	1.73	Tensile
DE	6.06	Tensile

SOLVED PROBLEM 12.10

A truss of span 12 m is loaded as shown in Figure 12.47. Determine the forces in the members using method of sections.

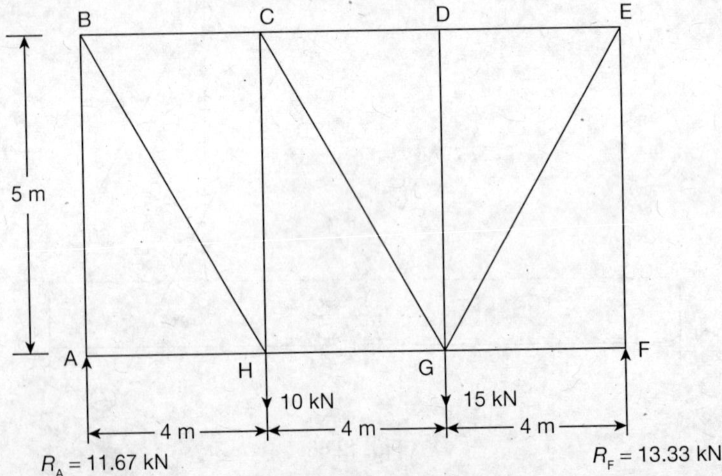

Fig. 12.47

Solution:

(i) Support reactions

The reactions have already beam determined in Problem 12.2, as $R_A = 11.67$ kN and $R_F = 13.33$ kN.

(ii) Forces in members

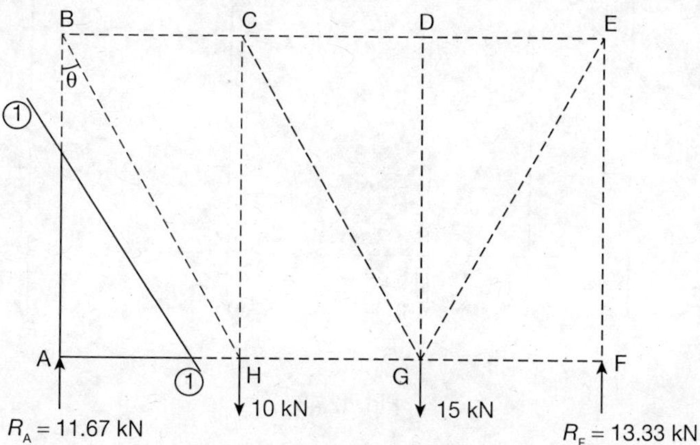

Fig. 12.48

A section (1) – (1) through AB and AH cutting the truss into two parts is considered. The forces F_{AB} and F_{AH} are assumed as shown in Figure 12.48.

The left part of the truss is considered.

Taking moment of forces acting on the left part of the truss with respect to B.

$$F_{AH} \times AB = 0$$

$$F_{AH} \times 5 = 0$$

i.e.,

$$F_{AH} = 0$$

As already discussed, the forces in members F_{AH}, F_{FG} and F_{DG} be zero.

Taking moment of forces with respect to H

$$F_{AH} \times AH = R_A \times AH$$

i.e.,

$$F_{AH} = R_A = 11.67 \text{ kN} \quad \text{(compressive)}$$

A section (2) – (2) through BC, BH and AH cutting the truss into two parts is considered. Forces F_{BH} and F_{BC} are assumed as tensile.

Taking moment of forces acting on the left part of the truss with respect to C.

$$R_A \times BC = F_{BH} \times CC'$$

$$F_{BH} = \frac{R_A \times BC}{CC'} = \frac{11.67 \times 4}{3.125} \left(CC' = 5 \times \sin\theta = 5 \times \frac{4}{6.4} = 3.125 \right)$$

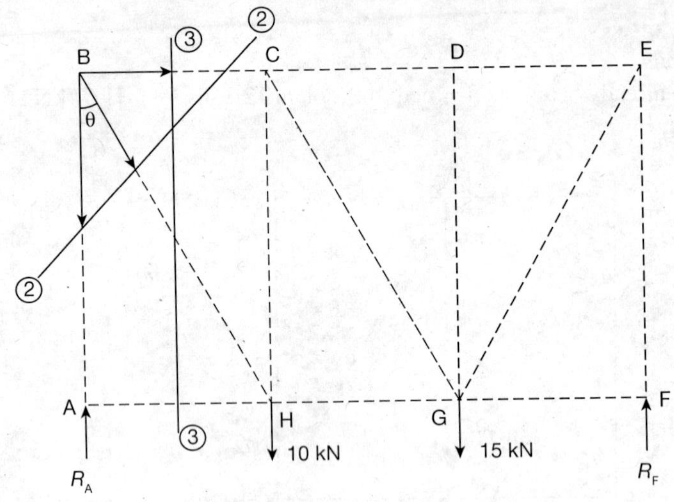

Fig. 12.49

i.e., $$F_{BH} = 14.96 \text{ kN} \quad \text{(tensile)}$$

Similarly, another section cutting BC, BH and AH are cut and the force in other members are found.

(iii) Summary

Forces in the members are tabulated below.

Member	Magnitude of forces in kN	Nature of force
AB	11.67	Compressive
AH	0	Nil
BC	9.35	Compressive
BH	14.96	Tensile
CD	10.69	Compressive
CG	2.14	Tensile
CH	1.67	Compressive
DE	10.69	Compressive
DG	0	Nil
EF	13.33	Compressive
EG	17.09	Tensile
FG	0	Nil
GH	9.35	Tensile

SOLVED UNIVERSITY QUESTIONS

SOLVED PROBLEM 12.11

Determine the forces in the truss shown in Figure 12.50. It carries a horizontal load of 16 kN and vertical load of 24 kN (Anna Univ., June 2007, ME).

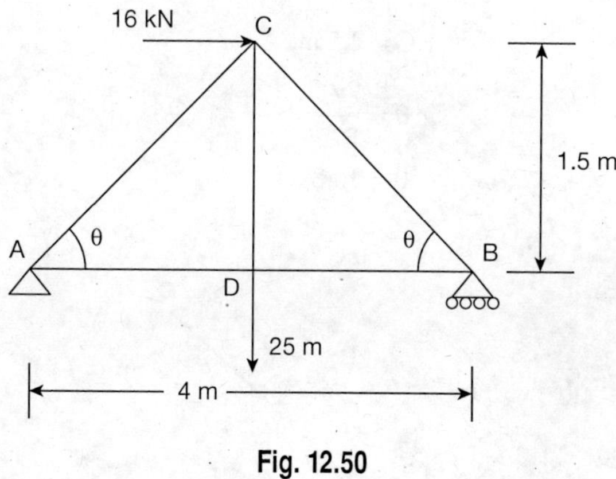

Fig. 12.50

Solution:

(i) Supported reactions

The truss is supported on rollers at B and hence the reaction at B (R_B) must be normal to the roller base, i.e., vertical (Figure 12.51).

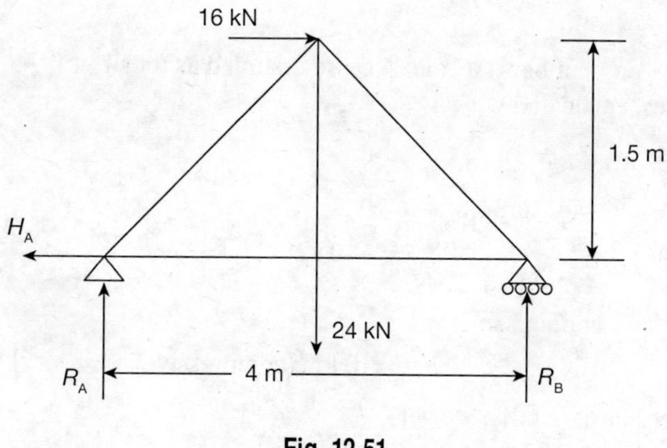

Fig. 12.51

The truss is hinged at A and hence the support reactions at the hinged end A will consist of a horizontal reaction H_A and a vertical reaction R_A.

Taking moment about A, we get

$$R_B \times 4 = 24 \times 2 + 16 \times 1.5$$

∴ $$R_B = 18 \text{ kN}$$

$$R_A = 24 - 18 = 6 \text{ kN}$$

And $$H_A = \text{Sum of all horizontal loads} = 12 \text{ kN}$$

In ΔBCD, $$BC^2 = CD^2 + B^2 = 1.5^2 + 2^2$$

i.e., $$BC = 2.5 \text{ m}$$

and $$\sin\theta = \frac{DC}{BC} = \frac{1.5}{2.5} = 0.6$$

∴ $$\theta = 36.8°$$

(ii) Joint A

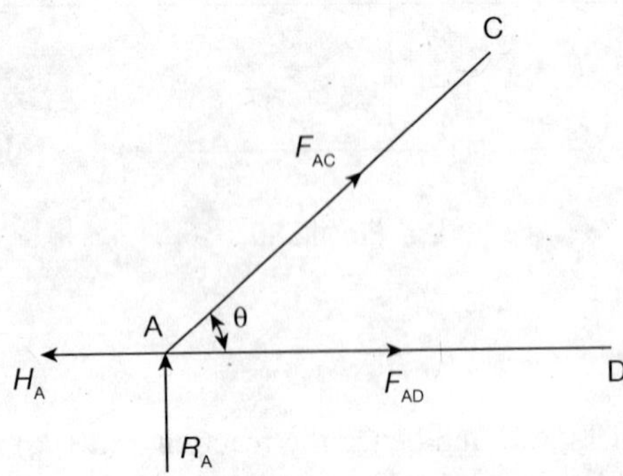

Fig. 12.52

The forces F_{AC} and F_{AD} on members AC and AD are assumed as tensile

Resolving the forces, vertically, we get

$$F_{AC}\sin\theta + R_A = 0$$

$$F_{AC}\sin 36.8° + 6 = 0$$

$$F_{AC} = -10 \text{ kN}$$

As it is negative, F_{AC} is compressive

i.e., $$F_{AC} = 10 \text{ kN (compressive)}$$

Resolving the forces horizontally, we get

$$-H_A + F_{AD} + F_{AC}\cos\theta = 0$$

i.e., $$-16 + F_{AD} - 10 \times \cos\ 36.8° = 0$$

Solving, $$F_{AD} \doteq 24 \text{ kN (tensile)}$$

(iii) Joint B

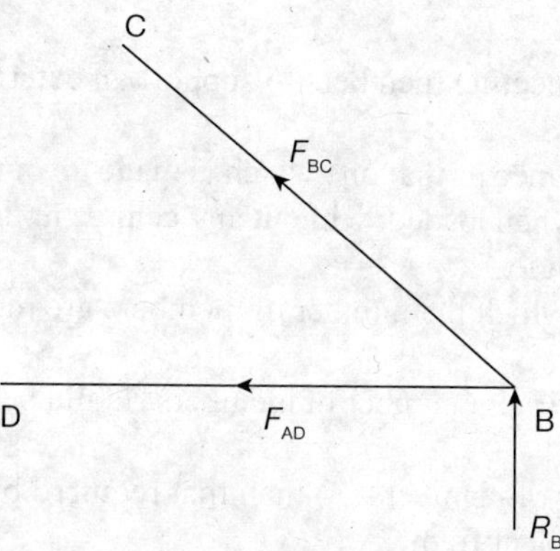

Fig. 12.53

Let the force F_{AD} and F_{BC} be assumed as tensile

Resolving the forces vertically, we get

$$R_B + F_{BC} \sin\theta = 0$$

i.e.,

$$18 + F_{BC} \sin 36.8° = 0$$

$$\therefore\ F_{BC} = -\frac{18}{\sin 36.8°} = -30 \text{ kN}$$

Hence, the assumption is not correct and the force F_{BC} is compressive

i.e.,

$$F_{BC} = 30 \text{ kN} \quad \text{(compressive)}$$

Resolving the forces horizontally we get

$$-F_{BC} \cos\theta - F_{BA} = 0$$

$$-F_{BC} \cos\theta = F_{BD}$$

i.e.,

$$-(-30) \cos 36.8° = F_{BD}$$

i.e.,

$$F_{BD} = 24 \text{ kN} \ \text{(tensile)}$$

(iv) Summary

Forces in the members are tabulated below.

Member	Magnitude of force in kN	Nature of force
AC	10	Compressive
AD	24	Tensile
BC	30	Compressive
BD	24	Tensile

SALIENT POINTS

- A structure formed by connecting members to support an external systems of forces is known as a frame.
- A perfect frame may be defined as that one which is made up of members just sufficient to keep the frame in equilibrium, when loaded, without any change in the shape. The simplest example of a perfect frame is a triangle.
- Imperfect frame is one in which the number of members are more or less than $(2j-3)$ where j is the member of joints.
- For a stable frame, the minimum number of members should be equal to two times the number of joints minus three.
- If a frame has less number of members (n) than that required by the relation $n=(2j-3)$ then the frame is known as deficient frame.
- A redundant frame is an imperfect frame, in which the number of members are more than $(2j-3)$.
- Assumption made in finding the forces in a frame are:
  (i) Frame is a perfect frame
  (ii) Frame carries loads only at the joints
  (iii) All the members are pin-jointed
  (iv) Self-weight of the members is neglected.

QUESTIONS

1. Distinguish between perfect and imperfect frames. How an imperfect frame is made as a perfect frame?
2. Define a redundant frame? How it is different from a deficient frame? Which one is stable?
3. Explain the method of joints.
4. How the method of sections is applied to analyze the forces in the members of a truss?
5. Determine the forces in all the members of a truss shown in Figure 12.54 by method of joints.

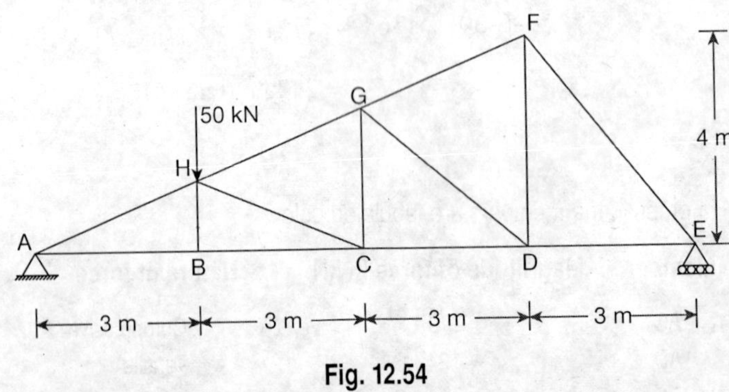

Fig. 12.54

6. Find the force in all members of the truss shown in Figure 12.55 by method of joints.

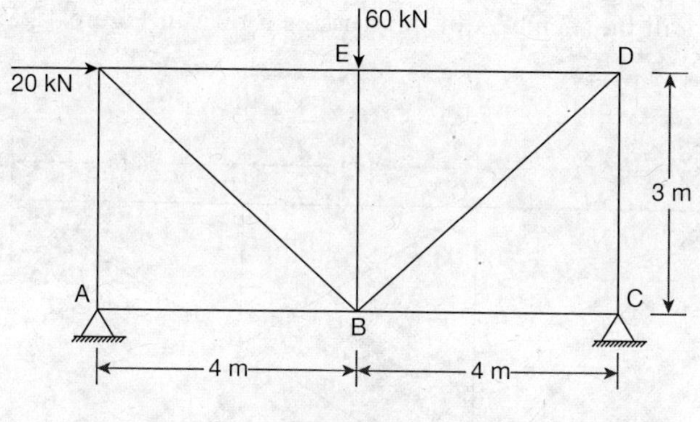

Fig. 12.55

7. A cantilever truss of Warner type is loaded as shown in Figure 12.56. Find the forces in the members of the frame by method of joints.

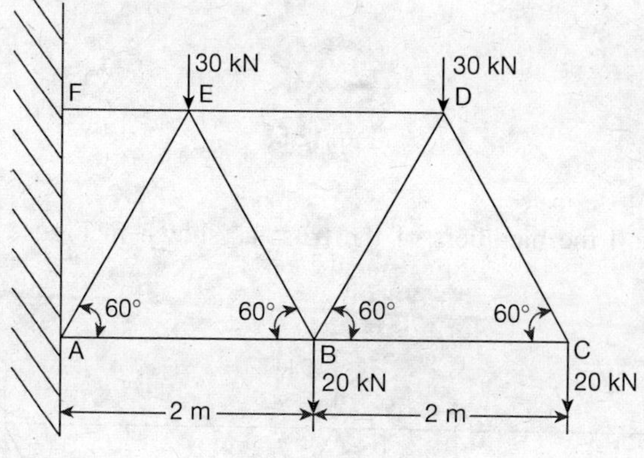

Fig. 12.56

8. Find the forces in all members of the truss shown in Figure 12.57.

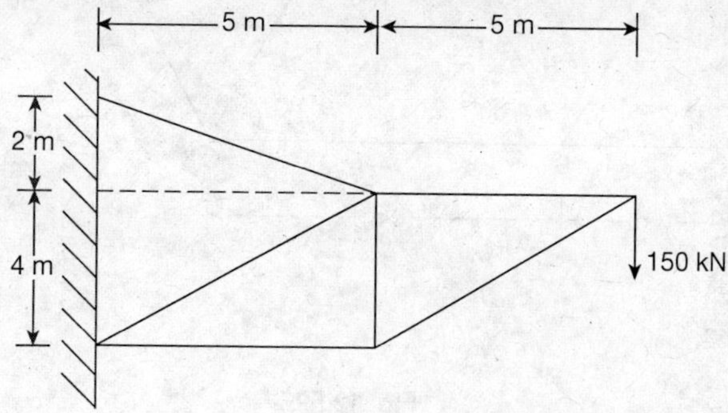

Fig. 12.57

Use method of sections.

9. Find the forces in all the members of the truss as shown in Figure 12.58. Adopt method of sections.

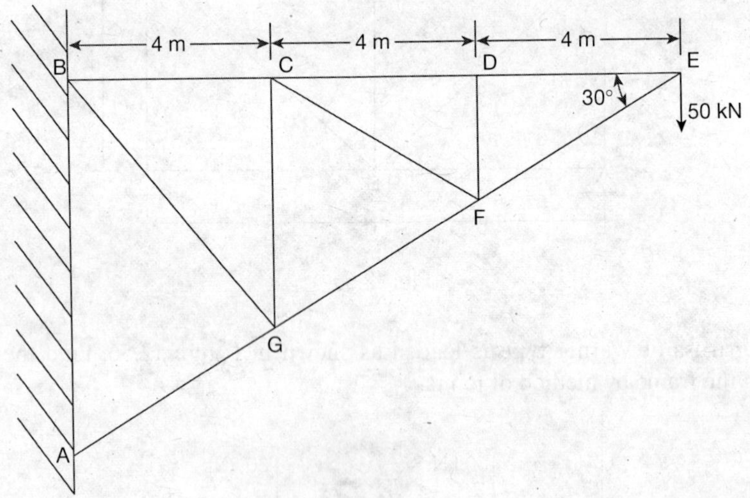

Fig. 12.58

10. Find the forces in all the members of the truss as shown in Figure 12.59. Apply a suitable method.

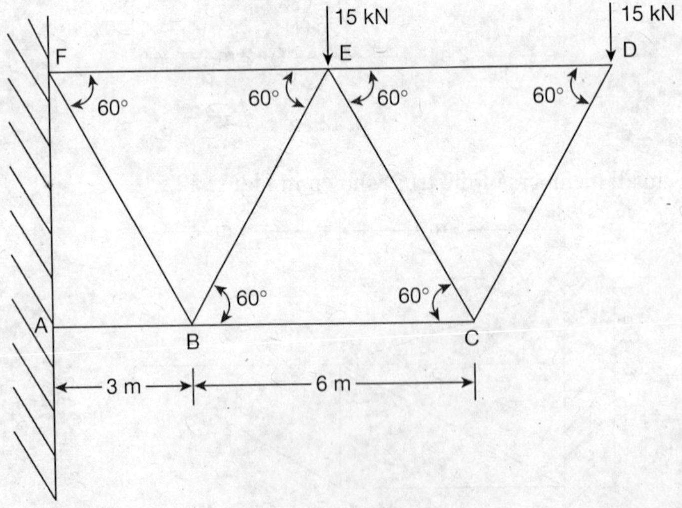

Fig. 12.59

Bibliography

Andrew, P., and Kiusalaas, J., *Strength of Materials*, Cartgage Learning India Pvt. Ltd, New Delhi, 2003.

Bansal, R. K., *Text Book of Strength of Materials*, Laxmi Publications, New Delhi, 2004.

Beer, F. P., and Johnson, R., *Mechanics of Materials*, Third Edition, McGraw Hill Book Co., New York, 2002.

Bhavikatti, S. S., *Strength of Materials*, Vikas Publishing House Pvt. Ltd, New Delhi, 1998.

Gambir, M. L., *Fundamentals of Solid Mechanics*, PHI Learning Private Limited, New Delhi, 2009.

Gere, J. M., and Goodno, B. J., *Strength of Materials*, Indian Edition, Cartgage Learning India Pvt. Ltd, New Delhi, 2009.

Hearn, E. J., *Mechanics of Materials*, Third Edition, Vol. 1, Elsevier, New York, 1997.

Hulse, R., Sherwin, K., and Cain, J., *Solid Mechanics*, Palgrave ANE Books, New Delhi, 2004.

Jindal, U. C., *Text Book of Strength of Materials*, Galgotia Publications, New Delhi, 1998.

Kazimi, S. M. A., *Solid Mechanics,* Tata McGraw Hill Publishing Co., New Delhi, 1981.

Morley, A., *Strength of Materials, ECBS*, Longman's Green & Company, London, 1962.

Nash, W. A., *Theory and Problems in Strength of Materials*, Schaum Outline Series, McGraw Hill Book Co., New York, 1995.

Nash, W. A., *Strength of Materials*, Fourth Edition, McGraw Hill, New York, 1998.

Negi, L. S., *Strength of Materials*, Tata McGraw Hill Education Pvt. Ltd, New Delhi, 2008.

Popov Egor, P., and Balan Toader, A., *Engineering Mechanics of Solids*, Second Edition, PHI Learning Pvt. Ltd, New Delhi, 2006.

Punmia, B. C., and Jain, A., *Strength of Materials*, Laxmi Publications, New Delhi, 2000.

Ryder, G. H., *Strength of Materials*, Third Edition, Macmillan, London, 2002.

Shames, I. H., and Pitarresi, J. M., *Introduction to Solid Mechanics*, Third Edition, PHI Learning Private Limited, New Delhi, 2009.

Singh, S., *Strength of Materials*, Khanna Publications, New Delhi, 2000.

Singh, A. K., *Mechanics of Solids*, PHI Learning Private Limited, New Delhi, 2007.

Srinath, L. S., *Mechanics of Solids*, PHI Learning Private Limited, New Delhi, 2007.

Srivastava, A. K., and Gope, P. C., *Strength of Materials*, PHI Learning Pvt. Ltd, New Delhi, 2009.

Subramanyan, R., *Strength of Materials*, Oxford University Press, New Delhi, 2005.

Timoshenko, S. P., *Strength of Materials*, Third Edition, Van Nostrand Reinhold, New York, 1955.

Timoshenko, S. P., and Young, D. H., Elements *of Strength of Materials*, Fifth Edition, Van Nostrand Reinhold, New York, 1968.

Index

U

V

W

Y